Paving the Way for 5G Through the Convergence of Wireless Systems

Ramona Trestian
Middlesex University, UK

Gabriel-Miro Muntean
Dublin City University, Ireland

A volume in the Advances in Wireless Technologies and Telecommunication (AWTT) Book Series

IGI Global
DISSEMINATOR OF KNOWLEDGE

Published in the United States of America by
IGI Global
Information Science Reference (an imprint of IGI Global)
701 E. Chocolate Avenue
Hershey PA, USA 17033
Tel: 717-533-8845
Fax: 717-533-8661
E-mail: cust@igi-global.com
Web site: http://www.igi-global.com

Library of Congress Cataloging-in-Publication Data

Names: Trestian, Ramona, 1983- editor. | Muntean, Gabriel-Miro, 1972- editor.
Title: Paving the way for 5G through the convergence of wireless systems / Ramona Trestian and Gabriel-Miro Muntean, editors.
Description: Hershey, PA : Information Science Reference, an imprint of IGI Global, [2019] | Includes bibliographical references and index.
Identifiers: LCCN 2018030361| ISBN 9781522575702 (hardcover) | ISBN 9781522575719 (ebook)
Subjects: LCSH: Wireless communication systems--Technological innovations. | Convergence (Telecommunication)
Classification: LCC TK5103.2 .P397 2019 | DDC 621.3845/6--dc23 LC record available at https://lccn.loc.gov/2018030361

This book is published in the IGI Global book series Advances in Wireless Technologies and Telecommunication (AWTT) (ISSN: 2327-3305; eISSN: 2327-3313)

British Cataloguing in Publication Data
A Cataloguing in Publication record for this book is available from the British Library.

For electronic access to this publication, please contact: eresources@igi-global.com.

Advances in Wireless Technologies and Telecommunication (AWTT) Book Series

Xiaoge Xu
University of Nottingham Ningbo China

ISSN:2327-3305
EISSN:2327-3313

Mission

The wireless computing industry is constantly evolving, redesigning the ways in which individuals share information. Wireless technology and telecommunication remain one of the most important technologies in business organizations. The utilization of these technologies has enhanced business efficiency by enabling dynamic resources in all aspects of society.

The **Advances in Wireless Technologies and Telecommunication Book Series** aims to provide researchers and academic communities with quality research on the concepts and developments in the wireless technology fields. Developers, engineers, students, research strategists, and IT managers will find this series useful to gain insight into next generation wireless technologies and telecommunication.

Coverage

- Broadcasting
- Radio Communication
- Wireless Broadband
- Mobile Web Services
- Mobile Technology
- Wireless Sensor Networks
- Network Management
- Grid Communications
- Mobile Communications
- Telecommunications

IGI Global is currently accepting manuscripts for publication within this series. To submit a proposal for a volume in this series, please contact our Acquisition Editors at Acquisitions@igi-global.com or visit: http://www.igi-global.com/publish/.

The Advances in Wireless Technologies and Telecommunication (AWTT) Book Series (ISSN 2327-3305) is published by IGI Global, 701 E. Chocolate Avenue, Hershey, PA 17033-1240, USA, www.igi-global.com. This series is composed of titles available for purchase individually; each title is edited to be contextually exclusive from any other title within the series. For pricing and ordering information please visit http://www.igi-global.com/book-series/advances-wireless-technologies-telecommunication/73684. Postmaster: Send all address changes to above address.

Titles in this Series

For a list of additional titles in this series, please visit: www.igi-global.com/book-series

Emerging Capabilities and Applications of Wireless Power Transfer
Alicia Triviño-Cabrera (University of Málaga, Spain) and José A. Aguado (University of Málaga, Spain)
Information Science Reference • copyright 2019 • 383pp • H/C (ISBN: 9781522558705) • US $195.00 (our price)

Sensing Techniques for Next Generation Cognitive Radio Networks
Ashish Bagwari (Uttrakhand Technical University, India) Jyotshana Bagwari (Uttrakhand Technical University, India) and Geetam Singh Tomar (THDC Institute of Hydropower Engineering and Technology, India)
Information Science Reference • copyright 2019 • 356pp • H/C (ISBN: 9781522553540) • US $215.00 (our price)

Mobile Applications and Solutions for Social Inclusion
Sara Paiva (Instituto Politécnico de Viana do Castelo, Portugal)
Information Science Reference • copyright 2018 • 354pp • H/C (ISBN: 9781522552703) • US $195.00 (our price)

Optimizing E-Participation Initiatives Through Social Media
Laura Alcaide-Muñoz (University of Granada, Spain) and Francisco José Alcaraz-Quiles (University of Granada, Spain)
Information Science Reference • copyright 2018 • 307pp • H/C (ISBN: 9781522553267) • US $195.00 (our price)

Centrality Metrics for Complex Network Analysis Emerging Research and Opportunities
Natarajan Meghanathan (Jackson State University, USA)
Information Science Reference • copyright 2018 • 183pp • H/C (ISBN: 9781522538028) • US $175.00 (our price)

Mobile Technologies and Socio-Economic Development in Emerging Nations
Fredrick Japhet Mtenzi (Dublin Institute of Technology, Ireland & Aga Khan University, Tanzania) George S. Oreku (Open University of Tanzania (OUT), Tanzania) Dennis M. Lupiana (Institute of Finance Management, Tanzania) and Jim James Yonazi (Tanzania Standard (Newspapers) Limited, Tanzania)
Information Science Reference • copyright 2018 • 366pp • H/C (ISBN: 9781522540298) • US $195.00 (our price)

Affordability Issues Surrounding the Use of ICT for Development and Poverty Reduction
Sam Takavarasha Jr. (University of Fort Hare, South Africa & University of Zimbabwe, Zimbabwe) and Carl Adams (University of Portsmouth, UK)
Information Science Reference • copyright 2018 • 319pp • H/C (ISBN: 9781522531791) • US $195.00 (our price)

Handbook of Research on Environmental Policies for Emergency Management and Public Safety
Augustine Nduka Eneanya (University of Lagos, Nigeria)
Engineering Science Reference • copyright 2018 • 393pp • H/C (ISBN: 9781522531944) • US $295.00 (our price)

IGI Global
DISSEMINATOR OF KNOWLEDGE

701 East Chocolate Avenue, Hershey, PA 17033, USA
Tel: 717-533-8845 x100 • Fax: 717-533-8661
E-Mail: cust@igi-global.com • www.igi-global.com

Table of Contents

Detailed Table of Contents

or reputation. In addition, the authors highlight open problems, such as the lack of proper models for dynamic and incomplete information scenarios. These upcoming scenarios are associated to computing and storage at the network edge, as well as, the deployment of large-scale IoT systems. The chapter finalizes by discussing a business model for future networks.

With the emerging mobile applications and needs of ever-increasing bandwidth, it is anticipated that the next-generation passive optical network (NG-PON) with much higher bandwidth is a natural path forward to satisfy these demands and to develop valuable converged fiber-wireless access networks for wireless network operators. NG-PON systems present optical access infrastructures to support various applications of many service providers. Hybrid passive optical networks (HPON) present a necessary phase to future PON networks utilized the optical transmission medium – the optical fiber. For developing hybrid passive optical networks, there exist various architectures and directions. They are specified with emphasis on their basic characteristics. For proposing reliable and survivable architectures, traffic protection schemes must be implemented. For converging Fi-Wi passive optical networks, an integration of optical and wireless technologies into common broadband access network must be considered. Finally, the HPON network configurator as the interactive software tool is introduced.

With increase in demand of data traffic with no compromise on the underlying quality of service (QoS), the coexistence problem arises due to high electricity consumption by the network architecture which results in a huge CO2 emission and thereby causing various health hazards. Efficient utilization of the resources can reduce the cost of power consumption which will increase the economy-characteristics of the network. The resource consumption can be reduced under an intelligent technology-neutral policies which optimizes the deployment of the network architecture along with their transmit power paving the way for fifth generation (5G) in green wireless communications. On another front, the ultra-dense deployment of the small cells can increase the frequency reuse factor as well as help in reducing the energy consumption. This chapter designs the energy efficient networks while satisfying the underlying QoS by joint optimization of available resources depending on the interoperability challenges in terrestrial, underwater acoustic, and free space optical (FSO) communications.

This chapter discusses the ongoing work around hybrid access and network convergence, with particular emphasis on recent works on ATSSS in 3GPP. Three main aspects are analyzed: policy enforcement, integration with 5G QoS framework, interaction with underlying multi-path transport protocol. The chapter also provides some preliminary testbed results showing the benefits of ATSSS in the management of multiple accesses analyzing some primary performance indicators such as achievable data rates, link utilization for aggregated traffic, and session setup latency. The chapter also provides some results by considering two examples of realization of ATSSS policies to avoid inefficiency in link utilization and to allow the fulfillment of data rate requirements.

With the rapid increase in data traffic and high data rate demands from cellular users, conventional cellular networks are becoming insufficient to fulfill these requirements. Femto cells are integrated in macro cellular network to increase the capacity, coverage, and to fulfill the increasing demands of the users. Time required for handoff process between the cells became more sensitive and complex with the introduction of femto cells in the network. Public internet which connect the femto base station with the mobile core network induces higher latency if conventional handoff procedures are also employed in macro-femto cell network. So, handoff process will become slower and network operation will become insufficient. Some standards, procedures, and protocols should be defined for macro-femto cell network rather than using existing protocols. This chapter presents a comprehensive survey of handoff process, types of handoff in macro-femto cell network, and proposed methods and schemes for frequent and unnecessary handoff reduction for efficient network operation.

This chapter discusses the most recent solutions for accelerating mobile data delivery, focusing on the different issues for achieving this goal. These issues include the efficiency of mobile data offloading and in-network caching. By characterizing different offloading options from the aspect offloading efficiency, dynamically offloading cellular traffic onto cost-effective options is able to achieve great performance gains. On the other hand, with concerns on tightly coupled dependency on original data sources, mobile users could easily suffer from the scarcity of content object resources. Researchers further account for the possibility of decoupling content availability dependency from original content sources. And a key enabling factor could be the recently emerged technology (i.e., in-network caching) whereby the emerging information-centric networking (ICN) performs its critical functionalities. Coupling ICN with mobile cellular network could be an attractive solution in order to resolve the tension between the tremendous growth of mobile content traffic and backhaul bottleneck.

5G mobile systems can be comprehended as highly flexible and programmable E2E networking infrastructures that provide increased performance in terms of capacity, latency, reliability, and energy efficiency while meeting a plethora of diverse requirements from multiple services. Network slicing is emerging as a prospective paradigm to meet these requirements with reduced operating cost and improved time and space functionality. A network slice is the way to provide better resource isolation and increased statistical multiplexing. With dynamic slicing, 5G will operate on flexible zone of the network, permitting varying, adaptable levels or bandwidth and reliability. In this chapter, a comprehensive survey of network slicing is presented from an E2E perspective, detailing its origination and current standardization efforts, principal concepts, enabling technologies, as well as applicable solutions. In particular, it provides specific slicing solutions for each part of the 5G systems, encompassing orchestration and management in the radio access and the core network domains.

The next generation of mobile wireless communications represents a heterogeneous environment which integrates variety of network generation like third generation (3G), fourth generation (4G), and fifth generation (5G). The major challenge in this heterogeneous environment is to decide which access point to use when multiple networks are available. Process of roaming mobile user from one technology to anther different is called vertical handover. In this chapter, the authors propose a new mechanism based on graph theory and cost function in order to determine the best path for the end user in terms of quality of service (QoS) when the vertical handover process is needed. Then, they investigate the impact of some existing weighting methods in order to determine the suitable method which can be coupled with the cost function. The experiments evaluation by using Mininet emulator demonstrate that the proposed approach can achieve a significant improvement concerning four QoS metrics: throughput, packet lost, packet delay, and packer jitter for two services FTP and video streaming.

Preface

Looking at the ever-evolving telecommunication industry it can be seen that as smart mobile computing devices have become increasingly affordable and powerful there is also a significant growth in the number of advanced mobile users as well as their bandwidth demands. People can enjoy now ubiquitous connectivity to the Internet even when they are on the move (e.g., pedestrians, in the car, on the bus, etc.) or stationary (e.g., at home/office/airport/coffee shops, etc.). Depending on the user location, the Internet connectivity is enabled either via wireline or wireless solutions provided by different Radio Access Technologies (RATs) such as: cellular networks, wireless broadband networks, wireless personal area networks, etc. These access networks provide coverage over various geographical locations and usually in an overlapping manner. Moreover, these RATs have different characteristics and differ in capacity, coverage area, monetary cost, connection speed, and can be deployed by one or more network operators.

The 'Always Best Connected' and 'Always Best Experienced' visions are looking at a scenario where mobile users seamlessly roam within a multi-operator multi-technology multi-terminal multi-application environment supported by the next generation of wireless networks. This heterogeneous environment enables mobile users to be equipped with multi-mode wireless mobile devices that allows them to get online anytime and from anywhere. The ubiquitous connectivity provided by this heterogeneous environment creates flexibility around the mobile user's activities, enabling them to use their personal or professional e-mails, take part in video conferencing, use video streaming services, mobile TV, entertainment, download or online listening/watching music/videos, transfer/receive files to/from business contacts or friends, online shopping, use various applications. Among these, the use of social networking applications (e.g., Instagram, Twitter, Facebook, Linkedin, etc.) has become a part of one's daily life and can be used either for business (e.g., to post a profile, or look for a job), or just to connect to people and share social media data (e.g., news, personal experience, videos, photos). Moreover, these applications can be accessed by any device with Internet connectivity. Because of the advances in technology, the high-end mobile devices have become more and more affordable, easy to use, and powerful, leading to an increase in the mobile users' expectations of rich media services at high quality levels.

However, the increase in the number of mobile users as well as their application demands led to an exponential increase in the mobile broadband traffic putting significant pressure on the wireless networks capacity. One single network technology cannot cope with this explosion of mobile broadband data traffic, making the network convergence one promising solution for the future 5G networks.

THE CHALLENGES

Within the rapidly-evolving Information Communications Technology (ICT) industry a stealth, but very powerful service-oriented revolution takes place step-by-step.

In the first instance, the high-end mobile computing devices have become powerful and affordable leading to a significant growth in their number, as well as their users and their processing, communication and display capabilities (e.g., improved CPU, graphics, display, etc.). This led to the mass-market adoption of these multi-mode high-end devices - smartphones, iPhones, netbooks, tablets, and laptops, which in turn led to a new challenge for the mobile operators that are now confronting with a massive traffic growth.

The second instance is represented by the wide range and increasing popularity of various services including those offered by video sharing sites such as YouTube, social websites, mobile TV, banking, gaming and other entertainment services. In this regard, Cisco noted that the global IP traffic will increase threefold by 2021, so it estimates that more than 278 Exabytes of data per month will be transferred, out of more than 63% will be exchanged wireless and around 82% will be rich media-based (Cisco, 2017). Some of these services (e.g. live Internet video, virtual reality and augmented reality traffic, 360°video, etc.) put important pressure on both content processing as well as delivery.

In the third instance, the Internet connectivity to gain access to various services from anywhere at anytime, is enabled by a wide range of technologies either while on the move or stationary. These technologies include broadcast access (e.g. DVB-T2, DVB-H, etc.), broadband access (e.g. IEEE 802.11g, IEEE 802.11n, etc.) or cellular (e.g. LTE, LTE-Advanced etc.) access technologies. Multiple coexisting radio access technologies will be fundamental for the next generation of wireless networks in order to handle the volume and diversity of user traffic demands.

In the fourth instance, we can identify a definite trend towards using mobile cloud computing or mobile edge computing for the next generation business processing and communication needs, offering both robustness and flexibility. By making use of cloud computing a mass market adoption of the mobile devices has been observed with the advantage of enabling the ubiquitous availability of mobile broadband. This requires data exchange and processing both in the foreground and background. Moreover, there is an important increase in inter-communicating objects or things, including machines, appliances, sensors and vehicles.

The last instance but not the least, there are on-going efforts put in energy consumption optimization and in finding sustainable solutions in diverse areas, including ICT, and networking in particular. The increased competition in the networked-based services and the current world-wide economic situation determines increased efforts for reducing the costs by making use of optimization and/or finding innovative and more efficient research and development directions.

Many years ICT has focused on providing solutions for supporting a wide range of service types and then on increasing user quality of experience only. However, in recent years a rising concern on finding sustainable and energy-efficient solutions appeared. This was driven by both the evolved societal interests and user conscience being backed by EU and world-wide policies related to the reduction of the gas emissions. Moreover, the current and future economic situation has determined all the important market players to work at reducing costs and to make their businesses more efficient. However, these cost and energy-efficiency targets contrast severely with both the user demand expectations of high quality levels and service providers' need to introduce new various interactive rich media services. All these new rich media-based data services are well known as being bandwidth and power-hungry applications.

To address all these challenges, within this multi-service multi-technology multi-provider multi-device multi-user environment there is a need for the development of a coherent framework and a set of new solutions to support cost and energy consumption optimization while offering high quality of experience levels to the mobile users while they avail from diverse services.

SEARCHING FOR A SOLUTION

It is estimated that networked delivered digital media, especially over a heterogeneous environment to mobile customers will become one of main economic driving forces in the coming years. However, this will only be possible by having the necessary infrastructure to accommodate the increasing number of mobile users and supporting their expected high quality of experience levels. There are many Internet Service Providers which provide excellent wired connectivity to residential and business users alike. For local network access and service support wireless technologies such as WLAN (wireless local area network) and WPAN (wireless personal area network), can be deployed. Additionally, full wireless connectivity is supported via different radio access technologies, offered by different operators. These operators can provide connectivity by making use of different technologies such as: LTE, LTE-Advanced, etc. One operator can cover various regions using different access technologies. However, no single network technology and no single network provider will be capable to deal with this explosion of mobile broadband traffic and the diverse services demands, making the optimized innovative use of the co-existing broadband, broadcast and cellular network technologies a key solution. In order to deal with this explosion of mobile broadband data, network operators have started deploying different radio access technologies in overlapping areas. This solution enables them to accommodate more mobile users and to keep up with the traffic demands. In this context, the new challenge that the network operators are facing is to ensure seamless rich media service delivery experience at high quality levels to the end-user in a heterogeneous environment.

This multi-user multi-technology multi-application multi-provider environment requires the development of new technologies and standards that seek to ensure the quality of experience for the global end-users. While wireless technologies had a spectacular evolution over the past years, the present trend is to adopt a global network of shared standards which comes to meet user applications' requirements. For example, the evolution of wireless network architecture with the 4G/LTE densification and the future 5G wireless networks represents the convergence of the physical assets of fixed networks and wireless access points with the aim to increase flexibility, reliability and cost savings. The current Internet-centric environment and the need for ubiquitous connectivity pave the path towards network convergence. This means that the network of the future will enable the delivery of high data rate traffic through a high capacity or *broadband* network that mixes multiple communication modes over a single network enabling convenience and flexibility.

Other solutions adopted by network operators when dealing with the increasing amount of mobile broadband data traffic is the use of WLAN offload. WLANs have had an important impact in the area of mobile communications and their use has grown significantly in recent years (e.g., extended coverage, low-latency, power-efficient connection, reduced loads, etc.). The Wireless-Fidelity (Wi-Fi) offload solution is already adopted by many service providers, (e.g., Deutsche Telekom and iPass launched WiFi Mobilize). This solution enables the transfer of some of the mobile broadband traffic from the core cellular network to the WiFi network at peak times. In this way users can avail of a wider service offering. However, the overall user experience is still far from optimal as providing high quality mobile video services with Quality of Service (QoS) provisioning over resource-constrained wireless networks remains a challenge. Moreover user mobility, as well as the heterogeneity of mobile devices (e.g., different operating systems, display size, CPU capabilities, battery limitations, etc.), and the wide range of the video-centric applications (e.g., VoD (Video On Demand), video games, live video streaming, video conferences, surveillance, etc.) opens up the demand for user-centric solutions that adapt the application to the underlying network conditions and device characteristics. In the current environment, mobile users want to be connected to the best value network that best satisfies their preferences for their current application(s) requirements. On the other hand, the network operators want to maximize their revenue by efficiently using their networks to satisfy and retain the most customers possible. Challenges for the operators include network optimization especially for video traffic, if it is to represent two-thirds of the overall wireless traffic. Uninterrupted, continuous, and smooth video streaming, minimal delay, jitter, and packet loss, must be provided in order to avoid degradation in the application quality and user experience. The main challenge for the users is to select the best available Radio Access Network (RAN). For example, there is a need for an efficient solution for selecting the best value network for the user, considering the user preferences, application requirements, and network conditions. The network selection decision is a complex one, with the challenge of trading-off different decision criteria, (e.g. service class type, user's preferences, mobile device being used, battery level, network load, time of day, price, etc.). This is further complicated by the combination of static and dynamic information involved, the accuracy of the information available, and the effort in collecting all of this information with a battery, memory, and processor limited device. This selection decision needs to be made once for connection initiation and subsequently as part of all handover decisions.

Another challenge is the rich media service delivery with QoS provisioning over wireless networks. This is due to the constraints of wireless links, and the user mobility. In this context, it is essential to provide QoS mechanisms to cater for the rich media throughput, delay, and jitter constraints, especially within the wireless environment where connections are prone to interference, high data loss rates, and/or disconnection. The aim of these mechanisms is to maintain an acceptable user perceived quality and make efficient use of the wireless network resources.

The energy consumption is another key component that consumers care highly about. Handsets are used as mobile work and entertainment centres, e.g. for communications, listening to music/radio, taking photos, GPS services, playing games, and for rich media playback/streaming. It is known that real-time applications, and in particular those which are based on multimedia, have strict QoS requirements, but they are also the most power-hungry. In this context, one of the impediments of progress is the battery lifetime of the mobile device. With advances in technology, the mobile user has now a wide choice of high capability mobile devices, from laptop computers and netbooks to PDAs, tablets and smart phones. However the batteries have not evolved as much as processors and memory, and their capability is very much limited. This deficiency in battery power and the need for reduced energy consumption provides motivation for the development of more energy efficient solutions while enabling always best connectivity and always best experience to mobile users.

This continuing growth of video content creates challenges for the network service providers in ensuring seamless multimedia experience at high end-user perceived quality levels, given the device characteristics and network resources. Technologies improvements alone would not be able to keep up with this explosion of mobile broadband traffic. This makes network convergence and the coexistence of different radio access network technologies to be seen as a key solution that could contribute to the economical, social and technological benefits of the future 5G networks.

ORGANIZATION OF THE BOOK

The book will cover important aspects of the emerging technologies in this multi-technology multi-application multi-terminal multi-user environment with the main focus on the convergence problems and quality of service provisioning. It presents various convergence approaches and adaptive techniques identifying the main research issues and challenges and presents a survey of the proposed solutions in the literature.

The manuscript is structured in two main sections. The first section on "Convergent Evolution" describes the current wireless multi-access environment which leads towards the next generation of wireless networks. It includes important chapters written by researchers from prestigious laboratories from Australia, Ireland, India, Italy, United Kingdom, Portugal, and Slovakia which present the current state of the art in the heterogeneous wireless communication environment including economical aspects, mobility challenges, resource allocation and management, ensuring always best experience, smart cities and vehicular networks, energy management, interference mitigation, wireless and fiber optics convergence, etc. The Convergence Evolution section consists of six chapters. A brief description of each of the chapters in this section is given below:

Chapter 1 identifies the economical aspects of the future 5G networks in terms of user risk attitude and degree of billing plans usage. The popularity and the increase usage of multimedia content by mobile users on daily basis burdens the network operators in terms of network resources, which might lead to network congestion. In order to provide a better control and reduce the network traffic, the network operators have capped their billing plans which might require the mobile user to pay extra money when exceeding the quota. However, this might results in undesirably high bills for the mobile user which could lead to a significant drop in users' satisfaction and increased churn rate. The chapter introduces a model that involves personalising the multimedia content based on the user attitude towards risk. It makes use of a model that assesses the user attitude towards risk by considering the user age, gender and risk attitude self-assessment. The research presented in this chapter adds to the state of the art by taking into account the user context (e.g. whether the user uses the roaming service or not, the current data consumption) as well as the user input. The improved model may be used to provide a more accurate trade-off: quality vs. price resulting in an increase of user satisfaction and a better service quality within the future 5G networks.

Chapter 2 highlights the importance of Vehicular Ad-hoc Networks (VANETs) in the context of 5G-enabled smarter cities and roads. Currently VANET attracts significant academic, industrial and governmental planning, research, and development efforts. A very promising avenue in VANET is to bring together multiple wireless technologies in the architectural design in order to build an intelligent public transportation systems based on real-time information, traffic management systems for congestion avoidance, safety applications (e.g. collision avoidance) and green applications (e.g. intelligent routing aiming to reduce fuel consumption, gas emissions or energy consumption) as well as enable autonomous vehicles. One of the solutions that can be employed in designing such a 5G-VANET architecture that successfully uses different technologies is clustering. Clustering will have an important role in the desired vehicular connectivity in the future cities and roads as it addresses some of 5G-VANETs' major challenges such as scalability and stability. This chapter presents a comprehensive survey of clustering schemes in the 5G-VANET research area, covering aspects that have never been addressed before in a structured manner. Moreover, the survey provides a general classification of the clustering algorithms, presents some of the most advanced and latest algorithms in 5G-VANETs, and in addition reviews the performance assessment of clustering algorithms.

Chapter 3 discusses the implementation details (i.e. architecture and network components), issues associated with heterogeneous broadband wireless access networks (BWANs) (i.e. handovers, network selection and base station placement), and the various resource allocation schemes (i.e. shared resource allocation in split handover and inter-RAT self-organizing networks) that can improve the performance of the system by maximizing the capacity and accommodating more users. The chapter looks at the existing solutions that have been proposed in the literature in order to overcome several issues faced by the future heterogeneous networks in terms of network selection, inter/intra-RAT handovers, BS placement, and resource allocation. These issues can be very complex mainly because of the heterogeneity of radio access networks.

Chapter 4 analyzes the possible challenges that might exist when supporting multicast services over Long Term Evolution (LTE), with particular attention to resource management, considered as the key aspect for an effective provisioning of Multimedia Broadcast Multicast Service (MBMS) services over cellular networks. Recently, the mobile operators observed a growing demand of multicast services over their radio cellular networks. Since multicast applications are expected to be massively exchanged over the forthcoming Fifth Generation (5G) systems, the Third Generation Partnership Project (3GPP) defined the MBMS standard that offers support for multicast services over LTE, the 4G wireless technology able to provide high quality services in mobile environments. Nevertheless, several issues related to the management of MBMS services and the challenges faced within the future 5G networks are still open and are discussed in this chapter.

Chapter 5 looks into the use of Game Theory for collaboration in future next generation networks. It has been shown that cooperative strategies have the great potential of improving network performance and spectrum utilization in future networking environments. This represents a new paradigm in terms of network management. However, this requires a novel design and analysis framework targeting a highly flexible networking solution with a distributed architecture. Game Theory is a comprehensive mathematical tool suitable for modeling highly complex interactions among distributed and intelligent decision makers. In this way, the more convenient management policies for the diverse players, e.g. content providers, cloud providers, edge providers, brokers, network providers or users, should be found to optimize the performance of the overall network infrastructure. This chapter discusses relevant theoretical models for enabling collaboration among players in different ways, namely pricing or reputation. Additionally the chapter highlights the open problems, such as the lack of proper models for dynamic and incomplete information scenarios and discusses a business model for future networks.

Chapter 6 explores the use of the hybrid passive optical networks (HPON) configurator in designing a converged Fi-Wi PON network based on the HPON network integrating optical fiber and wireless technologies. Looking at the emerging applications and the needs of ever increasing bandwidth demands, it is anticipated NG-PON will require higher bandwidth to satisfy all the end-user demands and for the network operators to develop valuable access networks. Therefore, this chapter identifies the challenges of integrating optical and wireless technologies into common broadband access networks. As the HPON present a necessary phase of the future transition towards optical fiber, the specific requirements of the HPON networks are also presented. The various architectures of the HPON with their characteristics and distinctions are discusses and the HPON network configurator is introduced with the main aim on helping users, professional workers, network operators and system analysts to design, configure, analyze and compare various variations of possible hybrid passive optical networks and the converged fiber-wireless access networks.

The second section entitled "QoS Provisioning Solutions" presents various solutions proposed by world known researchers in different areas of wireless communications. It includes important chapters written by researchers from prestigious laboratories from India, United Kingdom, Sweden, Pakistan, China, Japan, Serbia, and Morocco presenting results in the area of quality of service aware green communication, handover and network selection within heterogeneous environments, access traffic steering, switching and splitting in 3GPP, offload and in-network caching, etc. The QoS Provisioning Solutions section consists of six chapters with a brief description given below:

Chapter 7 investigates Quality of Service-aware green communication strategies for optimal utilization of resources within the future 5G networks. Efficient utilization of resources can reduce the power consumption cost which will increase the economy-characteristics of networks. The resource consumption can be reduced under an intelligent technology-neutral policies which optimizes the deployment of the network architecture along with their transmit power paving the way for fifth generation (5G) in green wireless communications. On another front, the ultra-dense deployment of the small cells can increase the frequency reuse factor as well as help in reducing the energy consumption. This chapter designs the energy efficient networks while satisfying the underlying QoS by joint optimization of available resources depending on the interoperability challenges in terrestrial, underwater acoustic, and free space optical (FSO) communications.

Chapter 8 discusses the ongoing work around hybrid access and network convergence, with particular emphasis on recent works on access traffic steering, switching and splitting (ATSSS) in 3GPP. Three main aspects are analyzed: policy enforcement, integration with 5G QoS framework, interaction with underlying multi-path transport protocol. Preliminary testbed results showing the benefits of ATSSS in the management of multiple accesses analyzing some primary performance indicators such as achievable data rates, link utilization for aggregated traffic and session setup latency are presented. Moreover, two examples of realization of ATSSS policies are considered, to avoid inefficiency in link utilization and to allow the fulfillment of data rate requirements.

Chapter 9 presents a comprehensive survey of various handoff strategies, handoff types in macro-femto cell networks and existing solutions for frequent and unnecessary handoffs. Femto cells are integrated in macro cellular network to increase the capacity, coverage and to fulfill the increasing demands of the users. Time required for handoff process between the cells became more sensitive and complex with the introduction of femto cells in the network, introducing higher latency. Consequently, the handoff process will becomes slower and network operation will become inefficient. Some standards, procedures and protocols should be defined for macro-femto cell network rather than using existing protocols.

Chapter 10 discusses the most recent solutions for accelerating mobile data delivery, focusing on the different issues for achieving this goal. These issues include the efficiency of mobile data offloading and in-network caching. By characterizing different offloading options from the aspect offloading efficiency, dynamically offloading cellular traffic onto cost-effective options is able to achieve great performance gains. On the other hand, with concerns on tightly coupled dependency on original data sources, mobile users could easily suffer from the scarcity of content object resources. Researchers further account for the possibility of decoupling content availability dependency from original content sources. A key enabling factor could be the recently emerged technology, i.e., in-network caching, whereby the emerging Information-Centric Networking (ICN) performs its critical functionalities. Coupling ICN with mobile cellular network could be an attractive solution, in order to resolve the tension between the tremendous growth of mobile content traffic and backhaul bottleneck.

Chapter 11 presents a comprehensive survey of network slicing from an end to end perspective, detailing its origination and current standardization efforts, principal concepts, enabling technologies, as well as applicable solutions. In particular, it also provides specific slicing solutions for each part of the 5G systems, encompassing orchestration and management in the radio access network and the core network domains. The vision of 5G mobile systems can be seen as a highly flexible and programmable end to end networking infrastructure that provides increased performance in terms of capacity, latency, reliability, and energy efficiency, while meeting plethora of diverse requirements from multiple services. Network slicing is emerging as a prospective paradigm to meet these requirements with reduced operating cost and improved time and space functionality. A network slice is referred as the way for providing better resource isolation and increased statistical multiplexing. With dynamic slicing, 5G will operate on flexible zones of the network, permitting varying, adaptable levels of bandwidth and reliability.

Chapter 12 presents an improved modeling for network selection based on graph theory and cost function in heterogeneous wireless systems. The next generation of mobile wireless communications represents a heterogeneous environment which integrates a variety of access technologies. The major challenge in this heterogeneous environment is to decide which access point to use when multiple networks are available. This chapter proposes a new mechanism based on graph theory and cost function in order to determine the best path for the end user in terms of quality of service when the vertical handover process is needed. The impact of some existing weighting methods in order to determine the suitable method which can be coupled with the cost function is also investigated. The experiments evaluation by using Mininet emulator demonstrate that the proposed approach can achieve a significant improvement concerning four QoS metrics: throughput, packet lost, packet delay, and packer jitter for two service types like FTP and video streaming.

The prospective audience of this book are mainly the undergraduate students, postgraduate students, and researchers who are interested in learning more about the latest developments in the area of mobile and wireless communications. It also targets industry professionals who are working or are interested in this area, providing them with a reference of the latest efforts which bring the research further by addressing some of the shortcomings of the existing solutions.

The *Paving the Way for 5G Through the Convergence of Wireless Systems* book is published by the IGI Global (formerly Idea Group Inc.), publisher of the "Information Science Reference" (formerly Idea Group Reference), "Medical Information Science Reference," "Business Science Reference," and "Engineering Science Reference" imprints. For additional information regarding the publisher, please visit www.igi-global.com.

The editors wish you a pleasant reading.

Ramona Trestian
Middlesex University, UK

Gabriel-Miro Muntean
Dublin City University, Ireland

REFERENCES

Cisco. (2017). *Cisco Visual Networking Index: Global Mobile Data Traffic Forecast Update.* Cisco.

Acknowledgment

The *Paving the Way for 5G Through the Convergence of Wireless Systems* book would not exist without the effort of many people whose names may not appear on the cover of the book. However, their hard work, cooperation, friendship and understanding were very important to the preparation and production of the book.

The editors would like to sincerely thank the entire team at IGI Global for their support and help in the publication of this book. In particular, we wish to thank Ms. Jordan Tepper the Assistant Development Editor at IGI Global. Her patience was amazing and appreciated, while her guidance was always timely, useful, and valued.

As the success of this book depends on the high-quality chapters submitted, the editors wish to express their deep appreciation to the authors of each chapter for their significant contribution. Without their insights and expertise this book would not be possible.

As in general the work associated with the chapter review is underestimated and forgotten, we would like to thank the team of reviewers for their generous commitment of time and effort they have put in the reviewing process and for providing their expertise to ensure a high-quality review process. We also thank the Editorial Advisory Board members for their implication in this project.

Last but not least we would like to thank our families for being the 'center of gravity' and for their continuous support along the way. In particular Gabriel-Miro would like to thank his dear parents Dora and Ivo for being an immense source of inspiration throughout his life. His special gratitude is for his wife Cristina, for her immense love, patience and support. His lovely kids: his smart son Daniel-Sasha and beautiful daughter Alexandra for making him such a proud dad and making his life worth living. Ramona would like to thank her wonderful and loving parents Maria and Vasile, for their unconditional love and care that represents a source of inspiration in everything she is doing. A special gratitude goes to her dear brother Ionut and his lovely wife Andreea. A special thank you goes towards her husband Kumar, for his love, patience and continued support that were essential to this project as well as towards her little bundle of joy Noah Anthony for bringing light into her life.

Ramona Trestian
Middlesex University, UK

Gabriel-Miro Muntean
Dublin City University, Ireland

Section 1
Convergent Evolution

Chapter 1
User-Based Adaptive Multimedia Delivery Over 5G Network

Andreea Molnar
Swinburne University of Technology, Australia

Cristina Hava Muntean
National College of Ireland, Ireland

ABSTRACT

The increase in multimedia traffic is likely to pose challenges for the 5G networks. The data bundles are likely to increase but also open the door to new applications which will increase the data consumption. This chapter discusses how multimedia adaptation could be improved in this context by considering the user and the contextual factors for multimedia adaptation. This could be used to deliver a more accurate trade-off between quality and price. As a result, this could reduce the network consumptions, reduce the cost the user has to pay, and improve user satisfaction with the multimedia services.

INTRODUCTION

The continuous increase in the number of mobile devices and users on one hand and advent of ubiquitous communication technologies, development of innovative networking applications and user demand for high quality rich media content on the other hand are behind the latest massive amounts of data generated and exchanged, whose volume continues to grow at an exponential pace. For instance, the data digital universe is experiencing a two-fold expansion every two years since 2012 so that the annual global IP traffic is expected to reach 2.3 zettabytes per year by 2020 (Cisco, 2017a). By 2020, wired communications will be responsible for 34% of IP traffic (down from 52% in 2015), whereas Wi-Fi and mobile devices will account for 66% of IP traffic. At the same time mobile video traffic is estimated to reach 77% of the total traffic by 2020, up from 55% in 2015 (Cisco, 2017b). These trends show that both the number of high-end mobile devices and user consumption of high-bandwidth services are increasing

DOI: 10.4018/978-1-5225-7570-2.ch001

(e.g. a tablet exchanges rich media traffic as much as 113 phones with basic features). The average number of devices per capita is also increasing, with classic PCs and new smart devices being lately complemented by wearable devices and device-to-device (D2D) communicating machines. Additionally, network technologies have also diversified and technologies including the latest ones such as IEEE 802.11ac for wireless broadband, LTE and LTE-A for cellular and DVB-T2 for broadcast are supporting high speed wireless information exchange.

In this context, no single network technology and no single network provider will be equipped to deal with the explosion of data and diverse services as described by UE views. It is also not realistic a scenario where the potentialities of the new 5G will be materialized based exclusively on completely brand new technologies/standards, that will substitute current network infrastructures. The proponents believe that the 5G services can only be realistically implemented making innovative use of the co-existing broadband, broadcast and cellular network technologies (Andrews et al., 2014; The European Commission, n.d.). Cross-network optimisation will be a key component of the 5G solution. In this context there should be a global effort towards the creation of a convergent service quality-oriented network delivery environment. This environment should be composed of inter-connected networks over which services are offered at best quality level, given service requirements, network technology availability, network delivery conditions, device characteristics and user profiles.

There are currently different views of the implications of the emerging 5G networks, services and applications. There is the mainstream push towards building a landscape with a very high increase in data volume and large number of network-connected devices, offering access to an increasing number of rich media services. At the same time, efforts are put to support quality-orientation for these rich media services and to enable operation cost reduction and energy-awareness.

In parallel, these technical offerings foster exploitation of media and content convergence opportunities. These opportunities appear in form of content aggregation, annotation, personalization, media-sharing for the diverse services, targeting increased user experience (e.g. hyper-personalized, real-time storytelling, free viewpoint, and other advanced contents and formats).

As 5G standard was completed in June 2018, the network operators have started to update their software using equipment that complies with the 5G standard. It is expected that the first 5G capable smartphones to be released on the marked early 2019. In parallel a number of the network operators have announced the launch of 5G service in 2019 and 2020. For example, AT&T plans to be the first carrier to launch a standards-based mobile 5G service in a dozen cities in the United States by late 2018 (Alleven, 2018). While T-Mobile is aiming for a 2019 launch followed by nationwide 5G coverage in 2020. United States, South Korea, Japan and China will be the leaders in the deployment of 5G technology by 2019 while Europe will start the deployment slowly in 2020. However, according to a 5G subscriber forecast presented in the Ericsson Mobility Report, it is expected 1 billion 5G subscribers by 2023, and more than 20% of the world's population will has access to the 5G technology.

The rapid growth of devices that can connect in the network and the huge amount of data that will be delivered will pose a challenge for the 5G networks, and the connections will be more demanding and intensive than what exists now. Various studies forecast a huge amount of money to be invested by network operators to build a 5G infrastructure. However, there are no words yet about how much the customers will pay for a 5G service. We estimate that initially, 5G subscriptions might cost more than 4G and billing plans that will differentiate between different services will exist. Low cost billing plans will have a cap on mobile data traffic while more expensive plans will provide unlimited data access.

Similar billing models that exist now for the 4G service will be introduced by the network operators for the 5G service in an effort to increase their revenues and impede congestion.

At the moment, mobile billing plans currently have "very strict quotas" (Raj et al., 2013) that can lead to high bills as the user has to pay extra for exceeding the quota or having their connection speed reduced (Jordan, 2017; Molnar & Muntean, 2013a, 2013b). This aspect creates problems for the users that are not willing/do not afford to pay the required price but still want to access rich media content (e.g. multimedia clips) over the Internet.

A price reduction for the user, as well as reduced bandwidth consumption, can be obtained as a trade-off in multimedia quality (Molnar & Muntean, 2013b; Oeldorf-Hirsch et al., 2012). However, not all users are affected in the same way by the content delivery cost. Some people prefer to pay as much as it is required for an excellent multimedia quality while others being sensitive to a price increase and they prefer to trade-off for quality (Molnar, 2016a; Molnar & Muntean, 2010). It has been shown that the difference in the attitude of people when choosing a certain video quality over the price to be paid might be explained by people's attitude towards risk (Molnar & Muntean, 2015). A method that classifies people either in willing to pay for high video quality (risk seekers) or not willing to pay (risk averse) based on their age, gender and their risk attitude self-assessment was proposed in (Molnar & Muntean, 2015). This classification may be used to personalise the content. This chapter investigates the applicability of the model presented (Molnar & Muntean, 2014) in the context of 5G by considering the user risk attitude and degree of billing plans usage.

The rest of this chapter is organised as follows. The next section provides an overview of the existing research focusing on the following aspects: personalisation, research on user risk attitude and types of user models. The personalization is divided into user profile based personalization, device based personalization and context based personalization. The related work section is followed by the context aware based extension proposed for the user risk attitude model presented in Molnar & Muntean (2014). An exemplification of the extended model is also provided. The chapter continues with a discussion of the implications and possible applications of the proposed model and ends with conclusions.

RELATED WORK

As this research is transdisciplinary, this section will start by presenting the existing work on personalisation, continue with research done on user risk attitude and concludes with the types of user models.

Adaptation and Personalisation

Personalisation Based on the User Profile

User profile based personalisation builds a profile that stores information about the user such as the user's knowledge, background, past interaction, preferred media type, etc. This type of personalization has been used a lot in customizing and adapting the user content provided to the consumers. It has been also used in personalizing the content aimed at students. Two projects in this area are JPELAS2 (Yin et al., 2010) and EducaMovil (Molnar & Frias-Martinez, 2015).

JPELAS2 stores information about the user such as name, gender, number of years in school, friends and relatives. This information is introduced by the user when s/he first enters into the system. In addition, the system automatically detects the user comprehension. The aim of the JPELAS2 is to facilitate learning of Japanese and its politeness rules.

EducaMovil (Molnar et al. 2015) considers the required time for a user to answer a question, the number of correctly answered questions and the number of errors done in order to determine the educational content to be shown to the user.

When personalising multimedia content, several adaptation algorithms have considered adapting the multimedia content based on the user region of interest (Ciubotaru, et al., 2014). This approach assumes that there are certain regions in a video clip that are of a higher interest to the user than the other regions where a more aggressive degradation of quality can be applied without affecting as much the user perceived user video quality as degrading equally the whole video clip.

Personalisation Based on the Device

The diversity of devices a user may have when accessing a multimedia clip, and the device characteristics have always been considered a challenge when it comes to personalisation. Devices can have different screen size, screen colour depth (bits/pixel), screen mode (it refers to whether the screen has portrait or landscape mode and if it supports switching between them), capabilities (i.e. whether the device is capable of displaying multimedia, audio or images), memory and type of supported network connectivity (WiMax, 3G, 802.11, etc.). For example, Karadeniz (2011) designed a mobile application that takes into account the screen size of the user's mobile phone when delivering multimedia; Huang et al. (2012) proposed a system that adapts the content regardless of the used device by taking into account the device resolution and supported colours; Louhab et al. (2018) propose a mechanism to adapt the multimedia courses based on the capabilities and characteristics of the device, considering the software, connectivity and hardware (in this case screen size, resolution and battery). Adaptation with the aim to deliver the multimedia content in an energy efficient manner has been proposed also by other researchers such as Ghergulescu et al. (2014) and Zou et al. (2018).

Personalisation Based on the Context

One of the main difference between the content accessed from the mobile devices as opposed to the one accessed through desktop is that this content can be access anytime, anywhere. This implies that the user could be in different contexts when accessing the context, and these contexts could influence what information they need. Examples of context aware personalised systems are JAPELAS2 - Japanese Polite-Expressions Learning Assisting System (Yin et al., 2010) and Dandelion (Varela et al., 2015). JAPELAS2 aims at suggesting which Japanese polite expression may be used in a given context. The users have a PDA equipped with IR (infrared), RFID (Radio Frequency Identification) tag, GPS and wireless LAN. Information about the user location is obtained through the RFID tag attached to the doors when the user is located inside the building, or by using GPS when the user is located outside. IR is used for simplifying the communication targets; instead of entering the interlocutor name, the user just points to him. The users are required to introduce information about them when they use the system for the first time. Based on this information, the system suggests the appropriate "level of politeness". Four "levels of politeness" were considered: casual, basic, formal, more formal. The level of politeness the user has to use changes according to hyponymy (e.g. age, position, etc.), social distance (e.g. family, colleagues, etc.), and formality of the situation (e.g. meeting). Dandelion is an user interface framework which makes use of ambient intelligent in order to adapt to different user scenarios in order to provide the most natural user interaction (Varela et al., 2015). The research presented in this chapter considers how much the user has consumed from the available bundle, in order to provide a better user risk attitude model.

Risk Attitude

One of the dispositions that drive economic decisions is the human attitude towards risk. Risk is considered "the pivotal element in consumer behaviour" (Taylor, 1974), a central element to economics and finances (Bucciol & Zarri, 2013). It is most often associated with uncertainty (e.g. missing information – making decisions about an unfamiliar brand, however, "full information" contexts do not imply missing uncertainty).

It has been found that risk attitude has a great impact on many of the person's decisions such as those involving investment, educational attainment, ownership of a home and occupational choices (Dohmen et al., 2011a). Understanding people's attitude towards risk is linked to predicting the consumer behaviour (Dohmen et al., 2011b).

Purchase decisions are one of the areas which involve perceived risk. Five kinds of risks that affect purchase decisions are presented in Solomon et al. (2010): monetary risk, functional risk, physical risk, social risk and psychological risk. Items which are perceived as having a high price are more susceptible to monetary risk. Functional risk refers to the performance of the product. The physical risk is greater when the accident is bigger. Social risk involves the pressure put by the peers when buying visible goods (e.g. when the choice is visible to other peers, there is a risk of embarrassment if the work decision is made). Psychological risk refers to how satisfied the consumer will be when acquiring a certain product. Risk is perceived differently by different consumers. For example, a self-confident consumer will not be so much affected by peer choices; person's wealth can also influence how monetary risk is perceived. As price is an issue in the capped billing plans provided by the mobile network operators, monetary risk is of a greater importance for the research community. However, by adopting an open user model the user is able to modify its profile and have an impact if, for example, a peer manifest has an influence on his/her choices.

Based on people's attitude towards risk, previous studies have shown that certain categories of people tend to be more risk averse than others. For example, women are less willing to take risks than men in general (Ding et al., 2010; Grossman, 2013; Hügelschäfer & Achtziger, 2014; Jin et al., 2017) and the risk could different based on the contexts (Dohmen & Falk, 2011; Gerhardt et al., 2016; Karhunen & Ledyaeva, 2010; Hügelschäfer & Achtziger, 2014); and age influences the decision to take risks (Dohmen et al., 2011b; Jin et al., 2017).

Based on their attitude towards risk people can be classified into:

- **Risk Averse:** They prefer not to assume risks;
- **Risk Neutral:** Who are neutral to risk;
- **Risk Seekers:** They love risk.

However, these three categories are not always used as they are. The first and last one are the most used ones, risk neutral people being included in any of the previous two categories. Depends on the relevance of the three categories and if the system is able to determine the risk neutral category, not all three categories are used. Experimental studies involving lottery and hypothetical questions are typically able to determine a risk neutral person. Studies that ask people to assess on a scale their attitude towards risk in general or in a certain context, were not able to place on the scale a risk neutral person The three risk type categories were used to determine the user attitude towards risk across different domains such as health, financial matters, and career, in order to predict the economic behaviour or to explain different decisions taken by people (Hammitt & Haninger, 2010; Dohmen et al., 2011b).

A controversy concerning whether the person's risk attitudes is constant or not across different contexts exists in between neo-classical economics and psychology. Neo-classical economists believe that risk preference is the same across all contexts, and psychology challenges this idea. More recent studies show that a stable risk attitude exists but it may vary across the contexts (Dohmen et al., 2011b; Gerhardt et al., 2016; Lauriola et al., 2013), therefore, when taking into account people's attitude towards risk, the context should also be considered in order to obtain better results. The proposed context aware user risk model presented in this chapter aims to model better the user's risk attitude by enhancing an already existing model that does not take into account the contextual factors.

User Model

User models are created to store information which could be used by the system to personalise the content delivered to the user. However, some systems (Kopeinik et al., 2017; Wongchokprasitti et al., 2015) have adopted an 'open user model' that allows the users to see and even to modify the information saved in the model. This allows alleviatingsome of the issues with the cold start in the recommender systems (Wongchokprasitti et al., 2015). An open user model is used in this research because we wanted the users to be able to see and modify their individual profiles regarding their economic behaviour, as it has been shown that the risk behaviour can vary across contexts (Dohmen et al., 2011b; Gerhardt et al., 2016; Guiso, et al., 2018).

USER MODEL FOR MULTIMEDIA ADAPTATION

The aim of this section is to present the extension brought to the risk attitude model presented in (Molnar & Muntean, 2015). The proposed enhancement considers contextual factors (i.e. user data consumption). The proposed open risk attitude user model allows the users to directly adjust their risk attitude and any other user related information stored by the model.

Risk Attitude Model

A user model that assesses user risk attitude by taking into account the user age, gender and his answer to the general risk question has been proposed in (Molnar & Muntean, 2012). This model assumes that a user who is risk averse will be willing to switch to a lower multimedia quality if monetary cost benefits are obtained, whereas a user who is a risk seeker prefers to pay for maintaining the multimedia quality (Molnar & Muntean, 2012). For the purpose of this paper, the user risk value computed as in Equation (1) and presented in more details in (Molnar & Muntean, 2015) is named a general user risk model, as it can be applied regardless of the user context.

$$RV_{General} = w_1 * RV_{GeneralRiskQuestion} + w_2 * RV_{AgeGender} \tag{1}$$

where, w1 + w2 = 1.

$RV_{GeneralRiskQuestion}$ and $RV_{AgeGender}$ take values between 0 and 10 and therefore the computed RVGeneral value will also be in the range of 0 to 10. If the $RV_{General}$ is a value between zero and less than or equal to 5, the user is classified as a risk averse. If the value is greater than 5 is considered as risk seeker.

The value of the first parameter, $RV_{GeneralRiskQuestion}$, is computed based on the answer given by the user to the general risk question. The general risk question asks users to assess their attitude towards risk on a 0 (risk averse) to 10 (fully prepared to take risks) scale. Users who pick a value between 0 and 5 (inclusive) are classified as risk averse, and the ones that select a value higher than 5 are classified as risk seekers (Dohmen et al. 2011b). The general risk quests used was used as described below and the scale is presented in Figure 1.

The second parameter, $RV_{AgeGender}$ considers the age and the gender of the person and it is computed based on an analysis performed on the SOEP data set (SOEP v.26). SOEP is a survey performed by the German Institute. It is a longitudinal research study, the survey taking place each year since 1984 and it is used for economic research. In 2009, the SOEP survey was administered to 20,869 people, 52% were women, and 48% men and they age varied from 17 up to 100 years old. The survey required the participants to provide, among others, information about their gender, the year when they were born, and to answer the general risk question. As users' attitude towards risk differs with age and between the two genders we have analysed the results of the SOEP study. A general formula for the computation of a user risk value based on person's age and gender was proposed and presented in Equation (2).

$$RV_{AgeGender} = \sum_{i=0}^{10} RiskValue_i * ProbabilityRiskValue_i \quad (2)$$

For a given person, $RV_{AgeGender}$ is computed as the sum of probabilities of having a certain risk value *i* for a certain gender and certain age (*ProbabilityRiskValue*$_i$), multiplied by the risk value *i* (*RiskValue*$_i$), where *i* can have a value from 0 up to 10, as the risk values scale from the general risk question.

For example, for a 24 years old female, the percentage of subjects from the SOEP survey that have assigned a particular risk value is presented in Table 1. The first row represents the 0 to 10 risk value range, and the second column the percentage of 24 years old females that were assigned for each risk value based on the SOEP survey analysis. Similar tables were created based on the SOEP data analysis for each age between 17 and 100.

When $RV_{AgeGender}$ value is computed, *RiskValue*$_i$ represents a particular risk value from the first row of the Table 1, and *ProbabilityRisk Value*$_i$ represents the percentage of people being assigned to *RiskValue*$_i$. Therefore, for a 24 years old female, $RV_{AgeGender}$ is computed as presented in Equation (3).

Figure 1. User's Risk Attitude Scale

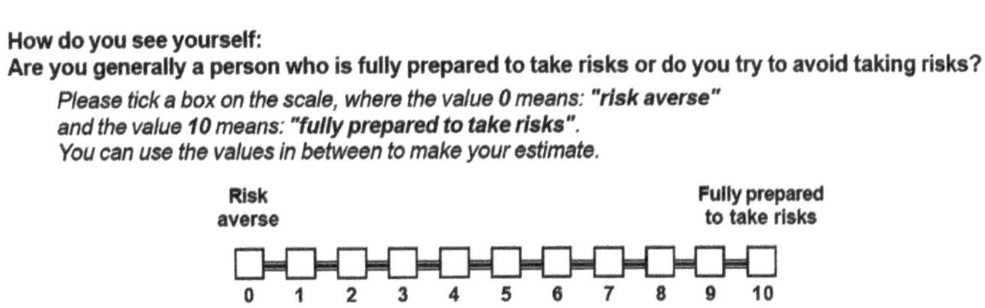

$RV_{24Female}$=0*0.021+1*0.0035+2*0.1049+3*0.1109+4*0.0909+5*0.2308

+6*0.1329+7*0.1678+8*0.0839+9*0.007+10*0.014

$$RV_{24Female} = 4.94 \tag{3}$$

The computed value is less than 5, therefore, all 24 years old female persons have $RV_{AgeGender}$ equal to 4.94 and they are classified as risk averse just based on their age and gender. $RV_{AgeGender}$ is combined with $RV_{GeneralRiskQuestion}$ and an individual risk attitude value is computed for a particular person. The general user risk model described above has been validated and results have been presented in (Molnar & Muntean, 2012). A multimedia delivery mechanism that uses the general user risk model has been already proposed in (Molnar & Muntean, 2013a).

Context Aware Extension for the Risk Attitude Model

It has been shown that the user attitude towards risk can vary across different contexts (Dohmen et al., 2011b; Lauriola et al., 2013). Among the factors that influence people decisions when it comes to paying for multimedia quality, is whether the user still has data available in the bundle and the price to be paid when exceeding the quota (Molnar & Muntean, 2012; Molnar, 2016a). Therefore, this research considers user billing plan usage as a contextual factor in the model.

Since a user could have multiple billing plans even in the home country (i.e. an increasing number of people use multiple SIMs to access the best tariff (Page et al., 2013), the risk attitude model presented in the next subsection computes a RV value for each of the billing plans the user has. The reason for this is that a different billing plan could influence the outcome and the user switching from one to another could lead to different attitudes. The user is however not restricted to two billing plans and can have multiple billing plans.

Open User Model

In this case, the user model allows the user to see and update his profile, depending on the context s/he is in, as we acknowledge that the current model is not able to consider all factors that may affect people's situation. The user may change directly the information found in the user model. When the user updates his profile, by changing the risk attitude value, with a new value named $RV_{UserUpdate}$, the new risk attitude value will be considered for the duration of the session, unless the user specifies otherwise.

Table 1. Percentages of 24 Years Old Females for each Risk Value (RV) on the 1-10 scale, based on SOEP Data Set

RV	0	1	2	3	4	5	6	7	8	9	10
%	2.10	3.50	10.49	11.19	9.09	23.08	13.29	16.78	8.39	0.70	1.40

Real Time Update of the User Risk Attitude Value

At a certain point, multiple factors can trigger a change of the user risk attitude such as user changes his risk value or billing plan data. As consequence, the computed risk value RV is updated. Two cases can be distinguished. If one of the factors is a user action that triggers the computation of the $RV_{UserUpdate}$ this value overwrites all the changes the other factors can have on the user model. If the user does not update his risk attitude value $RV_{UserUpdate}$, an average of the computed RV values corresponding to each activated factor is computed as in Equation (4). This approach has the advantage that multiple factors can be introduced in the model. The research presented in this paper considers another factor, except for $RV_{UserUpdate}$: RV_{Data} as in Equation (7). In the equation below RVFactor1… $RV_{FactorM}$ are risk attitude values computed for each activated factor.

$$RV_{RealTime} = Avg(RV_{Factor1}, RV_{Factor2}, \dots RV_{FactorM})$$

$$RV_{RealTime} = RV_{Factor1} \text{ when only one factor was activated} \tag{4}$$

The risk attitude value RV_i stored in the user profile is updated with a new value RV_{i+1} as in Equation (5) every time when $RV_{RealTime}$ is computed due to a re-computation of $RV_{UserUpdate}$, and/or RV_{Data}. The weight values, w'1 and w'2, are in this case 0.1 is 0.9. These values were taken in order to ensure a slow modification of the stored user risk attitude value over time, such that a certain change which might be triggered by an occasional change of context for the user, will not have a big effect on the user risk attitude value as a whole. A slow modification of the values is recommended when the initial value is a reasonable approximation (Rich, 1979). As previous research has shown (Molnar & Muntean, 2012) this also holds true for this case. The selected weight values were considered based on the TCP/IP model (RFC: 793, 1981). TC/IP model is used with the aim of ensuring that there are no sudden spikes in the traffic, and in this case is used to ensure that there are no sudden changes in the user model. The computed value will be used for the next session.

$$RV_1 = RV_{General}$$

$$RV_{i+1} = w'_1 * RV_{RealTime} + w'_2 * \frac{RV_i + RV_{General}}{2} \tag{5}$$

where

$$w'_1 + w'_2 = 1$$

RV_{Data} Computation

Bundle based billing plans (capped billing plans) are the most common billing plans currently used by mobile network operators (Molnar & Muntean, 2013a). A bundle billing plan requires the user to pay for a certain data quota in advance. When the data quota included in the bundle is exceeded, the user can be penalised in two ways: either to pay extra for the exceed quantity (most often at a higher price than the charge for the quantity of data contained in the bundle) or to throttle the user's bandwidth (Molnar, 2011). When the user has a bundle based billing plan (a capped billing plan) the personal consumption is monitored through the user risk attitude model. The aim is to maintain data consumption in the limits of the bundle quota. Therefore, when the user requests multimedia content, an estimation of the remaining data quantity from the bundle is computed as in Equation (6), where *DataQuota* represents the original data quota included in the bundle when the plan was bought; *TimeLimitInterval* the time interval for which the bundle is available, *CurrInterval* is the time period elapsed since the plan was activated. For example, if the bundle has an availability of 30 days, *CurrInterval* will be the number of the day when the multimedia was requested, counting as day one the day when the plan was activated. *DataCons* represents the quantity of data used so far from the bundle.

$$EstimatedRemainingData = \frac{DataQuota}{TimeLimitInterval} * CurrInterval - DataCons \tag{6}$$

If the *EstimatedRemainingData* has a negative value, the estimated user risk attitude value RV_{Data} is computed as in Equation (7), where RV_i is the last user risk attitude value saved in the user profile.

$$RV_{Data} = RV_i - RV_i * 10\% \tag{7}$$

Once the RV_{Data} was computed due to a multimedia clip request, the $RV_{RealTime}$ value will also be updated considering the new value of the RV_{Data}. As result a new user risk attitude value RV_{i+1} is computed and saved in the profile. This value becomes the new risk value, until the user requests data again, or particular contextual factor is activated. If the *EstimatedRemainingData* has a positive value, the risk value remains unchanged.

Exemplification of the User Risk Attitude Value Computation

A person living in Ireland has acquired a bundle plan with a 500MB data quota, valid for a 30 days period. The person has the current risk attitude value RV_i equal to 5.2 and the $RV_{General}$ value equal to 5. As the current risk attitude value is higher than 5 the person is classified as risk seeker. By the third day s/he has already consumed 100MB. The estimated remaining data in the bundle is computed as in Equation (9) and it is equal to -50, which means that the person has exceeded its quota for the first 3 days by 50MB. If a person exceeds the daily quota multiply by the number of days the bundle is already activated, the model assumes that by the end of the bundle period the user will be exceeding the bundle quota.

$$EstimatedRemainingData = \frac{500}{30} * 3 - 100 \tag{9}$$

$$EstimatedRemainingData = -50$$

As the estimated remaining data is negative the RV_{Data} value is computed and presented in Equation (10).

$$RV_{Data} = 5.2 - 5.2 * 10\% \quad (10)$$

$$RV_{Data} = 4.68$$

As the user has requested some data over the Internet the $RV_{RealTime}$ value is computed. Since there is only one factor considered in our algorithm, $RV_{RealTime}$ will be equal with $RV_{Data.}$

$$RV_{RealTime} = 4.68$$

As result of the $RV_{RealTime}$ computation, the user risk attitude value stored in the user profile is updated with a new value computed as in Equation (13).

$$RV_{i+1} = 0.1 * 4.68 + 0.9 * \frac{5.2 + 5}{2} \quad (13)$$

$$RV_{i+1} = 5.058$$

The newly computed risk attitude value is lower than the previous value but not lower than 5 so the model will maintain the same user risk attitude.

DISCUSSION

This chapter presented a user model which could be taken into account when delivering multimedia quality. The user model takes into account the user risk attitude and considering contextual factors such as the user changes the risk attitude value and bundle data quota.

Improvements to the Model

Further improvements can be made to the model by increasing its granularity: instead of stereotyping the people in two categories: risk averse and risk seekers other categories may be used to classify those persons that have a moderate opinion with regards to how much they are willing to trade-off. Another way of improving the model is by considering user preference for different multimedia content which could affect the user preference for video quality (Molnar, 2016b). Furthermore, the model might be improved by considering categories of multimedia clips for which the risk averse users are willing to pay for quality due to other extraneous benefits. For example, in the case of tele-monitoring glaucoma patients, it is a requirement of the doctor to have a very good (high definition) video image of the patient's eye in order to be able to consult the patient (Molnar & Weerakkody, 2016). In this case the patient is reducing the health related cost through other means such as avoiding the cost of travelling to the hospital, not queuing in the hospital anymore and/or the necessity to ask somebody to accompany her/him to the hospital, as the patients suffering from glaucoma are typically older – therefore giving more independence to the patient. A cost reduction for the hospital is also obtained due to the usage of tele-monitoring. A better patient management and hence a more efficient use of the ophthalmologists time may be obtained (Molnar & Weerakkody, 2016). Furthermore, the model is flexible enough to incorporate or remove factors. As the introduction of 5G might bring different billing plans or different ways of accessing the data by the users, new factors could be considered in the model to provide a better user profile. This might imply adding new factors or removing other factors which might not be relevant (e.g. limited billing plans).

Future Directions

As the network capacity is affected due to increase in traffic, especially in the mobile traffic, the proposed model could be used to incentivize a user to accept a lower multimedia quality that requires lower bandwidth thus allowing a critical application, such as ambulance service to make use of a higher bandwidth (Weerakkody et al., 2013). In the context of an emergency service, the ability to obtain fast and good quality data (e.g. videos, images) about the patient could improve patient health or even save the patient life. Although the introduction of 5G is likely to alleviate some of the bandwidth problems, the increased use in video content (Cisco, 2016), other types of multimedia such as 360 videos and virtual reality are likely to continue to put strains on the mobile networks. It is also likely that 5G will be first available in the metropolitan areas and the low bandwidth problems could still affect rural or remote areas until 5G will be introduced.

The context aware user risk attitude model may also be used by the Internet service providers as part of an adaptive mechanism with the aim to reduce the network congestion or by the service providers to reduce the traffic to the server/proxy. It could provide a better quality of experience model by improving the existing adaptation mechanisms which take into account network parameters and/or battery lifetime (Ciubotaru et al., 2014; Ciubotaru et al., 2017; Ghergulescu et al., 2014; Zou et al., 2018).

The model might be also used as an input for a handover mechanism. There are some studies that consider the cost of data delivery when performing a handover operation (Bakmaz, 2013) and risk attitude as a willingness to trade-off for handover related waiting time (Ormond et al., 2006). The risk averse users could prefer a network which has more bandwidth constraints as long as the price is reduced, whereas risk seekers could prefer a network that ensures high quality content delivery even if the delivery cost is high.

CONCLUSION

This research proposed an open user model that considers the user risk attitude and contextual factors, in our case the quantity of data consumed from the bundle plan and user's own update of the risk attitude value. Several examples of how the model works and a discussion on the different scenarios in which the model could be used as it is or enhanced had been also presented. The model could help reduce network congestion, costs of delivery for the user while trading off multimedia quality. However, further work is needed to validate the proposed model.

ACKNOWLEDGMENT

This work was partially supported by Irish Research Council for Science, Engineering and Technology.

REFERENCES

Alleven, M. (2018). *AT&T Plans to Launch Mobile 5G in a Dozen Cities by Late 2018*. Retrieved from https://www.fiercewireless.com/wireless/at-t-plans-to-launch-mobile-5g-dozen-cities-by-late-2018

Andrews, J. G., Buzzi, S., Choi, W., Hanly, S. V., Lozano, A., Soong, A. C. K., & Zhang, J. C. (2014). What Will 5G Be? *IEEE Journal on Selected Areas in Communications*, *23*(6), 1065–1082. doi:10.1109/JSAC.2014.2328098

Bakmaz, B. M. (2013). Network selection equilibrium in heterogeneous wireless environment. *Electronics and Electrical Engineering*, *19*(4), 91–96.

Bucciol, A., & Zarri, L. (2013). *Financial Risk Aversion and Personal Life History*. Retrieved June 19, 2018, from: http://leonardo3.dse.univr.it/home/workingpapers/risk_events_BZ.pdf

Cisco. (2016). *Cisco Virtual Networking Index: Forecast and Methodology, 2015-2020*. Retrieved from http://www.webvideomarketing.org/pdf/Cisco_Video_and_Visual_Networking_Index_Report_8.10.16.pdf

Cisco. (2017b). *Cisco Visual Networking Index: Forecast and Methodology, 2016-2021*. Retrieved from https://www.cisco.com/c/en/us/solutions/collateral/service-provider/visual-networking-index-vni/complete-white-paper-c11-481360.html

Cisco. (2017b). *The Zettabyte Era: Trends and Analysis*. Retrieved from https://www.cisco.com/c/en/us/solutions/collateral/service-provider/visual-networking-index-vni/vni-hyperconnectivity-wp.html

Ciubotaru, B., Ghinea, G., & Muntean, G. M. (2014). Subjective assessment of region of interest-aware adaptive multimedia streaming quality. *IEEE Transactions on Broadcasting*, *60*(1), 50–60. doi:10.1109/TBC.2013.2290238

Ciubotaru, B., Muntean, C. H., & Muntean, G. M. (2017). Mobile multi-source high quality multimedia delivery scheme. *IEEE Transactions on Broadcasting*, *63*(2), 391–403. doi:10.1109/TBC.2016.2635079

Ding, X., Hartog, J., & Sun, Y. (2010). *Can we measure individual risk attitudes in a survey?* IZA Discussion Paper No 4807. Retrieved June 19, 2018, from http://ftp.iza.org/dp4807.pdf

Dohmen, T., & Falk, A. (2011). Performance pay and multidimensional sorting - productivity, preferences and gender. *The American Economic Review*, *101*(2), 556–590. doi:10.1257/aer.101.2.556

Dohmen, T., Falk, A., Huffman, D., & Sunde, U. (2011a). The intergenerational transmission of risk and trust attitudes. *The Review of Economic Studies*, *79*(2), 645–677. doi:10.1093/restud/rdr027

Dohmen, T., Falk, A., Huffman, D., Sunde, U., Schupp, J., & Wagner, G. (2011b). Individual risk attitudes: Measurement, determinants and behavioural consequences. *Journal of the European Economic Association*, *9*(3), 522–550. doi:10.1111/j.1542-4774.2011.01015.x

Gerhardt, H., Biele, G. P., Heekeren, H. R., & Uhlig, H. (2016). *Cognitive load increases risk aversion* (No. 2016-011). SFB 649 Discussion Paper.

Ghergulescu, I., Moldovan, A. N., & Muntean, C. H. (2014, June). Energy-aware Adaptive Multimedia for Game-based E-learning. In *Broadband Multimedia Systems and Broadcasting (BMSB), 2014 IEEE International Symposium on* (pp. 1-6). IEEE. 10.1109/BMSB.2014.6873573

Grossman, P. J. (2013). Holding fast: The persistence and dominance of gender stereotypes. *Economic Inquiry*, *51*(1), 747–763. doi:10.1111/j.1465-7295.2012.00479.x

Guiso, L., Sapienza, P., & Zingales, L. (2018). Time varying risk aversion. *The National Bureau of Economic Research. Journal of Financial Economics*, *128*(3), 403–421. doi:10.1016/j.jfineco.2018.02.007

Hammitt, J., & Haninger, K. (2010). Valuing fatal risks to children and adults: Effects of disease, latency, and risk aversion. *Journal of Risk and Uncertainty*, *40*(1), 57–83. doi:10.100711166-009-9086-9

Huang, H. C., Wang, T. Y., & Hsieh, F. M. (2012). Constructing an adaptive mobile learning system for the support of personalized learning and device adaptation. *Procedia: Social and Behavioral Sciences*, *64*, 332–341. doi:10.1016/j.sbspro.2012.11.040

Hügelschäfer, S., & Achtziger, A. (2014). On confident men and rational women: It's all on your mind (set). *Journal of Economic Psychology*, *41*, 31–44. doi:10.1016/j.joep.2013.04.001

Jin, J., He, R., Gong, H., Xu, X., & He, C. (2017). Farmers' Risk Preferences in Rural China: Measurements and Determinants. *International Journal of Environmental Research and Public Health*, *14*(7), 713. doi:10.3390/ijerph14070713 PMID:28665332

Jordan, S. (2017). A critical survey of the literature on broadband data caps. *Telecommunications Policy*, *41*(9), 813–837. doi:10.1016/j.telpol.2017.09.001

Karadeniz, S. (2011). Effects of gender and test anxiety on student achievement in mobile based assessment. *Procedia: Social and Behavioral Sciences*, *15*, 3173–3178. doi:10.1016/j.sbspro.2011.04.267

Karhunen, P., & Ledyaeva, S. (2010). Determinants of entrepreneurial interest and risk tolerance among Russian university students: Empirical study. *Journal of Enterprising Culture*, *18*(3), 229–263. doi:10.1142/S0218495810000574

Kopeinik, S., Lex, E., Seitlinger, P., Albert, D., & Ley, T. (2017, March). Supporting collaborative learning with tag recommendations: a real-world study in an inquiry-based classroom project. In *Proceedings of the Seventh International Learning Analytics & Knowledge Conference* (pp. 409-418). ACM. 10.1145/3027385.3027421

Lauriola, M., Panno, A., Levin, I. P., & Lejuez, C. W. (2013). Individual differences in risky decision making: A meta-analysis of sensation seeking and impulsivity with the balloon analogue risk task. *Journal of Behavioral Decision Making*. DOI: doi:10.1002/bdm.1784

Louhab, F. E., Bahnasse, A., & Talea, M. (2018). Considering mobile device constraints and context-awareness in adaptive mobile learning for flipped classroom. *Education and Information Technologies*, 1–26.

Molnar, A. (2011). *Cost Efficient Educational Multimedia Delivery* (Doctoral thesis). National College of Ireland. Retrieved August 26, 2013, from http://trap.ncirl.ie/762/

Molnar, A. (2016a). Video quality vs. mobile data billing plans. *Telematics and Informatics*, *33*(1), 227–231. doi:10.1016/j.tele.2015.07.009

Molnar, A. (2016b). Factors affecting user preference for mobile video quality. In *Multimedia and Expo (ICME), 2016 IEEE International Conference on* (pp. 1-6). IEEE.

Molnar, A., & Muntean, C. H. (2015). What can people risk attitude tell us about people preference for multimedia quality? *Information Technology & People*, *28*(2), 383–404. doi:10.1108/ITP-11-2013-0199

Molnar, A., & Muntean, C. H. (2010). Educational content delivery: An experimental study assessing student preference for multimedia content when monetary cost is involved. *10th International Conference on Intelligent Systems Design and Applications*, 871-876.

Molnar, A., & Muntean, C. H. (2012). Mobile learning: An economic approach. Intelligent and Adaptive Learning Systems: Technology Enhanced Support for Learners and Teachers. IGI Global.

Molnar, A., & Muntean, C. H. (2013a). Cost oriented adaptive multimedia delivery. *IEEE Transactions on Broadcasting*, *59*(3), 484–499. doi:10.1109/TBC.2013.2244786

Molnar, A., & Muntean, C. H. (2013b). COMEDY: Viewer trade-off between multimedia quality and monetary benefits. *IEEE International Symposium on Broadband Multimedia Systems and Broadcasting*, 1-7.

Molnar, A., & Muntean, C. H. (2014). Economical Aspects of Heterogeneous Networks: When Money, Quality, and Context Matter. In Convergence of Broadband, Broadcast, and Cellular Network Technologies (pp. 1-15). IGI Global.

Molnar, A., Virseda, J., & Frias-Martinez, V. (2015). Insights from EducaMovil: Involving teachers in creating educational content for mobile learning games. *Journal of Interactive Learning Research*, *26*(2), 209–221.

Molnar, A., & Weerakkody, V. (2016). Clinicians' view of tele-glaucoma. *Health Policy and Technology*, *5*(2), 123–130. doi:10.1016/j.hlpt.2016.02.001

Oeldorf-Hirsch, A., Donner, J., & Cutrell, E. (2012). How bad is good enough? Exploring mobile video quality trade-offs for bandwidth-constrained consumers. *ACM 7th Nordic Conference on Human-Computer Interaction: Making Sense Through Design*, 49-58.

Ormond, O., Murphy, J., & Muntean, G. M. (2006). Utility-based intelligent network selection in beyond 3G systems. *IEEE International Conference on Communications- ICC'06*, 4, 1831-1836.

Page, M., Molina, M., & Jones, G. (2013). The mobile economy. *GSMA*. Retrieved June 18, 2018, from http://www.atkearney.com/documents/10192/760890/The_Mobile_Economy_2013.pdf/

Raj, H., Saroiu, S., Wolman, A., & Padhye, J. (2013). Splitting the bill for mobile data with SIMlets. *ACM Workshop on Mobile Computing Systems and Applications*, 1-6.

RFC 793. (1981). *Transmission Control Protocol*. Retrieved June 19, 2018, from ietf: http://www.ietf.org/rfc/rfc793.txt

Rich, E. (1979). User modelling via stereotypes. *Cognitive Science*, *3*(4), 329–354. doi:10.120715516709cog0304_3

Solomon, M. R., Bamossy, G., Askegaard, S., & Hogg, M. K. (2010). *Consumer Behaviour: A European Perspective*. Financial Times/Prentice Hall.

Taylor, J. W. (1974). The role of risk in consumer behaviour. *Journal of Marketing*, *38*(2), 54–60. doi:10.2307/1250198

The European Commission. (n.d.). *The 5G Infrastructure Public Private Partnership: The Next Generation of Communication Networks and Services*. Retrieved from https://5g-ppp.eu/wp-content/uploads/2015/02/5G-Vision-Brochure-v1.pdf

Varela, G., Paz-Lopez, A., Becerra, J. A., & Duro, R. J. (2015). Autonomous evaluation of interaction resource adequateness for ambient intelligence scenarios. In *International Conference on Ubiquitous Computing and Ambient Intelligence* (pp. 174-186). Springer.

Weerakkody, V., Molnar, A., Irani, Z., & El-Haddadeh, R. (2013). A research proposition for using high definition video in emergency medical services. *Health Policy and Technology*, *2*(3), 131–138. doi:10.1016/j.hlpt.2013.04.001

Wongchokprasitti, C., Peltonen, J., Ruotsalo, T., Bandyopadhyay, P., Jacucci, G., & Brusilovsky, P. (2015). User model in a box: Cross-system user model transfer for resolving cold start problems. In *International Conference on User Modeling, Adaptation, and Personalization* (pp. 289-301). Springer. 10.1007/978-3-319-20267-9_24

Yin, C., Ogata, H., Tabata, Y., & Yano, Y. (2010). JAPELAS2: Japanese polite expressions learning assisting system in ubiquitous environments. *International Journal of Mobile Learning and Organisation*, *4*(2), 214–234. doi:10.1504/IJMLO.2010.032637

Zou, L., Trestian, R., & Muntean, G. M. (2018). E 3 DOAS: Balancing QoE and Energy-Saving for Multi-Device Adaptation in Future Mobile Wireless Video Delivery. *IEEE Transactions on Broadcasting*, *64*(1), 26–40. doi:10.1109/TBC.2017.2722221

ADDITIONAL READING

Ding, R., & Muntean, G. M. (2013). Device characteristics-based differentiated energy-efficient adaptive solution for video delivery over heterogeneous wireless networks. In *IEEE Wireless Communications and Networking Conference*. Shanghai, China, 4442-4447.

Li, F., Shuang, F., Liu, Z., & Qian, X. (2018). A Cost-Constrained Video Quality Satisfaction Study on Mobile Devices. *IEEE Transactions on Multimedia*, *20*(5), 1154–1168. doi:10.1109/TMM.2017.2764329

Moldovan, A. N., & Muntean, C. H. (2017, June). QoE-aware video resolution thresholds computation for adaptive multimedia. In *Broadband Multimedia Systems and Broadcasting (BMSB), 2017 IEEE International Symposium on*(pp. 1-6). IEEE.

Molnar, A., & Muntean, C. H. (2015). Assessing Learning Achievements when Reducing Mobile Video Quality. *J. UCS*, *21*(7), 959–975.

Mondal, A., Trestian, I., Qin, Z., & Kuzmanovic, A. (2012). P2P as a CDN: A new service model for file sharing. *Computer Networks*, *56*(14), 3233–3246. doi:10.1016/j.comnet.2012.06.010

Sharma, S. K., Sarrab, M., & Al-Shihi, H. (2017). Development and validation of Mobile Learning Acceptance Measure. *Interactive Learning Environments*, *25*(7), 847–858. doi:10.1080/10494820.2016.1224250

Tal, I., Ibarrola, E., & Muntean, G. M. (2016, November). Quality and standardization in technology-enhanced learning. In ITU Kaleidoscope: ICTs for a Sustainable World (ITU WT), 2016 (pp. 1-8). IEEE. doi:10.1109/ITU-WT.2016.7805715

Traverso, S., Huguenin, K., Trestian, I., Erramilli, V., Laoutaris, N., & Papagiannaki, K. (2015). Social-aware replication in geo-diverse online systems. *IEEE Transactions on Parallel and Distributed Systems*, *26*(2), 584–593. doi:10.1109/TPDS.2014.2312197

Trestian, I., Ranjan, S., Kuzmanovic, A., & Nucci, A. (2012). Taming the mobile data deluge with drop zones. [TON]. *IEEE/ACM Transactions on Networking*, *20*(4), 1010–1023. doi:10.1109/TNET.2011.2172952

Trestian, R., Comsa, I. S., & Tuysuz, M. F. (2018). Seamless multimedia delivery within a heterogeneous wireless networks environment: Are we there yet? *IEEE Communications Surveys and Tutorials*, *20*(2), 945–977. doi:10.1109/COMST.2018.2789722

KEY TERMS AND DEFINITIONS

Bundle-Based Billing (Capped Billing Plan): A bundle-based billing is characterized by the fact that the user has to pay for a specific amount of data (bundle) in advance. That may be used over a given period of time. If the amount of data used is exceeded during the given period, the user may have the following options: (1) pays a different price for the exceeding quantity; (2) buys a new bundle at the same or different price; (3) the bandwidth is throttled.

Multimedia Adaptation: Adapting the multimedia properties (e.g., bit rate, frame rate) such that they satisfy certain constraints.

Open User Model: A user model that shows the user the information stored about him. Some open user models would also allow the user to modify the information stored.

Personalized Systems: Systems that adapt their results based on user characteristics, context, etc.

Risk-Averse Person: A person who prefers stability to risk.

Risk-Neutral Person: A person whose decision is not influenced by the uncertainty.

Risk-Seeking Person: A person who prefers to assume risks.

Chapter 2
Clustering and 5G-Enabled Smart Cities:
A Survey of Clustering Schemes in VANETs

Irina Tal
National College of Ireland, Ireland

Gabriel-Miro Muntean
Dublin City University, Ireland

ABSTRACT

This chapter highlights the importance of vehicular ad-hoc networks (VANETs) in the context of the 5G-enabled smarter cities and roads, a topic that attracts significant interest. In order for VANETs and its associated applications to become a reality, a very promising avenue is to bring together multiple wireless technologies in the architectural design. 5G is envisioned to have a heterogeneous network architecture. Clustering is employed in designing optimal VANET architectures that successfully use different technologies. Therefore, clustering has the potential to play an important role in the 5G-VANET-enabled solutions. This chapter presents a survey of clustering approaches in the VANET research area. The survey provides a general classification of the clustering algorithms, presents some of the most advanced and latest algorithms in VANETs, and it is among the fewest works in the literature that reviews the performance assessment of clustering algorithms.

INTRODUCTION

Nowadays, smart cities represent a very important research direction for academia, industry and governments that are eager to embrace various technologies, which will make cities "smarter". The main purpose of smart cities is to improve all the facilities provided in a city (e.g. buildings, infrastructure, transportation, energy distribution, etc.) in order to improve the citizens' quality of life, while creating a sustainable environment. Related to the transport, the declared aim of smart cities is to promote sustainable forms of transportation, to build intelligent public transportation systems based on real-time information, traffic management systems for congestion avoidance, safety applications (e.g. collision

DOI: 10.4018/978-1-5225-7570-2.ch002

avoidance) and green applications (e.g. intelligent routing aiming to reduce fuel consumption, gas emissions or energy consumption). Moreover, self-driving cars play an important role in the context of smart cities, due to their potential of improving citizen's life by improving their comfortability. Various statistics demonstrate that people are spending a lot of their time in the vehicles, in traffic (INRIX, n.d.). Self-driving cars would allow people to use this extra time in a more efficient way.

In this context, Vehicular Ad-hoc Networks (VANETs) or simply vehicular networks represent a hot research topic both for academia and industry due to their high potential to create not only smarter cities, but also smarter roads. This potential relies in the *on the wheels connectivity* provided by VANETs that can also meet the *always connected* need of drivers and passengers as they are spending much of their daily time in their vehicles. Moreover, VANET has a crucial role in the context of self-driving vehicles (Ydenberg, Heir & Gill, 2018). VANETs are based on "smart" vehicles that are able to communicate to each other and to the infrastructure via vehicle-to-vehicle (V2V) and vehicle-to-infrastructure (V2I) communications, known under the generic term of V2X communications, but also via other wireless communications technologies (e.g. cellular, WLAN).

V2X communications are considered the dedicated enabling technology of VANETs. They have exclusively dedicated spectrum that is of high importance particularly for safety applications. As this technology has a low penetration rate and also some limitations (i.e. short-lived and intermittent connectivity), in some architectures, other access technologies are employed as well, in order to support the diversity of VANET applications (i.e. safety, traffic management and infotainment applications). Due to the importance of VANET, cellular technologies that have a high market penetration have considered to accommodate this type of communications starting from 4G. More improvements and developments are planned in 5G. LTE for instance was mainly considered for the communication between vehicle and infrastructure because according to the studies performed, it seemed to be unable to support the huge amount of messages exchanged by vehicles during rush hours. The general consensus is that VANETs and their diversity of applications cannot rely on a single type of access technologies. Thus there is a need of bringing together multiple technologies, V2X communications, cellular technologies and WLAN, in order to enable support for a wide range of VANET applications.

In this context, clustering can play a very important role in the design of VANET architectures: on one hand clustering addresses some of the V2X communications limitations such as sparse deployment of the infrastructure, and intermittent connections and on the other hand it optimizes the communication via cellular access technology. In addition, clustering algorithms in VANET address some of the main VANET challenges: scalability and stability, and have been integrated in a various range of applications. This chapter presents a thorough survey of clustering algorithms in VANETs.

There are some reviews in the literature dedicated to clustering in VANET (Vodopivec, Bester & Kos, 2012; Cooper, Franklin, Ros, Safaei & Abolhasan, 2017), but these do not provide any classification of the performance assessment of these algorithms. This review is an extension of the one presented in (Tal&Muntean, 2014) and similar to this is trying to address a gap that there is in the literature, namely the lack of a well-structured analysis of the performance assessment of clustering algorithms in VANET, while considering new and significant stages in this research field.

The structure of the chapter is as follows. In the first sections, an overview of vehicular networks, their enabling technologies, applications and challenges is presented. The following sections are dedicated to clustering: general concepts of clustering, survey of clustering in VANETs – application, classification, performance assessment and representative algorithms. The chapter ends with future directions and conclusions.

BACKGROUND

Introduction to VANETs

Vehicular Ad-hoc Networks (VANETs) or simply vehicular networks are a specific class of mobile ad-hoc networks (MANETs), where the mobile nodes are represented by vehicles. Although they are a class of MANETs, they have specific characteristics that differentiate them, characteristics which will be discussed in a dedicated section. VANETs are mostly based on the communication between vehicle-to-vehicle (V2V), vehicle-to-infrastructure (V2I) or infrastructure-to-vehicle (I2V), generally referred to as V2X communications. This type of communications is mainly supported by a specific type of wireless access called Wireless Access for Vehicular Environment (WAVE). WAVE contains the standards dedicated to vehicular environment (Uzcategui & Acosta-Marum, 2009): IEEE 802.11p and IEEE P1609.x standards. IEEE 802.11p, developed to provide wireless access in vehicles, is a new amendment of IEEE 802.11 standard body (IEEE 802.11p, 2010). This is a justified decision in the context of the wide adoption and subsequently the low cost of IEEE 802.11 technologies. Both, IEEE 802.11p and IEEE 1609.x standards, are based upon the allocation of Dedicated Short Range Communications (DSRC) spectrum band. This initiative, started in USA in 1999, allocated dedicated spectrum of frequency to be used exclusively by V2X communications. In Europe, spectrum allocation was harder to achieve, as each country has different regulations, but agreement was eventually made on a spectrum similar to the USA. Seven channels of 10MHz in the 5.9GHz range are allocated for use in DSRC/IEEE 802.11p standard. Out of the 7 channels, 6 are service channels (SCH), while the one left is the control channel (CCH). CCH is reserved for system control and safety messages, an SCH channel is dedicated to safety messages as well, whereas the rest of SCHs are mainly used to exchange non-safety and larger data.

While IEEE 802.11p covers the Physical and MAC layers, IEEE P1609.x covers the entire VANET scope of services from application down to the MAC layer.

- IEEE P1609.1 (IEEE P1609.1, 2006) is the WAVE Resource Manager standard, defining the interfaces and services of WAVE applications and the format of data messages.
- IEEE P1609.2 (IEEE P1609.2, 2006) is the WAVE Security Services for Applications and Management Messages standard that defines the WAVE security: anonymity, authenticity and confidentiality and also the exchange of messages.
- IEEE P1609.3 (IEEE P1609.3, 2007) is the WAVE Networking Services that defines routing and transport services. It provides description and management to the protocol stack, network configuration management and also provides the transmission and reception of WAVE short messages.
- IEEE P1609.4 (IEEE P1609.4, 2006) is the WAVE Multi-channel Operations that provides the DSRC frequency band coordination and management.

In addition to V2X communications, other types of technologies are also used in supporting vehicular applications. Depending on how these VANET enabling technologies are employed in the vehicular applications, three types of VANET architectures are defined: pure ad-hoc, pure WLAN/cellular and hybrid (K.C. Lee, U. Lee & Gerla, 2010).

In the ad-hoc architecture, there is V2V communication only, without any infrastructure support. This scenario is feasible since the infrastructure and wireless access points are not everywhere and their deployment is limited by the cost or geography. Information exchanged between vehicles can be of extreme value, especially in difficult conditions or special circumstances (e.g. an icy road section previously detected by another car or an accident blocking the road).

In WLAN/cellular architecture, cellular base stations and WLAN access points facilitate vehicles' connection to the Internet and provide support for vehicular communications-based applications. Initially, in this type of architecture, the vehicles did not have support for directly communication with each other in a distributed manner with few exceptions. In this context, clustering can be successfully employed to limit for instance the cellular network communications. However, starting with 4G important steps have been made towards direct communication between vehicles. From Release 12 (Rel.12) a new feature known as Proximity Services was specified within 3GPP[1]. Proximity Services Direct Discovery and Proximity Services Direct Communication enable Device-to-Device (D2D) communications (Lin, Andrews, Ghosh & Ratasuk, 2014). However, this release of the Proximity Services specification has not considered the requirements of V2X communications as it provides low mobility support. Therefore, D2D communications specified in Rel.12 is not really suitable for V2V communications, especially not in highway scenarios characterized by high speeds. Cellular V2X communications, also called C-V2X was standardized in 3GPP based on LTE Release 14 (3GPP, n.d.), but this supports basic safety scenarios only. Improvements of these specifications aiming to support the diversity of VANET applications are planned in 5G (5G Americas Whitepaper, 2018).

In the hybrid architecture all types of communications are present. Vehicles can talk to each other and exchange information (V2V communications), but also can communicate with fixed infrastructure that is deployed alongside the road also referred to as *road side unit* (RSU) (V2I) or with access points, or wireless towers (WLAN, cellular). This is the most complex architecture and provides support for more complex applications. Especially infotainment applications which require richer content are based on this type of architecture, but also complex traffic management systems. Similar to the previous architecture, clustering can be a viable solution for an optimal communication between vehicle to infrastructure, if this is done via cellular technologies. In the case of V2I communications, clustering can be the response to limited range of communications, sparse deployment of RSUs and intermittent communication.

In VANET architectures the communication capabilities of a vehicle are provided by an in-vehicle component referred to as the *on-board unit* (OBU) that can have multiple network interfaces (V2X, UMTS, LTE, etc.). Note that this component was envisioned to be integrated in the cars by the car manufactures, but in the latter VANET solutions OBU can stand for different devices with wireless capabilities such as the driver's smartphone. OBU also supports intra-vehicle communication needed to collect the data from the vehicle's sensors and devices, data that is then used in the applications enabled by VANET. Most VANET applications assume that the position of the vehicle is known, so a GPS or other positioning system is considered to be integrated in OBU (or co-exist with OBU).

Figure 1. VANET Architectures

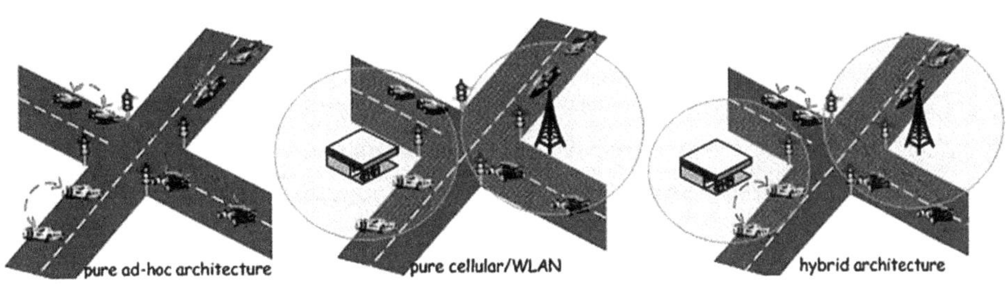

VANET Applications

A large plethora of applications have been envisioned and proposed for VANETs. These can be categorized in three big classes (Karagiannis et al., 2011): active road safety applications, traffic efficiency and management applications and infotainment applications. Moreover, VANET is considered one of the main enabling technology of self-driving cars (Ydenberg, Heir & Gill, 2018).

Active road safety applications aim to provide a safer driving environment by reducing the probability of accidents and preventing the loss of lives. Such applications are traffic signal violation warning, emergency electronic brake light, pre-crash sensing, lane change warning, cooperative forward collision warning, etc. These are mainly pro-active approaches that are trying to avoid accidents. Reactive safety approach based on VANETs can be developed in the context of emergency systems. "Green" routes for emergency vehicles can lead to saving many human lives. In their survey, Martinez, Toh, Cano, Calafate, & Manzoni (2010) emphasized on both the great potential of V2I/V2V communications in enhancing the emergency services and the need of designing systems based on this type of communications that ensure efficient emergency service delivery. The architecture and principles of a complete solution, a VANETs-based traffic management system ensuring "green" routes for emergency vehicles has been proposed in (Djahel, Salehie, Tal, & Jamshidi, 2013).

Traffic efficiency and management applications' goal is to improve the overall efficiency of transportation by managing the navigation of the vehicles via cooperative co-ordination (e.g. cooperative adaptive cruise control (Chang, Tsai & Liang, 2017)). Also, they aim to improve not only the overall efficiency, but the efficiency per vehicle via speed management applications (e.g. avoiding stopping to the intersections (Rakha & Kamalanathsharma, 2011)). This type of applications is situated somewhere at the border between safety and infotainment applications.

Infotainment applications are applications that are not directly related to traffic safety or efficiency, but they are designed for the needs and comfort of the users. These applications can be split into two big classes: entertainment applications and driver assistance applications.

Entertainment applications include solutions for different service delivery such as for instance live video streaming or multimedia delivery over VANETs (Lobato, Rosario, Gerla & Villas, 2017).

Driver assistance applications comprise countless VANETs-based solutions. This type of applications provide driver with useful information in driving process, but not only (e.g. applications that provide valuable information for driver, such as price of fuel or closest charging station, etc. are also included). Example of such applications are routing applications (Doolan & Muntean, 2017), free parking discovery applications (Lu, Lin, Zhu & Shen, 2009), applications that give driving/riding advices based on certain criteria: e.g. how to drive/ride in certain conditions in order to reduce gas emissions, fuel (Rakha el al., 2011) or energy consumption in the case of electric cars (Tielert, Rieger, Hartenstein, Luz & Hausberger, 2012) or electric bicycles (Tal, Zhu & Muntean, 2013; Tal, Ciubotaru & Muntean, 2016), etc.

VANET Characteristics and Challenges

VANETs have specific characteristics that differentiate them from any other type of ad-hoc networks. Some of these characteristics are very attractive for the researchers, while the others are creating new technical challenges that need to be addressed. The following features are among the attractive ones:

- **Theoretical Unlimited Power:** Is considered due to the fact that any vehicle-node is capable of generating power while moving. In the case of classic MANET mobile nodes, power is a very serious issue. However, this VANET characteristic is not applicable to the case of electric vehicles (EVs), where energy preservation is vital for increasing the travel range.
- **High Computational and Storage Capabilities:** Unlike the handheld devices in classic MANETs, vehicles can afford significant computational, storage and communication capabilities. This capability is partially made possible by the previously mentioned characteristic.
- **Predictable Mobility:** Is possible in VANETs due to the fact that vehicle movement is constrained by the roads, traffic regulations and driver behavior. So, given parameters such as the current position, current speed, route, average speed and/or learning about driver behavior, it is possible to predict the next position of the vehicle. On the contrary, the node mobility in classic MANETs is very hard to predict.

The challenging set of VANET features includes:

- **High Mobility:** Vehicle-nodes have very high speed compared to the nodes from MANETs. In highway scenarios speeds of up to 300km/h may occur, while in city scenarios speeds of up to 70km/h.
- **Rapidly Changing Topology:** The aforementioned high node mobility in VANETs leads to a frequent link disconnection between the vehicle-nodes and consequently to a rapidly changing network topology.
- **Diversity of Conditions:** Mainly refers to the diversity of the network density that can be very sparse or on the contrary, very dense. In a city scenario, especially during rush hours, the network is extremely dense, while in a highway scenario the network can be very sparse.
- **Frequent Disconnections:** In the network; mainly caused by the two previously mentioned characteristics. Road dead-ends is another factor that can produce frequent disconnections in VANETs.
- **Potentially Large Scale:** VANETs are networks with a potential high number of nodes. There is no limitation in terms of number of nodes, as it is in the case of other networks, so vehicle-nodes can potentially expand over the entire road network.
- **Diversity of Applications:** As presented in the previous section, a large plethora of applications have been envisioned for VANETs in the areas of traffic safety, traffic management and efficiency, and infotainment ranging from multimedia applications to driver assistance services. The requirements of these applications are as diverse as their range is. Consequently, much VANET- dedicated technology needs to be designed so these networks can cope with all this diversity of applications.

CLUSTERING AND VANETs

Introduction to Clustering

Clustering is a division technique that creates groups of similar objects (Wanner, 2009) mainly with the purpose of dealing with scalability. The similarity between objects is built upon one or more clustering metrics that are extremely varied and highly dependent on the context clustering is applied in. Clustering is widely used in data analysis, data mining, statistics, text mining, information retrieval, etc. Clustering has been widely adopted in MANETs, as it provides support for good system performance, good management and stability of the networks in the presence of mobility and large number of terminals (Yu & Chong, 2005). Thus, clustering helps solve some of the main issues in MANETs: scalability and stability (Wanner, 2009).

In MANETs, clustering involves dividing the nodes into virtual groups based on some rules that establish if a node is suitable to be within a cluster or not. These rules are defined based on clustering metrics that in MANETs can be node type, battery energy level, mobility pattern, etc.

In general, a clustering scheme considers that a node can be in one the following situations (Yu & Chong, 2005), based on the node membership and task associated to the node. If node situations are associated with states, one could consider the following as possible node states:

- *Unclustered*, also known as non-clustered or independent, when it does not pertain to any cluster
- *Cluster member* or clustered when the node is within a cluster
- *Cluster head* (CH) when the node has extra-responsibilities in a cluster. Usually, CH is the main controller of the cluster, the main coordinator of the communication within the cluster (i.e. intra-cluster communication) and has a main role in the functionality that is supposed to be provided by the cluster.
- *Gateway node* is the node that ensures the communication between the clusters, also called inter-cluster communications.

A general classification of MANET clustering schemes is based upon the following criterion: *CH-based clustering*, if there is a CH in the clusters created or *non-CH-based clustering*, if there is no CH in the cluster created. Note that in CH-based clustering, the performance of clustering is highly dependent on CH election as this node has the main responsibilities in its cluster. Therefore in this type of clustering algorithms the focus is mainly on CH selection algorithms. Another general classification of clustering is based upon the number of hops between node pairs in the cluster: *1-hop clustering* or *multi-hop clustering*.

Successfully applied in MANETs to address stability and scalability, clustering was adopted in VANETs, where these issues are even more augmented. At the beginning, MANET clustering algorithms were adopted and directly applied to VANETs without any modifications, but as this research direction evolved, new clustering algorithms dedicated to VANETs were designed to address their specific characteristics.

Figure 2. Illustration of node states in MANET clustering

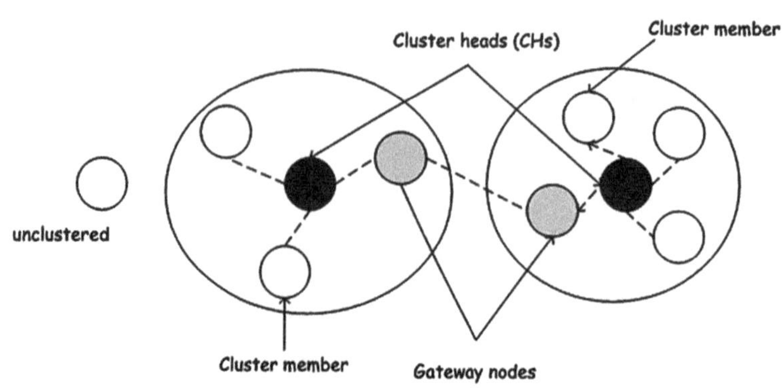

Clustering in VANETs

The clustering concepts presented in the context of MANETs are valid in VANETs context as well, especially given the fact that clustering in VANETs has evolved from MANETs. There are only some additional aspects that need to be mentioned and that derive from the adaptation of clustering to VANET-specific conditions.

Clustering metrics were adapted not only to address VANET challenges imposed by their specific characteristics such as high mobility, rapidly changing topology and diversity of conditions, but also to take advantage of some of these characteristics, such as predictability of their movement. Therefore, clustering in VANETs is based upon more metrics than in MANETs that need to describe the complexity of VANET environment. Among the most common metrics in VANET clustering are direction, vehicle's relative speed in comparison to other neighbouring vehicles, vehicle's relative position, but also traffic flow, the lane in urban scenarios (e.g. right lane, left lane, and ahead lane), predicted future speed and position, density of vehicles (sparse or dense), etc.

In the context of node states, as already described, additional node states have to be added in VANETs in order to address its more dynamic environment. These intermediate states include the *candidate node* and *CH backup* or *CH candidate states*. The candidate state was introduced by some approaches in order to obtain a better stability of the cluster. A node is not immediately given the cluster member state; it goes into the candidate state until it proves that it has certain stability in the cluster. The CH backup/CH candidate (quasi-CH in other approaches) state was introduced to make faster and smoother the process of changing the CH. Clustering in VANETs can be represented as a state machine, where the machine is the vehicle-node that can be in one of the following states: *unclustered*, *cluster member*, *CH* (in the case of CH-based clustering) and, optionally, in an intermediate state *candidate* and *CH backup/candidate* as previously defined (Figure 3).

Once adopted in VANETs, clustering gained popularity mostly due to its efficiency in addressing network stability issues. Clustering algorithms were implemented in the design of a large variety of VANET solutions: MAC protocols, routing protocols, data aggregation, security solutions, inter-vehicle

Figure 3. State machine representing CH-based clustering in VANETs

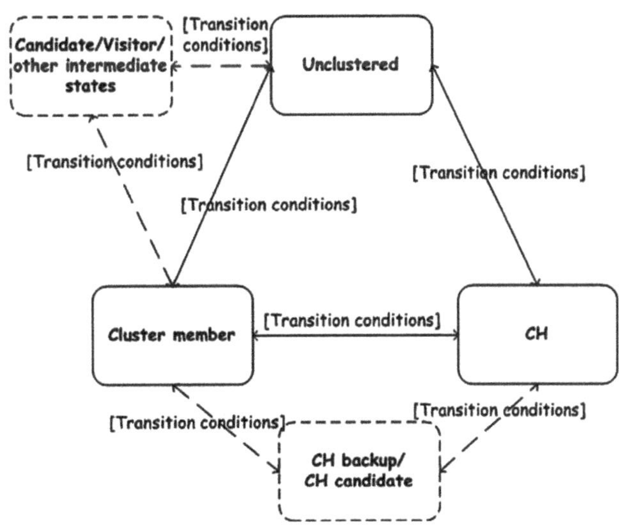

communication, and data and infotainment dissemination solutions and various architectures such as cluster-based heterogeneous networks architectures or vehicular cloud-based architectures. In addition, various generic clustering algorithms were defined for VANETs. A classification of the clustering algorithms based on the application criterion can be seen in Table 1.

Independent of the type of VANET solution the clustering algorithm is designed for, one of the main purpose of clustering is to achieve network stability. Therefore, the clustering metrics are focusing mainly on this aspect and they relate to VANET's dynamic environment. Thus independently of the context in which clustering is applied (i.e. MAC protocols, routing protocols, etc), clustering metrics focus on the same issues and they are similar to each other. They are only dependent on the ingeniously modeling of the VANET environment and they are different from solution to solution as researchers are experimenting in trying to find the best clustering metrics to express the dynamicity of the VANETs. Similarly, in clustering performance assessment, usually first the network stability achieved is measured and then, the overall assessment of the clustering solution is performed (the overall solution where clustering is integrated; e.g. MAC protocol, data aggregation, etc). All these considerations allows for a uniform

Table 1. Clustering Algorithms Classification: Application Criterion

Class	References
Generic Clustering Algorithms	Fan et al., 2005 Cherif et al., 2009 Kuklinski &Wolny, 2009 Rawashdesh &Mahmed, 2009 Shea et al., 2009 Almalag & Weigle, 2010 Dror et al., 2011 Maslekar et al., 2011 Dror et al, 2013 Harikrishnan & He, 2013 Tal & Muntean, 2013a Ucar et al, 2013 Khan, Abolhasan & Ni, 2018
Cluster-based routing algorithms	Goonewardene et al., 2009 Teshima et al., 2011 Wahab et al., 2013
Cluster-based data aggregation	Shoaib et al., 2012
Cluster-based MAC algorithms	Su & Zhang, 2007 Hafeez et al., 2011
Cluster-based architectures	Sivaraj et al., 2011; Taleb & Benslimane, 2010; Benslimane et al., 2011, Tung et al., 2013 El Mouna, Tabbane, Labiod & Tabbane, 2015 Arkian, Atani, Diyanat & Pourkhalili, 2015 Duan, Liu & Wang, 2017 Amad et al., 2018
Cluster-based data & infotainment dissemination	Huang et al., 2011 Tal & Muntean, 2012
Cluster-based security solutions	Gazdar et al., 2010 Gazdar et al., 2011 Sharma & Kaul, 2018

analysis of clustering algorithms in VANETs, independent of the type of solution/application in which they are integrated.

Although there is a considerable number of clustering solutions in VANETs, this research direction is still not mature. A closer analysis of the existent solutions in the literature reveals some major issues that relate to the performance assessment of clustering solutions in VANETs. So far, no analysis on this topic was provided in the literature and this is reflected by the fact that the existing clustering solutions use intuitively-defined performance assessment metrics or re-defined metrics similar with already existing ones mostly because researchers were not aware that such metrics have already been proposed in the literature. This resulted in metrics having various name versions. In particular, this is the case for the metrics used to measure the stability of the clusters, which contributes to network stability. These metrics are a direct measure of the performance of clustering algorithms in the context of VANETs, where the performance of clustering algorithms is reflected in how well clustering algorithms perform in achieving good network stability. It can be therefore concluded that the aforementioned major issues directly relate to the metrics used to evaluate the performance of clustering algorithms in VANETs.

In the absence of a study on performance assessment metrics of VANETs clustering solutions and in the absence of the standardized metrics, we performed a survey of the performance evaluation of clustering solutions in VANETs and of clustering algorithms designed for these solutions. This survey resulted in the identification and comprehensive definition, including in mathematical terms, of generic metrics that can be used in the evaluation of clustering algorithms in VANETs. Next sections describe in details the results of the study that can be considered an invaluable guide for the performance assessment of VANETs cluster-based solutions in general and VANETs clustering algorithms in particular and could support the standardization effort of these general metrics. This standardization is highly needed in order to avoid metrics being "re-invented" or intuitive assessment of the clustering solutions in VANETs to be performed. Moreover, evaluating the performance of clustering algorithms via these general metrics can greatly facilitate the comparison between clustering algorithms, independent from their type (i.e. generic algorithms or solution-specific).

Performance Assessment of Clustering in VANETs

The study conducted aimed to exhaustively analyze clustering solutions in VANETs. First, the focus was on the performance assessment of clustering solutions in general. Then the focus was moved on the evaluation of clustering algorithms designed for these solutions. As a result of the analysis performed, three major classes of performance assessment metrics for clustering solutions were identified and they are illustrated in Figure 4: network-specific metrics, application-specific metrics and topology-based metrics.

Network-specific metrics are well-known metrics applied in network communications, evaluating the performance of the clustered network mainly in terms of data transfer: throughput, loss, delay, data delivery ratio, overhead, etc.

Application-specific metrics depend on the type of the cluster-based solution employed. As emphasized above, clustering algorithms were implemented in the design of a large variety of VANET solutions: data aggregation solutions, MAC and routing protocols, security, etc. Therefore, this class includes a large variety of metrics as well. For instance, a data aggregation cluster-based solution is evaluated by measuring the size of data that needs to be disseminated, as the goal of a data aggregation scheme is to reduce the size of the data that needs to be disseminated. Note that these metrics and network-specific metrics are evaluating the performance of the overall solution based on clustering.

Figure 4. Classification of the performance assessment metrics used in VANETs clustering solutions

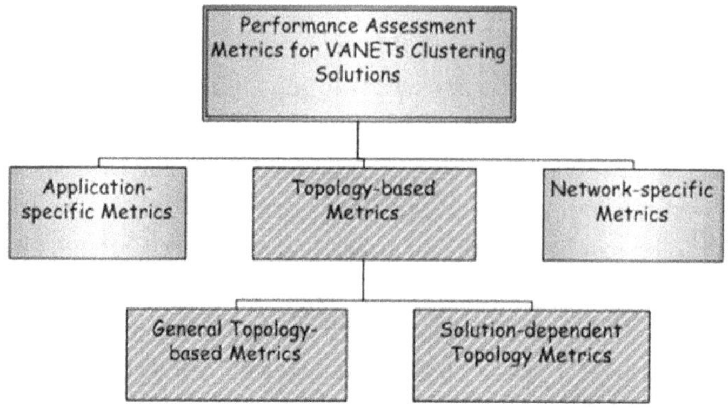

Topology-based metrics (hashed in Figure 4) evaluate the stability and robustness of the resulted clusters. Cluster stability translates into network stability, thus topology-based metrics are measuring the network stability. Network stability is emphasized as an important issue in VANETs due to their rapidly changing topology. Therefore, topology-related metrics are of great importance, fact acknowledged by researchers: the majority of proposed clustering solutions are using topology-based metrics in the performance assessment.

Independent of the type of VANET solution the clustering algorithm is designed for, the general aim of a clustering algorithm is to achieve network stability. In this context, the performance of clustering algorithms is not seen from a computational point of view (e.g. complexity of the algorithms). The focus is on how well clustering algorithms perform in achieving good network stability. Based on these considerations it can be said that in the context of VANETs, topology-based metrics are a measure of clustering algorithms performance.

TOPOLOGY-BASED METRICS FOR CLUSTERING ALGORITHMS ASSESSMENT IN VANETS

In the previous section, topology-based metrics were identified as the metrics suitable for the evaluation of clustering algorithms in VANETs. Therefore, this section contains an in-depth analysis of the topology-based metrics. The comprehensive definition of the performance metrics subscribing to the general topology-based class from the solution-dependent topology-based class is one of the purposes of this work. The first class contains the metrics we called as being general metrics for the evaluation of clustering algorithms in VANETs.

Next the topology-based metrics identified during the analysis of VANET clustering algorithms are presented. Several aspects regarding the form of presentation need to be mentioned before getting to the presentation of the topology-based metrics. These aspects relate to several issues revealed during the study performed on the clustering solutions in VANETs.

An important issue is the inconsistency in naming the metrics. As already mentioned some of the metrics have been re-defined and consequently bare different names. In some cases of re-defining, general metrics are constrained by particular conditions/characteristics of the algorithm they are used to assess: general metrics are defined either using particular conditions/assumptions of the algorithm or using particular parameters of the current algorithm. This is not necessary wrong as it is perfectly applicable for that particular algorithm, but can lead to misconceptions and misusing in case of other clustering algorithms. In addition, there are a considerable number of metrics that are not provided with a mathematical definition at all, being described in words only. So far, a single work in the literature (Su & Zhang, 2007) has provided mathematical definitions for some general metrics used in the evaluation of clustering.

Due to the aforementioned aspects, when presenting the metrics in the next sections, all naming versions are provided. The most representative name was chosen based on either the popularity or the degree of match with the metric description. In addition, general mathematical definitions for each of the general metrics were provided. The mathematical formulas were proposed based on the textual definitions of the metrics and the in-depth analysis of the results. The general mathematical definitions provided to some of the general metrics in by Su & Zhang (2007) were taken in the same form as presented in their work. Together with these metrics were taken the notations used in their definitions (i.e. first 6 notations from Table 2).

General Topology-Based Metrics or General Metrics for the Evaluation of Clustering Algorithms in VANETs

- **Average CH Lifetime:** $\overline{CHL}$ (average CH duration, CH time) is one of the most popular topology-based metrics. It is applicable to the CH-based clustering algorithms only. It was used for the evaluation of VANET clustering algorithms in a considerable number of works (see Table 3). The popularity of this metric is explainable: the importance of CHs lifetime is crucial as, usually, the CH is the main controller and content forwarder in the CH-based clustered networks. Smaller CH lifetime affects the overall performance of the clustering algorithm. Higher CH lifetime implies more stable cluster topologies are found, leading to a decrease in number of re-clusterings, and consequently avoiding the waste of system resources and excessive use of computation time.

Table 2. Notations used in topology-based metrics definition

Notation	Explanation
$\lvert V(t)\rvert$	Total number of vehicles at time t
$\lvert C(t)\rvert$	Numbers of clusters at time t
$\lvert CH_i(t)\rvert$	Number of cluster members in cluster i at time t
$\overrightarrow{SP_i}(t)$	Velocity vector of vehicle i at time t
$\lvert\overrightarrow{SP_i}(t)\rvert$	Speed of vehicle i at time t
S	Simulation time
$CHL_i(t)$	Time period between time t, when vehicle i becomes a CH, and the moment of time vehicle i changes to another state.
$CL_i(t)$	Time period between time t, when cluster i is formed and the moment of time cluster i is dismissed.
$CMT_j^i(t)$	Time period between time t, when vehicle j becomes a cluster-member of cluster i, and the moment of time vehicle j leaves cluster i.

Table 3. General Metrics for the Evaluation of Clustering Algorithms in VANETs: Summary

Metric	Mathematical Definition	Popularity	Restriction
Average CH Lifetime ($\overline{CHL}$)	$\overline{CHL} = \sum_{t=0}^{S} \frac{\sum_{i=1}^{\lvert V(t)\rvert} CHL_i(t)}{\sum_{i=1}^{\lvert V(t)\rvert} BC_i(t)}$, where $BC_i(t) = \begin{cases} 1, \text{if vehicle } i \text{ changes to cluster head at time } t \\ 0, \text{otherwise} \end{cases}$	Blum et al., 2003 Gunteret al., 2007 Su & Zhang, 2007 Rawashdesh & Mahmed, 2009 Shea et al., 2009 Huang et al., 2011 Lai et al., 2011 Tal & Muntean, 2012 Harikrishnan & He, 2013 Tal & Muntean, 2013a Ucar et al., 2013 Arkian, Atani, Diyanat & Pourkhalili, 2015	CH-based algorithms only
Average number of clusters ($\overline{NoC}$)	$\overline{NoC} = \frac{1}{S}\sum_{t=0}^{S} \lvert C(t) \rvert$	Wolny, 2008 Rawashdesh &Mahmed, 2009 Shea et al., 2009 Gazdar et al., 2010 Dror et al., 2011 Maslekar et al., 2011 Shoaib et al., 2012 Dror et. al, 2013 Tal & Muntean, 2013a Harikrishnan & He, 2013	-
Average cluster size ($\overline{CS}$)	$\overline{CS} = \frac{1}{S}\sum_{t=0}^{S}\sum_{i\in C(t)} \lvert CH_i(t) \rvert$	Fan et al., 2005 Su & Zhang, 2007 Kuklinski &Wolny, 2009 Hafeez et al., 2011 Teshima et al., 2011 Harikrishnan & He, 2013	-
Average CH change rate ($\overline{CHCR}$)	$\overline{CHCR} = \frac{1}{S}\sum_{t=0}^{S}\sum_{i=1}^{\lvert v(t)\rvert} CHD_i(t)$, where $CHD_i(t) = \begin{cases} 1, \text{ if vehicle i changes from cluster head to another type of node at time } t \\ 0, \text{ otherwise} \end{cases}$	Fan et al., 2005 Kuklinski &Wolny, 2009 Shea et al., 2009 Almalag & Weigle, 2010 Ucar et al., 2013 Khan, Abolhasan & Ni, 2018	CH-based algorithms only
Average cluster member lifetime ($\overline{CML}$)	$\overline{CML} = \frac{1}{\lvert C(t) \rvert}\sum_{t=0}^{S}\sum_{i\in C(t)} \frac{\sum_{j\in CH_i(t)} CMT_j^i(t) CMD_j^i(t)}{\sum_{k=0}^{S}\sum_{j\in CH_i(t)} CMD_j^i(k)}$ where $CMD_j^i(t) = \begin{cases} 1, & \text{if vehicle } j\text{,cluster member of cluster } i \text{ is dissmissed from the cluster at time } t \\ 0, & \text{otherwise} \end{cases}$	Goonewardene et al., 2009 Shea et al., 2009 Huang et al., 2011 Ucar et al., 2013	-
Cluster changes per node ($\overline{CC}$)	$\overline{CC} = \frac{1}{S}\sum_{t=0}^{S}\sum_{i\in C(t), j\in CH_i(t)} CMD_j^i(t)$, where *CMD* is defined as above	Wolny, 2008 Dror et al., 2011 Dror et al, 2013	-
Average cluster lifetime ($\overline{CL}$)	$\overline{CL} = \frac{\sum_{t=0}^{S}\sum_{i\in C(t)} CL_i(t)}{\sum_{t=0}^{S} \lvert C(t) \rvert}$	Cherif et al., 2009 Rawashdesh &Mahmed, 2009	-

continued on following page

Table 3. Continued

Metric	Mathematical Definition	Popularity	Restriction
Cluster reconfiguration rate ($\overline{CRR}$)	$\overline{CRR} = \frac{1}{S}\sum_{t=0}^{S}\sum_{i=1}^{\lvert C(t)\rvert} CD_i(t)$ Where $CD_i(t) = \begin{cases} 1, \text{ if cluster } i \text{ was dissmissed at moment } t \\ 0, \text{ otherwise} \end{cases}$	Wang et al., 2008 Huang et al., 2011	-
Average relative speed compared to the CH within a cluster ($\overline{RSWC}$)	$\overline{RSWC} = \frac{1}{\lvert C(t)\rvert S}\sum_{t=0}^{S}\sum_{i\in C(t)} \frac{\sum_{j\in CH_i(t)} \lvert \overrightarrow{SP_i}(t) - \overrightarrow{SP_j}(t)\rvert}{\lvert CH_i(t)\rvert}$	Su & Zhang, 2007 Hafeez et al., 2011	CH-based algorithms only
Average relative speed among CHs ($\overline{RSCH}$)	$\overline{RSCH} = \frac{1}{S}\sum_{t=0}^{S} \frac{\sum_{i,j\in C(t)\wedge i\neq j} \lvert \overrightarrow{SP_i}(t) - \overrightarrow{SP_j}(t)\rvert}{\lvert C(t)\rvert^2}$	Su & Zhang, 2007	CH-based algorithms only

The mathematical definition of $(\overline{CHL})$, as given by Su & Zhang (2007), is shown in equation (1).

$$\overline{CHL} = \sum_{t=0}^{S} \frac{\sum_{i=1}^{\lvert V(t)\rvert} CHL_i(t)}{\sum_{i=1}^{\lvert V(t)\rvert} BC_i(t)} \quad (1)$$

In equation (1), $BC_i(t)$ represents a function that defines the transition of a node (vehicle *i*) to the CH state as described by equation (2).

$$BC_i(t) = \begin{cases} 1, \text{if vehicle } i \text{ changes to cluster head at time } t \\ 0, \text{otherwise} \end{cases} \quad (2)$$

$\overline{CHL}$ is one of the metrics that were re-defined in the literature. In the work proposed by Rawashdesh & Mahmud (2009), this metric is given a mathematical definition dependent on the particularity of the clustering algorithm (equation (3)).

$$\overline{CHL} = \frac{1}{L}\sum_{i=1}^{L} C_i^{life} \text{ (3),}$$

where L is the number of clusters created throughout the session and C_i^{life} has the same meaning as $CHL_i(t)$. This is not a general definition, as the clusters are dismissed when the CH is changed. There are clustering algorithms (Hafeez et al., 2011) where clusters have back-up CHs. Often, when the vehicle CH changes its state, its role is taken by its back-up and the cluster is not dismissed. In this situation, equation (3) cannot be used in all CH-based clustering algorithms evaluation.

- **Average Number of Clusters:** ($\overline{NoC}$) is a very popular metric used to assess the performance of a large number of clustering algorithms (see Table 3). $\overline{NoC}$ is a general metric that can be applied to assess the performance of all the categories of clustering algorithms. $\overline{NoC}$ is a measure of network stability. When there are fewer clusters, better network stability is obtained. Equation (4) is proposed as a general mathematical definition of $\overline{NoC}$.

$$\overline{NoC} = \frac{1}{S}\sum_{t=0}^{S} | C(t) | \tag{4}$$

It can be stated that the same metric is used by Maslekar et al. (2011), although instead of number of clusters, the number of CHs is measured. This is due to the fact that in this particular solution, a cluster is dismissed when its CH is dismissed. In consequence, the number of CHs is equal to the number of clusters created.

- **Average Cluster Size:** ($\overline{CS}$) metric (average number of cluster members) measures the average size of the clusters throughout the session. This size of a cluster is considered to be determined by the number of vehicles in the cluster. $\overline{CS}$ is a metric applicable to all clustering algorithms and was highly used in the literature (see Table 3). Its general definition, as given by Su & Zhang (2007), is shown in equation (5).

$$\overline{CS} = \frac{1}{S}\sum_{t=0}^{S}\sum_{i \in C(t)} | CH_i(t) | \tag{5}$$

- **Average CH Change Rate:** ($\overline{CHCR}$) is another popular metric (see Table 3) and it is seen as a measure of cluster stability. In more stable clusters, nodes in general and CHs in particular change their cluster membership or their state less often. It is a metric that can be used for the performance assessment of CH-based algorithms. Equation (6) is proposed as the general mathematical definition of $\overline{CHCR}$.

$$\overline{CHCR} = \frac{1}{S}\sum_{t=0}^{S}\sum_{i=1}^{|v(t)|} CHD_i(t) \tag{6}$$

where $CHD_i(t)$ is the CH dismissed function of vehicle *i*, that was defined in equation (7) to express the transitions from CH to another type of node.

$$CHD_i(t) = \begin{cases} 1, & \text{if vehicle i changes from cluster head to another type of node at time } t \\ 0, & \text{otherwise} \end{cases} \quad (7)$$

- **Average Cluster Member Lifetime:** ($\overline{CML}$) (*average cluster member duration, cluster residence time*) was used as a performance metric by Goonewardene et al. (2009); Shea et al. (2009) and Huang et al. (2011). It is a general topology-based metric measuring the overall stability of clustering. $\overline{CML}$ is a similar metric to the average CH lifetime just that the lifetime is computed for all the nodes in the cluster members not only for the CH. Same as for $\overline{CHL}$, longer average cluster member lifetime indicates a more stable clustering topology. A general mathematical definition of $\overline{CML}$ is provided in equation (8).

$$\overline{CML} = \frac{1}{|C(t)|} \sum_{t=0}^{S} \sum_{i \in C(t)} \frac{\sum_{j \in CH_i(t)} CMT_j^i(t) CMD_j^i(t)}{\sum_{k=0}^{S} \sum_{j \in CH_i(t)} CMD_j^i(k)} \quad (8),$$

Where $CMD_j^i(t)$ is the cluster member dismissed function of vehicle *j* from cluster *i* being defined in equation (9).

$$CMD_j^i(t) = \begin{cases} 1, & \text{if vehicle } j\text{,cluster member of cluster } i \text{ is dissmissed from the cluster at time } t \\ 0, & \text{otherwise} \end{cases} \quad (9)$$

- **Cluster Changes Per Node:** ($\overline{CC}$) *(average number of cluster switches per node)* measures the number of transitions of a vehicle between clusters and is used in order to measure cluster stability. The less number of transitions indicates better cluster stability. In the work proposed by Dror et al. (2011), the metric is called improperly cluster stability and was measured through average number of cluster switches per node.

Based on the descriptions provided by Shea et al. (2009) and Wolny (2008) and on the evaluation of the results, equation (10) is proposed as the general mathematical formula for $\overline{CC}$, applicable to all clustering algorithms.

$$\overline{CC} = \frac{1}{S} \sum_{t=0}^{S} \sum_{i \in C(t), j \in CH_i(t)} CMD_j^i(t) \quad (10)$$

where $CMD_j^i(t)$ was defined in equation (9).

- **Cluster Stability:** Is stated as a metric by Wolny (2008), Dror et al. (2011) and Wahab et al. (2013). In the latter work, it is described and measured in terms of the average number of cluster switches per node during the simulation. As stated before, this metric is actually $\overline{CC}$. Wolny (2008) states that the cluster stability metric depends on the change rate of the cluster and provide a formula that is not generic, but solution-dependent and uses some undefined terms. A solution-dependent description is provided by Wahab et al. (2013) for what they call network stability metric. However, we argue that no metric can be called *cluster stability* until it is not able to comprise all the metrics that were defined so far for its measurement. In addition, cluster stability is a property of the VANETs which is assessed by most of the topology-related metrics defined and not only a metric. In consequence, cluster stability is not proposed and considered as a general metric in this work.

- **Average Cluster Lifetime:** ($\overline{CL}$) (Rawashdesh & Mahmud, 2009) *(average cluster lifecycle (Cherif et al. 2009))* is another general metric that can be used for assessing the performance of any type of VANETs clustering algorithm. It is a measure of cluster stability: larger average cluster lifetime translates into more stable clusters, thus a more stable network. Rawashdesh & Mahmud (2009) consider the average cluster lifetime equal to the average CH lifetime, as the clusters are dismissed whenever their CH changes. The authors define $\overline{CL}$ through equation (2) which was considered not to be an appropriate general formula for CH lifetime. It is obvious that equation (2) cannot be also considered a general formula for the $\overline{CL}$ as the cluster duration is not always dependent on the CH lifetime. This is valid in non-CH schemes and even in CH-based schemes. Equation (11) is proposed as a general mathematical definition for $\overline{CL}$.

$$\overline{CL} = \frac{\sum_{t=0}^{S} \sum_{i \in C(t)} CL_i(t)}{\sum_{t=0}^{S} | C(t) |} \tag{11}$$

- **Cluster Reconfiguration Rate:** ($\overline{CRR}$) (Wang et al. 2008) *(number of re-clusterings (Huang et al., 2011))* is a metric defined to measure cluster stability based on the fact that a good clustering algorithm should be stable and it should not change the cluster configuration too drastically when few nodes are moving and the topology changes rapidly. This metric does not have a mathematical definition in none of the solutions that use it. Moreover, these solutions are both CH-based algorithms and they state that the re-clustering/reconfiguration happens when CH changes. Described like this, the metric becomes identical to $\overline{CHCR}$. However, equation (12) proposed a general mathematical description for the $\overline{CRR}$ applicable to all clustering algorithms.

$$\overline{CRR} = \frac{1}{S} \sum_{t=0}^{S} \sum_{i=1}^{|C(t)|} CD_i(t) \text{ (12)},$$

where $CD_i(t)$ is the cluster dismissed (*CD*) function for cluster *i* as defined in equation (13).

$$CD_i(t) = \begin{cases} 1, \text{ if cluster } i \text{ was dissmissed at moment } t \\ 0, \text{ otherwise} \end{cases} \tag{13}$$

- **Average Relative Speed Compared to the CH Within a Cluster:** ($\overline{RSWC}$) (Su & Zhang, 2007) *(average cluster stability factor (Hafeez et al., 2011))* measures the topology stability of clusters. It is a general topology-based metric for all the CH-based algorithms. In all CH-based algorithms, a smaller average speed of the cluster member compared to that of the CH is translated into an increased stability of the cluster. However, $\overline{RSWC}$ is not a very common metric. It was defined and used as in equation (14) by Su & Zhang (2007). Hafeez et al. (2011) re-defined this metric as the *average cluster stability factor* and it is described depending on some particular parameters of the solution. However, a deeper analysis reveals that average cluster stability factor is identical to $\overline{RSWC}$.

$$\overline{RSWC} = \frac{1}{|C(t)|\,S}\sum_{t=0}^{S}\sum_{i\in C(t)}\frac{\sum_{j\in CH_i(t)}|\overrightarrow{SP_i}(t)-\overrightarrow{SP_j}(t)|}{|CH_i(t)|} \tag{14}$$

- **Average Relative Speed Among CHs:** ($\overline{RSCH}$) (Su & Zhang, 2007) is a general topology-based metric for CH-based algorithms. It measures the global topology of the network. Equation (15) represents the general mathematical definition of this metric as given by Su & Zhang (2007).

$$\overline{RSCH} = \frac{1}{S}\sum_{t=0}^{S}\frac{\sum_{i,j\in C(t)\hat{}i\neq j}|\overrightarrow{SP_i}(t)-\overrightarrow{SP_j}(t)|}{|C(t)|^2} \tag{15}$$

$\overline{RSCH}$ and $\overline{RSWC}$ are more complex metrics for CH-based algorithms that are indicators of both CH and network stability, as this type of metrics are measuring better the cluster stability in general. This statement also sustained by Fan et al. (2005), that fist assessed the clustering algorithms defined using $\overline{CS}$ and $\overline{CHCR}$ and then using relative measure ($\overline{CS}\,/\,\overline{CHCR}$) arguing that this is a better measurement of cluster stability.

All the general metrics presented and defined (except the cluster stability which is not considered a metric) are summarized in Table 3. This section and the table-based summary provided represent an invaluable guide for the performance assessment of VANETs clustering algorithms in particular and of VANETs cluster-based solutions in general. Evaluating the performance of clustering algorithms via these general metrics can greatly facilitate the comparison between the clustering algorithms, independent from their type: generic algorithms or integrated in a specific solution (e.g. clustering algorithm implemented in a MAC protocol).

Solution-Dependent Topology-Based Metrics

Node re-clustering time (Goonewardene et al., 2009)/ Re-affiliation frequency (Blum, 2003) are metrics very differently named, but both described as being the time between cluster associations for a given node. Solutions using this metric consider that this is a measure of the stability of a cluster membership and shorter node re-clustering time/re-affiliation frequency means an increased stability of a cluster membership. However, we consider that average cluster membership lifetime and average CH lifetime (in case of CH-based algorithms) are a better indicator of the stability of cluster membership and we base this statement on the following considerations. Particularly in VANET clustering, as emphasized before, the nodes can be in intermediate states. They have different states like candidate, visitor and they are not clustered until they demonstrate their future stability in the cluster, meaning a long lifetime as a cluster member. This translates into longer period of re-clustering/re-affiliation frequency, but this does not mean that the topology is less stable. These considerations represent also the motivation of including this metric in the class of solution-dependent topology-based clustering metrics. In this class, we also include the average percentage of clustered nodes metric (Kuklinski & Wolny, 2009). A bigger average percentage of clustered nodes it is usually translated into a better stability of the network topology. However, in the aforementioned particular clustering algorithms, this metric is not applicable because at some moments of time there can be a considerable number of nodes not-clustered (candidates or visitors).

CLUSTERING ALGORITHMS IN VANETs

Initial approaches of clustering in VANETs used clustering algorithms designed for MANETs. Lowest Id (Yu & Chong, 2005) is a state-of-the-art clustering algorithm in ad-hoc networks and was borrowed in VANETs from MANETs. Its principle is very simple. The nodes have assigned a unique fixed id which is broadcasted periodically in the network. The clusters are formed around the node with the lowest id among them, which is chosen as CH. Although its principle is very simple, it is a very efficient algorithm, more efficient than other clustering schemes, such as Highest-Degree (Yu & Chong, 2005), that take into consideration more factors (Fan et al., 2005). Highest-Degree is another state-of-the-art clustering algorithm in the area of ad-hoc networks. Its principle is similar to the Lowest Id algorithm, but the clusters are formed around the node with the highest number of neighbors. These two algorithms, as state-of-the-art algorithms in the area of ad-hoc networks, are very often used in the comparison-based assessment of the VANETs clustering algorithms and served as source of inspiration for many VANETs clustering approaches.

As emphasized before, although VANETs represent an instantiation of MANETs, they have unique features that need to be considered in order to design appropriate clustering algorithms for vehicular networks. On one hand some of the VANET's characteristics need to be overcome by the clustering schemes, such as their rapidly changing topology, high mobility and scalability, while on the other hand clustering schemes can make use of other characteristics such as predictable mobility due to the road topology, traffic regulations and driver's behavior. Researchers acknowledged these facts and VANET-dedicated clustering solutions have been proposed. After an overview of VANET clustering solutions in the literature, a very broad classification is provided here and several approaches are presented for each class for exemplification. The classification is made based on the cluster formation criterion: is the cluster formation dependent on some fixed structures such as road segments, grids, etc, or is it independent on any kind of structure and it is just following the traffic flow, vehicle's movement? In the first case, vehicles from the same structure (road segment, grid, etc) are grouped into a cluster. Thus static clusters are created bounded by this structure. Therefore, we called this type of VANET clustering algorithm under the generic name of static clustering algorithms. In the second case, cluster formation does not depend on any type of structures. Clusters are created by following the movement of the vehicles: vehicles with similar mobility patterns such as neighboring vehicles are grouped into clusters through exchange of clustering messages. In this type of approaches there is usually a beaconing message (a periodically broadcasted message in the network) sent either by the unclustered vehicle, either by a CH or a node with extra-responsibilities in the cluster. In the absence of predefined structures, this is necessary in order to announce the availability of joining the cluster or the availability of a cluster in zone so that a vehicle can join a cluster. The clusters created following this approach are mobile clusters, following the mobility of the vehicles and therefore we name this class of VANET clustering algorithms, mobile clustering algorithms.

Static Clustering Algorithms

Cherif *et al.* (2009) propose a CH-based clustering algorithm where the cluster formation is depended on fixed road segments. The communication area where vehicles can be reached by RSU via multi-hop communication is called extended communication area. This area is split into fixed length segments, vehicles located into the same segment forming a cluster. Beside CH and simple cluster member, nodes can have another status inside a cluster, called super-member. This is a node that has been a CH and is yielding the job to another node. Inside the cluster, a main area of interest is conceptually partitioned in the centre of the segment. This area is called central zone and has the radius equal to the transmission range. Central zone has an important role in the distributed election of the CH. Initially, each member in the cluster estimates the time period it is going to spend in the central zone. The main principle behind CH election algorithm is to choose as CH the vehicle with the highest probability to spend the longest duration in the central zone. The speed and the position of the vehicle are also taken into consideration. All these parameters are used in the computation of each vehicle's electing factor, based on which the CH is selected. After that, each vehicle periodically examines its status and, by using the laws of uniform motion from Physics $(distance = speed \times time)$ predicts its future position in the immediate next moment of time. If a CH determines that it will be leaving the central zone in this moment of time, it will resign as CH, and a new CH is elected following the same procedure.

The proposed algorithm takes into consideration the high mobility of VANET nodes and movement predictability. Algorithm's assessment is performed both via general topology-based metrics – $\overline{CL}$ – and network metrics – overhead, end to end delay and delivery ratio. These are evaluated in relation to network density, but it is to be mentioned as a limitation the fact that the solution is not compared against any other clustering scheme.

Luo *et al.* (2010) propose a CH-based clustering algorithm where the cluster's formation is based on square grids. The geographical area is divided into a subset of square grids. All the vehicles pertaining to a grid form a cluster. The vehicle having the closest position to the centre of the grid is elected as CH. This clustering scheme is implemented in a cluster and position-based routing protocol dedicated to VANETs and claims to reduce the overhead and packet delivery delay. CHs are the main data forwarders, a packet is sent from CH to CH until it gets to the CH that governs in the cluster where the destination node is positioned. The performance assessment is not very thoroughly, the authors presenting just a small analysis where they make some observation about their algorithm in comparison with state-of-the-art routing algorithms. Moreover, the clustering scheme neither tries to address any of VANETs challenging characteristics nor does it take advantage of any of VANETs characteristics. Thus, the clustering scheme, only by itself is not VANETs dedicated, but the routing protocol is taking advantage of the vehicle's knowledge about their own positioning via the GPS integrated in their OBU.

Ramakrishnan *et al.* (2011) adopt a similar approach in their proposed CH-based clustering algorithm to the one previously discussed: cluster formation is based on road segments called clustering areas. However, these clustering areas are not assigned with a fixed length value. Their size varies depending on the average speed of the vehicles within them. If the average speed is small then the cluster size is smaller, otherwise bigger. However, it is not mathematically described what smaller or bigger means. If an RSU is inside a cluster, then this is elected as CH. Otherwise, CH election is based on a single metric that is the velocity. As the clusters are static, the vehicle with the lowest speed in the cluster is going to spend the more time inside the cluster. Thus this vehicle is elected as CH. However, although the CHCR is reduced, is not clear how the fact that the position of CH related to the other cluster members is not taken into consideration is affecting the communication between CH and cluster members.

Performance assessment is done via topology metrics only, which are quite different than the ones typically used. Instead of measuring directly the rate of changes in CH or clusters, the times of creation of clusters or the time of electing CH is measured. However, these are not good measurements of the stability in clusters; instead these assess the initial performance of clustering.

There are some clustering solutions that represents a bridge between the two main identified types or a combination. Such is for instance the solution proposed by Tung et al. (2013). This is a clustering algorithm designed in the context of an intersection collision avoidance service. This clustering algorithm is employed in the design of a novel VANET WLAN-cellular architecture. This architecture is based on a heterogeneous network: LTE and WiFi. The communication messages inside the cluster are done via WiFi and they are called beacons, while CHs only are using the LTE interface for communicating with the base stations. As aforementioned, the proposed algorithm uses both static and mobile approach. On one hand, the clustering is bounded by the so called service region, region that is placed in the nearby of the intersection, but on the other hand it follows the mobility of the vehicles taking into account their direction. The proposed clustering algorithm is very specific to the solution built within. However, it indicates an efficient modality of bringing LTE in the vehicular networking context, as at this moment it appears to be more likely that LTE cannot handle the multiple messages that can be generated in VANET, especially during rush hours and in the traffic collision related applications when a huge number of messages can be generated. This solution was preceded by (Sivaraj et al., 2011) that employed a clustering algorithm to design a LTE-WAVE network architecture dedicated to multimedia delivery. This latest work uses a similar principle as in (Taleb & Benslimane, 2010; Benslimane et al., 2011), where a generic VANET UMTS-WAVE architecture based on clustering is designed, but instead of 3G brings 4G in the VANET context. The principle of these three works differ from (Tung et al., 2013) by delegating the responsibility of communicating to infrastructure (via 3G or 4G) to another node, a gateway node, while CH is the main forwarder of messages inside the cluster. Multiple metrics are involved in both selection procedures: CH and gateway node as both states are of great importance. Independently of the type of node that has the responsibility of communicating with the infrastructure, CH or gateway, there is a single node in each cluster that is accessing the cellular network interface. This leads to an optimized architecture that it is also proven to be reliable even for applications that require a rich content such as multimedia applications.

In (Tung et al., 2013), the procedure of selecting the CH is based on a single metric: the proximity to the base station. The algorithm is evaluated in the context of the overall solution using solution-dependent metrics. Although WiFi standard is chosen for the inter-vehicle communications, the authors suggest that this can be replaced with V2V communication (IEEE 802.11p). Such architecture is used in the clustering solution proposed by Harikrishnan & He (2013): IEEE 802.11p – V2V communication – for intra-cluster communication and LTE for the communication between CH and base station. This algorithm is a general CH-based clustering algorithm for VANETs. The clustering metrics are not clearly stated, but the CH is selected following the same policy as above: minimum distance to the base station. The algorithm is evaluated in terms of both network-specific performance metrics and topology-based performance metrics, namely: $\overline{CHL}$, $\overline{CS}$ and $\overline{NoC}$. Another clustering approach where clusters are bounded to an intersection region was proposed in (Chen et al., 2016). Unlike (Tung et al., 2013) the architecture of the clustered network is a pure ad-hoc architecture, based on V2V communication only. The clustering is done on the geographical location basis only, but the novelty of the approach is the employment of game theory in order to determine the CHs to decide on the aggregated transmission power and packet generation rate. Game theory has a secondary role in the context of the cluster-based solution proposed for congestion control. However, game theory started to be considered as the main decisional framework in the context of clustering algorithms. Some of these approaches will be discussed in the next chapter.

Mobile Clustering Algorithms

Su & Zhang (2007) proposed a CH-based clustering algorithm in the design of a dedicated VANET MAC protocol. The cluster formation is based on beaconing messages (an initial message periodically broadcast in the network either by a vehicle recently entered in the network, either by CHs) and other cluster messages among the same-direction neighbours. Thus in cluster formation the main criterion considered is the direction of the vehicles based on the assumption that vehicles flowing in the same direction have similar speeds and moving patterns that are regulated by the traffic rules. Another criterion considered in cluster formation is signal strength and its role is revealed in the next paragraph.

The possible states of a vehicle-node in this clustering algorithm are: CH, quasi-CH, cluster member and quasi-cluster member. Each vehicle is seen from the moment of entering on the road a potential CH, so it receives the quasi-CH state. If after a predefined period of time it does not receive any valid *invite-to-join* beaconing message from a CH, the vehicle elects himself as a CH, otherwise the vehicle joins the cluster and its state changes to cluster member. Note that valid *invite-to-join* message must have the signal strength greater than a predefined threshold. Thus the size of the cluster is determined by the signal strength threshold.

This algorithm is among the first mobile VANETs clustering algorithms. Its principle is simple, the only clustering metrics considered are direction and signal strength and the CH election is very simple, no decision process based on multiple metrics is involved. However, it is the first that considered direction metric in clustering the vehicles. In addition, this is the first approach in the literature that thoroughly defined some of the most popular general topology metrics in VANETs: $\overline{CHL}$ and $\overline{CS}$. Also, they defined 2 relative topology metrics, previously discussed: $\overline{RSWC}$ and $\overline{RSCH}$. These metrics are used to illustrate the performances of the clustering algorithm, but no other clustering algorithm is used as reference. The focus of the authors is on testing the MAC protocol where the clustering solution has been integrated. Tests show that this MAC protocol outperforms the standard IEEE 802.11p.

Kuklinski & Wolny (2009) propose a mobile clustering algorithm where mobile clusters are formed by the neighbouring vehicles through beaconing and other messages exchange. Multiple clustering metrics are considered in creating stable clusters such as: connectivity level that is actually measuring the density, link quality estimated by SNR, relative nodes position and the prediction of this position in the future (based on speed and position) and nodes reputation built upon the history of node connections. The prediction of vehicle positions aims on one hand to avoid situations like clustering the vehicles that are moving in different directions with high speed. On the other hand, it allows for clustering the vehicles that are moving in different directions but with a low speed (e.g. vehicles in traffic jam). This approach leads to a greater stability of the clusters. Moreover, in order to avoid a high rate of re-clusterings, a node is given three possible states, excepting the CH state: member, candidate and visitor. Vehicles must prove they are potentially stable members of the clusters before they can join. First, a vehicle is in the visitor change, then after a time threshold is given the candidate state and only after applying the other clustering metric (connectivity, future position, etc), its state is changed into a member. Candidate and visitor nodes do not have the same rights as members do. They are not provided with the services that are provided in the cluster and they only have the right to exchange clustering messages. CH election algorithm is not described, although in each cluster a vehicle is assigned with this role. In addition, it is not clear what the CH responsibilities are.

The proposed solution is compared against the state-of-the-art algorithm, Highest Degree and proves better performances in terms of $\overline{CS}$ and $\overline{CHCR}$ topology metrics.

Almalag & Weigle (2010) introduced a CH-based clustering algorithm designed mainly for urban scenarios that uses traffic flow in cluster formation. The authors focus on the CH election algorithm as it is a well-known fact that stable CHs conduct to stable clusters. This algorithm is based on multiple clustering parameters: density, distance between vehicles, speed and the lane of travelling. This last parameter is a new parameter considered so far in the clustering schemes and the key novelty of the algorithm. The rationale behind considering this parameter is that CH should be selected from a lane that the majority of vehicles are travelling in. Each vehicle first determines its own lane. Then each lane, referred as traffic flow, is given a weight. It is not explained what is the rationale behind weights' assignment for each traffic flow. Then for each vehicle it is determined on one hand the number of vehicles it is connected to (density), the comparison of its speed compared to others within its range and the comparison of its distance from all other vehicles within its range and on the other hand all these parameters but within their own traffic flows. The first group of parameters are multiplied with the traffic flow weights and then added to the second group in order to obtain the CH level of each vehicle. The vehicle with the highest CHL is selected as CH.

The proposed algorithm is compared against other three algorithms: the well-known Lowest Id, Highest Degree and against what authors generic named the Utility Function algorithm for VANETs. The latter clustering approach was proposed by Fan et al. (2005) having as models Lowest Id and Highest Degree and it is probably the first clustering scheme proposed for VANETs. The focus in this scheme is fully on the CH election that is suggested to be chosen for VANETs as the vehicle having the speed closest to the average and the distance between vehicles closest to the average. Although the authors do not provide details about what closest to the average means, they state that simulation results show better performance of their approach compared to Lowest-Id and Highest Degree. In the performance assessment of the traffic flow based algorithm, the authors use their own understanding of what closest to the average means for both speed and distance parameters. This is the same understanding that they used for implementing their own algorithm with respect to speed and distance metrics. The traffic flow-based algorithm outperforms all three algorithms (i.e. Lowest Id, Highest Degree and Utility Function) in terms of the topology metric used, $\overline{CHCR}$.

Shea et al. (2009) proposed another mobility-based clustering algorithm for VANETs with focus on the stability of the resulted clustered network. The novelty of the algorithm consists in employing affinity propagation (Frey & Dueck, 2007), a clustering technique that is borrowed from data clustering field. Same pattern for clustering formation is followed as in the other structure-free discussed algorithms: exchange of clustering messages between vehicles in 1-hop neighbourhood. Direction is the first parameter considered in clustering formation: the vehicles form clusters with their 1-hop same-direction neighbours. The focus is again on the CH election algorithm where the affinity propagation technique applies. This technique is based upon a similarity function that is tailored for VANETs. Thus it is based on the Euclidean distance between the position of the node and the positions of its same-direction neighbours and the Euclidean distance between the next position of the node and the next positions of its same-direction neighbours. The efficiency of the algorithm is demonstrated against the previously discussed clustering algorithm proposed by Su & Zhang (2007) by applying the most popular topology-based metrics: $\overline{CHL}$, $\overline{CML}$, $\overline{NoC}$ and $\overline{CHCR}$.

Goonewardene et al. (2009) proposed a mobile clustering algorithm based on exchange of clustering messages between 1-hop neighbours designed with a robust adaptability to mobility – RMAC (i.e .robust mobility adaptive clustering). The algorithm is designed to support geographic routing, although no routing protocol is proposed. An unclusterd node first makes a list of its 1-hop neighbours that answer to its beaconing messages with a message containing their speed, location and direction of travelling. Based on these metrics, the list is then sorted so that the most appropriate neighbour of the unclustered node to be selected as its CH. The appropriateness is decided as follows. First the position parameter is considered. Based on this the Euclidean distance is computed between the node and its neighbours. If the distances are comparable, then the next parameters, speed and location are considered. Based on these two parameters the next locations of the node and its neighbours are computed. The first neighbour in the list, the most appropriate to become the CH, is the one closer in the current moment of time and in the next one. This is quite a new approach in the literature, as usually a CH is elected in the cluster based on some values (id, computed weight using different techniques) that applies globally. The clustering algorithm proposed here is node-oriented – node precedence algorithm – as each node elects its own CH. If the first node in its 1-hop neighbours list is already a CH then the unclustered node becomes a member of its CH cluster. Otherwise, the vehicle selects this node as its CH and e new cluster is formed. Thus, beside cluster member and CH, a node can be in a dual state that is when it is a CH of a cluster and a member of another cluster. This leads to overlapping neighbouring clusters and no message overhead in case of a cluster member transition to a neighbouring cluster.

Another novel concept introduced by this algorithm is the zone of interest that enables each vehicle to keep an updated table of its neighbours that goes beyond their transmission range. Zone of interest' radius is established as two times their transmission range. Thus vehicles have prior knowledge about the network while they are travelling into the neighbourhood which it's translated into an optimized and smoother process of re-clustering.

The algorithm is compared against an algorithm proposed by Basagni (1999) that is shortly called DMAC (Distributed and Mobility Adaptive Clustering). DMAC is a generalised clustering algorithm designed for MANETs where the CH election is done globally and is not node-oriented. Each vehicle has a weight associated. The clustering process begins with each node examining the weights of all nodes within its own transmission range. The node with the highest weight becomes the CH. This algorithm can be tailored for VANETS where the weight of a vehicle is calculated using metrics such as distance/

speed/acceleration. RMAC outperforms DMAC in terms of $\overline{CML}$, and in terms of another topology metric: node re-clustering time.

Fuzzy Logic was employed in the decision making process in the context of several clustering approaches (Hafeez et al., 2011; Tal & Muntean, 2013a; Arkian, Atani, Diyanat & Pourkhalili, 2015; El Mouna, Tabbane, Labiod & Tabbane, 2015, Sharma & Kaul, 2018). Fuzzy Logic is an excellent mathematical framework for dealing with imprecision and multiple parameters. This is what needs to be modelled in VANET clustering: imprecision – it is impossible to define precisely how each of the clustering metrics influences the stability of CH in particular and clusters in general – and multiple clustering metrics that are imposed by the dynamicity and complex vehicular networking environment (Tal&Muntean, 2017). Moreover, Fuzzy Logic is widely used in prediction and detection systems.

Hafeez et al. (2011) introduced a clustering algorithm in the context of a new MAC protocol. Vehicles are organized in clusters on the basis of the beaconing and clustering messages they exchange in their neighbourhood. The focus is again on the CH election as CH is assigned with the main organizing and communication roles inside its cluster. The vehicles can have 5 different states: lone (not clustered), member, temporal CH, backup CH and CH. Temporal CH and backup CH roles aim to provide on one hand a stable CH in the cluster, a temporal CH must prove that it is the most stable selection, and on the other hand to ensure a smoother CH re-election, backup CH is ready to take over the CH role. CH election algorithm is based on a weighted stability factor that is built upon the exponential-weighted moving average of the previous stability factors. Stability factor is computed for each vehicle and it is based on the relative movement between the neighbouring vehicles reflected in the average speed difference between the vehicle speed and its neighbours' speed. The novelty of this clustering scheme consists in the technique implemented in order to provide a smoother CH re-election and consequently to improve the cluster stability. Basically, this technique states how backup CH is taking over the CH role. The implementation of this technique is based on a Fuzzy Logic system that aims to predict and learn driver's behaviour. Based on this the next position and speed of the vehicles are computed. If in this next moment of time if not all the member of its cluster are in its range anymore, but they are still in the range of the backup CH, then CH hands over its role to the backup CH.

The proposed clustering solution is assessed using a large variety of metrics from all the classes presented. Thus, the MAC protocol integrating the clustering solution is assessed as well. However as our interest relays in the clustering schemes we mention the topology-based metrics employed in assessment: $\overline{CHL}$, $\overline{CML}$, $\overline{CS}$. The proposed clustering algorithm is demonstrated to overcome the performances of another cluster algorithm designed for a VANETs MAC protocol that was previously discussed (Su & Zhang, 2007). Moreover, the MAC protocol based on the proposed clustering solution proves better performances than the protocol used as comparison.

Tal & Muntean (2013a) proposed a new CH-based scheme that has as main novelty the employment of Fuzzy Logic as decisional framework in selecting the CH. The clustering metrics considered are the average relative distance, average relative velocity, direction of travelling and the average relative compatibility. This later parameter was introduced as a novelty by Tal & Muntean (2012) and measures the compatibility in the users (vehicles' drivers/passengers) preferences in certain data/content. The aim is to increase the probability of users being provided with data/content of their interest inside the cluster. Thus both clustering schemes aim to provide a cluster-based architecture for disseminating data/content of users' interest inside the cluster. In addition, these two approaches emphasize on the capability of clustering of designing efficient and optimized VANET architectures where the communication with infrastructure is limited (only CH is communicating with RSU) and in the same time the communication range can be extended via multi-hop communication inside the cluster in case of a multi-hop clustering.

In an assessment that compares these both algorithms and in addition the Lowest Id it is shown that the Fuzzy Logic-based clustering algorithm performs better than others two. The performance metrics used were $\overline{CHL}$, $\overline{CS}$ and a solution–dependent relative topology metric.

Arkian, Atani, Diyanat & Pourkhalili (2015) introduced a new cluster-based vehicular cloud architecture. The purpose of this architecture is to provide a better management of the limited resources in vehicular networks. Fuzzy Logic is employed in the CH selection algorithm that is proven to perform better than the one proposed in (Tal & Muntean, 2013a) in terms of $\overline{CHL}$ performance metric and other solution-dependent performance metrics. The clustering metrics used in the algorithm are: neighborhood degree, average speed, and RSU link quality. The selection of the CH subscribes to the following policy: a CH should have a high neighborhood degree and the RSU link quality should also be high.

A newer trend in VN clustering is the employment of clustering algorithms in designing reliable and efficient VN architectures that bring together multiple access technologies. Such cluster-based hybrid architectures were proposed in some of the aforementioned clustering solutions (Sivaraj et al., 2011; Taleb & Benslimane, 2010; Benslimane et al., 2011, Tung et al., 2013). El Mouna, Tabbane, Labiod & Tabbane, 2015 proposed such an architecture, with Fuzzy Logic playing a central role. The vehicles in the cluster communicate via V2V communications based on the IEEE 802.11p standard, while a GW in the cluster is chosen for connection to the LTE Advanced infrastructure. Fuzzy Logic is employed in the GW selection. The selection takes into consideration multiple criteria: QoS classes, connectivity strength between the vehicle and infrastructure, connectivity strength between the CH and infrastructure, CH load and link connectivity between the vehicle and CH that encompasses the mobility of the vehicle. The algorithm is demonstrated to perform better than other clustering algorithms, and also the benefits of the clustered network are demonstrated against the non-clustered/flat network. The performance metrics were mainly network-specific metrics, including the overhead imposed by the clustering messages/control.

Security is a hot research topic in the context of 5G and VANET. Sharma & Kaul (2018) proposed a Fuzzy Logic cluster-based swarm optimized intrusion detection system for VANET, where Fuzzy Logic is employed as the main player in the CH selection. The main role of the clustering in the context of the proposed solution is to deal with the stability of the ever-changing VANET network.

Wahab et al. (2013) is a two-hop CH-based clustering scheme for VANETs that is also incorporating computational intelligence. This is one of the proposed clustering algorithm novelties: the employment of Ant Colony Optimization in selecting the nodes having a state called multi point relay. These nodes are selected by CH for inter-cluster communication. The other novelty of this approach consist in building 5 new clustering metrics models with the focus on QoS. The most complex one combines bandwidth, connectivity and mobility metrics specific to VANETs (i.e. relative speed and distance). The output of the model is a QoS factor that is computed for each vehicle. Each of these models can be employed in further QoS-oriented clustering algorithms for VANETs, the authors describing each model's recommended scenario. This QoS factor is used to elect the most suitable CH and the multi point relays, while the clustering formation is done only on the basis of 2-hop neighbouring.

The clustering algorithm is designed in the context of a new routing protocol dedicated for VANETs that derivates from a MANET routing protocol designed to improve QoS, QoS-OLSR that on its turns derives from the state-of-the-art routing protocol for MANETs, QLSR (Optimized Link State Routing). Thus the performance assessment aims to demonstrate on one hand the efficiency of the clustering algorithm in terms of network stability and on the other hand to prove the efficiency of this algorithm in the designed routing protocol. In the first case, the performance is tested using a solution-dependent topology metric defined by the authors that is generic entitled stability. The metric is highly dependent on the clustering parameters. The authors show the performances of the algorithm in terms of stability and network metrics (e.g. packet delivery ratio) in 5 different cases corresponding to the 5 different clustering metrics models. No comparison is done with other clustering schemes. The comparison-based assessment is done only when showing the performances of the overall solution. The cluster-based routing protocol outperforms QoS-OLSR and OLSR both.

Game theory started to be explored as a decision making tool in the context of clustering approaches in the more generic context of mobile networks (Massin, Le Martret & Ciblat, 2017), but also in the specific context of VANET clustering (Khan, Abolhasan & Ni, 2018). In (Khan, Abolhasan & Ni, 2018) game theory is employed in the decision making process for both cluster head selection and vehicle clustering. An Evolutionary Game Theoretic (EGT) framework aiming to create stable clusters and to select stable CHs is proposed. The decision criteria are throughput, link capacity between the RSU and CH and cluster size. Initially, the vehicles are randomly clustered and the cluster reformation and CH selection is triggered through the EGT framework. The clustering algorithm is evaluated in terms of both network-specific performance metrics and topology-based performance metrics, namely: $\overline{CHCR}$. The latter metric is the one used for measuring the stability of the proposed approach. Game theory was employed in another VANET clustering algorithm, but not as a decisional tool (Ahmad et al., 2018). In this V2X-LTE cluster-based solution, clustering is used in order to minimize LTE usage, while game theory is used to enforce a cooperation policy between the cluster members in order to avoid bottlenecks in sharing data.

FUTURE RESEARCH DIRECTIONS

In a roadmap of clustering algorithms in the VANET research area, its start is recorded in 2005 when the first studies have been performed by employing Lowest Id and Highest Degree in VANETs scenarios and by suggesting new approaches that relate to these (e.g. Fan et al. (2005)). Actually, most of the VANET mobile clustering algorithms are using the basic principles that fundament Lowest Id and Highest Degree algorithms. Back in 2005 some of the main clustering challenges in VANETs have been outlined as well: rapidly changing topology of VANETs, scalability, multiple services to be provided with different requirements (real-time traffic, non-real time traffic) (Reumerman, Roggero & Ruffini, 2005).

Since then, the clustering algorithms evolved, many approaches have been proposed to tackle especially the rapidly changing topology of VANETs. More and more mobility parameters have been considered in clustering: direction, lanes, speed, position, predicted speed and position and combined with other parameters such bandwidth, connectivity (or density of the vehicles) and signal strength. These parameters are mainly considered in selecting the nodes with extra-responsibilities in the cluster, especially CH. CH election algorithms are of high importance and some of the researchers are actually focusing on this aspect of clustering only. Usually CH has the main responsibilities in the cluster and therefore a stable CH is required. In addition, the stability of CH is highly influencing the stability of the cluster itself, as most of the times when CH is re-elected a cluster reconfiguration is required, too. Therefore, researchers employed all kind of techniques to combine the clustering parameters and to decide which is the most suitable CH. The predominant techniques are utility functions and weight-based techniques, but the more these algorithms evolved, more and more innovatory techniques such as Affinity Propagation, Fuzzy Logic, game theory, etc. have been employed. As a matter of fact, one of the latest trends in VANET clustering algorithms is the employment of computational intelligence. Initially, these techniques were employed in secondary roles, but afterwards they were employed as main players in CH selection/clustering algorithms. As such, Fuzzy Logic (Hafeez et al., 2011) and Ant Colony Optimization (Wahab et al., 2013) were employed for instance in some secondary roles in clustering such as predicting the future positions of the vehicles. Fuzzy Logic decisional systems, known as very powerful decisional systems, have started to be employed as main players in the context of CH selection algorithms (Tal & Muntean, 2013a; Arkian, Atani, Diyanat & Pourkhalili, 2015; El Mouna, Tabbane, Labiod & Tabbane, 2015). Fuzzy Logic is the perfect mathematical framework for dealing with imprecise information such as the one used in clustering (it is impossible to precisely define how each of the clustering parameters influence the stability of CH) and with multiple parameters. Similarly, game theory was employed as the main decision making tool in the context of another clustering scheme for both CH selection and clustering algorithms.

Another trend in VANET clustering is the employment of clustering algorithms in designing reliable and efficient VANET architectures that bring together multiple access technologies. The need to converge multiple types of technologies in VANET context in order to enable the diversity of vehicular applications is underlined more and more in the literature and is also enforced by the low penetration rate of the WAVE technology. Moreover, 5G will be characterized by a heterogeneous network environment. Thus, design techniques of VANET architectures that bring together multiple access technologies are of high interest. First, VANET cluster-based 3G-WAVE architectures were envisioned and more recently, VANET cluster-based architectures using 4G together with other technologies (WLAN or WAVE) were proposed. Very recently, an SDN enabled 5G-VANET clustered architecture was proposed (Duan, Liu & Wang, 2017), where the one of the main role of clustering is again to optimize cellular communication – reduce from the communication burden: every CM would communicate with the BS via CH only, while the intra-cluster communication would rely on another type of communication (e.g. IEEE802.11p).

As a conclusion to the aforementioned aspects, the future directions in this research area of clustering in VANETs can be summarized as follows:

- More mathematical frameworks incorporating computational intelligence should be experimented in trying to find the most appropriate method of combining the clustering parameters in order to obtain stable CHs and stable networks. Machine learning techniques and their application are gaining momentum in the context of VANET (Ye et al., 2018) and we argue that this would be a research direction to explore in the specific context of VANET clustering as well considering that clustering algorithms in general have their own particular role in the context of machine learning.
- Although this chapter does some significant steps forward in the context of performance assessment of clustering algorithms in VANETs, this remains still an open challenge. The analysis of the VANET clustering algorithms conducted in this work, leads to the idea that there is a need of standardizing clustering performance metrics, especially the general topology metrics. In addition, there is a huge need of traffic and mobility models for performing the testing of VANETs clustering algorithms.

- A direction to be followed is the one that relates to clustering capabilities of conducting to a reliable and optimized design of VANET architecture based on multiple access technologies that is able to support the diversity of VANET applications. As 5G will be characterized by a heterogeneous network environment this research direction is of particular interest.
- Moreover, VANET security in particular and 5G security in general is a hot research topic (Hussein, Elhajj, Chehab, & Kayssi, 2017; Hasrouny, Samhat, Bassil & Laouiti, 2017; Sharma & Kaul, 2018b). The potential of clustering to be used in the context of security solutions was highlighted in (Sharma & Kaul, 2018a; Sharma & Kaul, 2018b) where clustering is suggested to have an important role in overcoming security issues that relate to the network scalability and stability control. As clustering is a well-known solution to these later issues, this potential should be further exploited in future security solutions.

CONCLUSION

This chapter has emphasized the high importance of VANETs in the context of future 5G-enabled smart cities and roads. Initially, VANET was mainly associated with WAVE/DSRC technologies, but this work has underlined the need for the convergence of multiple wireless technologies for support of a reliable VANET architecture that is able to provide a variety of services and to cope with the multiple challenges of VANETs. Clustering can be employed in designing such a VANET architecture that successfully uses different wireless communications technologies in an optimal manner. Thus, clustering algorithms may play an important role in the future 5G heterogeneous networks. Moreover, clustering addresses some of VANETs major challenges such as scalability and stability.

This work presents a comprehensive survey of clustering schemes in the VANET research area covering aspects that were not really addressed before in a structured manner. The survey presented in this chapter provides a general classification of the clustering algorithms, presents some of the most advanced and latest algorithms in VANETs, and in addition, this is among the few works in the literature that also reviewed the performance assessment of clustering algorithms. In this chapter, we discussed the performance assessment metrics used in clustering in VANETs, provided a classification, identified and defined general performance metrics to be used in the evaluation of clustering algorithms in VANETs.

REFERENCES

Ahmad, I., Noor, R. M., Ahmedy, I., Shah, S. A. A., Yaqoob, I., Ahmed, E., & Imran, M. (2018). VANET–LTE based Heterogeneous Vehicular Clustering for Driving Assistance and Route Planning Applications. *Computer Networks*, *145*, 128–140. doi:10.1016/j.comnet.2018.08.018

Almalag, M. S., & Weigle, M. C. (2010, October). Using traffic flow for cluster formation in vehicular ad-hoc networks. In *Local Computer Networks (LCN), 2010 IEEE 35th Conference on* (pp. 631-636). IEEE.

5G. Americas Whitepaper. (2018). *Cellular V2XCommunications Towards 5G*. Retrieved from http://www.5gamericas.org/files/9615/2096/4441/2018_5G_Americas_White_Paper_Cellular_V2X_Communications_Towards_5G__Final_for_Distribution.pdf

Arkian, H. R., Atani, R. E., Diyanat, A., & Pourkhalili, A. (2015). A cluster-based vehicular cloud architecture with learning-based resource management. *The Journal of Supercomputing*, *71*(4), 1401–1426. doi:10.100711227-014-1370-z

Basagni, S. (1999). Distributed clustering for ad hoc networks. In Parallel Architectures, Algorithms, and Networks, 1999.(I-SPAN'99) Proceedings. Fourth InternationalSymposium on (pp. 310-315). IEEE. doi:10.1109/ISPAN.1999.778957

Benslimane, A., Taleb, T., & Sivaraj, R. (2011). Dynamic clustering-based adaptive mobile gateway management in integrated VANET—3G Heterogeneous Wireless Networks. *Selected Areas in Communications. IEEE Journal on*, *29*(3), 559–570.

Blum, J., Eskandarian, A., & Hoffman, L. (2003). Mobility management in IVC networks. In *Intelligent Vehicles Symposium, 2003. Proceedings. IEEE* (pp. 150-155). IEEE.

Chang, B. J., Tsai, Y. L., & Liang, Y. H. (2017). Platoon-based cooperative adaptive cruise control for achieving active safe driving through mobile vehicular cloud computing. *Wireless Personal Communications*, *97*(4), 5455–5481. doi:10.100711277-017-4789-8

Chen, C., Hu, J., Zhang, J., Sun, C., Zhao, L., & Ren, Z. (2016). Information Congestion Control on Intersections in VANETs: A Bargaining Game Approach. In *Vehicular Technology Conference (VTC Spring), 2016 IEEE 83rd* (pp. 1-5). IEEE.

Cherif, M. O., Senouci, S. M., & Ducourthial, B. (2009). A new framework of self-organization of vehicular networks. In *Information Infrastructure Symposium, 2009. GIIS'09. Global* (pp. 1-6). IEEE.

Cooper, C., Franklin, D., Ros, M., Safaei, F., & Abolhasan, M. (2017). A comparative survey of VANET clustering techniques. *IEEE Communications Surveys and Tutorials*, *19*(1), 657–681. doi:10.1109/COMST.2016.2611524

Djahel, S., Salehie, M., Tal, I., & Jamshidi, P. (2013). Adaptive Traffic Management for Secure and Efficient Emergency Services in Smart Cities. *Pervasive Computing and Communications (PerCom WIP track)*, 340-343.

Doolan, R., & Muntean, G. M. (2017). EcoTrec—a novel VANET-based approach to reducing vehicle emissions. *IEEE Transactions on Intelligent Transportation Systems*, *18*(3), 608–620. doi:10.1109/TITS.2016.2585925

Dror, E., Avin, C., & Lotker, Z. (2011). Fast randomized algorithm for hierarchical clustering in vehicular ad-hoc networks. In *Ad Hoc Networking Workshop (Med-Hoc-Net), 2011 The 10th IFIP Annual Mediterranean* (pp. 1-8). IEEE.

Dror, E., Avin, C., & Lotker, Z. (2013). Fast randomized algorithm for 2-hops clustering in vehicular ad-hoc networks. *Ad Hoc Networks*, *11*(7), 2002–2015. doi:10.1016/j.adhoc.2012.02.006

Duan, X., Liu, Y., & Wang, X. (2017). SDN enabled 5G-VANET: Adaptive vehicle clustering and beamformed transmission for aggregated traffic. *IEEE Communications Magazine*, *55*(7), 120–127. doi:10.1109/MCOM.2017.1601160

El Mouna Zhioua, G., Tabbane, N., Labiod, H., & Tabbane, S. (2015). A fuzzy multi-metric QoS-balancing gateway selection algorithm in a clustered VANET to LTE advanced hybrid cellular network. *IEEE Transactions on Vehicular Technology*, *64*(2), 804–817. doi:10.1109/TVT.2014.2323693

Fan, P., Haran, J. G., Dillenburg, J., & Nelson, P. C. (2005). Cluster-based framework in vehicular ad-hoc networks. In Ad-Hoc, Mobile, and Wireless Networks (pp. 32-42). Springer Berlin Heidelberg. doi:10.1007/11561354_5

Frey, B. J., & Dueck, D. (2007). Clustering by passing messages between data points. *Science, 315*(5814), 972-976.

Gazdar, T., Belghith, A., & Benslimane, A. (2010). A cluster based secure architecture for vehicular ad hoc networks. In *Computer Systems and Applications (AICCSA), 2010 IEEE/ACS International Conference on* (pp. 1-8). IEEE.

Gazdar, T., Benslimane, A., & Belghith, A. (2011). Secure clustering scheme based keys management in VANETs. In *Vehicular Technology Conference (VTC Spring), 2011 IEEE 73rd* (pp. 1-5). IEEE.

Girinath, D. R., & Selvan, S. (2010). A novel cluster based routing algorithm for hybrid mobility model in VANET. *International Journal of Computers and Applications*, *1*(15), 35–42. doi:10.5120/326-495

Goldsmith, A. (2012). Beyond 4G: What Lies Ahead for Cellular System Design. In *Wireless Communications and Networking Conference*. IEEE.

Goonewardene, R. T., Ali, F. H., & Stipidis, E. (2009). Robust mobility adaptive clustering scheme with support for geographic routing for vehicular ad hoc networks. *Intelligent Transport Systems, IET*, *3*(2), 148–158. doi:10.1049/iet-its:20070052

3. GPP. (n.d.). *Release 14*. Retrieved from http://www.3gpp.org/release-14

Gunter, Y., Wiegel, B., & Großmann, H. P. (2007). Medium access concept for VANETs based on clustering. In *Vehicular Technology Conference, 2007. VTC-2007 Fall. 2007 IEEE 66th* (pp. 2189-2193). IEEE.

Hafeez, K. A., Zhao, L., Liao, Z., & Ma, B. N. W. (2011). Clustering and OFDMA-based MAC protocol (COMAC) for vehicular ad hoc networks. *EURASIP Journal on Wireless Communications and Networking*, (1): 1–16.

Harikrishnan, Y., & He, J. (2013). Clustering algorithm based on minimal path loss ratio for vehicular communication. In *Computing, Networking and Communications (ICNC), 2013 International Conference on* (pp. 745-749). IEEE.

Hasrouny, H., Samhat, A. E., Bassil, C., & Laouiti, A. (2017). VANet security challenges and solutions: A survey. *Vehicular Communications*, *7*, 7–20. doi:10.1016/j.vehcom.2017.01.002

Huang, C. J., Lin, C. F., Li, C. Y., Lee, C. Y., Chen, H. M., Shen, H. Y., & Chen, I. F. (2011). Service-oriented routing and clustering strategies for vehicle infotainment dissemination. *International Journal of Innovative Computing, Information, & Control*, *7*(3), 1467–1480.

Hussein, A., Elhajj, I. H., Chehab, A., & Kayssi, A. (2017, May). SDN VANETs in 5G: An architecture for resilient security services. In *Software Defined Systems (SDS), 2017 Fourth International Conference on* (pp. 67-74). IEEE.

INRIX. (n.d.). *INRIX Global Traffic Scorecard*. Retrieved from http://inrix.com/scorecard/

Karagiannis, G., Altintas, O., Ekici, E., Heijenk, G., Jarupan, B., Lin, K., & Weil, T. (2011). Vehicular networking: A survey and tutorial on requirements, architectures, challenges, standards and solutions. *IEEE Communications Surveys and Tutorials*, *13*(4), 584–616. doi:10.1109/SURV.2011.061411.00019

Kayis, O., & Acarman, T. (2007). Clustering formation for inter-vehicle communication. In *Intelligent Transportation Systems Conference, 2007. ITSC 2007. IEEE* (pp. 636-641). IEEE. 10.1109/ITSC.2007.4357779

Khan, A. A., Abolhasan, M., & Ni, W. (2018). An Evolutionary Game Theoretic Approach for Stable and Optimized Clustering in VANETs. *IEEE Transactions on Vehicular Technology*, *67*(5), 4501–4513. doi:10.1109/TVT.2018.2790391

Kuklinski, S., & Wolny, G. (2009). Density based clustering algorithm for VANETs. In *Testbeds and Research Infrastructures for the Development of Networks & Communities and Workshops, 2009. TridentCom 2009. 5th International Conference on* (pp. 1-6). IEEE. 10.1109/TRIDENTCOM.2009.4976256

Lee, K. C., Lee, U., & Gerla, M. (2009). Survey of routing protocols in vehicular ad hoc networks. Advances in Vehicular Ad-Hoc Networks: Developments and Challenges, 21.

Lin, X., Andrews, J., Ghosh, A., & Ratasuk, R. (2014). An overview of 3GPP device-to-device proximity services. *IEEE Communications Magazine*, *52*(4), 40–48. doi:10.1109/MCOM.2014.6807945

Lobato, W., Rosario, D., Gerla, M., & Villas, L. A. (2017, December). Platoon-based driving protocol based on game theory for multimedia transmission over vanet. In *GLOBECOM 2017-2017 IEEE Global Communications Conference* (pp. 1-6). IEEE.

Lu, R., Lin, X., Zhu, H., & Shen, X. (2009). SPARK: a new VANET-based smart parking scheme for large parking lots. In INFOCOM 2009, IEEE (pp. 1413-1421). IEEE.

Luo, Y., Zhang, W., & Hu, Y. (2010). A new cluster based routing protocol for VANET. In *Networks Security Wireless Communications and Trusted Computing (NSWCTC), 2010 Second International Conference on* (Vol. 1, pp. 176-180). IEEE.

Martinez, F. J., Toh, C. K., Cano, J. C., Calafate, C. T., & Manzoni, P. (2010). Emergency services in future intelligent transportation systems based on vehicular communication networks. *Intelligent Transportation Systems Magazine, IEEE*, *2*(2), 6–20. doi:10.1109/MITS.2010.938166

Maslekar, N., Boussedjra, M., Mouzna, J., & Houda, L. (2009). Direction based clustering algorithm for data dissemination in vehicular networks. In *Vehicular Networking Conference (VNC)*, 2009 *IEEE* (pp. 1-6). IEEE.

Maslekar, N., Boussedjra, M., Mouzna, J., & Labiod, H. (2011). A stable clustering algorithm for efficiency applications in VANETs. In *Wireless Communications and Mobile Computing Conference (IWCMC), 2011 7th International* (pp. 1188-1193). IEEE.

Maslekar, N., Mouzna, J., Labiod, H., Devisetty, M., & Pai, M. (2011). Modified C-DRIVE: Clustering based on direction in vehicular environment. In *Intelligent Vehicles Symposium (IV)*, 2011 *IEEE* (pp. 845-850). IEEE.

Massin, R., Le Martret, C. J., & Ciblat, P. (2017). A Coalition Formation Game for Distributed Node Clustering in Mobile Ad Hoc Networks. *IEEE Transactions on Wireless Communications*, *16*(6), 3940–3952. doi:10.1109/TWC.2017.2690419

Moustafa, H., Senouci, S. M., & Jerbi, M. (2009). Introduction to Vehicular Networks. In H. Moustafa & Y. Zhang (Eds.), *Vehicular Networks: Techniques, Standards and Applications*. CRC Press. doi:10.1201/9781420085723

Rakha, H., & Kamalanathsharma, R. K. (2011). Eco-driving at signalized intersections using V2I communication. In *Intelligent Transportation Systems (ITSC), 2011 14th International IEEE Conference on* (pp. 341-346). IEEE.

Ramakrishnan, B., Rajesh, R. S., & Shaji, R. S. (2011). CBVANET: A cluster based vehicular adhoc network model for simple highway communication. *J. Advanced Networking and Applications*, *2*(04), 755–761.

Reumerman, H. J., Roggero, M., & Ruffini, M. (2005). The application-based clustering concept and requirements for intervehicle networks. *Communications Magazine, IEEE*, *43*(4), 108–113. doi:10.1109/MCOM.2005.1421913

Santi, P. (2012). *Mobility Models for Next Generation Wireless Networks*. Sussex, UK: Wiley. doi:10.1002/9781118344774

Shea, C., Hassanabadi, B., & Valaee, S. (2009). Mobility-based clustering in VANETs using affinity propagation. In *Global Telecommunications Conference, 2009. GLOBECOM 2009. IEEE* (pp. 1-6). IEEE.

Shoaib, M., Song, W. C., & Kim, K. H. (2012). Cluster based data aggregation in vehicular adhoc network. In *Communication Technologies for Vehicles* (pp. 91–102). Springer Berlin Heidelberg. doi:10.1007/978-3-642-29667-3_8

Sivaraj, R., Gopalakrishna, A. K., Chandra, M. G., & Balamuralidhar, P. (2011). QoS-enabled group communication in integrated VANET-LTE heterogeneous wireless networks. In *Wireless and Mobile Computing, Networking and Communications (WiMob), 2011 IEEE 7th International Conference on* (pp. 17-24). IEEE.

Su, H., & Zhang, X. (2007). Clustering-Based Multichannel MAC Protocols for QoS Provisionings Over Vehicular. *Vehicular Technology. IEEE Transactions on*, *56*(6), 3309–3323.

Tal, I., Ciubotaru, B., & Muntean, G. M. (2016). Vehicular-Communications-Based Speed Advisory System for Electric Bicycles. *IEEE Transactions on Vehicular Technology*, *65*(6), 4129–4143. doi:10.1109/TVT.2015.2442338

Tal, I., & Muntean, G. M. (2012). User-oriented cluster-based solution for multimedia content delivery over VANETs. In *Broadband Multimedia Systems and Broadcasting (BMSB), 2012 IEEE International Symposium on* (pp. 1-5). IEEE.

Tal, I., & Muntean, G. M. (2013a). User-oriented Fuzzy Logic-based Clustering Scheme for Vehicular Ad-hoc Networks. In *Vehicular Technology Conference*, (pp. 1-5). IEEE.

Tal, I., & Muntean, G.-M. (2013b). V2X communication-based power saving strategy for electric bicycles. In *Global Telecommunications Conference Workshops, 2013. GlobeCom Workshops 2013.* (pp 1-6).

Tal, I., & Muntean, G. M. (2014). Towards smarter cities and roads: A survey of clustering algorithms in vanets. In Convergence of Broadband, Broadcast, and Cellular Network Technologies (pp. 16-50). IGI Global.

Tal, I., & Muntean, G. M. (2017). Towards Reasoning Vehicles: A Survey of Fuzzy Logic-Based Solutions in Vehicular Networks. *ACM Computing Surveys*, *50*(6), 80. doi:10.1145/3125640

Tal, I., Tianhua, Z., & Muntean, G.-M. (2013). On the Potential of V2X Communications in Helping Electric Bicycles Save Energy. *Vehicular Networking Conference*, 1-4.

Taleb, T., & Benslimane, A. (2010). Design guidelines for a network architecture integrating vanet with 3g & beyond networks. In *Global Telecommunications Conference (GLOBECOM 2010)*, 2010 *IEEE* (pp. 1-5). IEEE.

Teshima, S., Ohta, T., Kohno, E., & Kakuda, Y. (2011). A data transfer scheme using autonomous clustering in VANETs environment. In *Autonomous Decentralized Systems (ISADS), 2011 10th International Symposium on* (pp. 477-482). IEEE.

Tian, D., Wang, Y., Lu, G., & Yu, G. (2010). A VANETs routing algorithm based on Euclidean distance clustering. In *Future Computer and Communication (ICFCC), 2010 2nd International Conference on* (Vol. 1, pp. V1-183). IEEE.

Tielert, T., Rieger, D., Hartenstein, H., Luz, R., & Hausberger, S. (2012). Can V2X communication help electric vehicles save energy? In *ITS Telecommunications (ITST), 2012 12th International Conference on* (pp. 232-237). IEEE.

Trivedi, H., Veeraraghavan, P., Loke, S., Desai, A., & Singh, J. (2011). SmartVANET: The Case for a Cross-Layer Vehicular Network Architecture. In *Advanced Information Networking and Applications (WAINA), 2011 IEEE Workshops of International Conference on* (pp. 362-368). IEEE.

Tung, L.-C., Mena, J., Gerla, M., & Sommer, C. (2013). A Cluster Based Architecture for Intersection Collision Avoidance Using Heterogeneous Networks. In *Annual Mediterranean Ad Hoc Networking Workshop (Med-Hoc-Net 2013)*. IFIP/IEEE.

Ucar, S., Ergen, S. C., & Ozkasap, O. (2013). VMaSC: Vehicular multi-hop algorithm for stable clustering in Vehicular Ad Hoc Networks. In *Wireless Communications and Networking Conference (WCNC)*, 2013 *IEEE* (pp. 2381-2386). IEEE.

Uzcategui, R. A., & Acosta-Marum, G. (2009). Wave: A tutorial. *IEEE Communications Magazine*, *47*(5), 126–133. doi:10.1109/MCOM.2009.4939288

Vodopivec, S., Bester, J., & Kos, A. (2012). A survey on clustering algorithms for vehicular ad-hoc networks. In *Telecommunications and Signal Processing (TSP), 2012 35th International Conference on* (pp. 52-56). IEEE.

Wahab, O. A., Otrok, H., & Mourad, A. (2013). VANET QoS-OLSR: QoS-based clustering protocol for Vehicular Ad hoc Networks. *Computer Communications*, *36*(13), 1422–1435. doi:10.1016/j.comcom.2013.07.003

Wang, Z., Liu, L., Zhou, M., & Ansari, N. (2008). A position-based clustering technique for ad hoc intervehicle communication. *Systems, Man, and Cybernetics, Part C: Applications and Reviews. IEEE Transactions on*, *38*(2), 201–208.

Wanner, L. (2004). *Introduction to Clustering Techniques*. International Union of Local Authorities.

Wolny, G. (2008, October). Modified DMAC clustering algorithm for vanets. In *Systems and Networks Communications, 2008. ICSNC'08. 3rd International Conference on* (pp. 268-273). IEEE. 10.1109/ICSNC.2008.28

Yang, Q., Lim, A., & Agrawal, P. (2008). Connectivity aware routing in vehicular networks. In *Wireless Communications and Networking Conference, 2008. WCNC 2008. IEEE* (pp. 2218-2223). IEEE. 10.1109/WCNC.2008.392

Ydenberg, A., Heir, N., & Gill, B. (2018, January). Security, SDN, and VANET technology of driver-less cars. In *Computing and Communication Workshop and Conference (CCWC), 2018 IEEE 8th Annual* (pp. 313-316). IEEE.

Ye, H., Liang, L., Li, G. Y., Kim, J., Lu, L., & Wu, M. (2018). Machine Learning for Vehicular Networks: Recent Advances and Application Examples. *IEEE Vehicular Technology Magazine*, *13*(2), 94–101. doi:10.1109/MVT.2018.2811185

Yu, J. Y., & Chong, P. H. J. (2005). A survey of clustering schemes for mobile ad hoc networks. *IEEE Communications Surveys and Tutorials*, *7*(1-4), 32–48. doi:10.1109/COMST.2005.1423333

Zaydoun, Z. Y., & Mahmud, S. M. (2009). Toward Strongly Connected Clustering Structure in Vehicular Ad Hoc Networks. In *Conference on Vehicular Technology* (pp. 1-5). IEEE.

ADDITIONAL READING

5G Americas Whitepaper. (2018) Cellular V2X Communications Towards 5G, available online: http://www.5gamericas.org/files/9615/2096/4441/2018_5G_Americas_White_Paper_Cellular_V2X_Communications_Towards_5G__Final_for_Distribution.pdf

Amadeo, M., Campolo, C., & Molinaro, A. (2012). Enhancing IEEE 802.11p/WAVE to provide infotainment applications in VANETs. *Ad Hoc Networks*, *10*(2), 253–269. doi:10.1016/j.adhoc.2010.09.013

Campolo, C., Iera, A., Molinaro, A., Paratore, S. Y., & Ruggeri, G. (2012, November). SMaRTCaR: An integrated smartphone-based platform to support traffic management applications. In *Vehicular Traffic Management for Smart Cities (VTM), 2012 First International Workshop on* (pp. 1-6). IEEE.

Campolo, C., Molinaro, A., Iera, A., & Menichella, F. (2017). 5G network slicing for vehicle-to-everything services. *IEEE Wireless Communications*, *24*(6), 38–45. doi:10.1109/MWC.2017.1600408

Chen, S. Z., & Kang, S. L. (2018). A tutorial on 5G and the progress in China. *Frontiers of Information Technology & Electronic Engineering*, *19*(3), 309–321. doi:10.1631/FITEE.1800070

Djilali, Y. D., Bakhtil, Y., Kouninef, B., & Senouci, B. (2018, July). Performances Evaluation Study of VANET Communication Technologies for Smart and Autonomous Vehicles. In *2018 Tenth International Conference on Ubiquitous and Future Networks (ICUFN)* (pp. 79-84). IEEE. 10.1109/ICUFN.2018.8436661

Eltoweissy, M., Olariu, S., & Younis, M. (2010). Towards autonomous vehicular clouds. In *Ad Hoc Networks* (pp. 1–16). Springer Berlin Heidelberg. doi:10.1007/978-3-642-17994-5_1

Fan, P., Haran, J., Dillenburg, J., & Nelson, P. C. (2006, January). Traffic model for clustering algorithms in vehicular ad-hoc networks. In *Proc. of CCNC* (pp. 168-172).

Gerla, M. (2012). Vehicular Cloud Computing. In *Ad Hoc Networking Workshop (Med-Hoc-Net), 2012 The 11th Annual Mediterranean* (pp. 152-155). IEEE. 10.1109/MedHocNet.2012.6257116

Gerla, M. (2013). Safe driving in Crowded Spectrum: Cognitive radios for VANETs. Keynote presentation at *Network Protocols (ICNP), IEEE International Conference on, Workshop on Vehicular Communications and Applications*.

Gerla, M., & Kleinrock, L. (2011). Vehicular networks and the future of the mobile internet. *Computer Networks*, *55*(2), 457–469. doi:10.1016/j.comnet.2010.10.015

Gerla, M., & Tsai, J. T. C. (1995). Multicluster, mobile, multimedia radio network. *Wireless Networks*, *1*(3), 255–265. doi:10.1007/BF01200845

Hartenstein, H., & Laberteaux, K. P. (2008). A tutorial survey on vehicular ad hoc networks. *Communications Magazine, IEEE*, *46*(6), 164–171. doi:10.1109/MCOM.2008.4539481

Hossain, E., Chow, G., Leung, V., McLeod, R. D., Mišić, J., Wong, V. W., & Yang, O. (2010). Vehicular telematics over heterogeneous wireless networks: A survey. *Computer Communications*, *33*(7), 775–793. doi:10.1016/j.comcom.2009.12.010

Ide, C., Dusza, B., Putzke, M., & Wietfeld, C. (2012, June). Channel sensitive transmission scheme for V2I-based Floating Car Data collection via LTE. In *Communications (ICC), 2012 IEEE International Conference on* (pp. 7151-7156). IEEE.

Ide, C., Niehoefer, B., Knaup, T., Weber, D., Wietfeld, C., Habel, L., & Schreckenberg, M. (2012, September). Efficient Floating Car Data Transmission via LTE for Travel Time Estimation of Vehicles. In *Vehicular Technology Conference (VTC Fall), 2012 IEEE* (pp. 1-5). IEEE.

Kihl, M., Bur, K., Mahanta, P., & Coelingh, E. (2012, July). 3GPP LTE downlink scheduling strategies in vehicle-to-infrastructure communications for traffic safety applications. In *Computers and Communications (ISCC), 2012 IEEE Symposium on* (pp. 448-453). IEEE.

Lottermann, C., Botsov, M., Fertl, P., & Mullner, R. (2012, November). Performance evaluation of automotive off-board applications in LTE deployments. In *Vehicular Networking Conference (VNC)*, 2012 *IEEE* (pp. 211-218). IEEE.

Morgan, Y. L. (2010). Notes on DSRC & WAVE standards suite: Its architecture, design, and characteristics. *IEEE Communications Surveys and Tutorials*, *12*(4), 504–518. doi:10.1109/SURV.2010.033010.00024

Papathanassiou, A., & Khoryaev, A. (2016). Cellular V2X as the essential enabler of superior global connected transportation services. *IEEE 5G Tech Focus, 1*(2).

Phan, M. A., Rembarz, R., & Sories, S. (2011). A Capacity Analysis for the Transmission of Event and Cooperative Awareness Messages in LTE Networks. In *18th ITS World Congress*.

Rémy, G., Senouci, S., Jan, F., & Gourhant, Y. (2011, December). LTE4V2X: LTE for a Centralized VANET Organization. In *Global Telecommunications Conference (GLOBECOM 2011)*, 2011 *IEEE* (pp. 1-6). IEEE.

Remy, G., Senouci, S. M., Jan, F., & Gourhant, Y. (2012, June). LTE4V2X—Collection, dissemination and multi-hop forwarding. In *Communications (ICC), 2012 IEEE International Conference on* (pp. 120-125). IEEE.

Selvanesan, S., Thomas, R. R., Fehrenbach, T., Wirth, T., & Hellge, C. (2018, June). Towards Advanced V2X Multimedia Services for 5G Networks. In *2018 IEEE International Symposium on Broadband Multimedia Systems and Broadcasting (BMSB)* (pp. 1-6). IEEE.

Sharma, S., & Kaul, A. (2018a). Hybrid fuzzy multi-criteria decision making based multi cluster head dolphin swarm optimized IDS for VANET. *Vehicular Communications*, *12*, 23–38. doi:10.1016/j.vehcom.2017.12.003

Sharma, S., & Kaul, A. (2018b). *A survey on Intrusion Detection Systems and Honeypot based proactive security mechanisms in VANETs and VANET Cloud*. Vehicular Communications.

Toor, Y., Muhlethaler, P., Laouiti, A., & La Fortelle, A. (2008). Vehicle ad hoc networks: Applications and related technical issues. *IEEE Communications Surveys and Tutorials*, *10*(3), 74–88. doi:10.1109/COMST.2008.4625806

Vegni, A. M., Biagi, M., & Cusani, R. (2013). Smart Vehicles, Technologies and Main Applications in Vehicular Ad hoc. *Networks*. doi:10.5772/55492

Vinel, A. (2012). 3GPP LTE Versus IEEE802.11p/WAVE: Which Technology is Able to Support Cooperative Vehicular Safety Applications. *Communications Letters, IEEE*, *1*(2), 125–128. doi:10.1109/WCL.2012.022012.120073

Vukadinovic, V., Bakowski, K., Marsch, P., Garcia, I. D., Xu, H., Sybis, M., ... Thibault, I. (2018). 3GPP C-V2X and IEEE 802.11 p for Vehicle-to-Vehicle communications in highway platooning scenarios. *Ad Hoc Networks*, *74*, 17–29. doi:10.1016/j.adhoc.2018.03.004

Wu, X., Subramanian, S., Guha, R., White, R., Li, J., Lu, K., ... Zhang, T. (2013). Vehicular communications using DSRC: Challenges, enhancements, and evolution. *Selected Areas in Communications. IEEE Journal on*, *31*(9), 399–408.

KEY TERMS AND DEFINITIONS

3G/4G/5G – 3rd/4th/5th: Generation of mobile cellular communications.

Clustering: Technique of grouping similar objects used mainly to deal with scalability, but its aim depends on the context it is applied in.

Performance Assessment: Evaluation; in the context of this chapter, the evaluation of clustering algorithms in VANETs.

Smart Cites: Cities of the future that provide increased comfort to their citizens while creating a sustainable environment.

V2X Communications: The main enabling technology of VANETs.

VANET: Vehicular ad-hoc networks.

WAVE: Wireless access vehicular environment; V2X communications standardization.

WLAN: Wireless Local Area Network.

ENDNOTE

1 3GPP TS 23.303. (July 2015). Proximity-based services (ProSe); Stage 2.V12.5.0 (Rel-12).

Chapter 3
Resource Allocation in Heterogeneous Wireless Networks

Chetna Singhal
Indian Institute of Technology Kharagpur, India

Swades De
Indian Institute of Technology Delhi, India

ABSTRACT

The advent of heterogeneous broadband wireless access networks (BWANs) has been to support the ever-increasing cellular networks' data requirements by increasing capacity, spectrum efficiency, and network coverage. The focus of this chapter is to discuss the implementation details (i.e., architecture and network components), issues associated with heterogeneous BWANs (i.e., handovers, network selection, and base station placement), and also the various resource allocation schemes (i.e., shared resource allocation in split handover and inter-RAT self-organizing networks) that can improve the performance of the system by maximizing the network capacity.

INTRODUCTION

Emergence of BWAN as a popular alternative to the wire-line access infrastructure is primarily associated with steady increase in data rate support and has inherent advantages, such as: easy scalability; ease of use in the end system; and low deployment and maintenance cost. According to Dahlman et al. (2008), the 4th generation (4G) BWAN, like LTE-A (long term evolution-advanced) and WiMAX-Mobile (world-wide interoperability for microwave access-mobile), have a maximum data rate of approximately 1 Gbps in downlink and 300Mbps in uplink as per the IMT-Advanced (international mobile telecommunications-advanced) specification. According to the data published by Ericsson (2018) and Cisco (2017), the mobile broadband data traffic has been increasing exponentially every year and the increase was nearly 18-fold over the past five years. Traffic forecast update by Cisco (2017) also projects an increase in mobile devices and connections to 11.6 billion and a 24 percent increase in network connection speeds by 2021. Latouche et al. (2013) has suggested value-added services (based on end-user information and user

DOI: 10.4018/978-1-5225-7570-2.ch003

identification) and industry "coopetation" (simultaneous cooperation and competition between service providers) as feasible solutions for the predicted mobile data explosion. Mobile data offloading (such as Wi-Fi offloading) is a cost-effective solution for delivering cellular network data that eases network congestion, provides seamless connectivity, and offers higher bandwidth to end-users.

The future heterogeneous network would efficiently and effectively integrate the prevailing heterogeneities as discussed by Chan et al. (2011), such as: communication modalities, channel types, technology generations, protocol types, and QoS requirements. To address dynamic, distributed, and unpredictable nature of these networks, they need to have self-organisation properties that range from self-configuration during startup, to self-adaptation to dynamic changes in operating environment, to self-healing in presence of failures and losses, as has been discussed by Razzaque et al. (2010). Among the standardization groups in this domain, are next generation mobile networks (NGMN) alliance that aims to bring affordable mobile broadband services to the LTE and LTE-A end users and Small Cell Forum/Femto Forum that works towards adoption of small cell technologies to improve coverage, capacity, and services delivered by mobile networks.

Heterogeneous Networks have proved their worth in 3G/4G cellular systems and because of the innumerable advantages associated with them, they will play a big role in future 5G networks as well. In 3G and 4G cellular networks the aim of BS densification is to improve the wireless transmission rate in partial regions and to cover the dark spots arising due to various natural and artificial obstacles, as discussed by Ge et. al. (2016).

The prominent need to enhance network capacity, throughput and users' quality of service (QoS) has led to the advent of heterogeneous next generation networks, comprising of different RAN (radio access networks) connected to a single core network. Implementing these networks, while ensuring high data rates in the wireless environment, poses certain resource allocation challenges such as: positioning the base station (BS) transceiver, handovers and network selection etc. In this chapter, we discuss: the architecture of heterogeneous BWAN; the issues associated with their implementation such as handover, network selection and BS placement; shared resource allocation solution to maximize the users capacity; and the idea of self-organizing network (SON) based inter-RAT (radio access technology) MRO (mobility robustness optimization).

ARCHITECTURE

Heterogeneous network (het-net) architecture is a prominent low-cost approach where an operator can exploit the different BWANs to provide additional areal capacity gain, indoor coverage improvement, and improved quality of service (QoS) in the network, as per Yeh et al. (2011).

Deployment Scenarios

Multitier Architecture Network Components

A heterogeneous BWAN architecture consisting of hierarchical multitier multiple radio access technologies (RAT) deployments is shown in Figure 1. This multitier deployment improves capacity and coverage by enabling dense reuse of the spectrum and enhancing link quality. The role of tiers and the larger and smaller footprint devices is examined as follows:

Figure 1.

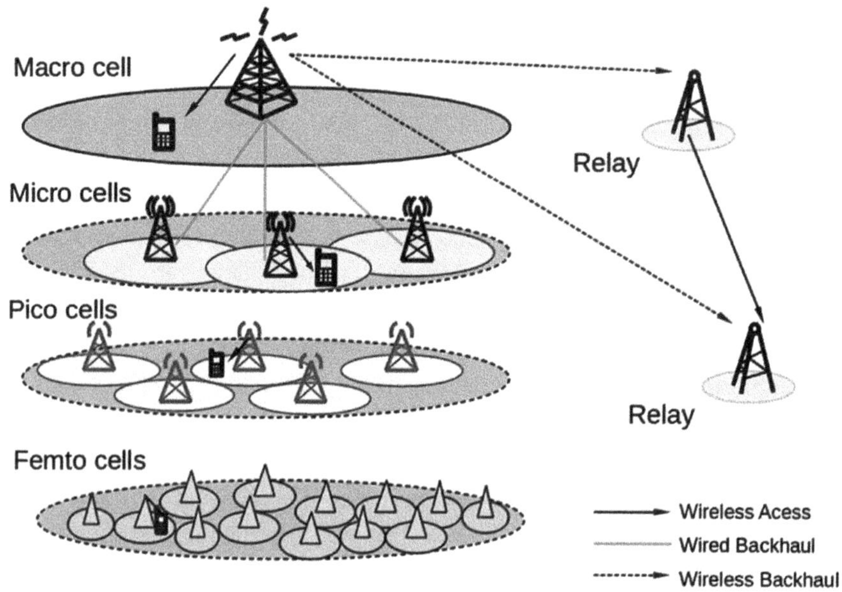

1. **Macrocells/Microcells:** Macrocells are deployed in rural or suburban areas with a coverage distance of more than 500m. The advantage of having large macrocells is that it provides support for high mobility users with reduced frequency of handovers. Both the macrocell and microcells provide essential coverage in het-nets.
2. **Picocells:** Pico BS (PBS) are deployed as hotspots to serve capacity hungry locations (example: airports, stadiums) where the serving area is smaller than micro-cells. PBSs are low cost, low power, simplified MBSs (macro/micro BS) that are accessible to all cellular users. MBSs and PBSs are planned and controlled by operators and horizontal optimization is beneficial across tiers. MBSs and PBSs are connected to the network through operator-owned backhaul.
3. **Femtocells:** Femto access point (FAP) covers a very small area (10-50m, in a house). These utilize residential backhaul links (digital subscriber line (DSL) or cable). FAPs are privately owned and deployed according to users' needs. FAPs are configured in closed subscriber groups (CSG), where FAP is accessible to only restricted users. CSG FAPs causes excessive interference to the network when the same spectrum is shared with other tiers. CSG FAP interference and network scalability to support large number of FAPs are important research challenges.
4. **Relays:** Relay stations serve similar sizes of footprints as PBSs. They provide coverage extension and throughput enhancement by forwarding an enhanced version of the received signal from BSs to mobile stations. Relays use wireless backhaul, so no landline resource is required; however, this reduces the amount of spectrum available for access. Operators may choose to implement infrastructural relays over PBSs at coverage holes where wired backhaul is unavailable or difficult to implement.
5. **Client Relay:** Client relay is the client cooperation (CC) tier that lies between clients comprising of very short range links. Utilizing good links between clients and BS, it improves the successful transmission probability for and virtual link quality for users in poor locations (i.e., cell edge users),

thereby reducing the amount of channel resources, battery power consumption, and interference caused to other cells. CC improves average network throughput.

It is efficient to deploy small cells in significant data generation sites (shopping mall, stadium, university campus, public transportation hub, etc.) and where subscribers spend considerable amount of time (home, office, etc.).

Handover

Handover is the process of transferring an active call or data session of a mobile user from one cell in a cellular network to another or from one channel in a cell to another. It plays a very vital role in mobile wireless networks, to ensure a seamless service to a mobile user. Because of high data rates in broadband wireless access networks, providing a seamless handover is a challenging issue. Handover is associated with loss of data resulting in a decreased throughput and increased latency, discussed by Vasudeva et. al. (2017).

Before performing handover, an appropriate BS candidate must be chosen and then the handover procedure should be continued based on the current technology and the specific application constraints of the MS (mobile station/subscriber). The exact procedures vary depending on the used technology, and usually within the technology several alternatives are available as well.

Classification and Associated Factors

The various handover schemes in het-nets are based on several factors that have been shown in Figure 2. The underlying factors for these handover schemes are enlisted as follows:

1. **Network Types Involved:** Based on the network types involved, handovers are classified as vertical handover and horizontal handover (depicted in Figure 2). Horizontal handover is the handover between two same radio access technology (RAT) networks and vertical handover is the handover between two different RATs. Vertical handovers can be further classified as upward vertical handover and downward vertical handover. In upward vertical handover the user device connects to a cell with larger coverage area and lower bandwidth than its source cell. In downward vertical handover the user device connects to a cell with lower coverage area and higher bandwidth. Downward handover is less critical as the device can stay connected to upper layer. In vertical handover, the decision of when to carry out handover itself is a complex problem since in vertical handover we cannot consider only the signal strength like we do in horizontal handover due to different physical techniques that are used by different overlay networks. In vertical handovers, the two key issues to be considered are network conditions for vertical handover and connection maintenance which have been discussed by Guo et al. (2004). Both the vertical and horizontal handover process constitute of three phases – handover decision, radio link transfer, and channel assignment, as discussed by Nasser (2006). These handover process phases are elaborately discussed later in the chapter.
2. **Frequencies Engaged:** Based on the frequencies engaged, handovers are classified as intra-frequency handover and inter-frequency handover. Intra-frequency handover is the handover between same frequency access points. Example: handovers in CDMA. Inter-frequency handover is the handover between different frequency access points. Example: handovers in GSM.

Figure 2. Handover classification

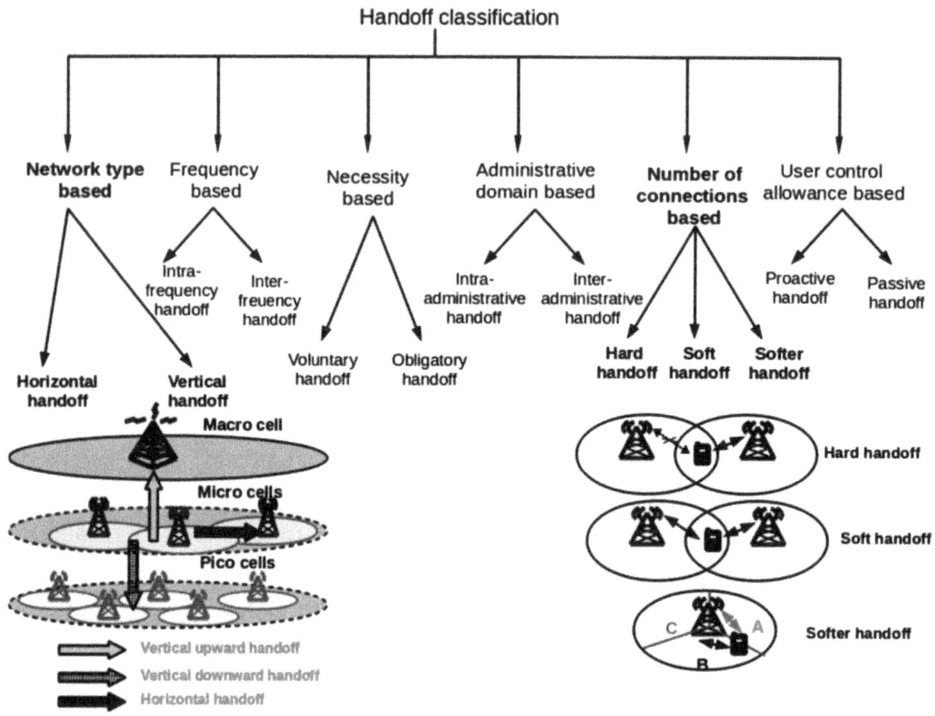

3. **Number of Connections Involved:** Based on the number of connections involved, handovers are classified as hard handover, soft handover, and softer handover. These are depicted in Figure 2. In a hard handover, the new connection will only be established after disconnecting from source base station. In hard handover the user device maintains only single connection at a time. In a soft handover, the user device maintains connection with atleast two base stations in an overlapping handover region. These types of handovers are possible when handover is between cells operating with same frequency. Softer handover is soft handover between sectors of a given cell. Different variants of hard handover and soft handover are semi-soft handover and fractional handover in OFDM based systems that have been discussed by Lee et al. (2011) and Chang et al. (2011), fast base station switching (FBSS) and macro-diversity handover (MDHO) specified in IEEE Standard for Local and Metropolitan Area Networks-Part 16. (2005). Soft and softer handovers improve forward link capacity and signal to interference ratio for mobile users near the cell and sector boundaries, as per Lee et al. (1998).
4. **Administrative Domains Involved:** Based on the administrative domains involved, handovers are classified as intra-administrative handover and inter-administrative handover. In intra-administrative handover the user device moves from one network to another which belongs to same operator. In Inter administrative handover the user device moves from one network to another which is operated by different operators.
5. **Necessity of Handover:** Based on the necessity of handover, handovers are classified as obligatory handover and voluntary handover. Obligatory handover is necessary to avoid connection termination. Voluntary handover is optional and mostly used in scenarios in order to improve QOS.

6. **User Control Allowance:** Based on the user control allowance, handovers are classified as proactive handover and passive handover. In proactive handovers, user can give preferences which will affect the handover decision. In passive handover, the handover decision is not controlled by the user.

As compared to soft/softer handover schemes, a further increase in capacity is achieved by using a split handover procedure as discussed by Singhal et al. (2014), which results in a capacity gain by employing a 2-dimensional shared resource allocation for cell edge users in orthogonal frequency division multiple access (OFDMA) networks. Here, the MSs at the cell-edge can maintain parallel connections with more than one BS when it is in their coverage area. A MS, before handover to a new BS, seeks to utilize additional resources from the other BSs if the BS through which its current session is registered is not able to satisfy its requirements. The BSs participate in split handover operation while guaranteeing that they are able to maintain QoS of the existing connections associated with them.

Handover Process Operations

The handover process constitutes of three phases (handover decision, radio link transfer, and channel assignment) and the following discussion is in context of hard and soft handover.

1. **Handover Decision:** In horizontal handover, the decision is mainly reflected by the QoS which in turn depends mainly on signal strength since the source and target base station are using the same physical techniques. So a horizontal handover is executed if the signal strength from the source BS is less than the threshold value and signal strength from neighboring BS is more than the threshold. Unlike in horizontal handover there are many network parameters that will affect the handover decision in vertical handover since it is between the BSs with different access technologies. These parameters are - quality of service, cost of service, power requirement, velocity, and proactive handover etc. Here the network selection which we will discuss subsequently comes into picture. Based on the above parameters, if a better network is available when the current one is not able to service then vertical handover will be executed. The handover parameters are briefly stated as follows:
   a. **Quality of Service (QOS):** Handing over to a better network with better QoS parameters such as - transmission rates, error rates, delay and connectivity etc.
   b. **Cost of Service:** It is one of the important parameter as some users are not willing to pay more than a certain limit and may want to connect to the cheapest network available with minimum requirements. Different access technologies have different costs, hence it is critical in vertical handover decision making.
   c. **Velocity:** It is also an important parameter because if the user is moving with higher velocity then it is better to connect him to a cell with larger radius (macro cell) rather than connecting him to a cell with smaller radius (micro or pico cell) which in turn increases the handovers.
   d. **Power Requirements:** The mobile devices have the limited battery power and when the battery level decreases it is better to handover to a network which consumes less battery power.
   e. **Proactive Handover:** By proactive handover, it means that the ultimate decision is with the user to satisfy their special requirement. If a user wants to handover to a network regardless of network conditions the vertical handover decision is executed.

2. **Radio Link Transfer:** There are two approaches for radio link transfer they are forward handover and backward handover. In forward handover the radio link transfer is initiated by the target BS unlike the backward handover in which the source BS initiates the radio link transfer. Both have their advantages and disadvantages, for instance the advantage of backward handover is that in this the initial signaling information is transmitted through the existing connection and thus does not need establishment of a new channel. However, the disadvantage of backward handover as per Shen & Zeng (2007) is that if the link quality of source BS is too poor, then the signaling information might be lost leading to unsuccessful handover. Backward handover is used in most cellular networks.
3. The final handover stage is the channel assignment stage, in which allocation of resources to the mobile node is done by the target BS. In this stage a call admission control (CAC) algorithm will take decision of accepting or rejecting the new mobile node based on the available amount of resources and the effect of existing connection on QoS that may have been caused due to a new connection.

NETWORK SELECTION

Network selection in het-nets is a challenge as the user device has to select from radio access networks (RANs) with different latency, coverage area, cost etc. These RANs may belong to different service providers who are competing with each other to maximize their own revenue or cooperating with each other for mutual benefit. The network selection may be needed mainly at two stages, while requesting for a new connection and in the handover process. Proper modeling of network selection by user device is important, since a bad network selection algorithm results in unnecessary handovers which in turn decreases the QoS of the users.

The network selection in het-nets is modeled mainly by using two different approaches that are: user oriented approach and network oriented approach. In network oriented approach, a central node runs the decision algorithm to assign the access networks to the users and users bind by the decision. The above approach mainly enhances the network capacity. It is very complex to implement and it creates extensive overhead in networks. On the contrary, in user oriented approach given by Charilas & Panagopoulous (2010), the users' mobile device executes the algorithms for network selection and implements them. This approach is more dynamic and it has less communication overhead and low complexity of implementation.

Steps Involved in Network Selection

Network selection for heterogeneous networks mainly involves the following five steps:

1. **Selection of Decision Criteria:** The user device is responsible for determining the set of decision criteria to be used. The decision criteria are classified into the following three categories:
    a. **Network Based Parameters:** Utilization, reliability, coverage, anonymity, security, seamless number of active users, etc.
    b. **Service Based Parameters:** Delay, jitter, packet loss, throughput average in kbps, bit error rate, signal-to-interference ratio in dB, etc.

Figure 3. User oriented network selection process

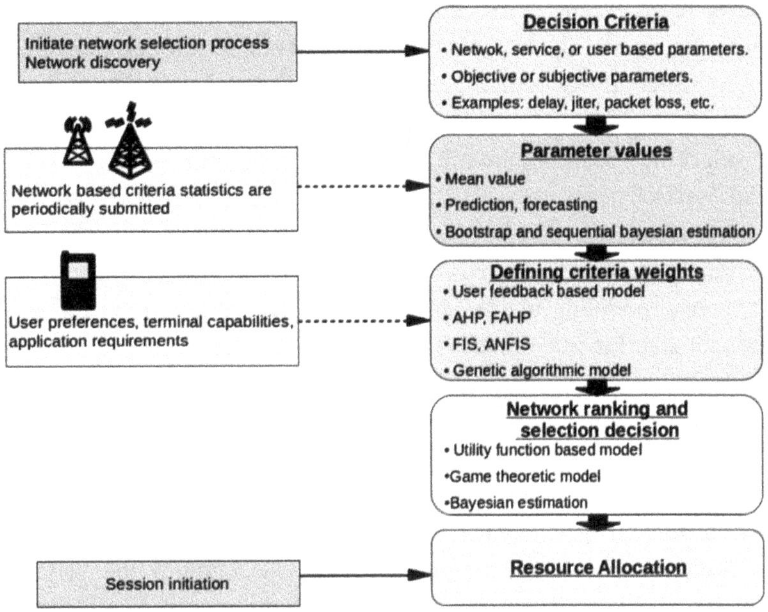

c. **User Based Parameters:** Sojourn time, forced termination probability for handover, blocking probability, etc.The above mentioned parameters can also be classified into two groups, as follows:
   i. **Objective Parameters:** Can be measured accurately and thus can be quantified.
   ii. **Subjective Parameters:** Can only be estimated usually in the form of fuzzy numbers (i.e., a number that refers to a set of possible values between 0 and 1), as given by Zhang (2004).

2. Parameters like jitter, packet loss are to be given low priority because they cannot be measured before establishing the connection to the network whereas parameters such as bandwidth, signal-to-interference ratio, and coverage area should be given comparatively higher priority since information about these can be known before connecting to the network.
3. **Collection of Values for Selected Criteria:** After listing of the available RANs, value of parameters for each RAN is collected, and monitoring of the network conditions is also done. The following four ways are used for collecting the parameter values:
   a. Collecting mean values from previous sessions. This is suitable when the network is not prone to abrupt changes.
   b. Collecting most recent values that reflect current status of the network.
   c. Predicting and forecasting the values of the parameters. The forecasting can be done by two techniques such as: direct estimation of parameters like load on the hot spot by using past values in weighted moving average or simple moving average methods; and predicting the probability of parameter (like SNR, blocking probability, termination probability) using markov chains for modeling, as discussed by Shen & Zeng (2007).
   d. Bootstrap and sequential Bayesian estimation of dynamic QoS parameters discussed by Ong & Khan (2008) and Zoubir & Boashash (1998). The bootstrap estimation generates probability distribution of parameters using samples collected from parameter values.

4. Defining criteria weights in the final outcome based on their importance and also user or application context based requirements. This can be performed by the following six models:
    a. User feedback based evaluation of weights i.e. obtained through questionnaire and provided to the device. Since this model depends on overall users' perception of the service, it cannot guarantee accurate results for objective criteria. However, in scenarios where user always wants to select the cheapest network, this model is helpful.
    b. Analytical hierarchy process (AHP), given by Pervaiz & Bigham (2009) is widely used in multi criteria models by dividing the decision problem into multi-level hierarchical model and determining weights by paired comparisons between parameters. When AHP is followed by game theory modeling for ranking, it results in achieving optimum network utility and customer's satisfaction in competing wireless networks.
    c. Fuzzy analytical hierarchy process (FAHP) is an extension of AHP that solves hierarchical fuzzy problems, since AHP cannot handle imprecise data represented in fuzzy numbers. The outputs of pair-wise comparison in FAHP are fuzzy numbers. Shen (2010) uses FAHP for access selection among various RATs and for optimizing system performance by considering multiple criterions of QoE to cater to user-preferences and experiences. Sgora (2010) combines two Multi Attribute Decision Making (MADM) methods, the AHP method to determine weights of the criteria and the fuzzy Total Order Preference by Similarity to the Ideal Solution (TOPSIS) method to obtain the final access network ranking.
    d. In fuzzy inference system (FIS), input is converted to linguistic variables by fuzzier; system is controlled using if-then rules (based on questionnaires, network measurement, and previous system knowledge) by the inference engine. The main disadvantage of this system is that it is not scalable, since the weights of all parameters are determined by all the combination of large number of rules in the inference engine.
    e. Adaptive network based FIS (ANFIS) is developed to add scalability and adaptability for FIS. The advantage of this system is that the user need not develop all the rules, instead a training data set is the input to ANFIS and the rules are automatically adapted. However, in FIS and ANFIS the number of rules increases exponentially with increase in the number of RANs. Jang (2002) presented FIS based architecture and learning procedure for adaptive networks.
    f. General algorithms (GA) can also be implemented to determine the weights. Unlike FIS and AFIS, genetic algorithms can be used when large numbers of RANs are involved. They can also be used to implement fuzzy logic for network selection, as have been discussed by Alkhawlani (2008).
5. **Ranking the Networks and Network Selection Decision:** The network selection decision is initiated by a new call connection or by handover of existing connection. The best network is selected by ranking of the available networks based on evaluation of the parameters collected in the previous step. The various procedures for ranking of networks are given as follows:
    a. **Utility Function:** This is an easy way to rank the available networks by maximizing the utility function value. Network with highest utility function value is selected. The networks are ranked in the decreasing order of utility function value. For example, according to Ormond et al. (2006), a possible utility function used for ranking can be consumer surplus (Datavolume -ServiceCost). A disadvantage of this method is that every user tries to maximize its own utility function irrespective of others.

b. **Game Theory Model:** It is the best tool to model the competitive environment where different players interact to maximize their utility functions, as has been discussed by Trestian et al. (2012). There are six different approaches of using game theory to model network selection:
   i. Users versus users cooperative approach, where users mutually cooperate with each other for mutual benefit.
   ii. Users versus users non-cooperative approach, where users compete with each other to maximize their own utility function.
   iii. Users versus networks cooperative approach, where users and networks mutually cooperate with each other for mutual benefit.
   iv. Users versus networks non-cooperative approach, where users and networks compete with each other for their own benefits.
   v. Networks versus networks cooperative approach, where networks mutually cooperate with each other for mutual benefit.
   vi. Networks versus networks non-cooperative approach where networks mutually compete with each other for their own revenue maximization.

c. Non-cooperative game model is mostly used as compared to cooperative (i.e. cooperative behavior between BSs) game model. In the above scenario, users try to maximize their utility function and networks try to maximize their revenue and capacity. The various game theory models like trading market game, auction game, repeated game, bargaining game and strategic game (prisoners dilemma) etc. can be used in different approaches. Watanabe et al. (2008) have used Evolutionary game for network selection and parameters considered for payoff function are loss rate, mean burst size and delay jitter; Zhu et al. (2010) have used Bayesian game and parameters considered for payoff function are bandwidth and price; Khan et al. (2010) and Khan et al. (2009) used Auction game for network selection and parameters considered are bandwidth, Delivery and response time, application requirements; Charilas (2009) used Prisoners dilemma game for network selection and parameters considered are delay, jitter, throughput, packet loss and cost; Niyato & Hossain (2008) have used Trading market game for network selection and parameters considered are bandwidth and network load. Bargaining game and repeated game are cooperative games while others are non-cooperative games.

d. **Bayesian Estimation:** It is the fundamental statistical approach for decision making in the presence of uncertain values. It is based on Bayes theorem. It maximizes the probability average values ($\leq$ corresponding thresholds) estimated and thus selects the network with highest probability.

6. **Resource Allocation at the New Base Station:** For network selection initiated by handover, after the network is selected, protocol is transferred from source BS to target BS and resource is allocated at target BS.

BASE STATION PLACEMENT

BS placement is also a very important challenge in heterogeneous networks because poor placement leads to wastage of resources or inefficient utilization which in turn decreases the capacity and revenue etc. Choosing the optimal places for BSs in heterogeneous networks is a complex problem because of its multi-layer structure (usage of macro, micro, pico and femto cells) and different radio access technologies with different coverage areas should coexist to enhance the effective coverage and increase spectrum efficiency. The parameters to be considered to find an optimal position for BSs for different radio access technologies are coverage area, capacity, available bandwidth, cell structure (antenna height), maintenance cost, and deployment cost etc.

There are several algorithms proposed in the literature and some of these are discussed here. Guruprasad (2011) proposed a Generalized Vornoi partition scheme. Vornoi partition also known as Dirchlet tessellation is mainly used in homogeneous scenario. So generalization of standard Vornoi partition is used in order to corporate the heterogeneity of BSs. Wen (2009) formulated the BS placement problem as an integer-nonlinear programming problem. In this the algorithms used to choose BS position are based on simulated annealing approaches and Lagrangian relaxation. The methods used by Wen (2009) mainly focused on decreasing the deployment cost of the heterogeneous network while satisfying users' QoS requirements.

Relay Based Heterogeneous Network

The relay based het-net improves the effective coverage area, and QoS, etc,and have been discussed in Peng et al. (2011). In relay based het-net, the relay nodes (relay stations) are introduced which will receive the signal from the BS, enhances it and retransmit it so that the cell edge users' QoS is improved. An iterative algorithm that has been discussed by Isalm et al. (2012), determines the optimum positions for the BSs and relay stations. Islam et al. (2012) showed that by using relays, the total transmitted power is significantly reduced and is thus preferred over increasing the number of BSs instead.

SHARED RESOURCE ALLOCATION IN HET-NETS

A significant improvement in capacity and QoS of the het-nets can be achieved by using shared resource allocation scheme (split handover) and network selection, wherein the resources of multiple RATs are pooled together so as to enhance the user experience in a mobile network environment. Here, we discuss the architecture, capacity enhancement analysis, and protocol functionalities in split handover with network selection facilitating differentiated QoS provisioning that accounts for MS speed, channel quality and load of different RAN involved in resource sharing.

System Model

With the advent of LTE and use of 2G and 3G being prevalent, the inter-RAT handover between LTE to 3G and from 3G to 2G, would gain significance due to limited LTE and 3G coverage since LTE is deployed more in areas having high traffic overlaying with the legacy 2G and 3G mobile systems. We consider BWA (broadband wireless access) system with LTE, 3G, and 2G mobile systems deployed. In order to incorporate the flexibility in the scheme Figure 4 depicts a new transport layer, two-level queueing system model, for minimized impact of user movement on the connection for active users. Queueing is applied at the data link control (DLC) layer and the transport control (TCP) layer, to distribute the traffic to the BSs which are participating in data transmission of the users in the shared region. This scheduling principle applies to both downlink and uplink (control message exchange related modifications) traf-

Figure 4. System architecture and downlink queuing model, with different queue structure for the shared and non-shared users

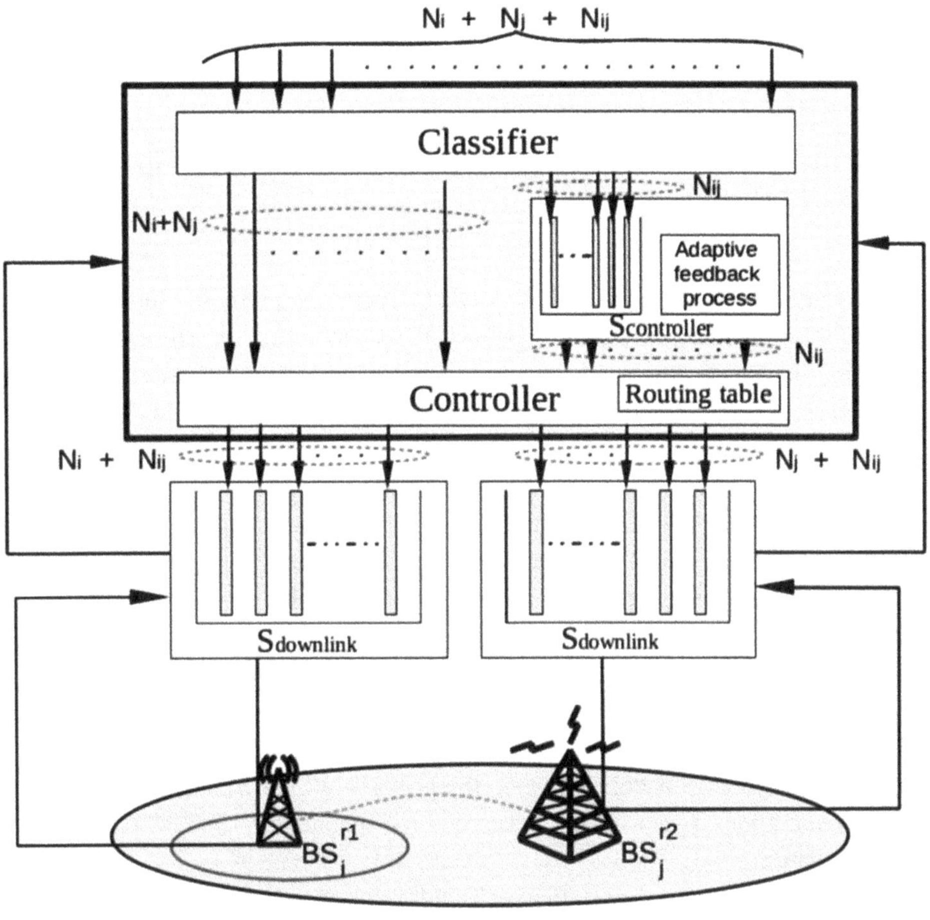

fic. In Figure 4, the node architecture and its interaction with other network entities is presented. This resource sharing between two BSs can be extended to three or more BSs.

Two BSs, BS_i^{r1} and BS_j^{r2} are shown in the figure belonging to different RAT *r1* and *r2* with an overlapping coverage. There are N_i users which are served only by BS_i^{r1}, N_j users served only by BS_j^{r2}, and N_{ij} users served by both BS_i^{r1} and BS_j^{r2} in shared mode. BSr1i maintains queues for Ni + Nij users and serves them using its $S_{downlink}$ scheduler. Likewise, BS_j^{r2} maintains queues for $N_j + N_{ij}$ users and serves them using $S_{downlink}$. It is assumed that one user has only one class of service at a time. If a single user maintains multiple parallel connections with different QoS requirements, then the scheduling can be easily handled by additional queues, called priority queues, at a BS. A controller directs the flows from the classifier according to the routing table maintained at the controller. Controller and classifier are the two logical entities which can be physically co-located. Based on the feedback from the BSs, the classifier is used to distinguish the incoming/outgoing flows if they are of a shared user - served by two BSs, or a non-shared user - served by only one BS. The controller also maintains the queues for all users which can be served by both BSs. Flow scheduling at the controller is according to the rule provided by $S_{controller}$. The parameters considered for splitting of traffic are fed back to the controller using feedback links.

Table 1. WiMAX multi-tier architecture base stations

Tier	Range	No. of Users per Sector	Typical RF Output Power	No. of Sectors
Macro	up to 5 km	>200	10W	3
Micro	up to 3km	<200	4W	3
Outdoor pico	1km	up to 100	1W	1
Indoor pico	100m	up to 50	500mW	1
Femto	20-30m	<10	100mW	1

Some of the advantages of the given architecture are: (i) centralized routing information maintenance for the subscribers to create multiple parallel connections when necessary; (ii) avoidance of packet duplication, by distributing packets for a cell-edge user across the BSs, thereby minimizing resource wastage; (iii) rule based splitting of traffic by using scheduler $S_{controller}$; (iv) possibility of resource allocation based on traffic classification.

The controller in split handover is connected to the BSs via high-speed wireline or wireless links. Beyond signal transmission-reception over the radio links, the BSs have a very little role to play. Functionality-wise, a controller will perform some extended tasks beyond a conventional BSC (base station controller) or a RNC (radio network controller). The specific activities of a controller in split handover are: (i) construction of universal DL-map and broadcasting to all BSs, and (ii) scheduling and traffic load balancing by accounting the CINR (carrier-to-interference- and-noise ratio) at the MS from the connected BS and the neighboring BSs, available resources of the neighboring BSs, and QoS requirements of subscribers. The participating BSs are assumed synchronized through the controller.

The split-handover of the MS occurs between- the primary BS (PBS) (the BS with which a MS exchanges the management messages as well as data and the secondary BS (SBS) (a BS with which the MS exchanges only data).

The WiMAX multi-tier architecture's BS specification for each tier is given in Table VI-A. For an inter-tier handover the PBS and SBS belong to different tiers. This concept is equally applicable to multi-tier architecture of other 3G and 4G wireless technologies.

Multi-Tier and Intra-RAT Split Handover Scheme for WiMAX

WiMAX standard based shared resource split-handover system functionalities for multi-tier architecture has been explained in the following discussion and an extension of this scheme is applicable to other RATs. Following the WiMAX standard notations for channel usage, the downlink interval usage code (DIUC) used by a MS with the PBS is denoted as DIUC1, and the DIUC used by a MS with the SBS is denoted as DIUC2. As indicated in the proposed system architecture (Section VI-A), traffic splitting is done at the transport layer. The controller stores the BS IDs and their associated loads. With respect to a particular MS, it stores the MS ID, its MAC address, PBS ID, DIUC1, priority calculated based on the service flow QoS parameters, and SBS ID and DIUC2 - in case the MS is in contact with two BSs. The timing diagram of a MS session and a split handover process is shown in Figure 5, in which only downlink data traffic is considered. The initialization procedure is similar as in a standard service flow set up. The MAC information and DIUC1 of the MS is passed to the controller at the network entry phase. During data and management message exchange with the PBS, the MS sends scanning request

Figure 5. Timing diagram of a MS session and split handoff process, where only the downlink traffic is considered

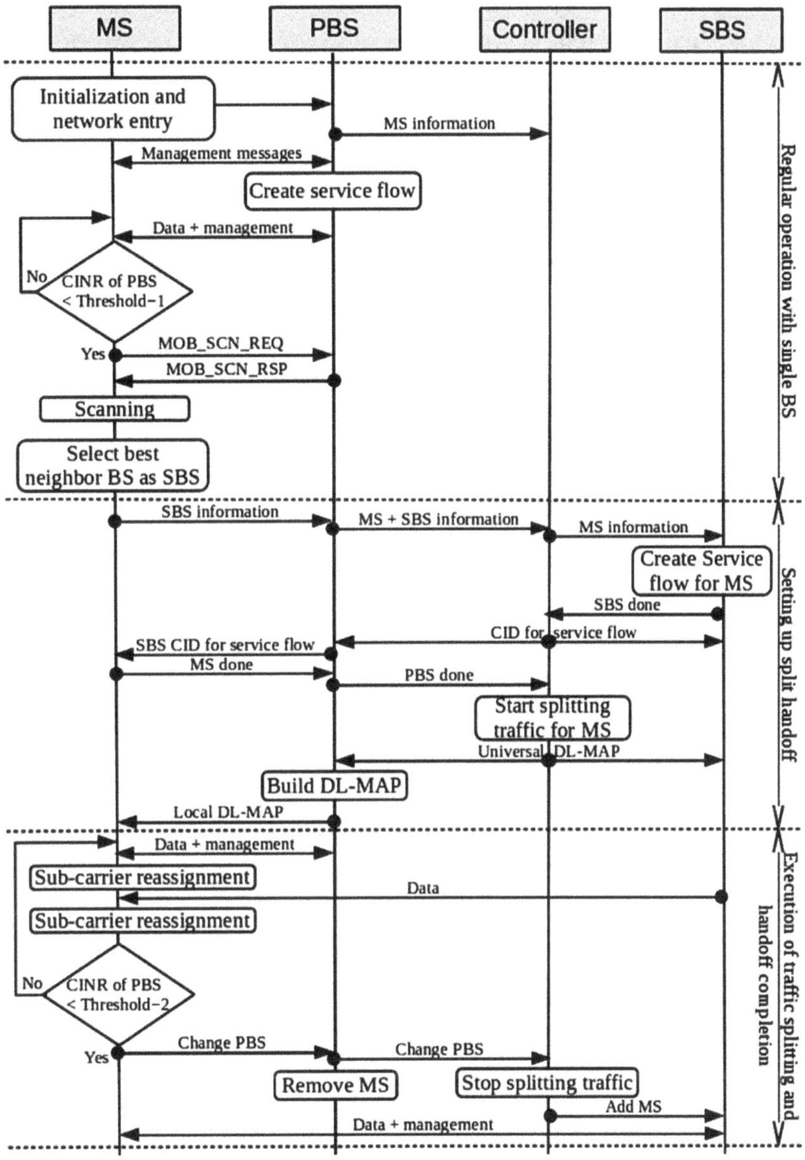

(MOB_SCN_REQ) to the PBS when the CINR from the PBS falls below Threshold-1 (which corresponds to a higher-than-the-lowest modulation and coding rate).

If the response (MOB_SCN_RSP) from the PBS is positive, the MS starts scanning for a SBS and synchronizes with one, initiating the split handover set up phase. Otherwise it continues its connection with the PBS only. During scanning, the MS sends (via the PBS) the SBS information (BS ID, DIUC2, traffic load) to the controller. Subsequently, a new CID for data connection with the SBS is created by the controller, which is forwarded to the MS. With the ongoing CID for the PBS and the new CID for the SBS, the user-level split handover procedure starts. The timing information of the bursts (slots) for connecting to the PBS and SBS is notified to the MS by the controller via a universal DL-map. To address user QoS and cell load imbalance, the controller accounts for the QoS priority, buffer status at the MS, and the BS traffic load at the time of burst scheduling. The burst timings for the PBS and SBS are separated within a frame such that the sub-carrier frequency reassignment latency of the MS is sufficiently accommodated. Note that, with frequency reuse factor = 1, it may be possible that the MS is connected to the BSs at different time slots over the same assignment of sub-carriers, in which case no reassignment is necessary. On the other hand, in case the adjacent BSs operate at different carrier frequencies, i.e., with frequency reuse factor < 1, split handover involves sub-carrier reassignment latency. Finally, when the CINR from the PBS falls below Threshold-2 (corresponding to the lowest allowable data rate), a PBS change request is sent to the controller. At that point the SBS assumes the responsibility of PBS for the MS. This process marks the *end of split handover*.

Analytical Model for Shared Resource Allocation in Het-Nets

In the following discussion an analytical framework for shared resource allocation is shown to result in capacity gain.

1. **Maximizing Capacity by Scheduling Shared Users:** Theoretically, QoS is defined by the maximum tolerable delay D_{max} for a user (traffic type) beyond which the delay violation probability exceeds a predefined threshold ε i.e. $\sup_t \Pr\{D(t) \geq D_{\max}\} \leq \varepsilon$. Also for a dynamic Also for a dynamic queuing system, where the arrival and service processes are stationary and ergodic, the probability that the delay at time t, $D(t)$ exceeds the threshold D_{max} can be accurately given as $\sup_t \Pr\{D(t) \geq D_{\max}\} \approx \gamma(\Omega)e^{\theta(\Omega)D_{\max}}$, where $\theta(\Omega)$ is a function of constant source rate Ω, D_{max} is a sufficiently large quantity, and $\gamma(\Omega)$ is the probability that the delay of a particular packet is nonzero, i.e., $\gamma(\Omega) = \Pr\{D(t) > 0\}$ at a randomly chosen time instant t. Here, $\theta(\Omega)>0$ is a parameter describing the exponential decay rate of probability of QoS violation. $\theta(\Omega)$ is referred as the QoS exponent. A large value of $\theta(\Omega)$ corresponds to stringent QoS requirement and a smaller value corresponds to a loose QoS requirement.

The effective capacity for a given QoS exponent $\theta(\Omega)$ specifies the maximum constant arrival rate that can be supported by the system at the link layer. Applying the effective capacity concept to wireless channels with arbitrary physical-layer characteristics, for a discrete-time stationary and ergodic service process with rate $\mu(n)$ and channel service rate $R(m) = \sum_{n=0}^{m} \mu(n)$, the effective capacity is given as $E_c(\theta) = \lim_{m\to\infty} \frac{1}{\theta_m} \ln E\{e^{-\theta R(m)}\}$ where m is the block length. For uncorrelated block fading channels, where the service process $\mu(n)$, $n=1,2,\cdots$ is also uncorrelated, the expression is reduced to $E_c(\theta) = \frac{1}{\theta_m} \ln E\{e^{-\theta\mu(n)}\}$, where the product T_fB is the total time-frequency resources available in one frame $S=T_fB$.

A saturation condition arises when the total resource demand is more than the available resources at the BS such that $\sum_{u=1}^{N_i} S^u = S_i$ where S_i is the total resource available per frame in BS_i and $(0 \le S_u \le S_i)$ is the resource allocated to the user u from BS_i, in order to maintain its QoS demand. It is assumed that the total resource available, at each BS in a cluster with FFR plan, is equal, i.e., $S_i = S \forall i \in N_c$.

The effective capacity of user u scheduled from BS_i, is $E^u_{C,i}(\theta^u) = -\frac{1}{\theta^u S} \ln E\{e^{\theta^u \mu^u_i}\}$ where θ^u is the QoS exponent of user u and $\mu^u_i = r^u_i S^u$ is the rate provided to user u from BS_i, with modulation index r_i^u. The *joint effective capacity* when the user u is scheduled from two BSs i.e. BS_i and Bs_j, is:

$$E^u_{C,joint}(\theta^u) = -\frac{1}{\theta^u S} \ln\left[E\{e^{\theta^u \mu^u_{i(1)}}\}E\{e^{\theta^u \mu^u_{i(2)}}\}\right],$$

such that

$$S^u = S^u_{i(1)} + S^u_{j(2)}, S^u > S^u_{i(1)}\ S^u > S^u_{j(2)} > 0, \gamma_i > \gamma_{th}$$

and $\gamma_j > \gamma_{th}$ i.e. joint resources from BS_i and BS_j are the same as before and the CINR of these BSs is above acceptable threshold. Considering $\Pr\{\gamma^u_i \le \gamma_{th}\}$ and $\Pr\{\gamma^u_j \le \gamma_{th}\}$. Assuming the same modulation index (i.e. r) for the user u from the two BSs then

$$E^u_{C,joint}(\theta^u) = -\frac{1}{\theta^u S} \ln\left[\left\{e^{-\theta^u r S^u_{i(1)}}(1-p_i) + p_i\right\} \bullet \left\{e^{-\theta^u r S^u_{j(1)}}(1-p_j) + p_j\right\}\right].$$

By substituting $S^u_{i(1)}$ with $S^u - S^u_{j(2)}$ in the expression for $E^u_{C,joint}(\theta^u)$ and differentiating with respect to $S^u_{j(2)}$ and equating it to zero, we have,

$$S^u_{j(2)} = \frac{S^u}{2} + \frac{S^u}{2\theta^u r} \ln\left[\frac{(1-p_j)/p_j}{(1-p_i)/p_i}\right], S^u_{i(1)} = \frac{S^u}{2} + \frac{S^u}{2\theta^u r} \ln\left[\frac{(1-p_i)/p_i}{(1-p_j)/p_j}\right]$$

that maximize the joint effective capacity. Hence for a MS in the coverage regions of two or more BSs and when the resource allocation is done from these BSs as per the iterated scheme increases the total effective capacity.

2. **Class Based Shared Resource Allocation for Mobile Users:** Class based resource allocation policy which is also influenced by the dynamic availability of bandwidth, cell load conditions, and shared BS resource usage. The usual standard service differentiation suggests division of user traffic into service classes. For differentiated shared resource usage, the following three classes categorize the user traffic of user: P_0 (voice packets), P_1 (video traffic), and P_2 (data traffic). P_0 is

most delay sensitive and needs a guaranteed bandwidth; P_1 requires a minimum bandwidth and has more delay flexibility. P_2 traffic does not have delay or bandwidth guarantee constraints. Since P_0 traffic is most delay sensitive, initially the resources are allocated to this traffic class, then the resources are allocated to P_1 traffic and the remaining resources of each BS are allocated to P_2 traffic.

For resource allocation to mobile users in a class P_c, $c \in \{0,1,2\}$ the user with the maximum scheduling function $\psi^n = \max\left\{\frac{\rho^u}{\tau^u}\right\}$ is successively selected. To maximize throughput the current bit rate of user u, ρ^u, ensures the user with best channel conditions is selected while the user throughput, τ^u, ensures none of users' experience starves. Fairness is ensured with the selection of the user with high ρ^u and /or low τ^u. The resource allocation for mobile users is influenced by statistical characteristics of the traffic arrival and channel behavior and backlog history usage. The resource required over a time frame interval T_f for the incoming traffic using a linear predictor is given as: $\tilde{S}^{u(v)}_{P_C}(n+1) = \sum_{l=0}^{L_{P_C}-1} \xi^{u(v)}_{P_{C,l}}(n-l)$ where the prediction order L_{P_C} is a function of the traffic type P_C, $\xi^u_{P_C,l}$ is the parameter that indicates the impact of actual resource requirements $\tilde{S}^{u(v)}_{P_C}(n-1)$ due to new arrivals in frame *(n-l)*. $\xi^u_{P_C,l}$ is updated by least mean square (LMS) algorithm as: $\xi^u_{P_C,l}(n+1) = \xi^u_{P_C,l}(n) + \eta^u_{P_C}(n)\frac{\varepsilon^u_{P_C}(n)}{S^{u(v)}_{P_C}(n)}$ where the prediction error $\varepsilon^u_{P_C}(n) = S^{u(v)}_{P_C}(n) - \tilde{S}^{u(v)}_{P_C}(n)$ and $\eta^u_{P_C}(n) = \frac{L_{P_C}}{\sum_{l=0}^{L_{P_C}-1}[S^{u(v)}_{P_C}(n-1)]^2}$ The requested resource for frame $(n+1)$ and P_c traffic type is given as: $S^{u(r)}_{P_C}(n+1) = S^{u(q)}_{P_C}(n) - \tilde{S}^{u(v)}_{P_C}(n)$ where $S^{u(q)}_{P_C}(n)$ is the resource required from nth frame due to user u's queued traffic and service type. For any given traffic type P_C, it is important to choose an important optimum number of taps that maximizes the quality of prediction using historical prediction. A higher value of L_{P_C} increases the prediction accuracy as it sharply predicts the burstiness of the traffic, but also causes increased lag in predicted traffic and detrimental to prediction quality. A small value of L_{P_C} tracks the traffic fast but is unable to track burstiness.

Shared Resource Allocation for Split Handover in Comparison to MDHO and FBSS

Prior to discussing the advantage that shared resource allocation has in comparison to the other soft handover scheme variants like MDHO and FBSS (shown in Figure 6), it is necessary to briefly discuss about these soft handover techniques.

- **Macro Diversity Handover (MDHO):** MS maintains a diversity set that constitutes a group of BSs that operate on same frequency channel and are synchronized on time and frame level. The BSs having a considerable CINR are selected for the diversity set by the MS by the exchange of

Figure 6. Soft handover scheme: MDHO and FBSS

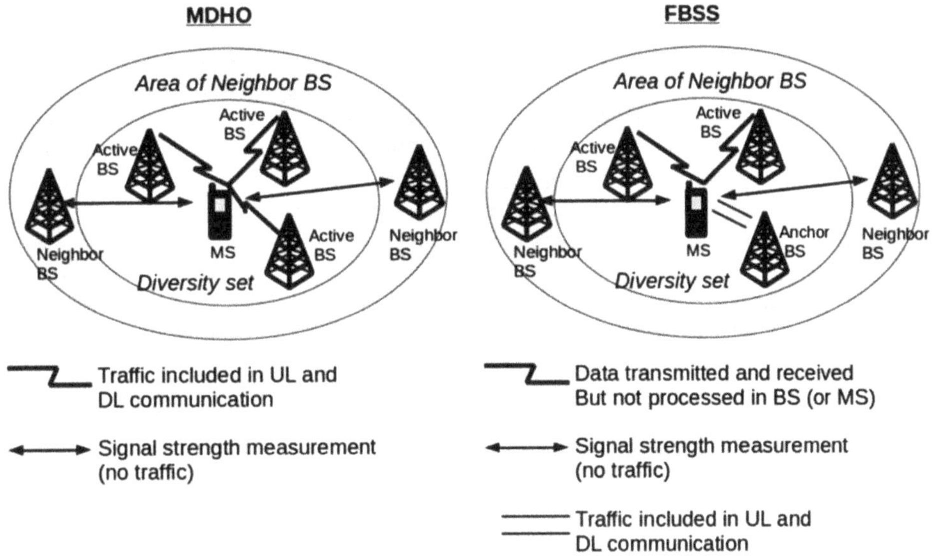

MAC context between BS and MS. All the BSs in the diversity set have the information of the MS and the MAC context and are said to be active BSs. The BS, with which the MS is registered or is handed over to, is the active BS. All the other BSs that are not a part of the diversity set and do not have any traffic exchange with MS but the MS is still able to perform signal strength measurement with these BSs, are said to be the neighbor BSs. The diversity set is updated when the CINR of the serving BS falls below a predefined threshold or when CINR of a BS is above another predefined threshold. In MDHO, during downlink communication MS receives data from all the active BSs in the diversity set and the MS combines the multiple signals by the virtue of diversity combining to obtain an improved data signal. During uplink communication the MS's data signal is received by different active BSs and the BS with the strongest signal finally makes the transmission after performing selection diversity.

- **Fast Base Station Switching (FBSS):** During handover the MS receives signal from all the BSs of diversity set and then from these BSs, MS selects one BS for exchanging uplink or downlink data. This BS is known as the anchor BS. MS registers and shares MAC context information with this anchor BS only. Anchor BS can be updated at a MS based on CINR levels

In the case of class based shared resource allocation, intelligent scheduling of MS by multiple BSs to increase users' capacity and QoS. In case of schemes like MDHO the diversity combining underutilizes the resources in the shared cell region, resulting in a poorer data rate for the users in this region as compared to the split-handover scheme. Since the class based traffic is scheduled by two different BSs (several BSs as the case may be), it results in increase in the effective throughput and data rate provided to the mobile user. These metrics are significantly higher than the hard handover or the other soft handover scheme variants (MDHO and FBSS). This is also reflected in the Figure 7, wherein the data rate of MS in shared cell region is highest by the split handover scheme, intermediate for MDHO soft handover scheme, and least for hard handover scheme.

Figure 7. Two cell scenario, comparison of data rate for the hard, soft (MDHO), and split handover schemes

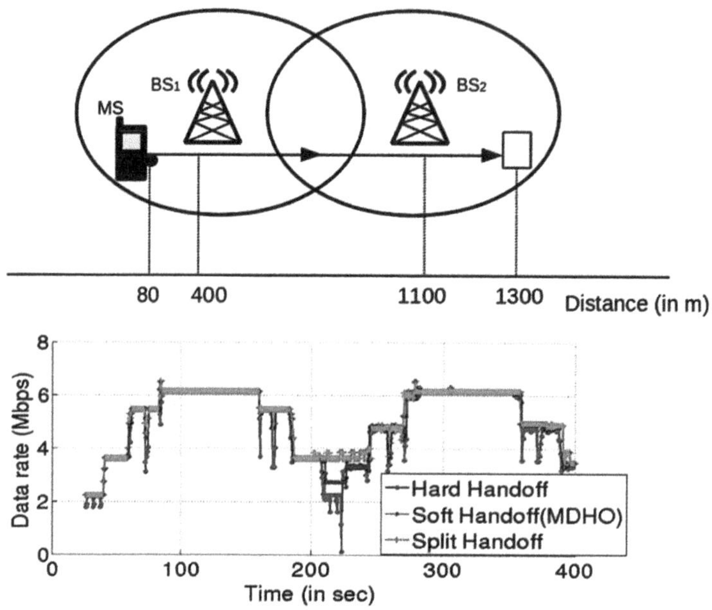

Shared Resource Allocation for Mobile Users in LTE-A Het-Net

One of the biggest challenges of a Heterogeneous NW is the frequent HO due to presence of a large number of small cells. Same are depicted in Figure 8 and are elaborated in Table 2.

A mobile UE, traversing across a macro-cell is bound to encounter a number of micro-cell in a heterogeneous network. This may result in decrease of throughput on two accounts:

- Loss of data during handover from macro-cell and micro-cell.
- Reduced resource availability at micro-cell to guarantee the ongoing services.

Table 2. Heterogeneous Architecture Parameters

S. No.	Parameter	3G/4G	5G
1	Density	Macrocell BS: 4-5 BS/Km2 Microcell BS: 8-10 BS/Km2 Femtocell AP	Small Cells: 40-50 BS/Km2
2	Planning	Macrocell: Properly planned Microcell: Semi planned Femtocell: Completely random	Small Cells:Properly planned
3	Coverage Area	Macrocell: 1.5-10 Km Microcell: 100-500m Femtocell: 10-25m	Small Cells: 100m
4	Challenges	Horizontal/ Vertical Handovers, Inter/ Intra Cell Interference Resource Allocation	Vertical Handovers, Backhaul Connectivity Resource Allocation

Both of these phenomenon take place in region I and II, shown in Figure 9.

For LTE-A Het-Net the shared resource allocation for Scenario shown in Figure 9 is simulated with parameters given in Table 3 for mobile users.

The simulation results are shown in Figure 10. It is seen from Figure 10 that 'Shared resource Allocation' achieves a higher throughput than the `actual throughput' without the shared resource allocation by compensating the deficient bandwidth by the macro cell 'resource Blocks'. The achieved throughput at the end of shared resource allocation is close to the ideal throughput, because of discrete nature of resource block and its allocation to mobile users.

Figure 8. Macrocell-Microcell Shared Resource Allocation

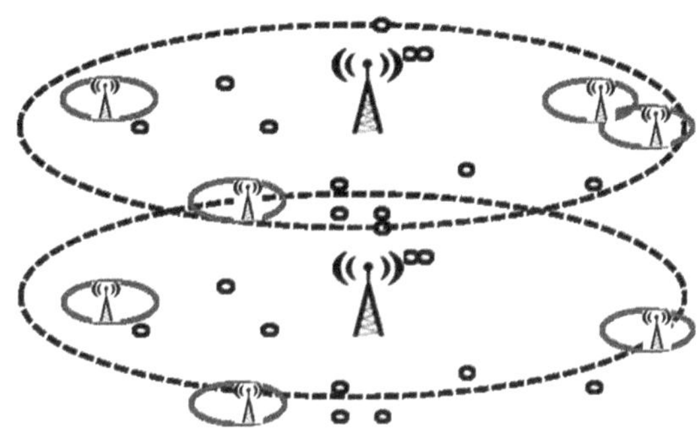

Figure 9. Reduction in Throughput at Cell Edge & Handover

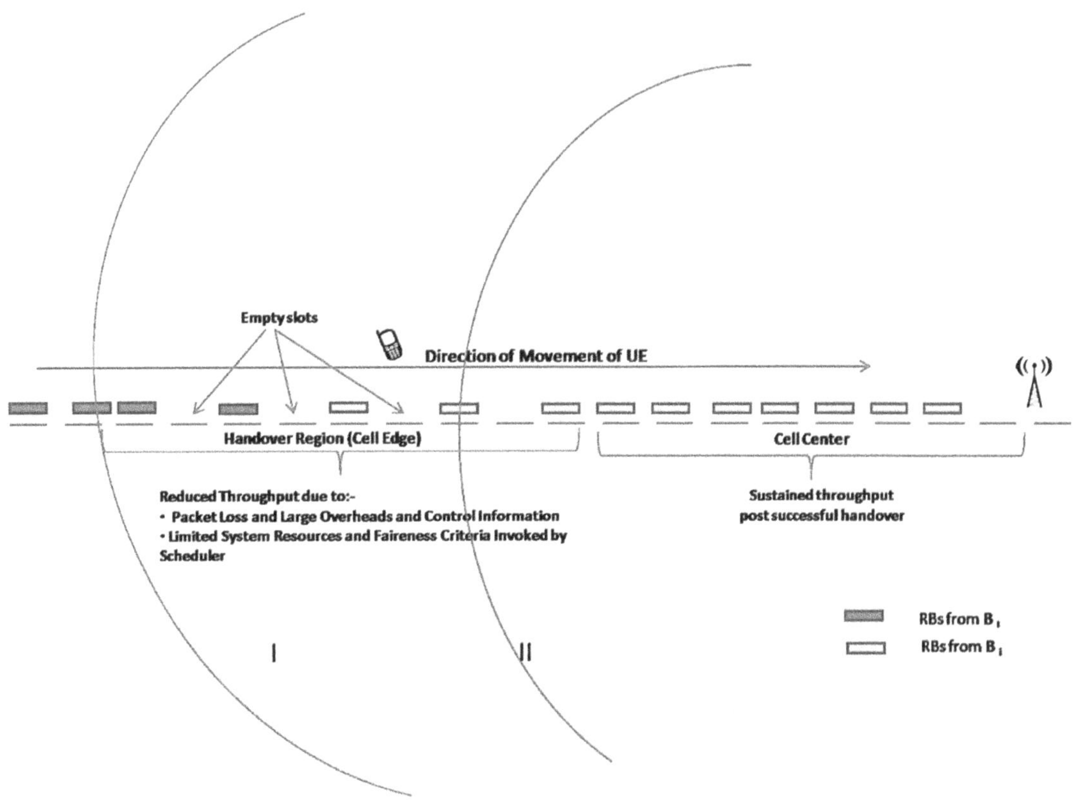

Table 3. Simulation Parameters for LTE-A Het-Net

S. No.	Parameter	Microcell	Macrocell
1	Cell Radius	150m	1000m
2	Distance between eNodeB	150m	850m
3	Downlink Bandwidth	10MHz	15 Mhz
4	NRB	10	As per scenario
5	AWGN PSD	-174 dB	-174 dB
6	Pathloss Coefficient	4	4
7	Transmit power	2W	39.4W
8	Log-normal Fading	8dB	8dB

Figure 10. Reduction in Throughput at Cell Edge & Handover

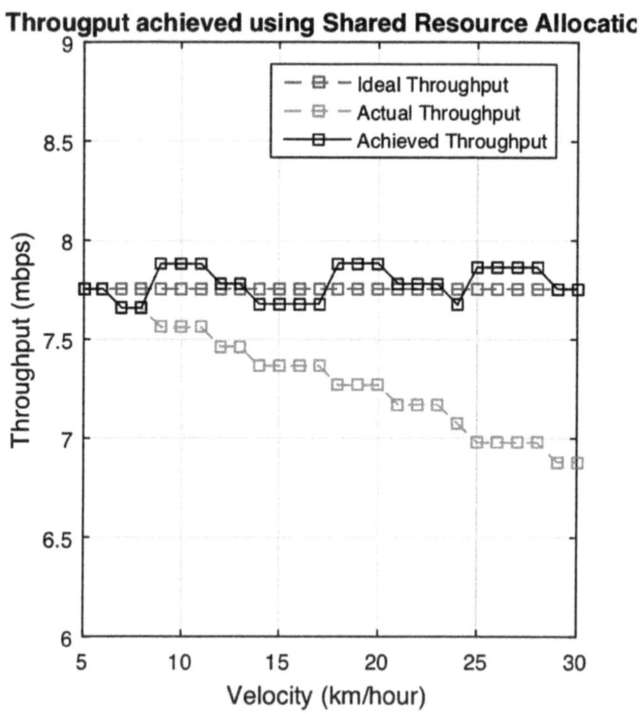

SON Based Inter-RAT MRO

The increasing demand for high speed communication services has led to the deployment of the new radio access technologies (RATs) (LTE overlaying legacy 3G and 2G mobile systems). Inter-RAT handover is necessary to fully utilize the capabilities of the various RATs to provide best QoS and meet users' demands. While discussing the heterogeneous networks, an important concept is that of SON (self-organizing network) and its important use case of MRO (mobility robustness optimization). SON is a technology that simplifies and speeds up the planning, management, configuration, healing and optimization of NGMN (Next Generation Mobile Networks) that comprises of several RATs deployed. An inter-RAT MRO algorithm given by Awada et al. (2013) automates the adjustment of handover threshold of each cell. Various measurement quantities such as SS (signal strength) and SQ (signal quality) form the basis for measurement event.

1. **Inter-RAT Handover Procedure:** At any given time the UE is served by an LTE, 3G or 2G cell. A UE u being served by a cell c experiences a radio link failure (RLF) when its SINR ($\gamma_{u,c}(t)$) falls below a threshold (Q_{out}) for a certain interval of time $T_{Q_{out}}$ i.e. $\gamma_{u,c}(t) < Q_{out}$ for $t_0 - T_{Q_{out}} < t < t_0$. The serving BS in LTE, 3G or 2G network configures the UE to perform signal strength measurement for intra or inter-RAT neighboring cells and the serving cell and sending a report to the serving BS periodically or by means of event trigger mechanism (report sent on fulfillment entering condition of measurement event). The handover is triggered by serving BS after receiving the measurement report. For LTE to 3G handover the serving LTE BS configures UE with measurement *event B2*. For a handover from 3G cell to LTE or 2G, the measurement event is *event 3A*. The measurements of the signal strength by UE in LTE cell is of received signal received power (RSRP), in 3G it is of received signal code power (RSCP). RSRP is the linear average of power contributions of the resource elements carrying cell-specific reference signals within the measurement frequency bandwidth considered, and RSCP is the received power measured on primary common pilot channel (CPICH). These measurements are inclusive of path loss, antenna gain, log-normal shadowing and fast fading. The phases before handover comprise of handover triggering condition, measurement by mobile UE of neighboring cells, and selection of the best candidate cell that fulfills the handover criteria condition.

The entering condition for the event 3A or B2 is fulfilled when serving cell's signal (i.e. $S_{u,c}(t)$) falls below threshold S_{thr} in dBm and the target cell's signal $T_{u,C_0}(t)$ greater than threshold (i.e., T_{thr}) in dBm. The entering condition needs to be fulfilled for time duration T_T for the measurement report to be sent by the UE i.e., $S_{u,c}(t) < S_{thr} \cap T_{u,C_0}(t) > T_{thr}$ for $-T_T < t < t_0$.

After receiving the measurement report from the UE *u*, the serving cell *c* sends a handover request to the target handover cell c_0 in order to prepare handover of UE. An additional delay termed as handover preparation time (i.e. T_{HP}) during which *c* waits for acknowledgement from c_0. Handover occurs after T_{HP} seconds after the measurement event trigger and the UE's SINR $\gamma_{u,c}(t)$ being greater than the threshold Q_{fail}. Thus, the execution of handover is based on the following conditions' fulfillment, i.e. $S_{u,c}(t) < S_{thr} \cap T_{u,C_0}(t) > T_{thr}$ for $t_{HO} - T_{HP} - T_T < t < t_{HO} - T_{HP}$ and $\gamma_{u,C}(t_{HO}) > Q_{fail}$.

The key performance indicators (KPI) for inter-RAT mobility are to capture the radio link failure (RLF) events and the costly inter-RAT handover like ping-pongs (PPs). There are three RLF mobility events: Too late inter-RAT handover (TLH), too early inter-RAT handover (TEH), and inter-RAT handover to wrong cell (HWC). There are two kinds of costly inter-RAT HOs: Inter-RAT PPs, and unnecessary HOs (UHs) from LTE to 3G. It is essential to minimize UHs to enable users to benefit from the newly deployed, higher priority LTE network. These KPIs are briefly described as follows:

- **TLH:** The UE is dropped before handover is executed or has been initiated from one RAT to another and subsequently the UE reconnects to a RAT that is different from the previous serving

Figure 11. Four cases of inter-RAT TLH

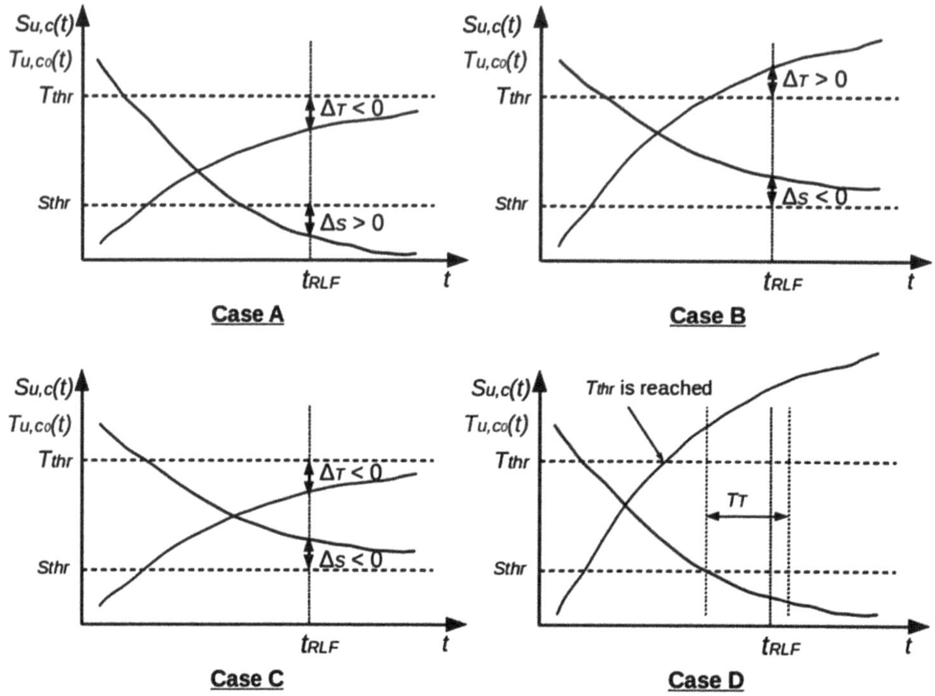

cell. This occurs when the entering condition of the measurement event is not fulfilled or the RLF occurs before the inter-RAT handover is executed in spite of the fulfillment of the entering condition of the measurement event. There are four cases (illustrated in Figure 11) in which TLH for inter-RAT handover scenario occurs. (i) Case A:

- **Misconfiguration of T_{thr} Is the Root Cause for TLH. (ii) Case B:** $S_{u,c}(t) > S_{thr} \cap T_{u,C_0}(t) > T_{thr}$ misconfiguration of S_{thr} is the root cause for TLH. (iii) Case C: $S_{u,c}(t) < S_{thr} \cap T_{u,C_0}(t) < T_{thr}$ misconfiguration of the threshold corresponding to the smallest values of Δ_S and Δ_T, is the root cause for TLH. (iv) Case D: Entering condition of measurement event is fulfilled but RLF occurs before inter-RAT HO, and the misconfiguration of one of the two thresholds that is reached later is the root cause for TLH.
- **TEH:** An RLF happens shortly after the UE is successfully handed over to a cell of a different RAT, and the UE reconnects to a previous RAT cell. Also when the UE fails to connect to c_0 using random access channel (RACH), the inter-RAT handover fails and is considered as a TEH. Misconfiguration of T_{thr} is the root cause of TEH. T_{thr} needs to be increased to ensure that signal of target cell of a different RAT i.e., c_0 is strong enough.
- **HWC:** Shortly after the UE is successfully handed over to a cell c_0 of a different RAT, an RLF happens and UE connects to a cell c_0 of the same RAt as c_0. Similar to TEH, misconfiguration of T_{thr} is the root cause for HWC. T_{thr} should be increased to guarantee that other RAT's cell signal is strong enough.
- **PP:** When the UE is handed over to a cell of a different RAT and then handed over back to the same cell or a different cell of the previous RAT within a time interval of T_{PP}. By delaying the

first inter-RAT handover by either increasing T_{thr} or decreasing S_{thr}, it is possible to resolve inter-RAT HO.

- **UH:** The UE is handover from a cell of high priority RAT to a cell of low priority RAT, even though the signal quality of the previous RAT is good enough. This can be evaluated in terms of reference signal received quality (RSRQ) of the previous LTE cell is higher than Q_{RSRQ} threshold for $T_{Q_{RSRQ}}$ time interval. To resolve UHs it is essential to decrease S_{thr} thereby increasing the LTE cell coverage.

The SON with inter-RAT MRO needs that the two thresholds (i.e. S_{thr} and T_{thr}) should be adjusted properly for the above mentioned KPIs, and for this purpose a feedback controller can be used to determine magnitude change in the handover threshold.

Although the above discussion is more focused on LTE to 3G cell inter-RAT HO, however this same analytical framework is equally applicable to 3G to 2G cell inter-RAT handover and had been discussed by Mohammed et al. (2007).

CONCLUSION AND FUTURE RESEARCH DIRECTIONS

According to ITUs vision of optimally connected anytime, anywhere, mobile users are free to roam between different RATs that are provided by different service providers, has resulted in advent of heterogeneous networks. Next generation networks are based on these heterogeneous networks with heterogeneity in terms of communication modalities, channel types, technology generations, protocol types, and QoS requirements. This chapter has discussed several issues faced in heterogeneous networks in network selection, inter/intra-RAT handovers, BS placement, and resource allocation. These issues can be very complex mainly because of heterogeneity in radio access networks. Selection of suitable scheme depends on many parameters which have been discussed. In different scenarios different parameters should be weighted with high priority, for which the various existing schemes in the literature have also been enlisted in this chapter.

Although, several aspects of next-generation heterogeneous networks have been studied by researchers in the recent times and are discussed in this chapter. However, still there are some other issues in heterogeneous wireless access networks that still needs to be focused upon, like pricing schemes, and several approaches where network operators or service providers will compete with each other to maximize their revenue and users will compete with each other to maximize their QoS, which will have effects in decision taken in network selection and handover. In addition to this, it is essential to also consider simultaneous cooperation and competition by the service providers in an optimized manner to integrate resources and provide users with improved service experience, and also consider user-preferences while configuring the network system in an automated manner. Even though any optimization is often based on tradeoffs and also since each of the solutions discussed previously look into a subset of the factors concerning heterogeneous network. Hence, a collated, well-balanced, and comprehensive solution is the need-of-hour, which would integrate the network heterogeneities to the advantage of users as well as the service providers, while utilizing the available resources in the best possible manner.

ACKNOWLEDGMENT

This work has been partly supported by the Department of Science and Technology (DST) under Grant DST/INSPIRE/04/2015/000793, SERB under Grant SB/S3/EECE/0248/2014, and IIT Kharagpur, ISIRD, under Grant IIT/SRIC/ECE/USW/2016-17.

REFERENCES

Alkhawlani, M. & Ayesh, A. (2008). Access network selection based on fuzzy logic and genetic algorithm. *Advances in Artificial Intelligence, 2008*(1), 1-12.

Awada, A., Wegmann, B., Viering, I., & Klein, A. (2013). A SON-based algorithm for the optimization of inter-RAT handover parameters. *IEEE Transactions on Vehicular Technology*, *62*(5), 1906–1923. doi:10.1109/TVT.2013.2251923

Chan, V., Chapin, J., Elson, P., Fisher, D., Frost, V., Jones, K., & Miller, G. (2011, March). Workshop on Future Heterogeneous Networks. In *FutureHetNets 2011*. Mountain View.

Chang, J., Li, Y., Feng, S., Wang, H., Sun, C., & Zhang, P. (2009,). A fractional soft handover scheme for 3GPP LTE-advanced system. *IEEE International Conference on Communications Workshops*, 1-5.

Charilas, D., & Panagopoulous, A. (2010). Multiaccess radio network enviroments. *IEEE Vehicular Technology Magazine*, *5*(4), 40–49. doi:10.1109/MVT.2010.939107

Charilas, D. E., Panagopoulos, A. D., Vlacheas, P., Markaki, O. I., & Constantinou, P. (2009). *Mobile Lightweight Wireless Systems Congestion Avoidance Control through Non-cooperative Games between Customers and Service Providers*. Berlin: Springer.

Cisco. (2017). *Cisco Visual Networking Index: Global Mobile Data Traffic Forecast Update, 2016–2021*. Author.

Cisco. (2017). *The Zettabyte Era: Trends and Analysis*. Cisco White Paper.

Dahlman, E., Parkvall, S., & Skold, J. (2008). *3G evolution: HSPA and LTE for mobile broadband*. Burlington: Academic Press.

Ericsson. (2018). *Mobility report*. Author.

Ge, X., Pan, L., Tu, P. S., Chen, H. H., & Wang, C. X. (2016). Wireless Backhaul Capacity of 5G Ultra-Dense Cellular Networks. In *IEEE Vehicular Technology Conference* (pp. 1-6). Montreal, Canada: IEEE.

Guo, C., Guo, Z., Zhang, Q., & Wenwu, Z. (2004). A seamless and proactive end-to-end mobility solution for roaming across heterogeneous wireless networks. *IEEE Journal on Selected Areas in Communications*, *22*(5), 834–848. doi:10.1109/JSAC.2004.826921

Guruprasad, K. R. (2011). Generalized Vornoi partition: A new tool for optimal placement of base stations. In *IEEE International Conference on Advanced Networks and Telecommunication Systems* (pp. 1-3). Bangalore: IEEE.

IEEE Standard for Local and Metropolitan Area Networks-Part 16. (2005). Air Interface for Fixed Broadband Wireless Access Systems (Incorporated into IEEE Standard 802.16e-2005 and IEEE Standard 802.16-2004/Cor 1-2005 E). *IEEE Standard P802.16/Cor1/D5.*

Islam, M. H., Dziong, Z., Sohraby, K., Daneshmand, M. F., & Jana, R. (2012). Capacity-Optimal Relay and Base Station Placement in Wireless Networks. *International Conference on Information Networking*, 358-363.

Jang, J.-S. R. (2002). ANFIS: Adaptive-network-based fuzzy inference system. *IEEE Transactions on Systems, Man, and Cybernetics*, *23*(3), 665–685. doi:10.1109/21.256541

Khan, M. A. & Sivrikaya, F. & Albayrak, S. & Mengaly, K. Q. (2009). Auction based interface selection in heterogeneous wireless networks. *Second IFIP Wireless Days*, 1-6.

Khan, M. A., Toseef, U., Marx, S., & Goerg, C. (2010). Auction based interface selection with Media Independent Handover services and flow management. *European Wireless Conference*, 429-436.

Khan, M. A., Toseef, U., Marx, S., & Goerg, C. (2010). Game-Theory Based User Centric Network Selection with Media Independent Handover Services and Flow Management. *Eighth Annual Communication Networks and Services Research Conference*, 248-255. 10.1109/CNSR.2010.40

Latouche, M., Rauschen, C., Oszabó, O., Creusat, J.-B., & Belmans, W. (2013). *Mobile Data Explosion How Mobile Service Providers Can Monetize the Growth in Mobile Data Through Value-Added Services.* Cisco White Paper: Cisco Internet Business Solutions Group (IBSG).

Lee, C. C., & Steele, R. (1998). Effect of soft and softer handoffs on CDMA system capacity. *IEEE Transactions on Vehicular Technology*, *47*(3), 830–841. doi:10.1109/25.704838

Lee, H., Son, H., & Lee, S. (2009). Semisoft handover gain analysis over OFDM-based broadband systems. *IEEE Transactions on Vehicular Technology*, *58*(3), 1443–1453. doi:10.1109/TVT.2008.927041

Mohammed, A., Kamal, H., & AbdelWahab, S. (2007). 2G/3G inter-RAT handover performance analysis. *The Second European Conference on Antennas and Propagation*, 1-8.

Nasser, N., Hasswa, A., & Hassanein, H. (2006). Handoffs in fourth generation heterogeneous networks. *IEEE Communications Magazine*, *44*(10), 96–103. doi:10.1109/MCOM.2006.1710420

Niyato, D., & Hossain, E. (2008). A Noncooperative Game-Theoretic Framework for Radio Resource Management in 4G Heterogeneous Wireless Access Networks. *IEEE Transactions on Mobile Computing*, *7*(3), 332–345. doi:10.1109/TMC.2007.70727

Ong, E. H., & Khan, J. (2008). Dynamic access network selection with QoS parameters estimation: A step closer to ABC. In *IEEE Vehicular Technology Conference* (pp. 2671-2676). Singapore: IEEE.

Ormond, O., Murphy, J., & Muntean, G. M. (2006). Utility-based intelligent network selection in beyond 3G systems. *IEEE International Conference on Communications*, 1831-1836.

Peng, M., Liu, Y., Wei, D., Wang, W., & Chen, H.-H. (2011). Hierarchical cooperative relay based heterogeneous networks. *IEEE Wireless Communications Magazine*, *18*(3), 48–56. doi:10.1109/MWC.2011.5876500

Pervaiz, H., & Bigham, J. (2009). Game theoretical formulation of network selection in competing wireless networks: An analytic hierarchy process model. *Third International Conference on Next Generation Mobile Applications, Services and Technologies*, 292–297.

Razzaque, M. A., Dobson, S., & Nixon, P. (2010). *Enhancement of Self-organisation in Wireless Networking through a Cross-layer Approach*. Berlin: Springer.

Sgora, A., Chatzimisios, P., & Vergados, D. D. (2010). *Mobile Lightweight Wireless Systems: Access Network Selection in a Heterogeneous Environment Using the AHP and Fuzzy TOPSIS Methods*. Berlin: Springer.

Shen, D. M. (2010) The QoE-oriented Heterogeneous Network Selection Based on Fuzzy AHP Methodology. *The Fourth International Conference on Mobile Ubiquitous Computing, Systems, Services and Technologies*, 275-280.

Shen, W., & Zeng, Q.-A. (2007). Cost-function-based network selection strategy in integrated wireless and mobile networks. *21st International Conference on Advanced Information Networking and Applications Workshops*, 2, 314-319. 10.1109/AINAW.2007.136

Singhal, C., Kumar, S., De, S., Panwar, N., Tonde, R., & De, P. (2014). Class-based shared resource allocation for cell-edge users in OFDMA networks. *IEEE Transactions on Mobile Computing*, *13*(1), 48–60. doi:10.1109/TMC.2012.210

Trestian, R., Ormond, O., & Muntean, G.-M. (2012). Game theory-based network selection: Solutions and challenges. *IEEE Communications Surveys and Tutorials*, *14*(4), 1212–1231. doi:10.1109/SURV.2012.010912.00081

Vasudeva, K., López-Pérez, D., & Güven, I. (2017). Analysis of Handover Failures in Heterogeneous Networks With Fading. *IEEE Transactions on Vehicular Technology*, *66*(7), 6060–6074. doi:10.1109/TVT.2016.2640310

Watanabe, E. H., Nenasche, D. S., de Souza, S., & Leao, R. M. M. (2008). Modeling Resource Sharing Dynamics of VoIP Users over a WLAN Using a Game-Theoretic Approach. *IEEE International Conference on Computer Communications*.

Wen, Y. F. (2009). Heterogeneous Base Station Placement for Wireless Networks. *IEEE Mobile WiMAX Symposium*, 87-92.

Yeh, S., Talwar, S., Wu, G., Himayat, N., & Johnsson, K. (2011). Capacity and coverage enhancement in heterogeneous networks. *IEEE Wireless Communications*, *18*(3), 32–38. doi:10.1109/MWC.2011.5876498

Zhang, W. (2004). Handover decision using fuzzy MADM in heterogeneous networks. *IEEE Wireless Communications and Networking Conference*, 2, 653-658.

Zhu, K., Niyato, D., & Wang, P. (2010). Network Selection in Heterogeneous Wireless Networks: Evolution with Incomplete Information. *IEEE Wireless Communications and Networking Conference*, 1-6.

Zoubir, A., & Boashash, B. (1998). The bootstrap and its application in signal processing. *IEEE Signal Processing Magazine*, *15*(1), 56–76. doi:10.1109/79.647043

KEY TERMS AND DEFINITIONS

Broadband Wireless Access (BWA) Networks: BWA technologies provide broadband data access over wireless LAN (Local Area Network), 3G, MAN (Metropolitan Area Network).

Differentiated QoS: It is an end-to-end, application-level, QoS-management technique that classifies application-traffic into multiple classes with different QoS parameters.

Effective Capacity: Effective capacity of a system is the maximum constant arrival rate that can be supported by the system at the link layer.

Mobility Robustness Optimization (MRO): It is an algorithmic optimization framework that automates the adjustment of handover threshold of each cell based on measurement quantities such as SS (signal strength) and SQ (signal quality).

Multi-Tier Handover: It is an intra-RAT handover technique where handover occurs between hierarchical tiers (macrocell, picocell, and femtocalls) of an infrastructural RAT deployment.

Network Selection: Network selection is a process in which the user devices (while requesting for a new connection or during the handover process) select from the radio access networks (RANs) that belong to competing service providers having different latency, coverage area, and cost.

Radio Access Technology (RAT): RAT is the underlying physical connection method for a radio based mobile communication network such as Bluetooth, Wi-Fi, 3G, 4G, and LTE.

Self-Organizing Network (SON): SON is a technology that simplifies and speeds up the planning, management, configuration, healing and optimization of NGMN (Next Generation Mobile Networks) that comprises of several RATs deployed.

Shared Resource Allocation: Resource allocation technique wherein resources of multiple RATs are pooled together to improve capacity and QoS in a mobile network environment.

Split Handover: Shared resource allocation technique for cell edge users in orthogonal frequency division multiple access (OFDMA) networks, where MSs maintain parallel connections with more than one BS.

Chapter 4
Resource Management of Mixed Unicast and Multicast Services Over LTE

Giuseppe Araniti
University Mediterranea of Reggio Calabria, Italy

Massimo Condoluci
King's College London, UK

Sara Pizzi
University Mediterranea of Reggio Calabria, Italy

Antonella Molinaro
University Mediterranea of Reggio Calabria, Italy

ABSTRACT

In recent years, mobile operators are observing a growing demand of multicast services over radio cellular networks. In this scenario, multicasting is the technology exploited to serve a group of users who simultaneously request the same data content. Since multicast applications are expected to be massively exchanged over the forthcoming fifth generation (5G) systems, the third-generation partnership project (3GPP) defined the multimedia broadcast multicast service (MBMS) standard. MBMS supports multicast services over long-term evolution (LTE), and the 4G wireless technology provides high quality services in mobile environments. Nevertheless, several issues related to the management of MBMS services together with more traditional unicast services are still open. The aim of this chapter is to analyze the main challenges in supporting heterogeneous traffic over LTE with particular attention to resource management, considered as the key aspect for an effective provisioning of mobile multimedia services over cellular networks.

DOI: 10.4018/978-1-5225-7570-2.ch004

INTRODUCTION

The increase in the content availability and the growing number of available devices with enhanced media capabilities (i.e., smartphones and tablets) enabled a tremendous growth in the demand of advanced services over mobile radio systems. Among those, *multicast* services are expected to be massively transmitted over Fifth Generation (5G) wireless systems and allow groups of users to simultaneously access services with high Quality of Service (QoS) (such as Mobile TV, news forecast, video calls, video conferencing, Internet video streaming). Since multicasting is considered as one of the main value-added services for 5G systems, standardization bodies and network providers are currently working to suitably support multicast services over Long Term Evolution (LTE), the most promising wireless technology that will lead the growth of mobile broadband services in the next years (Third Generation Partnership Project, 2012). With this aim, the Third Generation Partnership Project (3GPP) standardized the Multimedia Broadcast Multicast Service (MBMS). This standard, which defines all networks enhancements necessary to support the transmission of multicast services over LTE, introduces the *Point-to-Multipoint (PtM)* transmission mode and covers different functionalities related to the management of multicast services (e.g., service announcement, joining and leaving procedures, session setup and re-configuration). An example of PtM transmission mode is shown in Figure 1, where the main differences between PtM and the traditional unicast transmission mode, i.e., Point-to-Point (PtP), are highlighted. In particular, PtM simultaneously serves all users interested to a given multicast service through a shared channel, with the aim to improve the system capacity and "theoretically" serve an unlimited number of users per group (Lecompte & Gabin, 2012).

Although MBMS improves the capabilities of LTE in supporting multicast services, the main challenge in multicast environments is related to the Radio Resource Management (RRM), which includes all functionalities necessary to manage the radio resources available in the cellular system (Richard, Dadlani & Kim, 2013). In particular, the RRM is in charge of performing *link adaptation* procedures, i.e., the selection of the transmission parameters, such as the Modulation and Coding Scheme (MCS), for multicast content delivery according to the channel conditions experienced by the User Equipments (UEs). Indeed, in a multicast scenario, link adaptation must be accomplished on a per-group basis, i.e., by taking into account the channel state information of all terminals interested to a given multicast service.

Figure 1. A comparison of multicast service delivery through Point-to-Point (PtP) and Point-to-Multipoint (PtM) transmission modes

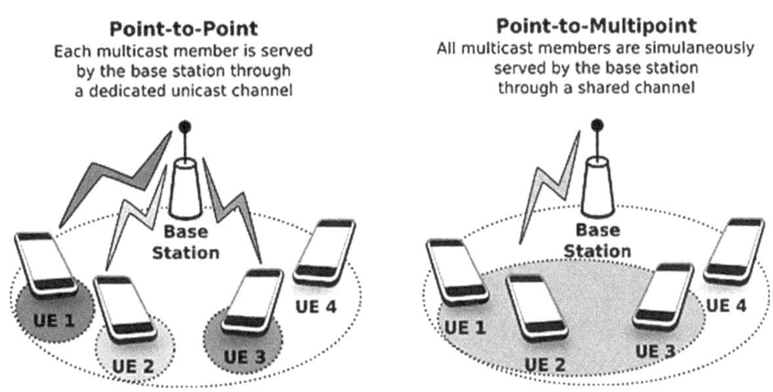

This may limit the session quality performance achieved by multicast members, due to the presence of cell-edge users which experience poor channel conditions and consequently cannot support high data rates. Moreover, the delivery of typical multicast applications (e.g., mobile TV) requires a large amount of radio resources and this further challenges the spectrum efficiency and the coexistence with other services (e.g., unicast flows) in the cell. In fact, while advantaging unicast user requests leads to a significant network capacity decrease, favoring multicast traffic causes an unfair bandwidth distribution between multicast and unicast traffic with negative consequences on unicast users' degree of satisfaction.

In this Chapter, we will focus on the challenges related to the RRM of multicast services alongside traditional unicast traffic over LTE. We firstly will present an overview on the most typical multicast applications, with the related requirements in terms of QoS, and we will discuss about the MBMS standard, by introducing the network architecture and the system features of LTE/MBMS. Then, we will address the structure of LTE packet scheduler, by underling the aspects related to the RRM. Finally, we will focus on the issues related to RRM in multicast environments over LTE, and we will present an overview of the different approaches defined in literature for the delivery of mixed unicast and multicast services.

MULTICAST SERVICES

A multicast service is defined as a service which is simultaneously transmitted towards a set of managed UEs, namely *multicast members*. Each multicast service transmitted in a given cell of a mobile radio system is associated to only one *multicast group*. The terminals can receive only those services related to the multicast group they belong to. A multicast service is announced by the base station through several advertising messages and, through this procedure, interested cellular users may join the multicast group. Once the multicast group is formed, the base station will send data content only towards those nodes that joined the multicast group. The multicast service delivery can be considered similar to a traditional broadcast transmission. Indeed, while at the physical layer the transmission of both broadcast and multicast services is the same, the difference between these two modes lies in the set of involved users: broadcast services are available to *all* users of a given mobile network (i.e., there is no need for subscriptions), whereas multicast services are restricted only to those UEs belonging to a multicast group.

In the following of this Section, we will discuss the multicast applications commonly exchanged over mobile radio systems.

Multicast Applications

Multicast services can cover several types of applications, each one characterized by different QoS requirements. One of the main multicast services expected to be massively exchanged over LTE is *audio and/or video streaming,* which can range over a wide set of applications such as standard video (e.g., Mobile TV) or audio streaming (e.g., web radio), news (e.g., weather forecasts) or advertisement message distribution, video or audio conference calls, and so on. The management of multicast audio/video streaming sessions poses several challenges since these services are characterized by strict QoS constraints. In particular, the most relevant QoS issues are related to the provisioning of a minimum data rate, a maximum packet transfer delay and tolerated jitter.

Another important multicast application is the *file downloading.* Multicast file downloading can be suitable for software update transmission (e.g., codec or plug-in), for image and text distribution and for multimedia content delivery (e.g., video or audio files) for off-line use. The management of these multicast services does not pose strict constraints in terms of delay and guaranteed bit rate are, but it is characterized by strict requirements in terms of packet loss ratio.

Recently, the transmission of *geographic information updates* is growing in importance in the mobile market scenario. Typical examples of such services are traffic reports, local news, weather forecast, stock prices and location-based advertisement. In these cases, data must be delivered only to those users located in a given area. The management of these services is quiet similar to the typical management of multicast file downloading, but additional issues must be considered in order to perform multicast group formation according to users' position. Moreover, additional procedures are required to dynamically update the members belonging to a given group by taking into account the users' direction and mobility speed.

The applications considered above can be divided into two categories. The *single-rate* applications require all multicast users to be served with the same session quality. Examples of single-rate services are file downloading and geographic information updates. On the contrary, for the *multi-rate* applications the original information is split into different levels of "quality", and the better the channel condition of a given multicast user the higher the quality experienced by such a user. Multi-rate services cover a large set of multicast applications, e.g., audio/video streaming, and they will be explored in the following of this Section.

Multi-Rate Services

Multi-rate applications foresee to split the original data stream into different sub-streams, where each sub-stream represents a "portion" of the overall information to be transmitted to multicast group members. The idea at the basis of multi-layer techniques is that the perceived session quality improves as users receive a higher number of sub-streams. As a consequence, a minimal service, namely *base sub-stream* (or base layer), is received by *all* multicast users, while terminals with higher channel gain (which potentially support less robust MCS and can accordingly attain higher data rates) can also receive additional *enhancement sub-streams* and, consequently, experience improved session quality.

Currently, two categories of multi-layer schemes are largely used: Multiple Description Coding (MDC) and Scalable Video Coding (SVC). In MDC, multimedia data is split into multiple descriptors, where each descriptor represents a given sub-stream of the original information. According to MDC, the higher the number of successfully received descriptors the higher the session quality experienced by a given user. It is worth noticing that the quality improvement is not related to the priority of each descriptor. Indeed, in MDC, all descriptors have equal priority and any combinations of the received descriptors can be decoded independently. Because of its sub-stream independence feature, MDC has received high attention from researchers, though MDC suffers of several limitations in case of real-time audio/video streaming applications. Concerning this latter type of multicast services, SVC represents a more attractive solution, which is based on the idea to dynamically adapt the quality of the streaming to the various needs or preferences of multicast users (as well as to varying capabilities of involved UEs or to different network conditions and loads) by removing one or more parts of the original audio/video data. Being similar to MDC, the term "scalable" refers to the fact that, also in the case one or more sub-streams of the original content are not successfully received by a UE, the received sub-stream (or sub-streams) represents a valid information for some target decoders. SVC applications enable three different types of scalability: temporal, spatial, and quality. Spatial and temporal scalabilities refer to the cases where the reception of only a subset of original sub-streams involves the reception of a video stream with a reduced picture size (i.e., spatial resolution) or frame rate (i.e., temporal resolution), respectively. With quality scalability, the sub-stream provides the same spatio-temporal resolution as the reception of the overall available sub-streams, but with a lower fidelity (where fidelity is often informally referred to as signal-to-noise ratio, i.e., SNR). The main difference between MDC and SVC is related to the order in the reception of the sub-streams. In SVC, a given sub-stream *n* can be considered as successfully received (i.e., it represents a valid information for the decoder) only if the sub-stream *n-1* has been previously received by the user. This involves that, in SVC applications, all available sub-streams are hierarchically ordered.

Multicasting Over LTE

The 3GPP, motivated by the increasing demand for high-quality services over mobile broadband networks, carried out several activities under the LTE and System Architecture Evolution (SAE) projects, finalized to respectively define the radio access and the core network for the next generation of mobile radio systems. The LTE system is one of the most promising wireless technologies able to support the growing demand of high-quality multicast services. LTE offers several benefits in terms of *(i)* high data rates in both downlink and uplink directions, *(ii)* low latency, *(iii)* low cost per bit, *(iv)* high spectrum efficiency even for cell-edge users, *(v)* high system capacity.

LTE is very appealing to network providers as a means to deliver high quality services with strict QoS constraints. With the aim to efficiently support multicast services, the 3GPP defined the MBMS standard, which will be the main topic addressed in this Section.

Multimedia Broadcast Multicast Service

The LTE/MBMS architecture, depicted in Figure 2, is based on a *flat all-IP* network infrastructure where both control and data information is transmitted through IP-based connections. This characteristic introduces great flexibility in the management of network devices and allows mobile operators to reduce installation and configuration costs.

The LTE/MBMS network is composed of:

- **eNodeB (i.e., the LTE Base Station):** The eNodeB is the entry point for UE to the LTE network. It is responsible for RRM procedures. In particular, it handles the configuration of transmission

Figure 2. The network architecture of LTE/MBMS

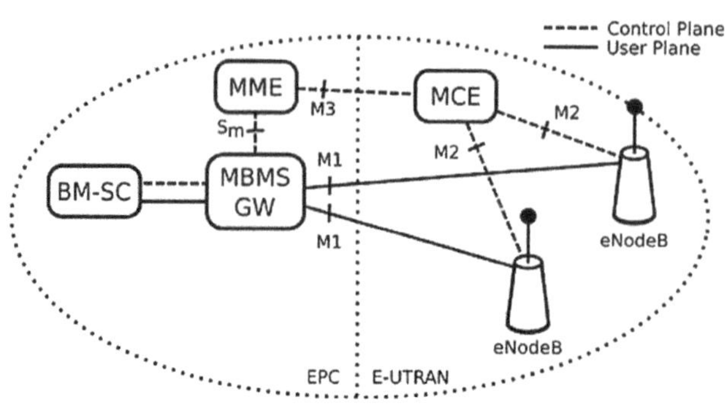

parameters (i.e., link adaptation) in single-cell MBMS mode (i.e., a MBMS service transmitted by only one eNodeB). The eNodeB is also in charge of collecting the channel state information of LTE subscribers and performs handover procedures.

- **MultiCell/Multicast Coordination Entity (MCE):** It is a logical entity (it can also be deployed within the eNodeB) involved in session control signaling towards multiple eNodeBs (the MCE does not perform signaling towards the MBMS receivers). One eNodeB is managed by one MCE while one MCE could manage multiple eNodeBs. The MCE covers different functionalities. Among those, the admission control and the allocation of the radio resources used by all the eNodeBs involved in a multi-cell MBMS transmission are the most important. This is typically the case of broadcast services simultaneously delivered by several base stations through single-frequency transmissions. In particular, in a multi-cell scenario, the MCE selects the transmission parameters (i.e., MCSs) to be simultaneously exploited by involved eNodeBs. The MCE performs the selection between multi- and single-cell transmission modes in the setup (or network reconfiguration) phase of the MBMS Service. The MCE also performs MBMS session suspension/ resumption based on the counting results for the corresponding MBMS service. The MCE is connected to the served eNodeB via the *M2* interface, used for MBMS session management and radio configuration signaling. The Stream Control Transmission Protocol (SCTP) is used over the *M2* interface to carry the control traffic.
- **Mobility Management Entity (MME):** It performs authentication and security procedures for subscriber's identification when a UE establishes a connection to the LTE network. The MME is also in charge for mobility management by storing the serving eNodeB for each terminal in *active*-mode while it stores the Tracking Area (which indicates a set of base stations relevant to a specific area) for UEs in *idle*-mode. Finally, the MME allocates the multicast IP address (in addition to the IP address relevant to the multicast source) to the eNodeB(s) that joined the MBMS session. The *M3* interface connects the MME and the MCE for conveying MBMS session management signaling. The SCTP is used over the *M3* interface to carry the control traffic.
- **MBMS-Gateway (MBMS-GW):** It is a logical entity (i.e., it may be a stand-alone device or it may be co-located with other network nodes) which accomplishes data content forwarding to the eNodeB(s) involved in the MBMS session. The MBMS-GW is in charge of maintaining the IP multicast groups. In particular, it allocates the multicast IP address to a multicast group. The eNodeB joins the IP multicast group to receive the data relevant to the MBMS user plane when the session starts, while it leaves the IP multicast group when the session stops. The *M1* interface connects the MBMS-GW and the eNodeB(s) in the user plane. The data delivery over the *M1* interface is performed by exploiting IP multicast transmissions (there is no uplink data over *M1* radio network layer). The MBMS synchronization protocol (SYNC) is used over the *M1* interface to allow content synchronization for MBMS service data transmission.
- **Broadcast Multicast-Service Center (BM-SC):** It is a functional entity which represents the entry point for content provider transmissions. It authorizes and initiates MBMS Bearer Services by performing several procedures relevant to service provisioning. Indeed, The BM-SC should be used for scheduling and delivery procedures relevant to a MBMS transmission. In addition, the BM-SC is able to accomplish service announcement and membership functions and performs functions such as security operations and content synchronization. Finally, the BM-SC is in charge for header compression in multi-cell MBMS services.

Multicast Session Procedures

The provisioning of a multicast service involves different procedures necessary for service announcement, multicast group formation and service delivery. In details, the phases to be accomplished for a MBMS multicast session set up are:

- **Subscription:** It is an agreement between the UE and the service provider for the reception of the MBMS service(s). The BM-SC stores the subscription information. At the end of this phase, the subscriber's terminal is updated with MBMS-specific information (MBMS service keys and user keys) useful to decipher the received data traffic.
- **Service Announcement:** It allows the MBMS members to collect service activation information (e.g., multicast IP address, service start time) about the available MBMS services. Service provider may consider different discovery mechanisms: *(i)* Point-to-Point or Cell Broadcast SMS; *(ii)* MBMS Broadcast mode to announce both multicast and broadcast services; *(iii)* MBMS Multicast mode for MBMS Multicast services advertising; *(iv)* URL (HTTP, FTP); *(v)* PUSH mechanisms (WAP, SMS-PP, MMS).
- **Joining:** The subscriber joins the multicast group. In this phase, the UE becomes a multicast group member and it initiates the relevant MBMS bearer services in order to receive the multicast data content.
- **Session Start:** The BM-SC performs the transmission of MBMS data content. A Session Start message is sent for each MBMS bearer service involved in the MBMS session. The Session Start message does not depend on the activation of the service by the user (i.e., the service activation can be performed by the user before or after the transmission of the Session Start message).
- **MBMS Notification:** In this phase multicast members are informed about forthcoming MBMS multicast data transfer.
- **Data Transfer:** In this phase the MBMS traffic is delivered to the multicast group members.
- **Session Stop:** The BM-SC stops data transmission and the MBMS bearers are accordingly released.
- **Leaving:** In this phase a subscriber leaves a multicast group.

Subscription, joining and leaving phases are performed on a *per-user* basis. On the contrary, other phases involve the whole set of users interested in the MBMS multicast session and are accordingly performed on a *per-group* basis.

MBMS Channels

The MBMS standard defines two logical channels to be exploited for multicast content delivery: the *Multicast Traffic Channel (MTCH)* and the *Multicast Control Channel (MCCH).* The MTCH is a PtM downlink channel for data transmission. Multiple MBMS services can therefore be transmitted using a single MTCH. The MTCH exploits the Radio Link Control (RLC) unacknowledged mode. According to this mode, no retransmissions at the RLC layer of LTE protocol stack are used, i.e., multicast terminals do not send any acknowledgement message to the eNodeB.

The MCCH is a PtM downlink channel for signaling traffic transmission. One MCCH carries information regarding one or several MTCHs, including the sub-frame allocation and the MCS relevant to each MTCH. The MCCH exploits the RLC unacknowledged mode.

In single-cell mode, these channels are mapped on the Downlink Shared Channel (DL-SCH). The design of the MCCH is a key challenge in order to efficiently manage the MBMS signaling traffic. The number of MCCHs varies according to the data transmission mode and, for single-cell MBMS deployments, one MCCH is activated for each MBMS service transmitted by the eNodeB.

RESOURCE ALLOCATION OF MIXED UNICAST AND MULTICAST SERVICES

The joint management of unicast and multicast traffic is a key design issue of future mobile networks. It is demonstrated by the intense research activity conducted by both standardization and academic bodies. Recently, also the 5G-PPP (5G Infrastructure Public Private Partnership) committee has indicated the full integration between emerging mobile group-based (i.e., multicasting) and more traditional unicast communications as one of the major research priorities in next-to-come 5G networks.

The main challenge in achieving this targeted goal is the design of effective resource allocation strategies, due to the use of different transmission modes for unicast and multicast services. Indeed, in case of unicast services, data is delivered to a single user; so the modulation and coding scheme (MCS) is selected according to *individual* channel conditions between the eNodeB and the user equipment (UE). Differently, in the multicast case, transmission parameters are selected by the eNodeB on a *per-group* basis. Multicast group members could measure different radio link qualities and support different MCSs: poor links need robust MCSs to face hostile propagation conditions, while users with good channels could receive data at higher bit rates. However, the multicast group member with the worst channel condition imposes the MCS selection for the entire group, and thus it becomes the bottleneck of the whole group; this severely affects the transmission efficiency.

In general, favoring multicast traffic allows a provider to achieve a high overall system capacity (measured as the number of users simultaneously served by the eNodeB). This is at the expense of an unfair bandwidth distribution between multicast and unicast traffic, with negative consequences on the degree of satisfaction of unicast users. On the contrary, by advantaging unicast users, the network capacity could significantly decrease, which is undesirable. This unbalance is due to the fact that multicast group(s) and unicast users have different sizes. It is clear that for the best balance between service quality and efficiency, there is a need for a combined unicast and multicast traffic management solution.

This Section presents a non-exhaustive review of current proposals in the literature that address: *(i)* support of multicast traffic in broadband wireless access networks, *(ii)* management of multiple multicast groups, and *(iii)* delivery of mixed unicast and multicast traffic.

Support of Multicast Traffic in Broadband Wireless Access Networks

In a context where the base station has to serve multiple destinations joined in the same multicast group, point-to-multipoint (PMP) transmissions represent an effective solution in order to efficiently exploit the broadcast nature of the radio channel by using a single transmission to feed the whole multicast group.

The main issue related to the PMP approach is that the link adaptation procedures, i.e., the selection of transmission parameters such as the MCS, are performed on a *per-group basis*. As a consequence, the presence of cell-edge users forces the base station to use more robust MCSs in order to guarantee an error free reception to the whole set of multicast receivers and this aspect dramatically influences the performance of subscriber stations with high channel quality, which could potentially attain higher data rate.

The Conventional Multicast Scheme (CMS) (Richard et al., 2013) performs the link adaptation procedures by adopting a conservative approach which foresees to assign the group data rate based on the user experiencing the *worst* channel conditions. Although this approach guarantees the highest fairness (i.e., all multicast members are served at the same data rate), it introduces severe inefficiencies in the spectrum management that is not fully exploited with a consequent poor spectral efficiency performance.

Several approaches have been proposed in order overcome the limitations of the CMS approach. For example, the opportunistic scheme (Low et al., 2009) foresees, during any given time slot, to serve only the "best" portion of multicast members in order to maximize the Quality of Service (QoS) of the served users.

A similar approach is proposed in (Hou et al., 2009), where is presented a cooperative transmission model: in the first phase the BS multicasts data at a high rate by exploiting an opportunistic scheme, whereas in the second phase users in good channel conditions help in relaying the received data to the remaining users.

Although the opportunistic multicasting introduces several improvements for high quality users, it could require the adoption of additional data coding, i.e, rateless codes, in order to make the service work properly (due to the fact that the portion of receivers could change slot by slot) and involves several issues for the cell-edge subscribers which will experience poor throughput performance and increased delivery latency.

An interesting approach that aims at enhancing the radio channel utilization in multicast environments by exploiting the multi-user diversity is the *subgrouping* (Araniti et al., 2012), depicted in Figure 3. It splits the multicast group members into several subgroups, where each subgroup collects subscribers which experience similar channel qualities. The goals of subgrouping are to serve the whole multicast group within every time slot while guaranteeing improvements in terms of session quality, user satisfaction and spectrum utilization. Subgrouping efficiently supports both single- and multi-rate multicast services.

The subgroup formation can be accomplished with different strategies. A basic approach is to split the multicast group into two subgroups and to define a single multicast transmission rate for such subgroups. For instance, the cell is divided into two QoS regions. The eNodeB transmits two data streams with different transmission parameters, i.e., MCS, and power level, according to the QoS definitions.

Figure 3. Multicast subgrouping

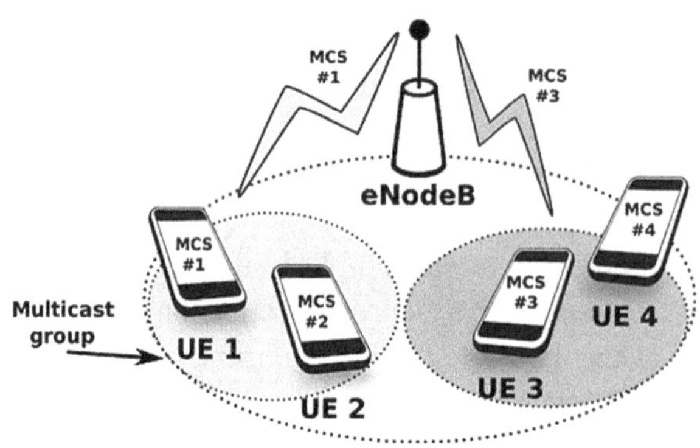

Each stream corresponds to a different QoS region. Users with high channel gain can receive both streams (i.e., they will achieve the highest quality session) while the users with poor channel qualities will receive only one stream (i.e., basic video quality).

Another strategy for multicast subgrouping is to design the subgroup formation according to an optimization problem. In this way, the most suitable subgroup configuration (i.e., number of subgroups with the related MCS, portion of users, assigned resources, and data rate for each enabled subgroup) is dynamically selected by the eNodeB based on an optimization problem which takes into account the users' CQI values and the QoS constraints of multicast session. The optimization problem can be formulated in order to achieve different goals, for example to maximize the system throughput or the spectral efficiency. Another strategy is to recast the subgroup creation phase with the aim to minimize the amount of RBs needed for multicast content delivery. Finally, another suitable approach is to enable the best subgroups in order to achieve proportional fairness allocation with the aim to attain both intra- and inter-group fairness.

Management of Multiple Multicast Groups

Several works in literature addressed the problem to efficiently deliver multimedia streaming content to different multicast groups simultaneously served by a given base station. Such works consider each video stream (or program) encoded into different layers and focus on the following problem (Deb et al., 2008): for any time slot within a given scheduling frame, which layer of which multicast group should be transmitted at what modulation coding scheme (MCS)? In order to reach this goal, the number of receivers per layer has to be dynamically adjusted according to the experienced channel conditions and the available network bandwidth, so as to maximize a given utility function.

Authors in (Kuo et al., 2007) propose a utility-based resource allocation scheme for layer-encoded IPTV multicast streaming service over IEEE 802.16 WiMAX networks. They prove that the problem of finding the best portion of users to serve per layer for each multicast group is NP-hard. Indeed in case of G multicast groups, K users per group and m MCSs, the number of possibilities to consider is equal to m^{KG}. Then, authors propose an approach that can run in polynomial time. Works (Deb et al., 2008) and (Li et al., 2009) follow an approach similar to (Kuo et al., 2007), by proposing near-optimal algorithms for subgroup management.

In particular, authors in (Deb et al., 2008) present a fast greedy algorithm, aimed at achieving proportional fairness resource allocation, that is provably within a constant approximation of the optimal solution (based on a metric that reflects video quality as perceived by the user).

The idea at the basis of work (Li et al., 2009) is to maximize the total system utility (defined as a generic non-negative and non-decreasing function of the received rate) through a two-step dynamic programming algorithm.

It is worth noting that, with the aim to maximize the total system utility of all video multicast sessions, (Li et al., 2009) assumes that the base layer should be not received by all users, i.e., some video sessions could be dropped or some users could not decode any layer.

Finally, (Sharangi et al., 2011) extended the referred problem by defining an algorithm for transmitting multicast data in bursts in order to conserve energy of mobile receivers.

The main issue related to these works is that they consider, within a scheduling frame, a given number of resources useful for supporting real-time services. They do not consider a practical case where such applications co-exist with other applications with heterogeneous QoS constraints in terms of guaranteed data rate and delay or jitter. A suitable allocation of resources across different class of applications characterized by different QoS requirements is the key issue we considered in our research work.

Delivery of Mixed Unicast and Multicast Traffic

Nowadays, broadband wireless communication systems are claimed to manage both multicast and unicast traffic. However, the design of effective algorithms for mixed unicast and multicast traffic is still an open research field.

In (Monserrat et al., 2012) the authors assess the joint performance of unicast and multicast services in LTE. In particular, they consider the transmission of both streaming and file delivery service when Multicast and Broadcast Single Frequency Network (MBSFN) operational mode is selected for the delivery of streaming services like mobile TV.

The system performance in terms of system capacity, worst average channel user's capacity, and outage probability for varying cell environments of the unicast and multicast transmission schemes are evaluated and compared in (Baek et al, 2009).

Furthermore, a transmission scheme in a mixed-traffic environment is presented. The proposed approach takes into account only the SNR threshold values and minimum performance guarantees are assured neither to unicast nor to multicast services. A comparison between unicasting and multicasting has been presented also in (Konrad et al., 2007), where authors determine the switching thresholds (based on the user density within the cell) able to make multicasting beneficial compared to unicasting for download or streaming services.

When focusing on the literature dealing with joint management of unicast and multicast traffic, three main trends can be identified: *Unicast Maximization* (UM), *Equal Sharing* (ES), and *Equal Competition* (EC). The first trend prioritizes unicast traffic; the second one aims to equally share the available resources between unicast and multicast users; and the third one aims to maximize the number of conveyed bits to improve the spectrum utilization.

Example of proposals which belong to the *UM* philosophy in multi-carrier Orthogonal Frequency Division Multiplexing (OFDM) systems are given in (Seo et al., 2007) and (Shen et al., 2009). The main idea behind these solutions is to guarantee a minimal required data rate to all multicast users, according to the CMS logic, and then to assign the remaining resources to unicast users following a maximum throughput approach. As a consequence, an increased throughput is offered to unicast destinations at the expense of multicast receivers, which experience significant limitations in QoS.

An example of the *ES* strategy is available in (Wu et al., 2012), which presents a power-saving scheduling algorithm that manages mixed unicast and multicast traffic. According to the *ES* philosophy, multicast and unicast services equally share the network capacity; the resources to assign to the two types of traffic (unicast and multicast) are statically split into two equal sets. This policy prevents a traffic class to utilize the resources assigned to the other class. Despite the approach is fair, its main inefficiency is due to the static bandwidth assignment that cannot adapt to the dynamic load variation of both unicast and multicast users, thus resulting in inefficient spectrum utilization and possible negative effects on the quality of one of the traffic classes.

Finally, a technique that follows the *EC* logic and aims to guarantee a minimum data rate of both unicast and multicast flows is designed in (Deng et al., 2012). The extra resources are assigned according to a maximum throughput scheme, which iteratively selects the service that delivers the highest number of bits to the destinations. Another example of the *EC* philosophy can be found in (Chen et al., 2015), which proposes a proportional fairness problem where unicast and multicast users equally compete to get the available resources (i.e., the available resources are assigned to either unicast or multicast users according to which ones are able to maximize the sum of logarithmic data rates). With the aim to maximize the system throughput, this technique intrinsically advantages multicast services, since multiple destinations can be satisfied with a single transmission. Therefore, multicast users have higher probability to be served with respect to unicast users, which may suffer from poor throughput performance.

FUTURE RESEARCH DIRECTIONS

From the discussion in the previous Sections, it clearly emerges that the design of an effective RRM policy for the management of multicast services over LTE is still an open issue. Among the addressed solutions, multicast subgrouping is the most promising scheme since it allows to improve the multicast session quality by efficiently exploiting multi-user diversity while guaranteeing full coverage. For this reason, future researches are required to definitively demonstrate the effectiveness of multicast subgrouping in several cell deployment scenarios and load conditions. Furthermore, since RRM procedures must be accomplished with strict time constraints (link adaptation is performed every 1ms-long TTI), a further challenge is related to the design of low-complexity subgrouping schemes which can perform subgroup formation without requiring high computational cost at the eNodeB.

Another important scenario to consider is related to the management of multiple multicast services transmitted in a cell. Indeed, since multicast services are considered as the value-added of 4G wireless systems, it is expected that several multicast sessions are simultaneously served by a given eNodeB. This scenario poses several challenges since the delivery of several multicast applications could meaningfully reduce the capacity of LTE systems. Finally, further investigations are still required to assess the impact of multicast subgrouping on the management of typical unicast services transmitted in a LTE cell.

CONCLUSION

This Chapter focused on the issues and the challenges related to the RRM of mixed unicast and multicast services over LTE systems. Multicasting is expected to be massively used over LTE together with more traditional unicast services, and effective solutions in terms of resource allocation and link adaptation procedures are required in order to allow unicast services to compete for resource distribution on an equal footing with multicast traffic. In this Chapter we analyzed the different strategies proposed in literature for the management of multiple multicast groups, and for the delivery of mixed unicast and multicast traffic.

It appears clear that providing satisfactory services to all users represents a challenging task and that further research is needed in order to design novel approaches for efficiently managing heterogeneous traffic.

REFERENCES

Baek, S. Y., & Hong, D. Y. J. (2009). Adaptive Transmission Scheme for Mixed Multicast and Unicast Traffic in Cellular Systems. *IEEE Transactions on Vehicular Technology*, *58*(6), 2899–2908. doi:10.1109/TVT.2008.2010515

Chen, J., Chiang, M., Erman, J., Li, G., Ramakrishnan, K. K., & Sinha, R. K. (2015). *Fair and optimal resource allocation for LTE multicast (eMBMS): Group partitioning and dynamics*. Paper presented at the IEEE Conference on Computer Communications.

Deb, S., Jaiswal, S., & Nagaraj, K. (2008). *Real-Time Video Multicast in WiMAX Networks*. Paper presented at the IEEE Conference on Computer Communications (INFOCOM).

Deng, Tao, & Lu. (2012). Qos-aware resource allocation for mixed multicast and unicast traffic in OFDMA networks. *EURASIP Journal on Wireless Communications and Networking*.

Hou, F., Cai, L. X., Ho, Pin-Han, Shen, X., & Zhang, J. (2009). A cooperative multicast scheduling scheme for multimedia services in IEEE 802.16 networks. *IEEE Trans. on Wireless Comm.*, *8*(3), 1508-1519.

Konrad, M., Koch, W., & Huschke, J. (2007). *Throughput Based Switching Threshold for Multicast and Unicast in Cellular OFDM Networks*. Paper presented at the IEE VTC-2007 Fall Conference.

Kuo, W. H., Liu, T., & Liao, W. (2007). *Utility-Based Resource Allocation for Layer-Encoded IPTV Multicast in IEEE 802.16 (WiMAX) Wireless Networks*. Paper presented at the IEEE International Conference on Communications (ICC).

Lecompte, D., & Gabin, F. (2012). Evolved Multimedia Broadcast/Multicast Service (eMBMS) in LTE-Advanced: Overview and Rel-11 Enhancements. *IEEE Communications Magazine*, *50*(11), 68–74. doi:10.1109/MCOM.2012.6353684

Li, P., Zhang, H., Zhao, B., & Rangarajan, S. (2009). *Scalable Video Multicast in Multi-carrier Wireless Data Systems*. Paper presented at the IEEE International Conference on Network Protocols (ICNP).

Low, T. P., Pun, M. O., Hong, Y. W. P., & Kuo, C. C. J. (2009). Optimized opportunistic multicast scheduling (OMS) over wireless cellular networks. *IEEE Transactions on Wireless Communications*, *9*(2), 791–801. doi:10.1109/TWC.2010.02.090387

Monserrat, J. F., Calabuig, J., Fernandez-Aguilella, A., & Gomez-Barquero, D. (2012). Joint Delivery of Unicast and E-MBMS Services in LTE Networks. *IEEE Transactions on Broadcasting*, *58*(2), 157–167. doi:10.1109/TBC.2012.2191030

Richard, A., Dadlani, A., & Kim, A. (2013). Multicast scheduling and resource allocation algorithms for OFDMA-based systems: A survey. *IEEE Communications Surveys and Tutorials*, *15*(1), 240–254. doi:10.1109/SURV.2012.013012.00074

Seo, H., Kwack, S., & Lee, B. G. (2007). *Channel Structuring and Subchannel Allocation for Efficient Multicast and Unicast Services Integration in Wireless OFDM Systems*. Paper presented at the IEEE Globecom Conference.

Sharangi, S., Krishnamurti, R., & Hefeeda, M. (2011). Energy-Efficient Multicasting of Scalable Video Streams Over WiMAX Networks. *IEEE Transactions on Multimedia, 13*(1), 102–115. doi:10.1109/TMM.2010.2076799

Shen, J., Yi, N., Wu, B., Jiang, W., & Xiang, H. (2009). *A greedy-based resource allocation algorithm for multicast and unicast services in OFDM system*. Paper presented at the WCSP Conference.

Third Generation Partnership Project. (2012, September). TS 36.300, Evolved Universal Terrestrial Radio Access (E-UTRA) and Evolved Universal Terrestrial Radio Access Network (E-UTRAN), Release 11.

Wu, C., Sun, X., & Zhang, T. (2012). *A power-saving scheduling algorithm for mixed multicast and unicast traffic in MBSFN*. Paper presented at the ComComAp Conference.

ADDITIONAL READING

Akyildiz, I. F., Gutierrez-Estevez, D. M., & Chavarria Reyes, E. (2010). The evolution to 4G cellular systems: LTE-Advanced. *Physical Communication, 3*(4), 217–244. doi:10.1016/j.phycom.2010.08.001

Alexious, A., Bouras, C., Kokkinos, V., Papazois, A., & Tsichritzis, G. (2010, October). *Efficient MCS selection for MBSFN transmissions over LTE networks*. Paper presented at the IFIP Wireless Days (WD), Venice, Italy. 10.1109/WD.2010.5657749

Araniti, G., Condoluci, M., Militano, L., & Iera, A. (2013). Adaptive resource allocation to multicast services in LTE systems. *IEEE Transactions on Broadcasting, 59*(4), 658–664. doi:10.1109/TBC.2013.2271387

Bakanoglu, K., Mingquan, W., Hang, L., & Saurabh, M. (2010, April). *Adaptive resource allocation in multicast OFDMA systems*. Paper presented at the IEEE Wireless Communications and Networking Conference, Sydney, Australia.

Belleschi, M., Gabor, F., & Abrardo, A. (2011, December). *Performance analysis of a distributed resource allocation scheme for D2D communications*. Paper presented at the IEEE Global Telecommunications Conference, Houston, Texas, USA.

Condoluci, M., Araniti, G., Molinaro, A., & Iera, A. (2013, September). *Exploiting Frequency-Selectivity in Real-Time Multicast Services over LTE Networks*. Paper presented at the IEEE 24rd International Symposium on Personal, Indoor and Mobile Radio Communications. London, UK.

Condoluci, M., Araniti, G., Molinaro, A., Iera, A., & Cosmas, J. (2013, June). *On the Impact of Frequency Selectivity on Multicast Subgroup Formation in 4G Networks*. Paper presented at the IEEE International Symposium on Broadband Multimedia Systems and Broadcasting, London, UK.

Gopala, P. K., & Gamal, H. E. (2004, November). *Opportunistic multicasting*. Paper presented at the Thirty-Eighth Asilomar Conference on Signals, Systems and Computers, Pacific Groove, California.

Huang, C. W., Huang, S. M., Wu, P. H., Lin, S. J., & Hwang, J. N. (2012). OLM: Opportunistic Layered Multicasting for Scalable IPTV over Mobile WiMAX. *IEEE Transactions on Mobile Computing, 11*(3), 453–463. doi:10.1109/TMC.2011.34

Kuo, W. H., Liu, T., & Liao, W. (2007, June). *Utility-based resource allocation for layer-encoded IPTV multicast in IEEE 802.16 (WiMAX) wireless networks*. Paper presented at the IEEE International Conference on Communications, Glasgow, Scotland.

Ma, Y., Letaief, K., Wang, Z., Murch, R., & Wu, Z. (2010, December). *Multiple description coding-based optimal resource allocation for OFDMA multicast service*. Paper presented at the IEEE Global Telecommunications Conference, Miami, Florida, USA.

Militano, L., Condoluci, M., Araniti, G., & Iera, A. (2013, June). *Multicast service delivery solutions in LTE-Advanced systems*. Paper presented at the IEEE International Conference on Communications, Budapest, Hungary.

Ngo, D. T., Tellambura, C., & Nguyen, H. H. (2009, January). *Efficient resource allocation for OFDMA multicast systems with fairness consideration*. Paper presented at the IEEE Radio and Wireless Symposium, San Diego, CA, USA. 10.1109/RWS.2009.4957364

Pyattaev, A., Johnsson, K., Andreev, S., & Koucheryavy, Y. (2013, June). *Proximity based data offloading via network assisted device-to-device communications*. Paper presented at the IEEE Vehicular Technology Conference, Dresden, Germany.

Rhee, W., & Cioffi, J. (2000, May). *Increase in capacity of multiuser OFDM systems using dynamic subchannel allocation*. Paper presented at the IEEE 51st Vehicular Technology Conference, Tokyo, Japan.

Schwarz, H., Marpe, D., & Wiegand, T. (2007). Overview of the Scalable Video Coding Extension of the H.264/AVC Standard. *IEEE Transactions on Circuits and Systems for Video Technology*, *17*(9), 1103–1120. doi:10.1109/TCSVT.2007.905532

Suh, C., & Mo, J. (2006, April). *Resource Allocation for Multicast Services in Multicarrier Wireless Communications*. Paper presented at the IEEE 25th International Conference on Computer Communications, Barcelona, Spain.

Third Generation Partnership Project (2006, October). *TR 25.814, Physical layer aspect for evolved Universal Terrestrial Radio Access (UTRA)*, Release 7.

Third Generation Partnership Project (2012, March). *TS 23.246, Multimedia Broadcast/Multicast Service (MBMS); Architecture and functional description*, Release 11.

Third Generation Partnership Project (2012, September). *TS 36.213, Evolved Universal Terrestrial Radio Access (E-UTRA); Physical layer procedures*, Release 11.

Third Generation Partnership Project (2012, December). *TS 36.201, LTE physical layer; General description*, Release 11.

Third Generation Partnership Project (2012, September). *TS 36.440, General aspects and principles for interfaces supporting Multimedia Broadcast Multicast Service (MBMS) within E-UTRAN*, Release 11.

Zhang, L., He, Z., Niu, K., Zhang, B., & Skov, P. (2009, September). *Optimization of coverage and throughput in single-cell E-MBMS*. Paper presented at the IEEE 70th Vehicular Technology Conference Fall, Anchorage, AK.

KEY TERMS AND DEFINITIONS

CQI: Channel state information transmitted by a UE to the base station.
FDPS: The unit responsible of link adaptation procedures over LTE.
Link Adaptation: Selection of the most suitable transmission parameters according to users' CQI.
LTE: Radio mobile system able to support high data rate even for users located at the cell-edge.
MBMS: Standard allowing to efficiently support multicast services over LTE.
Multicast: A service which is simultaneously transmitted towards multiple users.
RRM: Set of functionalities relevant to the management of radio resources over cellular networks.
UE: Mobile terminal connected to the LTE system.

Chapter 5
Game Theory for Cooperation in Multi-Access Edge Computing

Jose Moura
Instituto Universitário de Lisboa, Portugal

Rui Neto Marinheiro
Instituto Universitário de Lisboa, Portugal

Joao Carlos Silva
Instituto Universitário de Lisboa, Portugal

ABSTRACT

Cooperative strategies amongst network players can improve network performance and spectrum utilization in future networking environments. Game Theory is very suitable for these emerging scenarios, since it models high-complex interactions among distributed decision makers. It also finds the more convenient management policies for the diverse players (e.g., content providers, cloud providers, edge providers, brokers, network providers, or users). These management policies optimize the performance of the overall network infrastructure with a fair utilization of their resources. This chapter discusses relevant theoretical models that enable cooperation amongst the players in distinct ways through, namely, pricing or reputation. In addition, the authors highlight open problems, such as the lack of proper models for dynamic and incomplete information scenarios. These upcoming scenarios are associated to computing and storage at the network edge, as well as, the deployment of large-scale IoT systems. The chapter finalizes by discussing a business model for future networks.

1. INTRODUCTION

Game Theory (GT) techniques have recently emerged in many engineering applications, notably in communications and networking. With the emergence of cooperation as a new communication paradigm, alongside the need for self-organizing, decentralized, and autonomic networks, it has become imperative to seek suitable GT tools to analyze and study the behavior and interactions of nodes in Future Networks (FNs). The final goal is to find low-complexity distributed algorithms that can efficiently manage the

DOI: 10.4018/978-1-5225-7570-2.ch005

high-complexity future network environment formed by heterogeneous technologies, enhancing collaboration among players and punish selfish or misbehaving nodes. In addition, the new management solutions should reduce the unwanted effects of stale information (e.g. oscillation around a specific network status) by choosing the proper values, namely, for both sampling rate of network status and delay associated to the dissemination of status information amongst the network nodes. This chapter fills a hole in existing communications literature, by providing a comprehensive review about GT models/concepts that are highly relevant for enabling collaboration in FNs environments.

In FNs, distributed and intelligent management algorithms can manage (control) the network infrastructure. These algorithms create incentive mechanisms that force the players to cooperate instead of pursuing their own interest. This novel player's behavior enables the efficient usage of available (sometimes-constrained) network resources, satisfying the heterogeneous requirements of data flows. Broadly speaking, the current literature highlights two different ways to encourage cooperation (collaboration) among the players: one with a short-term control effect and the other with a long-term control effect. The first approach uses virtual payments (credit-based games) to relieve costs for relaying traffic, and the second approach enforces the creation of communities (or groups, clusters) to establish long-term relationships among the nodes (reputation-based games). The reputation-based games sustain cooperation among the players because defection against a specific node causes personal retaliation or sanction by others. In the limit, nodes that do not cooperate will not be able to use the network themselves. Effective corrective actions against cheating nodes are also required with either permanent or temporary measures. Other interesting perspective to investigate is the deployment of hybrid solutions combining credit-based and reputation-based methods to enhance collaboration amongst players.

There is a relatively new and a very interesting set of games designated by evolutionary coalitional games that can enable more intelligent, self-adjustable, and robust algorithms for the management of FNs. In addition, the social networks, like Facebook or Flickr, can rapidly disseminate the positive impact of collaborative actions among the users of FNs (Apicella, Marlowe, Fowler, & Christakis, 2012) (Bond et al., 2012). Furthermore, the deployment in large scale of vehicular and sensor networks supported by the convergent (Moura & Edwards, 2015) and heterogeneous (Moura & Edwards, 2016) wireless access can enable some collaborative behavior amongst players.

The current chapter reviews the literature to discuss the more promising GT proposals that can incentivize the collaboration among the diverse players, aiming to use more intelligently and efficiently the available resources of FNs. This chapter has the following structure. Section 2 introduces and discusses important GT aspects for FNs. Section 3 gives the background and highlights collaborative strategies in FNs. It also presents our vision about FNs. Then, section 4 describes how GT can enable and enhance collaboration in FNs. Section 5 offers a broad GT literature survey in wireless networking. Section 6 discusses some relevant research work about how GT addresses the more significant functional aspects we expect to be present in FN environments. In addition, Section 7 discusses the business perspective for FNs. Finally, Section 8 concludes with relevant GT open problems to support collaboration in FNs.

2. DISCUSSING GAME THEORY

The current section introduces and discusses relevant aspects of GT, which can be very useful to model the emergent network environments of FNs.

2.1. Roots and Scope

The earliest predecessors of GT are economic analysis of imperfectly competitive markets of the French economist Augustin Cournot in 1838 (Dutta, 1999). The next great advance is due to John Nash who, in 1950, introduced the Nash equilibrium (NE) which is the most widely used concept in modern GT. The NE consists on a game status where no rational actor playing that game has enough incentives to deviate from its current strategy. In fact, as any player would decide to use a different strategy from the one associated to the NE state then that player would be punished in the sense that his (her) reward is reduced. Nash´s initial work created a new branch in GT grouping all non-cooperative games. Further GT historical evolution is available in (Dutta, 1999).

GT is the study of multi-person decision problems (which differentiates it from the classical decision theory) in applications drawn from industrial organization, labor economics, macroeconomics, financial economics, and international economics (Gibbons, 1992). Alongside with previous applications in Economics and Finance, GT could be applied to other completely different real world cases (Dutta, 1999).

Classical GT essentially requires that all the specified players of a specific game make rational choices among a pre-defined set of static strategies. Therefore, it is fundamental in GT that each player must consider the strategic analysis that the players' opponents are making in determining that his (her) own static strategic choice is appropriate to receive the best payoff (reward) as possible. Otherwise, if other players do not influence a player's reward, then GT is not a proper tool. In this case, it is more convenient to use constrained optimization in the place of GT. Following, we discuss how GT can create a mathematical model (e.g. matrix form) that mimics real-life scenarios with conflict situations among the players, trying to solve those conflict situations.

2.2. Matrix Games

Matrix games are those in which the payoff to a player can be determined from a matrix of payoffs. The payoffs are assigned to each element of the matrix assuming that interactions among players are pairwise. One player chooses a row of the matrix and the other chooses a column of the matrix. The intersection between the row and the column points out a unique element of the matrix. As an example, if player A's strategy is to choose the third row and player B's strategy is to choose the first column, the resultant payoff to player A is the value in the third row and first column of the matrix. A consequence of this is that the number of strategies available to the players is finite and discrete.

The matrix games can be asymmetric or symmetric. On one hand, a game is asymmetric if players have different set of strategies and/or if players are distinctively rewarded from choosing a given strategy against an opponent with a particular strategy. A classic example of an asymmetric game is the battle of sexes that is modelled by two distinct payoff matrixes. On the other hand, a game is symmetric if players have the same set of strategies and experience the same reward of using a given strategy against an opponent with a particular strategy. A classic example of a symmetric game is the prisoner's dilemma, which can be modelled with a single matrix. Following, we discuss with further detail the prisoner's dilemma because is the classical GT approach to solve the dilemma of an individual choice between cooperate or defect (not cooperate) with others, which is the focus of the current chapter.

The prisoner's dilemma can be formulated in terms of a single payoff matrix with two players, each one with two possible strategies, as shown in Table 1. Suppose that two individuals are being held in a prison in isolated cells. In this game, regardless of what the other prisoner decides, each prisoner gets a

Table 1. Payoff matrix of prisoner's dilemma

		Prisoner B	
		Cooperate (Silent)	Defect (Betray)
Prisoner A	Cooperate (Silent)	1, 1	3, free
	Defect (Betray)	free, 3	2, 2

higher pay-off by betraying the other ("defecting"). The reasoning involves an argument by dilemma: B will either cooperate or defect. If B cooperates, A should defect, since going free is better than serving 1 year. If B defects, A should also defect, since serving 2 years is better than serving 3. Therefore, either way, A should defect. Parallel reasoning shows that B should also defect. As both players choose to defect, they will be serving 2 years. Yet both players choosing to cooperate obtain a higher payoff (serving only 1 year) than both players defecting! In this way, GT results in both players being worse off than if each chose to lessen the sentence of his accomplice at the cost of spending more time in jail himself. Later, in the current chapter, we use this game to show that the cooperation among network operators is very useful to all of them. In the following text, we discuss evolutionary game theory.

2.3. Evolutionary Game Theory

In opposition to the classical GT, Evolutionary GT (EGT), states that the players aren´t completely rational. The players have limited information about available choices and consequences and their strategies are not static. In fact, the players have a preferred strategy that continuously compare with other strategies, checking if they need to change their current strategy to get a better reward (fitness). The decision to change the preferred strategy can be also influenced by other neighboring players belonging to the same population (by observation and leaning). In this way, the strategy with the highest selection score inside a group of individuals forming a community will become the predominant strategy for that generation of individuals. Then, this strategy is transferred to the next generation of individuals (evolutionary aspect). Following, we discuss how EGT can model the upcoming scenarios of FNs. These future scenarios will be more complex and dynamic than current networking scenarios. Table 2 briefly compares traditional GT with EGT.

EGT has been developed as a mathematical framework to study the interaction among rational biological agents in a population. In evolutionary games, the agent revolves the chosen strategy based

Table 2. Comparison between traditional GT and EGT

Game Characteristic	Traditional GT	EGT
Pure strategies	Yes	No
Strategy adaption over time	No	Yes
Hyper rational behavior	Yes	No
Equilibria is always possible	No (in some scenarios due to restrictions on the strategy options)	Yes (i.e. at least it discovers an asymptotic equilibrium due to unrestricted strategy space)
Model dynamic and high complex game	No	Yes

on its payoff. In this way, both static and dynamic behavior of the game can be analyzed (Han, Niyato, Saad, Baar, & Hjrungnes, 2012). In this way, on one hand, evolutionary stable strategies (ESS) are used to study a static evolutionary game. On the other hand, replicator dynamics is used to study a dynamic evolutionary game.

EGT usually considers a set of players that interact within a game and then die, giving birth to a new player generation that fully inherits its ancestor's knowledge. The new player strategy is evaluated against the one of its ancestors and its current environmental context. Also, through mutation, a slightly distinct strategy may be selected by a set of players belonging to a specific generation, probably offering better payoffs. Next, each player competes with the other players within the evolutionary game using a strategy that increases its payoff. In this way, strategies with high payoffs will survive inside the system as more players will tend to choose them, while weak strategies will eventually disappear. Following, we present a tutorial in how EGT can be applied to wireless networks (Y. Zhang & Guizani, 2011).

Formally, we should consider within an evolutionary game an infinite population of individuals that react to changes of their environmental surroundings using a finite set of n pure strategies S = {s_1, s_2, …, s_n}. There is also a population profile, i.e. x = {x_1, x_2, …, x_n}, which denotes the popularity of each strategy $s_i \in S$ among the individuals. This means that x_i is the probability that a strategy s_i is played by the individuals. By this reason, x is also designated by the set of mixed strategies.

Consider an individual in a population with profile x. Its expected payoff when choosing to play strategy s_i is given by f (s_i, x). In a two-player game, if an individual chooses strategy s_i and its opponent responds with strategy s_j, the payoff of the former player is given by f (s_i, s_j). In a more generic way, the expected payoff of strategy s_i is evaluated by (1) $f_i = \sum_{j=1}^{n} x_j . f\left(s_i, s_j\right)$. whereas the average payoff is given by (2) $f_x = \sum_{i=1}^{n} x_i . f_i$.

The replicator dynamics is a differential equation that describes the dynamics of an evolutionary game without mutation (Y. Zhang & Guizani, 2011) (Taylor & Jonker, 1978). According to this differential equation, the rate of growth of a specific strategy is proportional to the difference between the expected payoff of that strategy and the overall average payoff of the population, as stated in (3) $\dot{x} = x_i . \left(f_i - f_x\right)$. Using this equation, if a strategy has a much better payoff than the average, the number of individuals from the population that tend to choose it increases. On the contrary, a strategy with a lower payoff than the average is preferred less and eventually is eliminated from the system set of strategies.

Considering now the mutation issue, suppose that a small group of mutants $m \in \left[0,1\right]$ with a profile $x' \neq x$ invades the previous population. The profile of the newly formed population is given by (4) $x_{final} = m.x' + \left(1 - m\right) .x$. Hence, the average payoff of non-mutants will be

$$f_{x_{final}}^{non-mutant} = f\left(x, x_{final}\right) = \sum_{j=1}^{n} x_j . f\left(j, x_{final}\right)$$

given by (5) and the average payoff of mutants will be given by (6)

$$f_{x_{final}}^{mutant} = f\left(x', x_{final}\right) = \sum_{j=1}^{n} x'_j . f\left(j, x_{final}\right).$$

In this context, a strategy x is called evolutionary stable strategy (ESS) if for any $x' \neq x, m_{mut} \in .[0,1]$ exists such that for all $m \in [0, m_{mut}]$, then equation (7) holds true. $f_{x_{final}}^{non-mutant} > f_{x_{final}}^{mutant}$. In this way, when an ESS is reached, the population is immune from being invaded by other groups with different population profiles. By other words, in this context the population is not affected by mutation issues.

$$f_i = \sum_{j=1}^{n} x_j . f\left(s_i, s_j\right) \tag{1}$$

$$f_x = \sum_{i=1}^{n} x_i \cdot f_i \tag{2}$$

$$\dot{x} = x_i \cdot \left(f_i - f_x\right) \tag{3}$$

$$x_{final} = m.x' + \left(1 - m\right) .x \tag{4}$$

$$f_{x_{final}}^{non-mutant} = f\left(x, x_{final}\right) = \sum_{j=1}^{n} x_j . f\left(j, x_{final}\right) \tag{5}$$

$$f_{x_{final}}^{mutant} = f\left(x', x_{final}\right) = \sum_{j=1}^{n} x'_j . f\left(j, x_{final}\right) \tag{6}$$

$$f_{x_{final}}^{non-mutant} > f_{x_{final}}^{mutant}. \tag{7}$$

EGT may be successfully applied to model a variety of network problems. The authors of (Y. Zhang & Guizani, 2011) review the literature concerning the applications of EGT to distinct network types such as wireless sensor networks, delay tolerant networks, peer-to-peer networks and wireless networks in general, including heterogeneous 4G networks and cloud environments. In addition, (Han et al., 2012) discusses selected applications of EGT in wireless communications and networking, including congestion control, contention-based (i.e. Aloha) protocol adaptation, power control in CDMA, routing, cooperative sensing in cognitive radio, TCP throughput adaptation, and service-provider network selection. By service-provider network selection, (Han et al., 2012) suggests EGT to study different scenarios:

- User churning behavior that impacts the revenue of service providers;

- User choice among candidate service providers of the access network that maximizes the perceived QoS for a service type.

In (Nazir, Bennis, Ghaboosi, MacKenzie, & Latva-aho, 2010), an evolutionary game based on replicator dynamics is formulated to model the dynamic competition in network selection among users. Each user can choose a service class from a certain service provider (i.e. available access network). They present two algorithms, namely, population evolution and reinforcement-learning for network selection. Although the network-selection algorithm based on population evolution can reach the evolutionary equilibrium faster, it requires a centralized controller to gather, process, and broadcast information about the users within the corresponding service area. In contrast, with reinforcement learning, a user can gradually learn (by interacting with the service provider) and adapt the decision on network selection (through a trial-and-error learning method) to reach evolutionary equilibrium without any interaction with other users.

Some work (Nazir et al., 2010) (Bennis, Guruacharya, & Niyato, 2011) investigated and compared the convergence behavior of Q-learning with EGT to enable a satisfactory performance of cellular networks with femtocells. The authors of (Nazir et al., 2010) introduce two mechanisms for interference mitigation supported by EGT and machine learning. In the first mechanism, stand-alone femtocells choose their strategies, observe the behavior of other players, and make the best decision based on their instantaneous payoff, as well as the average payoff of all other femtocells. They also formulate the interactions among selfish femtocells using evolutionary games and demonstrate how the system converges to equilibrium. By contrast, using the second mechanism (i.e. reinforcement learning), the information exchange among femtocells is no longer possible and hence each femtocell adapts its strategy and gradually learns by interacting with its environment (i.e., neighboring interferers). The femtocells can self-organize by relying only on local information, while mitigating interference inside the macrocell. In this way, the macrocell user can meet its Quality of Service requirements. They have concluded that the biologically inspired evolutionary approach converges more rapidly to the desired equilibrium as compared to the reinforcement learning and random approach. Nevertheless, this faster convergence requires more context information at the femtocells. The authors of (Bennis et al., 2011) reached equivalent results as (Nazir et al., 2010).

Further references that address EGT applications to the networking area are available for wireless (M. A. Khan, Tembine, & Vasilakos, 2012a) (M. A. Khan, Tembine, & Vasilakos, 2012b) and wireline (Eitan Altman, El-Azouzi, Hayel, & Tembine, 2009) networks. The impact of evolutionary games in future wireless networks is analyzed in (Tembine, Altman, El-Azouzi, & Hayel, 2010). Evolutionary models have been also proposed for hierarchical mobile (Semasinghe, Hossain, & Zhu, 2015) (Lin, Ni, Tian, & Liu, 2015) and vehicular (Shivshankar & Jamalipour, 2015) networks. In the text below, we discuss the Stackelberg game, which it is like a NC repeated game.

2.4. Stackelberg Game

Figure 1 shows the model of a Stackelberg game (SG). This game is like a Non-Cooperative (NC) game but instead of the players playing a single shot as a typical NC game, the players execute the SG game via a step-by-step way. In addition, a SG has a player, designated by a leader that has the highest priority to take the first action. However, before doing that, the leader observes other players' strategies. Then, the leader announces its preferred strategy to the remaining players, also designated by followers. The followers perceive the leader's action and adjust their strategies to minimize their own cost. After, the followers reveal their strategies again to the leader. In summary, the SG model is a sequential one with

Figure 1. Steps of a Stackelberg Game Theory

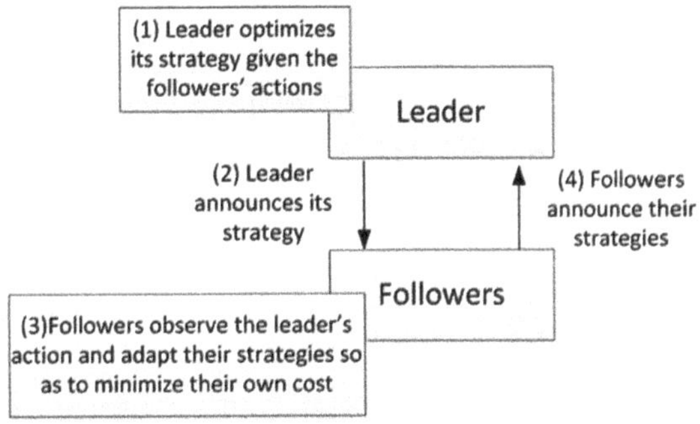

hierarchical decision-making that analyses the interaction between a leader (or leaders) and a set of followers to achieve a specific set of model goals. The final aim of a SG model is to discover the Stackelberg Equilibrium (SE), i.e. (Strategy_leader, Strategy_follower). We conclude that SE is an evolution from a NC game, where the former model adds two novel aspects: action observation and stage repetition.

Some applications of SE games are: Software Defined Networking (SDN) scenarios, where the SDN controller is the Leader; Femtocell power control (Han et al., 2012) in hierarchized mobile networks; and device-to-device (D2D) communication (Zhu & Hossain, 2015). The main advantage of using a SG model is to optimize diverse virtualized resources (e.g. computation, storage, and networking) of very complex topologies at the network edge under users' Quality of Experience. The main challenges the network designer should be aware of are as follows: i) implement a robust mechanism to ensure the correct and synchronous shift among leaders and followers; ii) the Stackelberg Equilibrium (SE) could give a worst result than NE due to the hierarchical decision-making process among leaders and followers (Han et al., 2012); and iii) a SE game requires complete and perfect information about other strategies and payoffs. In this situation, communication jitter among a leader and followers of a SG could disrupt the right control sequence and create instabilities on the control loop, affecting the obtained results from that model. In the next section, we discuss a model game that deals with a real problem that each player could have. It is related with the player uncertainty (full or partial) about the other players decisions. In this way, the players hardly predict how the pool of network resources shared among all them will be used.

2.5. Bayesian Game

In a Bayesian game (BG) the players have incomplete information about their environment (Y. Zhang & Guizani, 2011). This can occur due to some practical physical impairments that counteract the global dissemination among the nodes of useful information about the status of the system being studied, e.g. channel gain (Duong, 2016). Following Harsanyi's work (Harsanyi, 2004), a BG has a special player with random behaviour, i.e. 'Nature'. These games are called Bayesian because they require a probabilistic analysis. Players have initial beliefs about others' payoff functions. A belief is a probability distribution over the possible types for a player. Then, the initial beliefs might change based on the actions the players of the game have taken. As a game with incomplete information is repeated, the folk theorem (Fudenberg & Maskin, 1986) can find its social-optimum solution. The game also enables a distributed model to study the system. In this way, this game type can support user privacy as users do not need to disclose private data to an external centralized server or controller. However, it could be complicated to find the Bayesian NE, due to the dynamic characteristic of this game, where the players adjust their decisions based on their learning from the acquired information during the time the game is played (Han et al., 2012). The players' learning could be adversely affected also by jitter, security attacks, interference, errors, available battery energy to transmit, system unpredictability, etc. The reader could find in (Böge & Eisele, 1979) a comparison between a BG and a non-BG. In (Chawla & Sivan, 2014) a Bayesian mechanism design is also explained.

We have found in the literature some BGs for wireless networking environments. These games cover the following areas: hierarchical small cells (Bu, Yu, & Yanikomeroglu, 2015) (Z. Khan, Lehtomaki, DaSilva, Hossain, & Latva-aho, 2016) (Duong, Madhukumar, & Niyato, 2016); D2D communications (Kebriaei, Maham, & Niyato, 2015) (Xiao, Chen, Yuen, Han, & DaSilva, 2015) (Yan, Huang, & Wang, 2013)(Yan, 2013); vehicular scenarios (Duong et al., 2016) (Kumar, Misra, Rodrigues, & Obaidat, 2015) (Kumar, Zeadally, Chilamkurti, & Vinel, 2015); and wireless sensors (Kumar, Chilamkurti, & Misra, 2015) (La, Quek, Lee, Jin, & Zhu, 2016) (Zheng, Liu, & Qi, 2012).

2.6. Mechanism Design

There is a subfield of GT designated by Mechanism Design (MD) that allows a game designer to define initially the desired outcome and then specify the game rules to achieve that outcome (Han et al., 2012, 221-252). This is the opposite of game analysis, in which the game rules are predefined and then the outcome is investigated, as shown in Figure 2. That is why MD is also designated as reverse GT.

A very important result in MD is the Revelation Principle that states for any Bayesian Nash Equilibrium is associated a Bayesian game with the same equilibrium outcome but in which players truthfully report their choices (it could be a preference list), which simplifies the game analysis, eliminating the need to consider either strategic behavior or lying. So, no matter what the mechanism, a designer can confine attention to equilibrium in which players only report truthfully. To accomplish this, the model needs to consider incentives for players to truthfully cooperate among them, optimizing the game outcome.

3. BACKGROUND AND TRENDS IN FUTURE NETWORKS

According to the Cisco Global Forecast (CISCO, 2016) more than three-fourths of the world's mobile data traffic will be video by 2021. From the same source, sixty percent of total mobile data traffic was offloaded onto the fixed network through Wi-Fi or femtocell in 2016. This traffic offloading occurs

Figure 2. Game theory (GT) vs. mechanism design (MD)

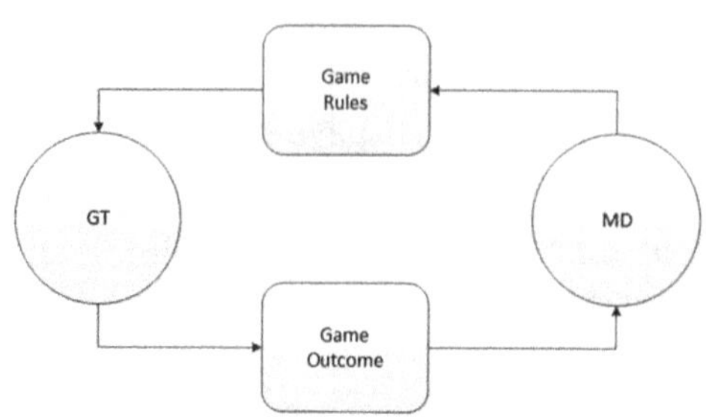

due to the lack of capacity in the mobile network infrastructure, originally dimensioned to support only voice and messages. The traffic offloading is one possible solution to mitigate congestion, avoiding the loss on the perceived quality by users' applications.

However, the first approach to the problem has been to perform an inter-technology handover between available technologies, with all the traffic routed through the most convenient access technology. A survey about mobility is available in (Fernandes & Karmouch, 2012). In our opinion, a better usage of available resources on the network-edge with a more fine-grained traffic management based on flows (e.g., Web traffic, VoIP) should alleviate the negative impact of network congestion, which has been reported very often essentially in the mobile broadband access. Multi-interface handheld terminals will soon have the battery autonomy and the capability to perform network access using simultaneous multi-radio access technologies (RAT). In addition, it is of particular interest the support of simultaneous data/multimedia flows through different access systems (LTE-A, WLAN, Wimax). Recent works (Yap et al., 2012) (Silva, Marinheiro, Moura, & Almeida, 2013) (Moura & Edwards, 2015) (Moura & Edwards, 2016) (Alves, Silva, Neto Marinheiro, & Moura, 2018) propose that mobile multimode terminals should use all the available connectivity options simultaneously. The mobile terminal should choose dynamically the most suitable network to each flow, obtaining faster connections by stitching flows over multiple networks, decrease the usage cost by choosing the most cost-effective network that meets application requisites, and reduce the energy consumption by selecting the technologies with the lowest energy-usage per byte. The management of the flows per network interface may not only be implemented independently by the terminal, but also be assisted transparently by the network (Alves et al., 2018).

This concept for FNs contributes to the perspective of integrating complementary access technologies with overlapping coverage to provide the expected ubiquitous coverage and to achieve the Always Best Connected (ABC) concept (Louta, Zournatzis, Kraounakis, Sarigiannidis, & Demetropoulos, 2011). This concept allows a flow to use at any time the most suitable access network/Network Attachment Point (NAP). This management of flows should be done in a distributed way with low complexity and reliable algorithms/protocols in networks formed by heterogeneous access technologies, where the most part of involved nodes should cooperate. Network brokers such as in (Moura & Edwards, 2016) follow on this idea. Brokerage systems, possibly implementing GT algorithms, can manage the network architecture, in which distributed nodes discover relevant context information to enhance the usage of local available connectivity resources (Mateus & Marinheiro, 2010). In this way, mobile operators can develop policies for IP flow mobility, and control which traffic is routed over different access technologies (Alves et al., 2018).

Another aspect to consider is that the Internet was initially designed to support communications between remote hosts. Since its early days, the Internet has evolved drastically, with a huge evolution in broadband access penetration and dissemination of mobile terminals with unforeseen capacities. This evolution has altered the Internet into a medium to connect people in multiple ways with content made available in completely new and complex modes through the entire network infrastructure. In fact, current users are more interested in searching for information over Google, watch videos on YouTube, and share files via Dropbox than to worry about connectivity to a particular host.

This content demand has catalyzed an exponential growth of Internet traffic volume and content distribution is increasingly becoming more centric in the Internet, and this is challenging and changing how the Internet is being organized.

Content delivery network (CDN) operators, content providers as well as ISPs are important players to consider in the typical content-centric cases of FNs. However, these players interact with a mix of technologies that are difficult to manage in a comprehensive and global ways.

Research efforts have been made to move the Internet away from its current reliance on purely point-to-point primitives and, to this end, have proposed detailed designs that make the Internet more data-oriented or content-centric (Jacobson et al., 2009)(L. Zhang et al., 2014). As such Information-centric networking has emerged as a new approach in the research community (Cheriton & Gritter, 2000) (Ahlgren, Dannewitz, Imbrenda, Kutscher, & Ohlman, 2012) to integrate content delivery as a native network feature and make networks natively content-aware.

Due to this, FNs most probably will sustain the next generation of the Internet infrastructure, interconnecting people and content through mobile cloud networks (as said before, the Internet is evolving from a node discovery to enable the discovery of specialized objects). These cloud networks will operate on an always best-connected scenario, where a person is allowed to choose the best available access technology (from small cells to standard base stations), access network and terminal device at any point in time. Generally, the idea is to enhance FNs to automatically interpret, process, and move content (information) independently of users' location. Additionally, the traditional approach, where resources are provided by remote clouds, is also not capable of giving adequate response to the fast-growing number of connected devises and their resource requirements. For all these reasons, new cloud architectures have been evolving, by migrating resources, such as services and data, closer to end users and devices (Figure 3).

With remote clouds, devices communicate directly with traditional distant resource-rich servers. These clouds can provide unlimited resources, but this approach does not easily scale, and long latency, bandwidth bottleneck, communication overhead, and location blindness is experienced. In face of this, it is necessary to bringing computing resources closer to end-users, to overcome the limitations of remote cloud computing (C. Li, Xue, Wang, Zhang, & Li, 2018). This is on the genesis of the edge computing paradigm, that allows more responsive cloud services, accomplished by extending the services from

Figure 3. Cloud architectures providing services to end devices

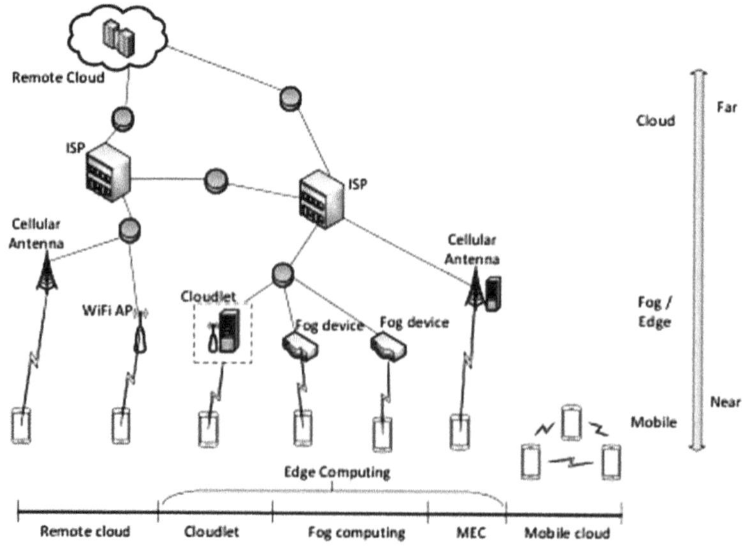

the core in cloud data centers to the edge of the network, by placing intermediate nodes between the cloud and the end user, which are responsible for better serving ubiquitous smart devices, fulfilling user resource requests.

Edge computing may follow different architecture implementations such as cloudlets, fog computing, or Multi-Access Edge Computing (MEC).

Cloudlets (Jararweh, Tawalbeh, Ababneh, Khreishah, & Dosari, 2014) are trusted devices or a cluster of devices with high capabilities. They are most often installed along with Access Points (AP) to allow mobile devices to access it, and in some cases both of the cloudlet and AP are integrated in one entity. Fog Computing is a term introduced by Cisco Systems (CISCO, 2015). Their rationale for coining this term is that a fog is nothing more than a cloud that is closer to the ground. Fog computing's main feature is that the fog system is deployed close to end users in a widely distributed manner (Yi, Li, & Li, 2015) (C. Li et al., 2018), in the form of fog nodes (Tordera et al., 2016), possibly at different levels and numbers (Balevi & Gitlin, 2018). The MEC paradigm (Taleb et al., 2017) was introduced by an industry lead initiative (ETSI 2014), to provide IT and cloud-computing capabilities within the Radio Access Network (RAN) in close proximity to mobile subscribers. Mobile network operators will allow the use of the access network, where low latency and high-bandwidth as well as direct access to real-time radio network information (such as subscriber location, cell load, etc.) is available. This can be used to allow content, services and applications to be accelerated, increasing server responsiveness from the edge. Additionally, MEC servers are context aware, as they manage information on end devices, such as their location and network information. Nevertheless, their capacity is limited, therefore deciding which and how resources can be managed at the edge can still be a trick endeavour (Gabry, Bioglio, & Land, 2016).

The clouds are migrating even closer to end users, with new computing architectures where mobile devices use their extra resources in a coordinated manner, to support cloud services. This contrasts with the previous edge implementations, where the mobile device's exclusive role in the cloud was that of a consumer. There is a myriad of proposals, either with centralised control, such as Hyrax (Marinelli, 2009) and FemtoClouds (Habak, Ammar, Harras, & Zegura, 2015), or a decentralised control, where nodes keep track of their own resources, such is the case with EECRS (Hu, Zhu, Xia, Chen, & Luo, 2012) (Lu et al., 2013), Phoenix (Panta, Jana, Cheng, Chen, & Vaishampayan, 2013), Mobile Host (Srirama & Paniagua, 2013) and (Monteiro, Silva, Lourenço, & Paulino, 2015).

This migration of clouds, to the proximity of users, in particular in the extreme case of clouds supported by autonomous mobile devices, brings new challenges regarding resource management. Once again this portrays a perfect scenario to apply game theoretic approaches, where conflicting interests have to be mediated.

Another trend gaining momentum for FNs is the Internet of Things (IoT). However, the IoT paradigm is not new (Corcoran, 2016), but building end-to-end IoT systems from scratch has always been a challenging and a risky enterprise, many times with ambiguous and uncertain business cases. To overcome this, a new trend in IoT, engaged by a surge of companies, is the building of complete solutions that encapsulate aspects of an end-to-end IoT system using building blocks that can be used in a repeatable and replicable way. These aggregated building blocks materialize many IoT platforms that allow companies to reduced development and deployment time and costs, and allow the creation of new business models, such as paying per use or fixed licensing. (Gluhak et al., 2016) provides an exhaustive review on different IoT platforms. In fact, IoT platforms have become so popular, which are present over 360 platforms on the market, with many more providers and consumers of this kind of platforms. The diversity of players at stake sometime have conflicting goals, and this challenge is an ideal use case for game theory approaches, such is the case with resource management (Semasinghe, Maghsudi, & Hossain, 2017).

But the IoT paradigm is progressing even further influencing new developments in various domains, such as the Internet of Mobile Things (IoMT), Autonomous Internet of Things (A-IoT), Autonomous System of Things (ASoT), Internet of Autonomous Things (IoAT), Internet of Things Clouds (IoT-C) and the Internet of Robotic Things (IoRT) (Vermesan et al., 2017), where new challenges are at stake. In fact, the initial tendency of centralized platforms, usually deployed at a remote cloud, following the classic centralized computing paradigm, faces several of such challenges such as high latency, low capacity and network failure. Because of these, the trend is now evolving to more distributed IoT platforms that can also, but not only, deployed at edge. This follows the same principles of fog computing to bring the cloud closer to IoT devices. The fog can provide IoT data processing and storage locally, instead of sending them to remote clouds, providing services with faster response and greater quality, enabling the IoT to provide efficient and secure services for many IoT users. (Atlam, Walters, & Wills, 2018) and (Mahmud, Kotagiri, & Buyya, 2018) reviewed pertinent state-of-the-art fog computing architectures and emerging IoT applications that will be improved by using the fog model, highlighting the benefits and implementation challenges. In these approaches, distributed resource management is usually more difficult to attain, and IoT devices are more than ever expected to act smart and resolve diverging goals. Once again, this is also a good used case for game theory approaches.

There are many applications for the IoT that include smart cities, like smart vehicles, surveillance systems, traffic monitoring, and smart parking, or homes and communities, like smart homes, wearable devices/mobile phones, healthcare, and hospitals, or the industry like business and production lines factories, or even agriculture, like automation and precision agriculture, and so on. This diversity of uses cases has also lead, regarding communication technologies, to the proliferation of a myriad of multi-radio access technologies for IoT, sometimes optimized to specific applications, to connect devices at the edge. (Vermesan et al., 2017). This has generated heterogeneous mobile networks that need complex configuration, management and maintenance, where it is important to have devices that play a more active role, at the edge of the network, making decisions and performing tasks without human intervention.

One of the major challenges for the the FN is how to achieve security in a growing networked world of distributed devices and services. To overcome this, blockchain and smart contracts have been a key technology to consider. The idea that supports blockchains, also referred to as distributed ledgers, is that distributed users maintain a public and identical dynamic digital register of all transactions that have taken place. The history of the recorded transactions alone determines the ownership, so it is imperative that transactions within this database are audited and agreed upon by consensus (Mingxiao, Xiaofeng, Zhe, Xiangwei, & Qijun, 2017). This decentralized method of keeping track of changes ensures the ledger cannot be practically controlled by any one entity. It also eliminates the possibility of single-points of failure and allows for the verification of transactions without the need for third-party intervention. The seminal paper for the Bitcoin protocol (Nakamoto, 2008) has triggered all this. With a blockchain in place, applications that could previously run only through a trusted intermediary, can now operate in a decentralized manner, without the need for a central authority, and achieve the same or better functionality with the same amount of certainty. This has prompted a new wave on security supported by blockchains and smart contracts, in several fields relevant for the FN such as the IoTs (Christidis & Devetsikiotis, 2016) and Wireless Mesh Networks (Selimi, Kabbinale, Ali, Navarro, & Sathiaseelan, 2018).

Blockchain technology can very well change and even disrupt future network (Mougayar, 2016) (Marsal-Llacuna, 2018) in several ways: it's reliable peer-to-peer communication model can lend to more effective IoT ecosystems; applications can be developed and hosted within decentralized storage environments, data bases can be connected using smart contracts; the overhead of managing and tracking large networks of devices without the need for a centralized controller could be reduced; network management could be further simplified using self-executing smart contracts, programmed to perform actions when certain requirements are met; transferring assets could be streamlined, in cloud-based architectures where edge devices are playing a greater role in networking; distributed and cooperative cloud storage environment over a peer-to-peer network could be possible.

Of course, the upcoming design of FNs (MEC/FC/IoT/Security) scheme to operate in a satisfactory way, a great number of very demanding requirements must be fulfilled, not only technical ones (e.g. autonomic self-x requisites with cognitive radios like self-learning) but also in terms of business relationships among operators and service providers, as well as, the handling of the service subscription.

The course of finding a solution that can satisfy all the involved entities in the high complex network environment of FNs, like content providers, cloud providers, home providers, brokers, network providers or users, can be found by means of GT (Moura & Hutchison, 2018). In this way, as the players define their strategies then the GT can find ways to build-up win-win situations for all of them. Cooperation between technologies and/or providers, alongside Machine to Machine (M2M) communications or Internet of Things deployment will require complex and dynamic management algorithms to maximize network efficiency, pricing, Quality of Service (QoS), Quality of Experience (QoE) and ultimately, profit.

Considering all previous facets, we foresee that FNs will have to form a network infrastructure with a collective intelligence, as shown in Figure 4. This intelligence is very pertinent in FNs to address emergent traffic requisites, the management complexity of the heterogeneous wireless access technologies, and the challenges faced by a more content and data centric network.

Figure 4. Collective intelligence in FNS to manage emergent traffic and functional requisites

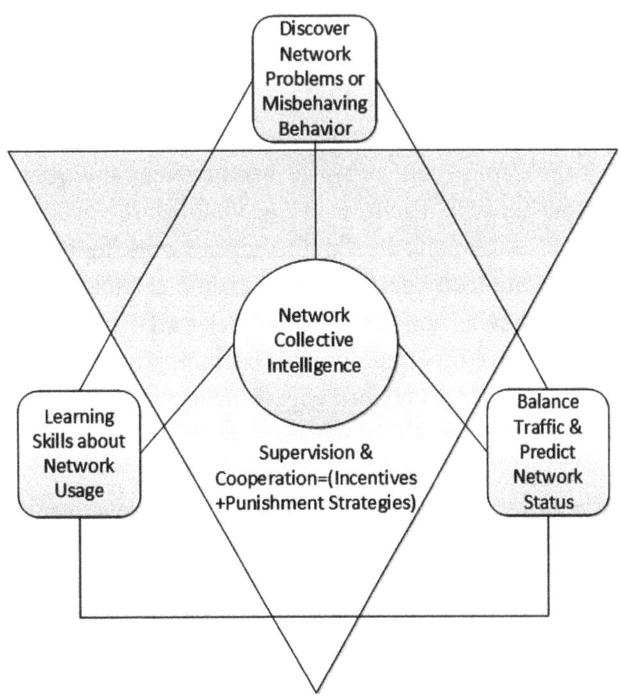

To enhance the network intelligence, the future network infrastructure needs to be supervised in order to enable learning processes on management algorithms when these control some network problems (e.g. congestion situation, node misbehaving behavior). In this way, the network intelligence will be enhanced, enabling the network infrastructure to manage the high complex future heterogeneous access infrastructure in a much more efficient way. As an example, the load could be balanced among the diverse wireless access technologies, reacting to a detected congestion situation to mitigate its negative effects. Alternatively, the load could be also balanced in a flash crowd scenario where a network problem is predicted and some policies are applied to the network to avoid the occurrence of that problem, e.g. offloading flows from the technology that soon could become disrupted to other available technologies with low levels of traffic load. In addition, congestion situations could be controlled by limiting the transmission rate of some users and freeing network resources to others. The one-billion-dollar question that remains to be answered is to find out the more efficient levels of aggressiveness of the algorithm that dynamically increases/decreases the rate transmission in a high complex networking scenario with diverse wireless access technologies and flow requirements.

To enable the network collective intelligence, we argue that it is important to obtain cooperation among the nodes. In this way, the network nodes need to be incentivized to cooperate, and the nodes that do not cooperate should be detected in a truthful way and be gradually penalized (e.g. their access rate is diminished). Eventually, uncooperative nodes that afterwards would change to a cooperative behavior, they could have their reputation values being restored to values that allow them to use again the network resources without any restriction on their access rate.

In practical terms, the FNs should require distributed management algorithms to support the network self-configuring feature. GT seems a very important area to model, analyze and decide how these distributed algorithms need to be deployed. In the following Section, we discuss some literature contributions that use theoretical games to enhance the cooperation among the diverse network players.

4. GAME THEORY CONTRIBUTIONS FOR ENHANCING NETWORK COOPERATION

FNs will be demanding for the deployment of novel management solutions aiming more efficiently and fairly usage of the available network resources. To accomplish the overall network goals, the nodes should collaborate or cooperate essentially in a multi-hop network topology, the typical scenario of future heterogeneous and high-complexity networks. For example, a terminal node should process both related and non-related traffic, whereas non-related classifies traffic not originated (not destined) from (to) that node. This new collaborative functionality will become possible at the physical layer in future multi-hop wireless networks because the network edge infrastructure will be vastly deployed by radio technologies, which allow the easy share of data messages among local terminals due to their broadcast transmission characteristic.

A very significant number of researchers have proposed GT models to encourage players (terminals and networks) to cooperate and enhance the overall network performance instead for acting selfishly to optimize their own performance. In this way, some additional incentives are required in FNs to enable collaboration among the nodes, defeating eventual misbehaving nodes like selfish or malicious ones. A selfish node may refuse to forward a non-related message to save its battery. In this way, this node needs a correct incentive to forward traffic, e.g. the network could increase the throughput of flows originated (destined) from (to) that node as a reward to previous collaboration in forwarding non-related traffic. Alternatively, a malicious node may try to disrupt the network functionality; in this case, the network could isolate that node from the network for a certain period as a punishment to that wrong procedure.

Broadly discussing, the right incentives to the nodes collaborate among them can be divided in two large groups: monetary-based and reputation-based. On one hand, the monetary-based solutions typically aim to achieve short/medium-term relationships among nodes. On the other hand, the reputation-based solutions typically aim to establish long-term relationships among nodes. This section will be highlighting some relevant work from these two groups, which is summarized in Figure 5.

The first group of contributions makes use of virtual payments for channel use and to incentive the collaboration among nodes in a multi-hop wireless network topology, as shown in Figure 6. Here, there are typically three types of nodes: the senders, the forwarders (intermediates) and the destination nodes. Some proposed credit-based systems suggest that distinct node types should be charged to cover the costs for packet forwarding. In fact, some proposals suggest that only the senders should be charged with a tariff initially specified (Zhong, Chen, & Yang, 2003) (L. Buttyan & Hubaux, 2000) (Buttyán & Hubaux, 2003) (Ileri, Siun-Chuon Mau, & Mandayam, 2005) (Shastry & Adve, 2006) (Chen, Yang, Wagener, & Nahrstedt, 2005) (T. Alpcan, Basar, Srikant, & Atman, 2001) (Saraydar, Mandayam, & Goodman, 2002) (Vassaki, Panagopoulos, Constantinou, & Vázquez-Castro, 2010). Alternatively, the destination nodes are charged (L. Buttyan & Hubaux, 2000) (Hua Liu & Krishnamachari, 2006) or destination and senders are both charged (Levente Buttyan & Hubaux, 2001) (Yanchao Zhang, Wenjing Lou, & Yuguang Fang, 2004). In addition, an incentive mechanism called bandwidth exchange was proposed in (D. Zhang, Ileri, & Mandayam, 2008), where a node can delegate a portion of its bandwidth to another node in exchange for relay cooperation. Finally, a different approach of credit-based schemes appear in (Chen & Nahrstedt, 2004) (Demir & Comaniciu, 2007), where auction-based incentive models are proposed. The basic idea of these schemes is that each intermediate node operates as a market; the users of the network put bids for their packets, the packets are accordingly scheduled to transmission and then charged after

Figure 5. Summary of game theory work supporting cooperation incentives

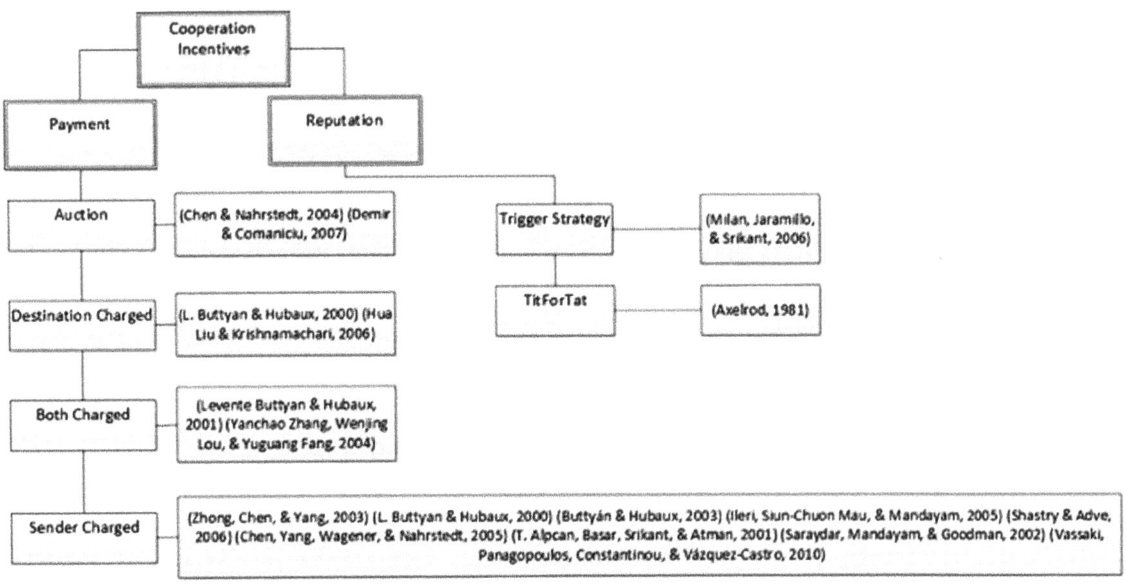

their transmission. The goals to achieve with auction models could be node truthful bidding and social network welfare maximization (Chen & Nahrstedt, 2004) or balancing residual battery energy and the current currency (credit) levels of the nodes in the network (Demir & Comaniciu, 2007).

The main advantage of credit-based approaches is that they succeed in large-scale networks to enforce a distributed cooperation mechanism among selfish nodes. Moreover, credits are useful when an action and its reward are not simultaneous. This is valid for multi-hop wireless networks: the action is packet forwarding and the reward occurs after sending their own packets. These approaches could be useful to discover the more convenient routing policies, solving very challenging dilemmas in multi-hop networks. For example, these approaches could help to choose the cheapest route between a source and a destination node either by minimizing the total number of hops (minimizing end-to-end flow delay) or by choosing the less-congested hops (increasing flow data rate). The drawbacks of credit-based proposals are extra overhead and complexity to charge users fairly and avoid cheating, turning these proposals hard to deploy.

In FNs, customers can be billed using a congestion-sensitive tariff, where prices are set in real time according to current load and taking full advantage of demand elasticity to maximize efficiency and fairness (Saraydar et al., 2002). The demand elasticity utilizes historical information about expected peak load periods. According to (Felegyhazi & Hubaux, 2006), an investigation area where pricing has practical relevance is service provisioning among operators (e.g., renting transmission capacity).

The second group of contributions makes use of reputation-based proposals (Trestian, Ormond, & Muntean, 2011) (Munjal & Singh, 2018) to incentivize the collaboration among nodes in a multi-hop wireless network topology. The reputation metric represents the amount of trust the network community has about a node. Figure 7 illustrates the typical phases of a reputation system to incentivize a correct node behavior.

During the initial phase, the reputation information of each node is collected to a central node connected to the wired network. After receiving the new reputation information, the central node updates a reputation matrix, which stores the reputation information from all the nodes (second phase). Then, in the next phase, management decisions are selected, which, during the fourth and last phase, are applied to the network infrastructure. In this way, as an example, members that have good reputation, because they helpfully contribute to the community welfare, can use the network resources; while nodes with a bad reputation, because they usually refuse to cooperate, are excluded from that community.

Figure 6. Credit-based incentive mechanism

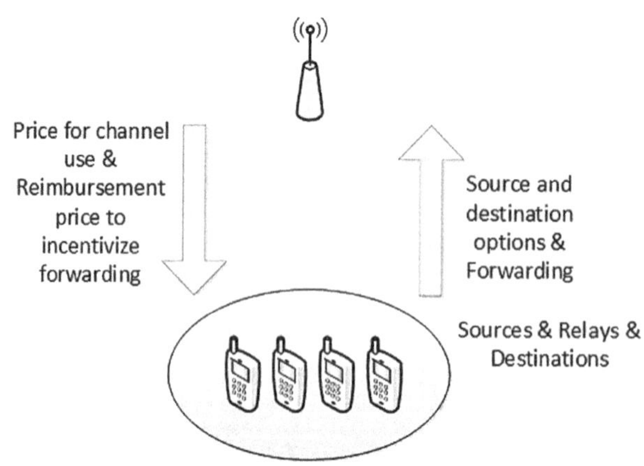

Figure 7. Reputation-based incentive mechanism

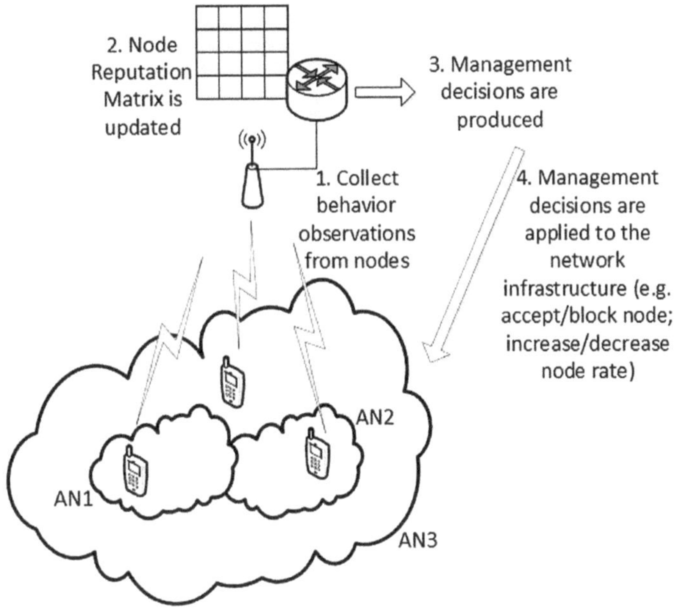

A very popular game-theoretic approach for reputation analysis is the repeated game because in this context it does not make sense that a game for reputation is based uniquely in its current (instantaneous) value; in fact, the reputation should be also evaluated through a historical term, normally with a higher weight than the one associated with the instantaneous value of reputation. In this way, it is possible to avoid false misbehavior detections due to temporary link communications failures. In addition, the uncertainty about the information that is available to other players and their decisions is normally modeled with Bayesian Game or Game with Incomplete Information (Harsanyi, 2004). Finally, to correctly model the robustness to changes on the behavior of the participants, auction games are preferred (Nurmi & Nurmi, 2006).

There are at least two different strategies on how the reputation could incentivize cooperation among nodes (or players). One of the ways is to develop a strategy such that the cooperation of a node is measured and if the fraction of packets it has dropped is above a threshold, it is considered selfish and is disconnected for a given amount of time. This strategy is known as a Trigger Strategy (Milan, Jaramillo, & Srikant, 2006). An alternative way is designated by Tit For Tat (TFT) (Axelrod, 1981). A player using this strategy will cooperate initially and then act regarding the opponent´s previous action: if the opponent previously was cooperative then the former player will be cooperative as well; otherwise, the former player will not cooperate. To illustrate the advantages of the TFT strategy being used by game players, a Finite Repeated Prisoner's Dilemma Game was simulated via Matlab (5000 iterations). The game is between two players. Each player tries to score the most number of points against each opponent player during each game. In this case, the player Operator1 can choose in each game's iteration between 'cooperate' or 'defect', like player Operator2. In each game's iteration, points are then awarded to both players based on the combination of their choices, following what is shown in Table 3.

The maximum number of points a player can win during a game's iteration is five. This maximum score only occurs if that player defects and the opponent cooperates. Nevertheless, the former player

Table 3. Points awarded to each player based on individual player's choices

		Operator2	
		Cooperate	Defect
Operator1	Cooperate	3, 3	0, 5
	Defect	5, 0	1, 1

scores one point instead five points if both players defect. As one can easily conclude, the main difficulty imposed to each player of the current game is to choose the option that maximizes his reward because he ignores the opponent's choice, as both players, during a game's iteration, perform their choices simultaneously. The previous difficulty in a player choosing the right option to maximize the reward points won by that player is perfectly evident from the simulation results presented in Figure 8. In fact, the random strategy used by each player to make a choice gives the worst performance. In opposition, TFT strategy shows a better performance.

Despite the good performance of TFT, it could reveal some drawbacks in a wireless scenario. As an example, TFT does not distinguish uncooperative behavior from a transmission failure due to a collision. In this way, TFT could penalize a collaborative player that had the bad luck of suffering a collision during a data transmission tentative. Consequently, a few TFT variants have been proposed (Milan et al., 2006) (Jaramillo & Srikant, 2007) (Q. Li, Zhu, & Cao, 2010) (Vedhavathy & Manikandan, 2018) (Ntemos, Plata-Chaves, Kolokotronis, Kalouptsidis, & Moonen, 2018) to correct that problem.

For a multi-hop wireless network, there is an interesting tradeoff between the amount of available information to evaluate a node´s behavior (reputation) and the protocol overhead/complexity used to disseminate the necessary information through the network. Some proposals are more concerned with all the nodes having access to the full information about node behavior (Buchegger & Le Boudec, 2002) (Jochen & Le, 2005) (Qi He, Dapeng Wu, & Khosla, 2004) to enhance the accuracy on how the reputation is evaluated. These proposals could have problems related with fake information disseminated

Figure 8. Outcomes of a finite repeated prisoner's dilemma game using two distinct strategies

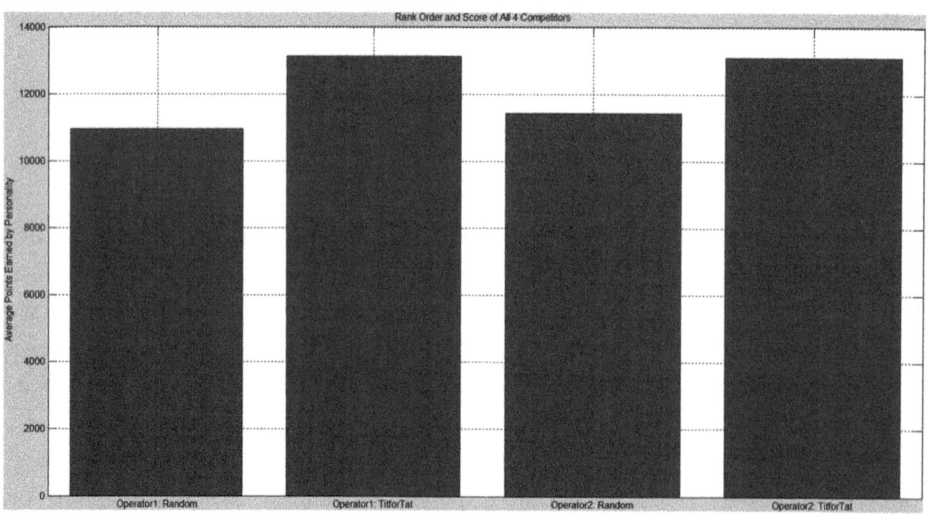

among the nodes that create wrong reputation values. To avoid these problems, the protocol used to disseminate the reputation values through the network must be enriched with additional authentication and trust functional features. Alternatively, to keep the protocol overhead low, each node should only disseminate the reputation values he directly measured to its neighbors (it only uses first-hand reputation changes) (Bansal & Baker, 2003).

Recent work proposed dynamic reputation-based incentives for cooperative relays present in a network topology formed by heterogeneous networks (Hwang, Shin, & Yoon, 2008) (Skraparlis, Sakarellos, Panagopoulos, & Kanellopoulos, 2009) (Z. Zhang, Long, Vasilakos, & Hanzo, 2016) (Kwon, Lim, Choi, & Hong, 2010). The incentive for cooperation among nodes can be given either by additional throughput (Hwang et al., 2008) or by additional time-slots for transmission (Skraparlis et al., 2009) (Z. Zhang et al., 2016) (Kwon et al., 2010).

Regarding strategies for penalizing misbehaving users, the research community has proposed several ways to perform it: isolate misbehaving users from the network (Buchegger & Le Boudec, 2002), reduce misbehaving users' bandwidth (Hwang et al., 2008) or reduce the transmission slots of misbehaving users (Skraparlis et al., 2009).

The main advantage of reputation-based proposals is that they rely on observations from multiple sources, turning it relatively resistant to the diffusion of false information from a small number of lying nodes. Some potential problems are the usage of additional bandwidth and battery energy to intensively monitor the behavior of each network node. In addition, some nodes could collude to cheat the reputation of other nodes by the dissemination of false information through the network about the latter nodes to the former nodes increase their benefits.

Game theory approaches have also been applied to blockchain security. The decentralized cooperative method of keeping track of blockchains, by miners, without the need for third-party intervention, is a relevant use case for many game models. (Nakamoto, 2008) already uses incentives in a simple, albeit insufficient, model. But unfortunately, distilling the essential game-theoretic properties of blockchain maintenance is far from trivial, and there have been many works that examines possible types of attacks against the blockchains and suggest adaptations of the protocol to ensure its security. A brief mention of some of these works follows.

In (Kroll, Davey, & Felten, 2013) the equilibria of the Bitcoin game are considered, and prove that any monotonic strategy is a Nash equilibrium (one of many). In (Eyal and Sirer, 2014), present a specific attack strategy called the "Selfish Mine" and examine when it is beneficial for a pool of miners. This is further exploited by (Sapirshtein, Sompolinsky, & Zohar, 2017), with a wider set of possible strategies, that includes the "Selfish-Mine" strategy, and explore this space computationally. In (Eyal, 2015) the author considers attacks performed between different pools where users are sent to infiltrate a competitive pool, giving raise to a pool game, the miner's dilemma, an instance of the iterative prisoner's dilemma. Further on, (Lewenberg, Bachrach, Sompolinsky, Zohar, & Rosenschein, 2015) has made a (cooperative) game theoretic analysis regarding pool mining. (Babaioff, Dobzinski, Oren, & Zohar, 2012) deals with Sybil attacks and propose a reward scheme which will make it in the best interest of a miner to propagate transactions. (Kiayias, Koutsoupias, Kyropoulou, & Tselekounis, 2016) have considered two simplified forms of a stochastic game, in which the miners have complete information: the Immediate-Release Game and the Strategic-Release Game.

The development of more suitable and fair schemes to incentivize cooperation in FNs is a challenging research direction. According to the authors of (Han et al., 2012) (Bouhaddi, Radjef, & Adi, 2018) (Ungureanu, 2018), hybrid schemes that combine both reputation and credit aspects are of particular interest to be further investigated. Lastly, by defining mechanisms of incentives for cooperation and disincentives against cheating or selfish behavior, and applying repeatedly both of these mechanisms, the cooperation among the players apparently becomes stronger in a distributed way without the need to sign a contract among the players (Trestian et al., 2011) (Munjal & Singh, 2018).

5. GAME THEORY FOR WIRELESS NETWORKING

In this section, we revise the literature in terms of how GT can be successfully applied to networking and wireless communications areas, including IoT.

(MacKenzie & DaSilva, 2006) describes ways in which GT can be applied to real applications in wireless communications and networking, such as: pricing, flow control, power control, medium access and interference avoidance. They also pointed out some appealing future applications of GT: cognitive networks and learning, mobility support and cooperation in wireless networks. (Y. Zhang & Guizani, 2011) explores applications of different economic approaches, including bargaining, auctions, cooperation incentives and dynamic coalition games for cooperation. (Han et al., 2012) discusses game-theoretic models in a wide range of wireless and communication applications such as cellular and broadband wireless access networks, wireless local area networks, multi-hop networks, cooperative networks, cognitive-radio networks, and Internet networks. In addition, some relevant Internet problems such as, congestion control, pricing, revenue sharing among Internet service providers, and incentive mechanisms to enable cooperation into peer-to-peer applications, are also discussed.

(Jianwei Huang & Zhu Han, 2010) presents several GT models/concepts that are highly relevant for spectrum sharing, including iterative water-filling, potential game, supermodular game, bargaining, auction, and correlated equilibrium. (Huang, 2013) outlines a taxonomy to systematically understand and tackle the issue of economic viability of cooperation in dynamic spectrum management. The framework divides the problem space according to four orthogonal dimensions, including complete/incomplete network information, loose/tight decision couplings, user/operator interactions, and static/dynamic decision processes. The vast majority of the key methodologies for each dimension involve GT. (Walid Saad, Han, & Hjørungnes, 2011) reviews coalitional GT for cooperative cellular wireless networks. (Marina, Saad, Han, & Hjørungnes, 2011) revises GT work about malicious behavior.

From the literature a significant number of surveys have been found about GT application in wireless communications and networking, as summarized in Figure 9. These surveys cover the following areas: wireless networks (Charilas & Panagopoulos, 2010) (Akkarajitsakul, Hossain, Niyato, & Kim, 2011) (Ghazvini, Movahedinia, Jamshidi, & Moghim, 2013) (Niyato & Hossain, 2007) (Trestian et al., 2011) (Larsson, Jorswieck, Lindblom, & Mochaourab, 2009) (M. A. Khan et al., 2012a); wireless Ad Hoc networks (Srivastava et al., 2005); wireless sensor networks (WSNs) (Machado & Tekinay, 2008) (Shen, Yue, Cao, & Yu, 2011) (Shi, Wang, Kwok, & Chen, 2012); MIMO systems (Scutari, Palomar, & Barbarossa, 2008); cognitive radio networks (Beibei Wang, Wu, & Liu, 2010) (B Wang, Wu, Liu, & Clancy, 2011); 4G networks (M. A. Khan et al., 2012b); smart grids (Fadlullah, Nozaki, Takeuchi, & Kato, 2011) (W Saad, Han, Poor, & Basar, 2012); telecommunications (E Altman, Boulogne, El-Azouzi, Jiménez, & Wynter, 2006); and Internet of Things (IoT) (Semasinghe et al., 2017).

Covering the area of wireless networks where GT is applied, we can explicit the following surveys: a significant number of GT proposals are discussed in a network-layered perspective (Charilas & Panagopoulos, 2010); multiple access games are analyzed in (Akkarajitsakul et al., 2011); games of random

Figure 9. Summary of game theory surveys

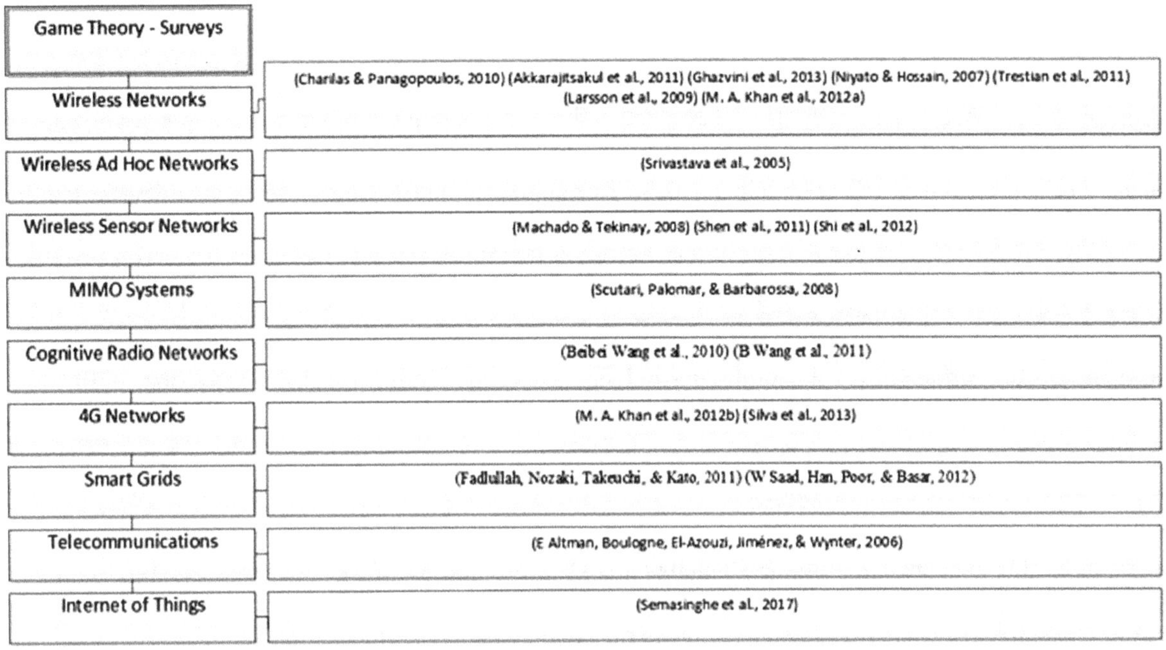

access with Carrier Sense Multiple Access (CSMA) are covered in (Ghazvini et al., 2013); games about resource management and admission control are addressed by (Niyato & Hossain, 2007); games for network selection and resource allocation are available in (Trestian et al., 2011); games of spectrum allocation, power control, interference are covered in (Larsson et al., 2009); and finally, evolutionary coalitional games for wireless networking and communications are available in (M. A. Khan et al., 2012a).

Since the application of GT to enhance cooperation in FNs, formed by heterogeneous wireless access networks, is the main focus of the present chapter, we particularize now some surveys related to Wireless Sensor Networks (WSNs), cognitive radio networks and 4G networks. (Machado & Tekinay, 2008) reviewed the literature about the usage of game-theoretic approaches to address problems related to security and energy efficiency in WSNs. (Shen et al., 2011) main concern was to revise GT approaches towards the enhancement of WSN security. Finally, (Shi et al., 2012) offered a more comprehensive survey than previous referred ones about GT applied to WSNs.

The games for cognitive radio networks are classified by (Beibei Wang et al., 2010) into four categories: non-cooperative spectrum sharing, spectrum trading and mechanism design, cooperative spectrum sharing, and stochastic spectrum sharing games. For each category, they explained the fundamental concepts and properties, and provided a detailed discussion about the methodologies on how to apply these games in spectrum sharing protocol design. They also discussed some research challenges and future research directions related to game theoretic modeling in cognitive radio networks.

Cognitive attackers may exist in a cognitive radio network, who can adapt their attacking strategy to the time-varying spectrum opportunities and secondary users' strategy. To alleviate the damage caused by cognitive attackers, a dynamic security mechanism is investigated in (B Wang et al., 2011) by a stochastic game modeling. The state of the anti-jamming game includes the spectrum availability, channel quality, and the status of jammed channels observed at the current time slot. The action of the secondary users reflects how many channels they should reserve for transmitting control and data messages and how to switch between the different channels. Since the secondary users and attackers have opposite goals, the antijamming game can be classified as a zero-sum game.

Regarding IoT, many challenges need to be addressed in order to efficiently manage available resources. Centralised resource management is however infeasible when a large number of entities is involved, not only because of the computational complexity involved but also due to information acquisition requirements. For this reason, in IoT there has been a trend in performing distributed resource management, in particular using game theoretic approaches such as proposed by (Al-Kashoash, Hafeez, & Kemp, 2017) (Borah, Dhurandher, Woungang, & Kumar, 2017) (Kim, 2016) (Sedjelmaci, Senouci, & Al-Bahri, 2016).

However, conventional game models are not always suitable for large-scale IoT systems, due to the massive information acquisition overhead, the slow convergence to equilibrium, the inefficiency of equilibrium, the extreme computational complexity, and the complexity required to characterize the equilibrium set (Semasinghe et al., 2017). Therefore, non-conventional game theoretic models are required to match the intrinsic characteristics of future large-scale IoT systems. These are characterized by having random deployments, scalability issues, limited fronthaul/backhaul, inhomogeneity, non-guaranteed energy supply, uncertain and incomplete information. Game models will inevitably have to overcome these challenges for an efficient distributed resource management. (Semasinghe et al., 2017) discusses several promising game models for IoT such as evolutionary games, mean field games, minority games, mean field bandit games, and mean field auctions. They describe the basics of each of these game models and access the potential IoT-related resource management problems that can be solved by using these models (Table 4)

The authors of (M. A. Khan et al., 2012b) study game dynamics and learning schemes for heterogeneous 4G networks. They propose a novel learning scheme called cost-to-learn that incorporates the cost to switch, the switching delay, and the cost of changing to a new action. Considering a dynamic and uncertain environment, where the users and operators have only a numerical value of their own payoffs as information, and strategy reinforcement learning (CODIPAS-RL) is used, they show the users are able to learn their own optimal payoff and their optimal strategy simultaneously. Using evolutionary game dynamics, they prove the convergence and stability properties in specific classes of dynamic robust games. They also provide various numerical and simulation results in the context of network selection in wireless local area networks (WLAN) and Long Term Evolution (LTE). In addition, (Silva et al., 2013) clearly shows the main advantages of cooperation among wireless access technologies. The following sections justify why the collaboration aspect should be very important in FNs and how GT can help to study the best ways to deploy this new functionality in a distributed way.

6. GUIDELINES TO APPLY GAME THEORY ON FUTURE NETWORKS

The current section discusses some relevant research work in how GT can be used to address the more significant operational or functional expected aspects of Future Network (FN) environments. The mostpart of the discussed scenarios belongs to the network edge of Internet. More specifically, these scenarios are concerned in how the heterogeneous wireless access infrastructure can be efficiently used by multimode terminals, as well as, to guarantee a reliable access to the Internet through wireless backhaul links. In this way, several possible functional/operational enhancements are envisioned to use efficiently the heterogeneous wireless access infrastructure in the following topics: network planning, multi-technology wireless networks, network management, Internet of Things (multi-hop reliable networks) and reliable

Table 4. Potential IoT applications for different game models

Game Theoretic Model	Potential Use Case for IoT
Evolutionary games	Power control, spectrum/subcarrier allocation, transmission mode/network selection
Mean field games	Energy/queue/channel-aware resource allocation, resource management under mobility
Minority games	Scheduling, transmission mode/network selection, interference management
Mean field bandit games	User association, scheduling, channel allocation
Mean field dynamic auctions	User association, scheduling, channel allocation

wireless backhaul. These should be hot research areas in FNs and are summarized in Table 5 together with references for relevant work that should be initially studied in order to find innovative ways to plan, control, manage and operate FNs.

6.1. Network Planning

Imperfect network coverage, especially in indoor locations is an important problem in existing cellular networks. To overcome this problem, the concept of Femtocell Access Points (FAPs) has recently been proposed as a means to overlay, on existing mobile networks, low-power and low-cost Base Stations (BSs). FAPs are connected by an IP backhaul through a local broadband connection such as DSL, cable or fiber.

Notably, various benefits of using FAP technology have been already identified:

- Enhances indoor coverage
- Provides high data rates
- Improves Quality-of-Service (QoS) to subscribers
- Ensures longer battery life for handheld terminals

Table 5. Relevant FN topics/areas where GT can be successfully applied

Topic/Area	Scenario/Game Type	Reference
Network planning	Stackelberg game to control power transmission in a network formed by macrocells and femtocells	(Guruacharya, Niyato, Hossain, & Kim, 2010)
Multi-technology wireless networks	Bayesian game to study vertical-handovers in which the users have distinct bandwidth requirements	(Zhu, Niyato, & Wang, 2010)
Network management	Evolutionary game to study rate selection for VoIP service; non-zero sum game for studying user admission control to avoid congestion	(Watanabe, Menasche, de Souza e Silva, & Leao, 2008) (Yu-Liang Kuo, Eric Hsiao-Kuang Wu, & Gen-Huey Chen, 2004)
Internet of things (multi-hop reliable networks)	Hop price-based routing game; auction theory to support truthfulness and security;	(Hua Liu & Krishnamachari, 2006) (Anderegg & Eidenbenz, 2003) (Eidenbenz, Resta, & Santi, 2008)
Reliable wireless backhaul	Evolutionary game to study traffic routing through multi-hop wireless backhaul links	(Anastasopoulos, Arapoglou, Kannan, & Cottis, 2008)
Multi-access edge computing	How game theoretical games should model wireless data communication networks to understand how to deploy in an efficient way upcoming edge technologies/services, such as the Internet of Things, user wearables, and virtual/augmented reality applications.	(Moura & Hutchison, 2018)

- Offloads traffic from the mobile operator's backhaul to the wired residential broadband connection, reducing the backhaul cost of the mobile operator.

When FAPs are deployed on top of an existing cellular system, and since FAPs operate on the same frequency bands as macrocell BSs, a new problem arises. This problem is related with the interference among channels that can impair the overall network performance. In such a network scenario, it is of interest to study the problem of transmit-power control in the downlink, minimizing the interference problem and ensuring an acceptable network performance.

In this section, we adopt the approach of (Guruacharya et al., 2010), also thoroughly discussed in (Han et al., 2012), for studying the transmit-power control in the downlink from a game-theory perspective. First, we model the scenario as a Stackelberg game. Then, we discuss the properties of the considered game and its solution. In the following text, we present a low-complexity algorithm to reach the desired outcome (Han et al., 2012).

6.1.1. Stackelberg Game to Control Transmission Power

In order to tackle the power-control problem using GT, a framework of a Stackelberg game has been used (Han et al., 2012). In the studied femtocell deployment model, it is considered that the macrocell BSs are the leaders and the FAPs are the followers in a Stackelberg game, as summarized in Table 6. In this multi-leader multi-follower Stackelberg game, there exists a competitive game among the leaders and a competitive game among the followers. The Stackelberg game keeps a distinct hierarchy among leaders and followers such that the leaders can anticipate, and take this into consideration, the behavior of the followers (the reciprocal is not true), before making their own decisions to maximize their data rate.

It was considered a Stackelberg game with complete and perfect information. As already mentioned, the leaders are the set of macrocell BS transceivers *M*, the followers are the set of FAPs *N*. Therefore, the total set of players in this game is *M U N*. The strategy space of the leaders is given by $P^{up} = \prod_{i \in M} P_i$, and any point in P^{up} is called a leader strategy. Let P_i denote the set of all feasible power vectors of transmitter *i*. The leaders compete with each other in a non-cooperative way to maximize their individual data rate, while always anticipating the strategic responses of the followers. This game among the leaders is referred as the upper subgame, and its equilibrium is referred as the upper subgame equilibrium. After the leaders apply their strategies, the followers make their moves in response to the leaders' strategies.

The strategy space of the followers is $P^{low} = \prod_{i \in N} P_i$, and any point in P^{low} is called a follower strategy. The followers also compete with each other in a non-cooperative way to maximize their own data rate,

Table 6. Summary of relevant characteristics of femtocell deployment game

Scenario	Game Type	Player	Player's Strategy	Payoff
Femtocell deployment	Stackelberg with complete and perfect information to control power	Base-stations (leaders)/ femtocell access points (followers)	Choose the maximum transmission power constrained by power constraints	Maximize Shannon data rate that each player can achieve

(Guruacharya et al., 2010)

and this competition among the followers is referred as the lower subgame. It is expected this game could offer an equilibrium state designated by the lower subgame equilibrium.

For any user $i \in \{M \cup N\}$, it is defined the best-response function as shown in (8).

$$p_i = \arg\max_{p_i} \left(p_i, p_{-i}\right) = b_i\left(p_{-i}, \overline{p_i}, \overline{m}_i\right) \tag{8}$$

where the notation $-i$ refers to all of the users in the set $\{M\ U\ N\}$ except user i; $\overline{p_i}$ is the total power constraint; $\overline{m_i}$ .is the individual power constraint, where $\overline{m_i}$ is chosen so as to maximize user i's capacity function subject to the power constraints.

6.1.2. Lower Subgame Equilibrium

It is defined the lower subgame equilibrium as any fixed point $p^{low^*} = \left(p_1^*, \cdots, p_N^*\right) \in P^{low}$ such that expression in (9) is satisfied.

$$p_i^* = b_i(p_{-i}^*, p^{up}, \overline{p_i m_i}) \tag{9}$$

where $p^{up} \in P^{up}$ is a fixed but arbitrary leader strategy for all the $i \in N$. Note that this definition is the same as a Nash Equilibrium (NE) of the lower subgame.

Following (Han et al., 2012), since every user participating in the lower subgame will maximize in a myopic way their individual data rate, the best response $b_i(.)$ of each user in the subgame can be given by the following water-filling game function (Lai & El Gamal, 2008), as shown in (10).

$$p_i = F\left(p_{-i}, \overline{p_i}, \overline{m_i}\right) = w_i\left(A_i\right)v_i + r_i\left(A_i, S_i\right) \tag{10}$$

where $W_i\ (A_i)$ is an $L_i\ x\ L_i$ symmetric matrix which contents is explained in more detail in (Han et al., 2012); $r_i\ (A_i,\ S_i)$ is an L_i-dimensional column vector detailed in (Han et al., 2012).

The main goal of a water-filling game is to identify a set of resource allocation strategies distributed among rational and selfish users (i.e. not interested in the overall system performance), who are interested in maximizing the utilities they obtain from the network (Lai & El Gamal, 2008).

By letting $b^{low} \equiv (b_i(.))^N_{i=1}$, it is possible to express the lower subgame equilibrium as any fixed point of the system-power space $p^* \in P$ such that $p^* = b^{low}\left(p^*\right)$.

Note that the function $b^{low}(.)$ does not impact the upper subgame strategy.

6.1.3. Upper Subgame Equilibrium

It is defined the upper subgame equilibrium as any fixed point $p^{up^*} = \left(p_1^*, \cdots, p_M^*\right) \in P^{up}$ such that the expression in (11) is satisfied.

$$p_i^* = b_i\left(p_{-i}^*, p^{low^*}, \overline{p_i}, \overline{m_i}\right) \tag{11}$$

where $p^{low^*} \in P^{low}$ is an equilibrium follower strategy conditioned on the upper subgame strategy, for all $i \in M$.

Equivalently, let $b^{up} \equiv (b_i(.))^M_{i=1}$; then the upper subgame equilibrium as the fixed point $p^{up^*} \in P^{up}$ such that (12) is a valid expression.

$$p^{up^*} = b^{up}\left(p^{up^*}, b^{low}\left(p^{low^*}, p^{up^*}\right)\right) \tag{12}$$

For convenience, the notation can be further simplified by writing the upper subgame equilibrium in terms of a system-power vector, i.e. as any fixed point $p^* \in P$ such that (13) is true.

$$p^* = b^{up}\left(b^{low}\left(p^*\right)\right) \tag{13}$$

Note that although the function $b^{up}(.)$ acts only on the upper subgame strategy, the lower subgame equilibrium strategy (the reaction of the followers) associated with each upper subgame strategy needs to be computed as well, since the leaders compute their strategies given their knowledge of what the followers might play.

6.1.4. Multi-Leader Multi-Follower Stackelberg Equilibrium

A suitable solution for the formulated hierarchical game between the base stations and the FAPs is the Stackelberg equilibrium. In such a multi-leader multi-follower game, the Stackelberg equilibrium is defined as any fixed-point $\left(p^{up^*}, p^{low^*}\right) = p^* \in P$ .that satisfies both conditions as shown in (14).

$$\begin{cases} p^* = b^{low}\left(p^*\right) \\ p^* = b^{up}\left(b^{low}\left(p^*\right)\right) \end{cases} \tag{14}$$

6.1.5. Algorithm for Reaching the Stackelberg Equilibrium

Finding, iteratively, the fixed point of the lower subgame using the water-filling algorithm usually yields an unstable system for a random channel gain matrix (Han et al., 2012). Therefore, it can be used a technique designated by Mann iterative methods, which allows a weaker stability criterion but it ensures that a stable system status point can be reached. To achieve this further discussion is available in (Han et al., 2012).

6.2. Multi-Technology Wireless Networks

The FN environment will be a heterogeneous network infrastructure composed by distinct wireless access technologies and several users/terminals aiming to monitor and select the best technology/Access Point (AP)/ Base Station (BS) to connect to, depending on their Quality-of-Service (QoS) requirements. One possible QoS requirement is the best throughput as possible each user can have through each AP/BS taking in consideration the overload imposed by the other attached users. Each user (the player of this network selection/vertical handover game) after its monitoring phase about all the available AP/BS connection possibilities should choose the one that ensures the maximum throughput value among all the options. Most of the existing work on vertical handover assumes that users have complete information on one another (Han et al., 2012). In FNs, the users will lack the ability to predict the behaviors of others based on past actions. In this case, it is more convenient to utilize a game with incomplete information, i.e. a Bayesian game, like the one adopted by (Zhu et al., 2010). Since the payoff (i.e. utility) for a mobile user is composed by private information (see Table 7), each user has to make a network selection given only the distribution of the preferences of other users (Han et al., 2012). In this game, it is very interesting to investigate the impact of different system parameters on the game performance itself using a practical setting, like the one composed by three different access technologies (Wifi, Wimax and cellular). The studied system parameters have been the convergence property of the aggregate best-response dynamics for the considered network selection game, the game adaptation for different handover costs (delay or packet loss), the impact of connection price on the equilibrium distribution and the impact of learning (i.e. user strategy adjustment) rate on game dynamics. The obtained results are discussed in (Han et al., 2012).

6.3. Network Management

The support of voice service in FNs will be a challenging task due the heterogeneity of both the network infrastructure and user requirements. A very interesting starting point to this problem is available in (Watanabe et al., 2008). It is proposed an analytical model based on Evolutionary Game Theory (EGT)

Table 7. Summary of relevant characteristics of network selection with incomplete information game

Scenario	Game Type	Player	Player's Strategy	Payoff
Network selection with incomplete information	Bayesian game	Users in a service area with K available access networks	Represents the probability of choosing an access network K and the minimum user bandwidth requirement (only the user knows about this, which turns this game an incomplete one)	User utility combines user achieved throughput above a minimum threshold (user private information) vs. price paid for the connection

(Zhu et al., 2010)

(see Table 8) to analyze the consequences of a situation in which all users are allowed to freely choose the transmission rate. They perform that by selecting the codec and Forward Error Correction (FEC) mode to maximize the voice quality (payoff), which can be experienced by them. They show that in a scenario where the users know only their own perceived voice quality, the system converges to a total transmission rate close to that of the effective cell's capacity. They concluded that each individual user's MOS, which is estimated by a Random Neural Network (RNN), can also be satisfied. Further, cell's congestion is avoided by local user adaptation (dynamically changing its codec/FEC to maximize its perceived quality) without any intervention from a centralized controller.

6.4. Internet of Things (Multi-Hop Reliable Networks)

FN environments will have a large-scale deployment of wireless networks, which consist of small, low-cost nodes with simple processing and networking capabilities. This emergent environment is commonly designated inside the research community as the Internet of Things. In order to reach the desired destination such as the data sink node, transmissions depending on multiple hops are necessary (Han et al., 2012). Because of this, the routing optimization is a pertinent problem that involves many aspects but the one more relevant for the current work is the nodes not willing to fully cooperate in the routing process through multiple wireless hops, forwarding traffic from other nodes, because relaying external traffic consumes their limited battery power. Hence, it is crucial to design a distributed –control mechanism encouraging cooperation among the nodes in the routing process (see Table 9). The literature describes two typical approaches to enforce cooperation. First, in a price-based approach, each hop has a price and the game outcome is controlled between the source-destination pair and the intermediate hops. Second, an auction-based approach is suggested to ensure that users reveal their information truthfully to others for network cooperation, because this strategy will bring them the best benefits.

6.5. Reliable Wireless Backhaul

In FN environments wireless multi-hop backhaul links are expected to be very popular deployments. In this case, the channel quality between relay stations can fluctuate because of fading. Therefore, the users (players) at the source node must be able to observe, learn, and change the routing strategy to achieve the most reliable path from source node to the Internet gateway, as summarized in Table 10.

Table 8. Summary of relevant characteristics of an evolutionary game to study rate selection for VoIP service

Scenario	Game Type	Player	Player's Strategy	Payoff
Study rate selection to guarantee the QoS offered to VoIP users	Evolutionary game	VoIP users in a service area	Each user selects the transmission rate through the codec and FEC mode	Voice quality experienced by the user and measured via a Mean Opinion Score (MOS) technique

(Watanabe et al., 2008)

Table 9. Summary of characteristics of games to incentivize cooperation among multi-hop nodes

Scenario	Game Type	Player	Player's Strategy	Payoff
Incentivize cooperation among nodes (Hua Liu & Krishnamachari, 2006)	Hop price-based reliable routing game	All the nodes except the destination one	A node to participate in this game should at least choose one next hop node in the path from the source to the destination; otherwise it is out of this game	The source's utility is the expected income (destination payment minus the payments to all of the intermediate nodes, times the probability that the packet will be delivered over the route) minus the link set-up cost for the first hop of the route; The utility for each intermediate routing node equals the expected payment that it obtains from the source node, times the ongoing route reliability minus the transmission cost per packet to its next-hop neighbor. If any node does not participate in the routing, it gains (and loses) nothing.
Incentivize cooperation among nodes (Eidenbenz et al., 2008)	Vickrey-Clarke-Groves (VCG) auction to prevent players from lying and to route messages along the most energy-efficient paths (as defined by the topology control protocol)	All the network nodes	A strategy is a combination of strategies from the following base space: 1. a node can declare any value for its type; 2. a node can drop control messages that it should forward; 3. a node can modify messages before forwarding, and 4. a node can create bogus messages.	Maximizing the node's utility. The sender's node utility is the difference between the amount of money it is willing to pay for the connection and the amount it effectively pays for that; the intermediate's node utility is the difference about the amount of money received from the sender and the total cost incurred by relaying the sender's packet.

Table 10. Summary of relevant characteristics of game to study traffic routing through multi-hop wireless backhaul links

Scenario	Game Type	Player	Player's Strategy	Payoff
Multi-hop Wireless Backhaul Links	Evolutionary game	Users	Users periodically and randomly sampling different wireless backhaul links to select a convenient path between a source node and an Internet gateway	Find a backhaul link that ensures the smallest number of packet errors due to rain attenuation

(Anastasopoulos et al., 2008)

6.6. Computing and Storage at the Network Edge

The paradigm of Multi-Access Edge Computing (MEC) (Yi et al., 2015) (Abbas, Zhang, Taherkordi, & Skeie, 2018) (Hang Liu et al., 2017) is finally possible to deploy because foundational technologies such as virtualization (e.g. docker, Linux Containers) and communications (e.g. 5G) are becoming a reality now more than ever. Edge computing aims to provide more compute and storage power at either Base Stations or Access Points. The potential benefits to the data traffic are: i) diminish the data access latency; ii) decrease the load on the backhaul links; iii) save users' cost because less traffic is exchanged with remote clouds. Nevertheless, this new paradigm can increase the battery consumption on mobile nodes. To overcome this potential problem, computation offloading from mobile devices to edge devices (APs or BSs) can be a viable solution.

The authors of (Nawab, Agrawal, & El Abbadi, 2018) propose extending edge computing technology with dynamic, mobile edge datacenters, which they designate as nomadic datacenters. Nomadic datacenters are small and portable edge datacenters that can be easily moved around according the traffic load needs. In this way, nomadic datacenters can replace a damaged communications infrastructure by a natural disaster. Alternatively, nomadic datacenters can temporarily extend the capacity of a mobile network in the case of a public event that concentrates several hundreds of thousands of people (e.g. musical concert within a stadium).

There is a huge number of recent literature contributions on MEC, namely covering the next topics: i) the communication perspective (Mao, You, Zhang, Huang, & Letaief, 2017); ii) computation offloading (Mach & Becvar, 2017); iii) convergence of computing, caching and communications (S. Wang et al., 2017); iv) emerging 5G network edge cloud architecture and orchestration (Taleb et al., 2017); v) software-defined networking (Kumar, Chilamkurti, et al., 2015); vi) architecture harmonization between cloud radio access networks and fog networks (Hung, Hsu, Lien, & Chen, 2015); vii) Internet of Things (Chiang & Zhang, 2016); and viii) latency control in software-defined mobile-edge vehicular networking (Deng, Lien, Lin, Hung, & Chen, 2017). None of the previous surveys comprehensively analyzes GT into MEC. Nevertheless, there is a very work (Moura & Hutchison, 2018) that tries to discuss in a comprehensive way those two pertinent areas, in the sense to understand how GT can address in a successful way the emerging requirements of MEC use cases. They also discuss GT research topics related to MEC, namely on wireless sensor networks, cognitive small cells, vehicular networks, and unmanned vehicles.

7. BUSINESS PERSPECTIVE

From a business viewpoint, collaboration may be positive or negative – if a certain company plans to take over the market, it would engage in an open competition with its competitors in the hope to conquer most of the market, but in doing so it would have to pay the price of being competitive; be it by lowering their prices or by increasing their quality. If the company would engage in collaboration with its competitors, they would assess the market needs together and become more efficient, operating at a point where all would maximize their profit. In this case, assuming a total of n companies, and taking Pm as the maximum price for the product that all the companies manufacture, we can build an inverse demand curve (the demand curve solves for Quantity, whereas the inverse for Price), where the price charged for one of those products is equal to the Pm minus the quantity Q, multiplied by a factor α (assuming a linear relationship for simplicity, as shown in (15); more detailed curves can be found in (O'Sullivan, Sheffrin, & Perez, 2003).

$$P = P_m - \alpha Q \tag{15}$$

In the case of n competing companies, we have the total quantity given by (16).

$$Q = Q_1 + Q_2 + \ldots + Q_n \tag{16}$$

The total revenue is given by the product price x quantity. The total revenue for all companies can thus be calculated as shown in (17).

$$TR = P \times Q = \left(P_m - \alpha Q\right) \times Q = P_m Q - \alpha Q^2 \tag{17}$$

The derivative of TR with respect to Q gives us the Marginal Revenue (MR), which in our simple linear model is given by (18).

$$MR = P_m - 2\alpha Q \tag{18}$$

In Figure 10, we have an illustration of the above formulas. Note that the MR curve represents the slope of the TR curve (which is drawn at a different scale, for practical reasons), and that maximum total revenue is obtain when MR=0.

Since the objective is to maximize profit, we must check at which point MR equals the Marginal Cost (MC), to find the ideal price and quantity to produce, as we evaluate below:

$$MR = MC \rightarrow P_m - 2\alpha Q - MC = 0 \Leftrightarrow$$

$$\Leftrightarrow Q = \frac{P_m - MC}{2\alpha} \Leftrightarrow$$

$$\Leftrightarrow P = \frac{P_m + MC}{2}$$

The Figure 11 illustrates the trend's point in which MC=MR (point b), and the corresponding Price and Quantity (P_1 and Q_1). From the previous figure, notice also that Q_0 denotes the free market equilibrium quantity, in which the marginal cost of producing a unit equals its price – in a hyper competition setting, companies may operate close to this point, ideally a bit more to the left to obtain some (minimal) profit. The difference between quantities Q_1 and Q_0, and between prices P_1 and C_0, can be regarded as

Figure 10. Inverse demand curve, total revenue and marginal revenue

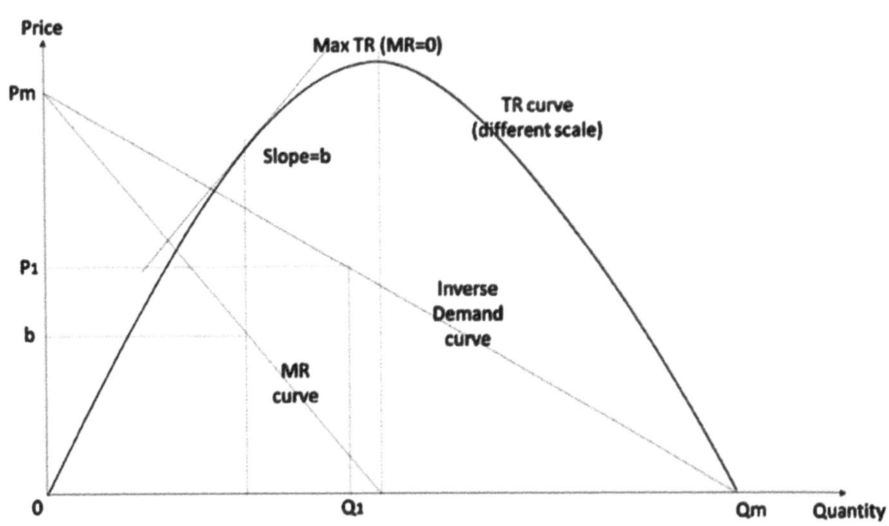

Figure 11. Collaboration and competition operating points

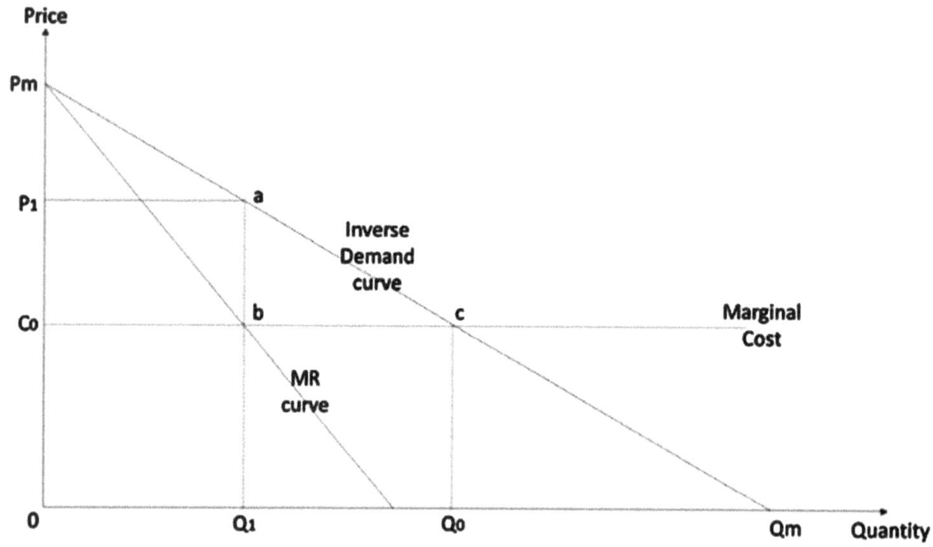

the difference between operating collaboratively (or in a monopoly) and competitively (the area of the triangle *abc* is also referred to as the deadweight loss of monopoly). It is up to each company's strategy if it should consolidate or invest/risk to (try to) conquer market (operating close to C_0 will lead many less cost-effective companies to file for bankruptcy, allowing for the bigger companies to take over their market-share).

Should the companies engage in competition, all would be losing out, operating with a higher quantity each, and subsequently lower prices and profit margins. There is a clear incentive to appeal to collaboration, although all companies want the best possible deal for themselves, which sometimes might hinder the collaboration attempts. Game theory models try to assess the best possible solution for all parties, quantifying the possible gains and losses predicted by the used model. Collaboration goes beyond revenue sharing (Bhaskaran & Krishnan, 2009); it may include the sharing of knowledge about the markets and technologies, setting the market standards, the sharing of facilities, etc. (Goyal & Joshi, 2003). To reap the full benefits of collaborating, one must analyze the collaborative conditions beforehand, hence negotiating the collaboration terms is of paramount importance. Important decisions need to be taken, such as the level of investment, profit sharing, knowledge and trust.

There has been some interesting work on the field to obtain win-win solutions for collaborative partners, and the authors would like to highlight (Arsenyan, Büyüközkan, & Feyzioğlu, 2015) for the proposed mathematical model integrating trust, coordination, co-learning and co-innovation for collaborative product development using Nash bargaining, as well as work in Hospital Information Exchange (HIE) networks that attempt to quantify the benefits and losses from the exchange of patient information both to the hospitals and patients (Martinez, Feijoo, Zayas-Castro, Levin, & Das, 2018). In the same field, Desai developed a game theoretical model to analyze the potential loss of competitive advantage due to HIE adoption (Desai, 2014).

8. CONCLUSION

8.1. Cooperation: Current Status and Open Issues

Cooperation is a revolutionary wireless communication paradigm that can achieve much higher network performance and spectrum utilization in future networking environments. Many technical challenges, however, need to be addressed to make this vision a reality. In particular, the distributed and dynamic nature of the sharing of information about node cooperation requires a new design and analysis framework. GT provides a very solid solution for this challenging task. In this book chapter, we describe several GT models that have been successfully used to solve various problems associated with node cooperation.

The most part of discussed models relies on the concept of Nash Equilibrium (NE) in games with complete information and static strategies. Although mathematically convenient, this may not be the most suitable GT model in practice. For example, the complete information assumption is difficult to be satisfied in practice, due to the dynamic and uncertain environment associated to FNs (or MEC / FC) formed by heterogeneous wireless access technologies and a huge variety of flow types. A model of incomplete games will be more suitable. Moreover, NE assumes rational players and static strategies but the players in FNs aren´t completely rational; the players have limited information about available choices and consequences of others; the game strategies are not static (in fact, the strategies are highly dynamic). A recent branch of GT - Evolutionary GT (EGT) seems a very promising alternative to the traditional GT to be applied in FNs. Some preliminary work has been reported along these directions (M. A. Khan et al., 2012a) (Nazir et al., 2010) (Bennis et al., 2011) (Eitan Altman et al., 2009) and definitely much more is required. As a pertinent example, evolutionary network models can provide useful guidelines

Table 11. Open issues in applying GT to future wireless networking scenarios

Scenario	Reference	Open Issue
Network planning	(Guruacharya et al., 2010)	Due to the notorious computational burden of estimating the Stackelberg equilibrium, a low complexity algorithm based on Lagrangian dual theory was chosen. However, the numerical results show that the adopted algorithm is suboptimal.
Multi-technology wireless networks	(Zhu et al., 2010)	Future work can study based on the Equilibrium distribution, how the service providers can adjust the system capacity and price accordingly to maximize the profits
Network management	(Watanabe et al., 2008)	The experiments were performed with small populations. Future work can devise more scalable experiments.
Internet of things (multi-hop reliable networks)	(Hua Liu & Krishnamachari, 2006)	Add the destination as a player; consider scenarios where the destination can choose from several source nodes for a given piece of information. This will allow for an auction to be held among the source nodes to optimize destination's payoff
Internet of things (multi-hop reliable networks)	(Eidenbenz et al., 2008)	Enhance previous protocol to be robust against malicious nodes and collusion
Reliable wireless backhaul	(Anastasopoulos et al., 2008)	Extend previous work in the direction of IEEE 802.11s (wireless mesh networking)
Computing and Storage at the Network Edge	(Moura & Hutchison, 2018)	Low data/service access latency; distributed offloading computing hierarchical mobile environments; proactive data caching at the network edge managed by data popularity, social links, and available node battery energy; low-power wireless communications and networking for IoT; fingerprinting localization technology to support indoor location-based services; and understand emerging security and privacy problems in cyberspace and potential solutions.

for upgrading protocols/algorithms to achieve stable infrastructure functionality around preferred status/ configuration in FNs. Other interesting models are Stackelberg and Bayesian ones. Finally, in Table 11, some relevant contributions found in the literature, which can be the foundations for new work in the FN area, are listed together with some associated open issues.

As a final conclusion, the current authors think that non-conventional game theoretic models are required to fulfill the demanding requisites of upcoming scenarios, such as: i) computing and storage at the network edge; ii) large-scale IoT systems; and iii) business models based on virtual money.

REFERENCES

Abbas, N., Zhang, Y., Taherkordi, A., & Skeie, T. (2018). Mobile Edge Computing: A Survey. *IEEE Internet of Things Journal*, *5*(1), 450–465. doi:10.1109/JIOT.2017.2750180

Ahlgren, B., Dannewitz, C., Imbrenda, C., Kutscher, D., & Ohlman, B. (2012). A survey of information-centric networking. *IEEE Communications Magazine*, *50*(7), 26–36. doi:10.1109/MCOM.2012.6231276

Akkarajitsakul, K., Hossain, E., Niyato, D., & Kim, D. I. (2011). Game Theoretic Approaches for Multiple Access in Wireless Networks: A Survey. *IEEE Communications Surveys and Tutorials*, *13*(3), 372–395. doi:10.1109/SURV.2011.122310.000119

Al-Kashoash, H. A. A., Hafeez, M., & Kemp, A. H. (2017). Congestion Control for 6LoWPAN Networks: A Game Theoretic Framework. *IEEE Internet of Things Journal*, *4*(3), 760–771. doi:10.1109/ JIOT.2017.2666269

Alpcan, T., Basar, T., Srikant, R., & Atman, E. (2001). CDMA uplink power control as a noncooperative game. In *Proceedings of the 40th IEEE Conference on Decision and Control (Cat. No.01CH37228)* (Vol. 1, pp. 197–202). IEEE. 10.1109/CDC.2001.980097

Altman, E., Boulogne, T., El-Azouzi, R., Jiménez, T., & Wynter, L. (2006). A Survey on Networking Games in Telecommunications. *Computers & Operations Research*, *33*(2), 286–311. doi:10.1016/j.cor.2004.06.005

Altman, E., El-Azouzi, R., Hayel, Y., & Tembine, H. (2009). The evolution of transport protocols: An evolutionary game perspective. *Computer Networks*, *53*(10), 1751–1759. doi:10.1016/j.comnet.2008.12.023

Alves, H., Silva, L. M., Neto Marinheiro, R., & Moura, J. (2018). PMIPv6 Integrated with MIH for Flow Mobility Management: A Real Testbed with Simultaneous Multi-Access in Heterogeneous Mobile Networks. *Wireless Personal Communications*, *98*(1), 1055–1082. doi:10.100711277-017-4908-6

Anastasopoulos, M., Arapoglou, P.-D., Kannan, R., & Cottis, P. (2008). Adaptive Routing Strategies in IEEE 802.16 Multi-Hop Wireless Backhaul Networks Based On Evolutionary Game Theory. *IEEE Journal on Selected Areas in Communications*, *26*(7), 1218–1225. doi:10.1109/JSAC.2008.080918

Anderegg, L., & Eidenbenz, S. (2003). Ad hoc-VCG Ad hoc-VCG: a truthful and cost-efficient routing protocol for mobile ad hoc networks with selfish agents. In *Proceedings of the 9th annual international conference on Mobile computing and networking - MobiCom '03* (p. 245). New York: ACM Press. 10.1145/938985.939011

Apicella, C. L., Marlowe, F. W., Fowler, J. H., & Christakis, N. A. (2012). Social networks and cooperation in hunter-gatherers. *Nature*, *481*(7382), 497–501. doi:10.1038/nature10736 PMID:22281599

Arsenyan, J., Büyüközkan, G., & Feyzioğlu, O. (2015). Modeling collaboration formation with a game theory approach. *Expert Systems with Applications*, *42*(4), 2073–2085. doi:10.1016/j.eswa.2014.10.010

Atlam, H. F., Walters, R. J., & Wills, G. B. (2018). Fog Computing and the Internet of Things: A Review. *Big Data and Cognitive Computing, 2*(2).

Axelrod, R. (1981). The Emergence of Cooperation among Egoists. *The American Political Science Review*, *75*(02), 306–318. doi:10.2307/1961366

Babaioff, M., Dobzinski, S., Oren, S., & Zohar, A. (2012). On bitcoin and red balloons. In *Proceedings of the 13th ACM Conference on Electronic Commerce - EC '12* (p. 56). New York: ACM Press. 10.1145/2229012.2229022

Balevi, E., & Gitlin, R. D. (2018). Optimizing the Number of Fog Nodes for Cloud-Fog-Thing Networks. *IEEE Access: Practical Innovations, Open Solutions*, *6*, 11173–11183. doi:10.1109/ACCESS.2018.2808598

Bansal, S., & Baker, M. (2003). Observation-based Cooperation Enforcement in Ad Hoc Networks. *CoRR,* cs.NI/0307.

Bennis, M., Guruacharya, S., & Niyato, D. (2011). Distributed Learning Strategies for Interference Mitigation in Femtocell Networks. In *Global Telecommunications Conference (GLOBECOM 2011)*, 2011 *IEEE* (pp. 1–5). IEEE. 10.1109/GLOCOM.2011.6134218

Bhaskaran, S. R., & Krishnan, V. (2009). Effort, Revenue, and Cost Sharing Mechanisms for Collaborative New Product Development. *Management Science*, *55*(7), 1152–1169. doi:10.1287/mnsc.1090.1010

Böge, W., & Eisele, T. (1979). On solutions of Bayesian games. *International Journal of Game Theory*, *8*(4), 193–215. doi:10.1007/BF01766706

Bond, R. M., Fariss, C. J., Jones, J. J., Kramer, A. D. I., Marlow, C., Settle, J. E., & Fowler, J. H. (2012). A 61-million-person experiment in social influence and political mobilization. *Nature*, *489*(7415), 295–298. doi:10.1038/nature11421 PMID:22972300

Borah, S. J., Dhurandher, S. K., Woungang, I., & Kumar, V. (2017). A game theoretic context-based routing protocol for opportunistic networks in an IoT scenario. *Computer Networks*, *129*, 572–584. doi:10.1016/j.comnet.2017.07.005

Bouhaddi, M., Radjef, M. S., & Adi, K. (2018). An efficient intrusion detection in resource-constrained mobile ad-hoc networks. *Computers & Security*, *76*, 156–177. doi:10.1016/j.cose.2018.02.018

Bu, S., Yu, F. R., & Yanikomeroglu, H. (2015). Interference-Aware Energy-Efficient Resource Allocation for OFDMA-Based Heterogeneous Networks With Incomplete Channel State Information. *IEEE Transactions on Vehicular Technology*, *64*(3), 1036–1050. doi:10.1109/TVT.2014.2325823

Buchegger, S., & Le Boudec, J.-Y. (2002). Performance analysis of the CONFIDANT protocol. In *Proceedings of the 3rd ACM international symposium on Mobile ad hoc networking & computing - MobiHoc '02* (p. 226). New York: ACM Press. 10.1145/513800.513828

Buttyan, L., & Hubaux, J.-P. (2000). Enforcing service availability in mobile ad-hoc WANs. In *2000 First Annual Workshop on Mobile and Ad Hoc Networking and Computing. MobiHOC (Cat. No.00EX444)* (pp. 87–96). IEEE. 10.1109/MOBHOC.2000.869216

Buttyan, L., & Hubaux, J.-P. (2001). Nuglets: A Virtual Currency to Stimulate Cooperation in Self-Organized Mobile. *Ad Hoc Networks.*

Buttyán, L., & Hubaux, J.-P. (2003). Stimulating Cooperation in Self-Organizing Mobile Ad Hoc Networks. *Mobile Networks and Applications*, *8*(5), 579–592. doi:10.1023/A:1025146013151

Charilas, D. E., & Panagopoulos, A. D. (2010). A Survey on Game Theory Applications in Wireless Networks. *Computer Networks*, *54*(18), 3421–3430. doi:10.1016/j.comnet.2010.06.020

Chawla, S., & Sivan, B. (2014). Bayesian Algorithmic Mechanism Design. *SIGecom Exch.*, *13*(1), 5–49. doi:10.1145/2692375.2692378

Chen, K., & Nahrstedt, K. (2004). iPass: an incentive compatible auction scheme to enable packet forwarding service in MANET. In *24th International Conference on Distributed Computing Systems, 2004. Proceedings.* (pp. 534–542). IEEE. 10.1109/ICDCS.2004.1281620

Chen, K., Yang, Z., Wagener, C., & Nahrstedt, K. (2005). Market models and pricing mechanisms in a multihop wireless hotspot network. In *The Second Annual International Conference on Mobile and Ubiquitous Systems: Networking and Services* (pp. 73–82). IEEE. 10.1109/MOBIQUITOUS.2005.36

Cheriton, D. R., & Gritter, M. (2000). *TRIAD: A New Next-Generation Internet Architecture.* Academic Press.

Chiang, M., & Zhang, T. (2016). Fog and IoT: An Overview of Research Opportunities. *IEEE Internet of Things Journal*, *3*(6), 854–864. doi:10.1109/JIOT.2016.2584538

Christidis, K., & Devetsikiotis, M. (2016). Blockchains and Smart Contracts for the Internet of Things. *IEEE Access: Practical Innovations, Open Solutions*, *4*, 2292–2303. doi:10.1109/ACCESS.2016.2566339

CISCO. (2015). *Cisco Fog Computing Solutions: Unleash the Power of the Internet of Things.* Retrieved July 26, 2018, from https://www.cisco.com/c/dam/en_us/solutions/trends/iot/docs/computing-solutions.pdf

CISCO. (2016). *VNI Global Fixed and Mobile Internet Traffic Forecasts (2016-2021).* Retrieved June 2, 2018, from https://www.cisco.com/c/en/us/solutions/service-provider/visual-networking-index-vni/index.html

Corcoran, P. (2016). The Internet of Things: Why now, and what's next? *IEEE Consumer Electronics Magazine*, *5*(1), 63–68. doi:10.1109/MCE.2015.2484659

Demir, C., & Comaniciu, C. (2007). An Auction based AODV Protocol for Mobile Ad Hoc Networks with Selfish Nodes. In *2007 IEEE International Conference on Communications* (pp. 3351–3356). IEEE. 10.1109/ICC.2007.555

Deng, D.-J., Lien, S.-Y., Lin, C.-C., Hung, S.-C., & Chen, W.-B. (2017). Latency Control in Software-Defined Mobile-Edge Vehicular Networking. *IEEE Communications Magazine*, *55*(8), 87–93. doi:10.1109/MCOM.2017.1601165

Desai, S. (2014). *Electronic Health Information Exchange, Switching Costs, and Network Effects*. SSRN Electronic Journal. doi:10.2139srn.2525084

Duong, N. D., Madhukumar, A. S., & Niyato, D. (2016). Stackelberg Bayesian Game for Power Allocation in Two-Tier Networks. *IEEE Transactions on Vehicular Technology*, *65*(4), 2341–2354. doi:10.1109/TVT.2015.2418297

Dutta, P. K. (1999). *Strategies and games: Theory and practice*. MIT Press.

Eidenbenz, S., Resta, G., & Santi, P. (2008). The COMMIT Protocol for Truthful and Cost-Efficient Routing in Ad Hoc Networks with Selfish Nodes. *IEEE Transactions on Mobile Computing*, *7*(1), 19–33. doi:10.1109/TMC.2007.1069

Eyal, I. (2015). The Miner's Dilemma. In *2015 IEEE Symposium on Security and Privacy* (pp. 89–103). IEEE. 10.1109/SP.2015.13

Fadlullah, Z. M., Nozaki, Y., Takeuchi, A., & Kato, N. (2011). A survey of game theoretic approaches in smart grid. In *Wireless Communications and Signal Processing (WCSP), 2011 International Conference on* (pp. 1–4). Academic Press. 10.1109/WCSP.2011.6096962

Felegyhazi, M., & Hubaux, J.-P. (2006). *Game Theory in Wireless Networks: A Tutorial*. Academic Press.

Fernandes, S., & Karmouch, A. (2012). Vertical Mobility Management Architectures in Wireless Networks: A Comprehensive Survey and Future Directions. *IEEE Communications Surveys and Tutorials*, *14*(1), 45–63. doi:10.1109/SURV.2011.082010.00099

Fudenberg, D., & Maskin, E. (1986). The Folk Theorem in Repeated Games with Discounting or with Incomplete Information. *Econometrica*, *54*(3), 533–554. doi:10.2307/1911307

Gabry, F., Bioglio, V., & Land, I. (2016). On Energy-Efficient Edge Caching in Heterogeneous Networks. *IEEE Journal on Selected Areas in Communications*, *34*(12), 3288–3298. doi:10.1109/JSAC.2016.2611845

Ghazvini, M., Movahedinia, N., Jamshidi, K., & Moghim, N. (2013). Game Theory Applications in CSMA Methods. *IEEE Communications Surveys and Tutorials*, *15*(3), 1062–1087. doi:10.1109/SURV.2012.111412.00167

Gibbons, R. (1992). *A primer in game theory*. Harvester.

Gluhak, A., Vermesan, O., Bahr, R., Clari, F., MacchiaMaria, T., Delgado, T., … Barchetti, V. (2016). *Report on IoT platform activities*. Retrieved July 26, 2018, from http://www.internet-of-things-research.eu/pdf/D03_01_WP03_H2020_UNIFY-IoT_Final.pdf

Goyal, S., & Joshi, S. (2003). Networks of collaboration in oligopoly. *Games and Economic Behavior*, *43*(1), 57–85. doi:10.1016/S0899-8256(02)00562-6

Guruacharya, S., Niyato, D., Hossain, E., & Kim, D. I. (2010). Hierarchical Competition in Femtocell-Based Cellular Networks. In *2010 IEEE Global Telecommunications Conference GLOBECOM 2010* (pp. 1–5). IEEE. 10.1109/GLOCOM.2010.5683278

Habak, K., Ammar, M., Harras, K. A., & Zegura, E. (2015). Femto Clouds: Leveraging Mobile Devices to Provide Cloud Service at the Edge. In *2015 IEEE 8th International Conference on Cloud Computing* (pp. 9–16). IEEE. Retrieved from https://www.cc.gatech.edu/%7B~%7Dkhabak3/papers/FemtoCloud-CLOUD%7B%25%7D2715.pdf

Hajir, M., Langar, R., & Gagnon, F. (2016). Coalitional Games for Joint Co-Tier and Cross-Tier Cooperative Spectrum Sharing in Dense Heterogeneous Networks. *IEEE Access: Practical Innovations, Open Solutions, 4*, 2450–2464. doi:10.1109/ACCESS.2016.2562498

Han, Z., Niyato, D., Saad, W., Baar, T., & Hjrungnes, A. (2012). *Game Theory in Wireless and Communication Networks: Theory, Models, and Applications.* Cambridge University Press.

Harsanyi, J. C. (2004). Games with Incomplete Information Played by "Bayesian" Players, I–III: Part I. The Basic Model. *Management Science, 50*(12S), 1804–1817. doi:10.1287/mnsc.1040.0270

He, Q., Wu, D., & Khosla, P. (2004). SORI: a secure and objective reputation-based incentive scheme for ad-hoc networks. In 2004 IEEE Wireless Communications and Networking Conference (IEEE Cat. No.04TH8733) (p. 825–830 Vol.2). IEEE. doi:10.1109/WCNC.2004.1311293

Hu, Y., Zhu, M., Xia, Y., Chen, K., & Luo, Y. (2012). GARDEN: Generic Addressing and Routing for Data Center Networks. In *Cloud Computing (CLOUD), 2012 IEEE 5th International Conference on* (pp. 107–114). IEEE. 10.1109/CLOUD.2012.9

Huang, J. (2013). Economic viability of dynamic spectrum management. In T. Alpcan, H. Boche, M. L. Honig, & H. V. Poor (Eds.), *Mechanisms and Games for Dynamic Spectrum Allocation* (pp. 396–432). Cambridge University Press; doi:10.1017/CBO9781139524421.018

Huang, J., & Han, Z. (2010). Game {Theory} for {Spectrum} {Sharing}. In Cognitive {Radio} {Networks} (pp. 291–317). CRC Press. Retrieved from doi:10.1201/EBK1420077759-c10

Hung, S.-C., Hsu, H., Lien, S.-Y., & Chen, K.-C. (2015). Architecture Harmonization Between Cloud Radio Access Networks and Fog Networks. *IEEE Access: Practical Innovations, Open Solutions, 3*, 3019–3034. doi:10.1109/ACCESS.2015.2509638

Hwang, J., Shin, A., & Yoon, H. (2008). Dynamic Reputation-based Incentive Mechanism Considering Heterogeneous Networks. In *Proceedings of the 3Nd ACM Workshop on Performance Monitoring and Measurement of Heterogeneous Wireless and Wired Networks* (pp. 137–144). New York, NY: ACM. 10.1145/1454630.1454651

Ileri, O., Siun-Chuon Mau, & Mandayam, N. B. (2005). Pricing for enabling forwarding in self-configuring ad hoc networks. *IEEE Journal on Selected Areas in Communications, 23*(1), 151–162. doi:10.1109/JSAC.2004.837356

Jacobson, V., Smetters, D. K., Thornton, J. D., Plass, M. F., Briggs, N. H., & Braynard, R. L. (2009). Networking Named Content. In *Proceedings of the 5th International Conference on Emerging Networking Experiments and Technologies* (pp. 1–12). New York: ACM. 10.1145/1658939.1658941

Jaramillo, J. J. & Srikant, R. (2007). DARWIN: Distributed and adaptive reputation mechanism for wireless ad-hoc networks. In *Proceedings of the 13th annual ACM international conference on Mobile computing and networking - MobiCom '07* (p. 87). New York: ACM Press. 10.1145/1287853.1287865

Jararweh, Y., Tawalbeh, L., Ababneh, F., Khreishah, A., & Dosari, F. (2014). Scalable Cloudlet-based Mobile Computing Model. *Procedia Computer Science*, *34*, 434–441. doi:10.1016/j.procs.2014.07.051

Jochen, M., & Le, B. J. (2005). Analysis of a robust reputation system for self-organised networks. *European Transactions on Telecommunications*, *16*(5), 375–384. doi:10.1002/ett.1066

Kebriaei, H., Maham, B., & Niyato, D. (2015). Double Sided Bandwidth-Auction Game for Cognitive Device-to-Device Communication in Cellular Networks. *IEEE Transactions on Vehicular Technology,* (99), 1–1. doi:10.1109/TVT.2015.2485304

Khan, M. A., Tembine, H., & Vasilakos, A. V. (2012a). Evolutionary coalitional games: Design and challenges in wireless networks. *IEEE Wireless Communications*, *19*(2), 50–56. doi:10.1109/MWC.2012.6189413

Khan, M. A., Tembine, H., & Vasilakos, A. V. (2012b). Game Dynamics and Cost of Learning in Heterogeneous 4G Networks. *IEEE Journal on Selected Areas in Communications*, *30*(1), 198–213. doi:10.1109/JSAC.2012.120118

Khan, Z., Lehtomaki, J. J., DaSilva, L. A., Hossain, E., & Latva-aho, M. (2016). Opportunistic Channel Selection by Cognitive Wireless Nodes Under Imperfect Observations and Limited Memory: A Repeated Game Model. *IEEE Transactions on Mobile Computing*, *15*(1), 173–187. doi:10.1109/TMC.2015.2412940

Kiayias, A., Koutsoupias, E., Kyropoulou, M., & Tselekounis, Y. (2016). Blockchain Mining Games. In *Proceedings of the 2016 ACM Conference on Economics and Computation - EC '16* (pp. 365–382). New York: ACM Press. 10.1145/2940716.2940773

Kim, S. (2016). Cognitive hierarchy thinking based behavioral game model for IoT power control algorithm. *Computer Networks*, *110*, 79–90. doi:10.1016/j.comnet.2016.09.020

Kroll, J. A., Davey, I. C., & Felten, E. W. (2013). The Economics of Bitcoin Mining, or Bitcoin in the Presence of Adversaries. In *Proceedings of WEIS* (p. 21). Academic Press.

Kumar, N., Chilamkurti, N., & Misra, S. (2015). Bayesian coalition game for the internet of things: An ambient intelligence-based evaluation. *IEEE Communications Magazine*, *53*(1), 48–55. doi:10.1109/MCOM.2015.7010515

Kumar, N., Misra, S., Rodrigues, J. J. P. C., & Obaidat, M. S. (2015). Coalition Games for Spatio-Temporal Big Data in Internet of Vehicles Environment: A Comparative Analysis. *IEEE Internet of Things Journal*, *2*(4), 310–320. doi:10.1109/JIOT.2015.2388588

Kumar, N., Zeadally, S., Chilamkurti, N., & Vinel, A. (2015). Performance analysis of Bayesian coalition game-based energy-aware virtual machine migration in vehicular mobile cloud. *IEEE Network*, *29*(2), 62–69. doi:10.1109/MNET.2015.7064905

Kuo, Y.-L., Wu, E. H.-K., & Chen, G.-H. (2004). Noncooperative admission control for differentiated services in IEEE 802.11 WLANs. In *IEEE Global Telecommunications Conference, 2004. GLOBECOM '04* (Vol. 5, pp. 2981–2986). IEEE. 10.1109/GLOCOM.2004.1378899

Kwon, T., Lim, S., Choi, S., & Hong, D. (2010). Optimal Duplex Mode for DF Relay in Terms of the Outage Probability. *IEEE Transactions on Vehicular Technology*, *59*(7), 3628–3634. doi:10.1109/TVT.2010.2050503

La, Q. D., Quek, T. Q. S., Lee, J., Jin, S., & Zhu, H. (2016). Deceptive Attack and Defense Game in Honeypot-Enabled Networks for the Internet of Things. *IEEE Internet of Things Journal*, *3*(6), 1025–1035. doi:10.1109/JIOT.2016.2547994

Lai, L., & El Gamal, H. (2008). The Water-Filling Game in Fading Multiple-Access Channels. *IEEE Transactions on Information Theory*, *54*(5), 2110–2122. doi:10.1109/TIT.2008.920340

Larsson, E. G., Jorswieck, E. A., Lindblom, J., & Mochaourab, R. (2009). Game theory and the flat-fading gaussian interference channel. *IEEE Signal Processing Magazine*, *26*(5), 18–27. doi:10.1109/MSP.2009.933370

Lewenberg, Y., Bachrach, Y., Sompolinsky, Y., Zohar, A., & Rosenschein, J. S. (2015). Bitcoin Mining Pools: A Cooperative Game Theoretic Analysis. In *Proceedings of the 2015 International Conference on Autonomous Agents and Multiagent Systems* (pp. 919–927). Richland, SC: International Foundation for Autonomous Agents and Multiagent Systems. Retrieved from http://dl.acm.org/citation.cfm?id=2772879.2773270

Li, C., Xue, Y., Wang, J., Zhang, W., & Li, T. (2018). Edge-Oriented Computing Paradigms: A Survey on Architecture Design and System Management. *ACM Computing Surveys*, *51*(2), 1–34. doi:10.1145/3154815

Li, Q., Zhu, S., & Cao, G. (2010). Routing in Socially Selfish Delay Tolerant Networks. In 2010 Proceedings IEEE INFOCOM (pp. 1–9). IEEE. doi:10.1109/INFCOM.2010.5462138

Lin, S., Ni, W., Tian, H., & Liu, R. P. (2015). An Evolutionary Game Theoretic Framework for Femtocell Radio Resource Management. *IEEE Transactions on Wireless Communications*, *14*(11), 6365–6376. doi:10.1109/TWC.2015.2453170

Liu, H., Eldarrat, F., Alqahtani, H., Reznik, A., de Foy, X., & Zhang, Y. (2017). Mobile Edge Cloud System: Architectures, Challenges, and Approaches. *IEEE Systems Journal*, 1–14. doi:10.1109/JSYST.2017.2654119

Liu, H., & Krishnamachari, B. (2006). A price-based reliable routing game in wireless networks. In *Proceeding from the 2006 workshop on Game theory for communications and networks - GameNets '06* (p. 7). New York: ACM Press. 10.1145/1190195.1190201

Louta, M., Zournatzis, P., Kraounakis, S., Sarigiannidis, P., & Demetropoulos, I. (2011). Towards realization of the ABC vision: A comparative survey of Access Network Selection. In *2011 IEEE Symposium on Computers and Communications (ISCC)* (pp. 472–477). IEEE. 10.1109/ISCC.2011.5983882

Lu, Y., Zhou, B., Tung, L.-C., Gerla, M., Ramesh, A., & Nagaraja, L. (2013). Energy-efficient content retrieval in mobile cloud. In *Proceedings of the second ACM SIGCOMM workshop on Mobile cloud computing - MCC '13* (p. 21). New York: ACM Press. 10.1145/2491266.2491271

Mach, P., & Becvar, Z. (2017). Mobile Edge Computing: A Survey on Architecture and Computation Offloading. *IEEE Communications Surveys and Tutorials*, *19*(3), 1628–1656. doi:10.1109/COMST.2017.2682318

Machado, R., & Tekinay, S. (2008). A survey of game-theoretic approaches in wireless sensor networks. *Computer Networks*, *52*(16), 3047–3061. doi:10.1016/j.gaceta.2008.07.003

MacKenzie, A. B., & DaSilva, L. A. (2006). Game Theory for Wireless Engineers. *Synthesis Lectures on Communications*, *1*(1), 1–86. doi:10.2200/S00014ED1V01Y200508COM001

Mahmud, R., Kotagiri, R., & Buyya, R. (2018). Fog Computing: A Taxonomy, Survey and Future Directions. In B. Di Martino, K.-C. Li, L. T. Yang, & A. Esposito (Eds.), *Internet of Everything: Algorithms, Methodologies, Technologies and Perspectives* (pp. 103–130). Singapore: Springer Singapore. doi:10.1007/978-981-10-5861-5_5

Mao, Y., You, C., Zhang, J., Huang, K., & Letaief, K. B. (2017). A Survey on Mobile Edge Computing: The Communication Perspective. *IEEE Communications Surveys and Tutorials*, *19*(4), 2322–2358. doi:10.1109/COMST.2017.2745201

Marina, N., Saad, W., Han, Z., & Hjørungnes, A. (2011). Modeling malicious behavior in cooperative cellular wireless networks. In E. Hossain, D. I. Kim, & V. K. Bhargava (Eds.), *Cooperative Cellular Wireless Networks* (pp. 382–422). Cambridge University Press. doi:10.1017/CBO9780511667008.015

Marinelli, E. E. (2009). *Hyrax: Cloud Computing on Mobile Devices using MapReduce* (MSc thesis). CMU. Retrieved from https://pdfs.semanticscholar.org/8cd2/11cc816952f036ed65a7022adba063486008.pdf

Marsal-Llacuna, M.-L. (2018). Future living framework: Is blockchain the next enabling network? *Technological Forecasting and Social Change*, *128*, 226–234. doi:10.1016/j.techfore.2017.12.005

Martinez, D. A., Feijoo, F., Zayas-Castro, J. L., Levin, S., & Das, T. K. (2018). A strategic gaming model for health information exchange markets. *Health Care Management Science*, *21*(1), 119–130. doi:10.100710729-016-9382-2 PMID:27600378

Mateus, A., & Marinheiro, R. N. (2010). A Media Independent Information Service Integration Architecture for Media Independent Handover. In *2010 Ninth International Conference on Networks* (pp. 173–178). IEEE. 10.1109/ICN.2010.71

Milan, F., Jaramillo, J. J., & Srikant, R. (2006). Achieving cooperation in multihop wireless networks of selfish nodes. In *Proceeding from the 2006 workshop on Game theory for communications and networks - GameNets '06* (p. 3). New York: ACM Press. 10.1145/1190195.1190197

Mingxiao, D., Xiaofeng, M., Zhe, Z., Xiangwei, W., & Qijun, C. (2017). A review on consensus algorithm of blockchain. In *2017 IEEE International Conference on Systems, Man, and Cybernetics (SMC)* (pp. 2567–2572). IEEE. 10.1109/SMC.2017.8123011

Monteiro, R., Silva, J., Lourenço, J., & Paulino, H. (2015). Decentralized Storage for Networks of Hand-held Devices. In *Proceedings of the 12th EAI International Conference on Mobile and Ubiquitous Systems: Computing, Networking and Services* (pp. 299–300). ACM. 10.4108/eai.22-7-2015.2260263

Mougayar, W. (2016). *The Business Blockchain: Promise, Practice, and Application of the Next Internet Technology*. John Wiley & Sons, Inc.

Moura, J., & Edwards, C. (2015). Future Trends and Challenges for Mobile and Convergent Networks. In A. Yarali (Ed.), *4G & Beyond: The Convergence of Networks, Devices and Services* (pp. 35–80). Nova Science Publishers. Retrieved from https://arxiv.org/abs/1601.06202v1

Moura, J., & Edwards, C. (2016). Efficient access of mobile flows to heterogeneous networks under flash crowds. *Computer Networks*, *107*(2), 163–177. doi:10.1016/j.comnet.2016.04.010

Moura, J., & Hutchison, D. (2018). Survey of Game Theory and Future Trends with Application to Emerging Wireless Data Communication Networks. *ArXiv E-Prints*, 46. Retrieved from https://arxiv.org/abs/1704.00323v1

Munjal, M., & Singh, N. P. (2018). Utility aware network selection in small cell. *Wireless Networks*, 1–14. doi:10.100711276-018-1676-5

Nakamoto, S. (2008). *Bitcoin: A peer-to-peer electronic cash system*. Retrieved July 27, 2018, from https://bitcoin.org/bitcoin.pdf

Nawab, F., Agrawal, D., & El Abbadi, A. (2018). Nomadic Datacenters at the Network Edge: Data Management Challenges for the Cloud with Mobile Infrastructure. In M. H. Böhlen, R. Pichler, N. May, E. Rahm, S.-H. Wu, & K. Hose (Eds.), *Proceedings of the 21th International Conference on Extending Database Technology, {EDBT} 2018, Vienna, Austria, March 26-29, 2018* (pp. 497–500). OpenProceedings.org. 10.5441/002/edbt.2018.56

Nazir, M., Bennis, M., Ghaboosi, K., MacKenzie, A. B., & Latva-aho, M. (2010). Learning based mechanisms for interference mitigation in self-organized femtocell networks. In *2010 Conference Record of the Forty Fourth Asilomar Conference on Signals, Systems and Computers* (pp. 1886–1890). Academic Press. 10.1109/ACSSC.2010.5757866

Niyato, D., & Hossain, E. (2007). Radio resource management games in wireless networks: An approach to bandwidth allocation and admission control for polling service in IEEE 802.16. *IEEE Wireless Communications*, *14*(1), 27–35. doi:10.1109/MWC.2007.314548

Ntemos, K., Plata-Chaves, J., Kolokotronis, N., Kalouptsidis, N., & Moonen, M. (2018). Secure Information Sharing in Adversarial Adaptive Diffusion Networks. *IEEE Transactions on Signal and Information Processing over Networks*, *4*(1), 111–124. doi:10.1109/TSIPN.2017.2787910

Nurmi, P., & Nurmi, P. (2006). A Bayesian framework for online reputation systems. In *Advanced International Conference on Telecommunications and International Conference on Internet and Web Applications and Services. IEEE.*

O'Sullivan, A., Sheffrin, S., & Perez, S. (2003). *Economics : principles in action*. Prentice Hall.

Panta, R. K., Jana, R., Cheng, F., Chen, Y.-F. R., & Vaishampayan, V. A. (2013). Phoenix: Storage Using an Autonomous Mobile Infrastructure. *IEEE Transactions on Parallel and Distributed Systems*, *24*(9), 1863–1873. doi:10.1109/TPDS.2013.84

Saad, W., Han, Z., & Hjørungnes, A. (2011). Coalitional games for cooperative cellular wireless networks. In E. Hossain, D. I. Kim, & V. K. Bhargava (Eds.), *Cooperative Cellular Wireless Networks* (pp. 347–381). Cambridge University Press. doi:10.1017/CBO9780511667008.014

Saad, W., Han, Z., Poor, H. V., & Basar, T. (2012). Game-Theoretic Methods for the Smart Grid: An Overview of Microgrid Systems, Demand-Side Management, and Smart Grid Communications. *IEEE Signal Processing Magazine*, *29*(5), 86–105. doi:10.1109/MSP.2012.2186410

Sapirshtein, A., Sompolinsky, Y., & Zohar, A. (2017). *Optimal Selfish Mining Strategies in Bitcoin*. Berlin: Springer; doi:10.1007/978-3-662-54970-4_30

Saraydar, C. U., Mandayam, N. B., & Goodman, D. J. (2002). Efficient power control via pricing in wireless data networks. *IEEE Transactions on Communications*, *50*(2), 291–303. doi:10.1109/26.983324

Scutari, G., Palomar, D. P., & Barbarossa, S. (2008). Competitive Design of Multiuser MIMO Systems Based on Game Theory: A Unified View. *IEEE Journal on Selected Areas in Communications*, *26*(7), 1089–1103. doi:10.1109/JSAC.2008.080907

Sedjelmaci, H., Senouci, S. M., & Al-Bahri, M. (2016). A lightweight anomaly detection technique for low-resource IoT devices: A game-theoretic methodology. In *2016 IEEE International Conference on Communications (ICC)* (pp. 1–6). IEEE. 10.1109/ICC.2016.7510811

Selimi, M., Kabbinale, A. R., Ali, A., Navarro, L., & Sathiaseelan, A. (2018). Towards Blockchain-enabled Wireless Mesh Networks. In *Proceedings of the 1st Workshop on Cryptocurrencies and Blockchains for Distributed Systems - CryBlock'18* (pp. 13–18). New York: ACM Press. 10.1145/3211933.3211936

Semasinghe, P., Hossain, E., & Zhu, K. (2015). An Evolutionary Game for Distributed Resource Allocation in Self-Organizing Small Cells. *IEEE Transactions on Mobile Computing*, *14*(2), 274–287. doi:10.1109/TMC.2014.2318700

Semasinghe, P., Maghsudi, S., & Hossain, E. (2017). Game Theoretic Mechanisms for Resource Management in Massive Wireless IoT Systems. *IEEE Communications Magazine*, *55*(2), 121–127. doi:10.1109/MCOM.2017.1600568CM

Shastry, N., & Adve, R. (2006). Stimulating Cooperative Diversity in Wireless Ad Hoc Networks through Pricing. In *2006 IEEE International Conference on Communications* (pp. 3747–3752). IEEE. 10.1109/ICC.2006.255655

Shen, S., Yue, G., Cao, Q., & Yu, F. (2011). A Survey of Game Theory in Wireless Sensor Networks Security. *JNW*, *6*(3), 521–532. doi:10.4304/jnw.6.3.521-532

Shi, H.-Y., Wang, W.-L., Kwok, N.-M., & Chen, S.-Y. (2012). Game Theory for Wireless Sensor Networks: A Survey. *Sensors (Basel), 12*(7), 9055–9097. doi:10.3390120709055 PMID:23012533

Shivshankar, S., & Jamalipour, A. (2015). An Evolutionary Game Theory-Based Approach to Cooperation in VANETs Under Different Network Conditions. *IEEE Transactions on Vehicular Technology, 64*(5), 2015–2022. doi:10.1109/TVT.2014.2334655

Silva, J., Marinheiro, R., Moura, J., & Almeida, J. (2013). Differentiated classes of service and flow management using an hybrid broker. *ACEEE International Journal on Communication, 4*(2), 13–22.

Skraparlis, D., Sakarellos, V., Panagopoulos, A., & Kanellopoulos, J. (2009). Outage Performance Analysis of Cooperative Diversity with MRC and SC in Correlated Lognormal Channels. *EURASIP Journal on Wireless Communications and Networking, 707839*(1), 707839. doi:10.1155/2009/707839

Srirama, S., & Paniagua, C. (2013). Mobile web service provisioning and discovery in android days. In *Conference on Mobile Services* (pp. 15–22). Washington, DC: IEEE Computer Society. 10.1109/MS.2013.14

Srivastava, V., Neel, J., Mackenzie, A. B., Menon, R., Dasilva, L. A., Hicks, J. E., ... Gilles, R. P. (2005). Using game theory to analyze wireless ad hoc networks. *IEEE Communications Surveys and Tutorials, 7*(4), 46–56. doi:10.1109/COMST.2005.1593279

Taleb, T., Samdanis, K., Mada, B., Flinck, H., Dutta, S., & Sabella, D. (2017). On Multi-Access Edge Computing: A Survey of the Emerging 5G Network Edge Architecture & Orchestration. *IEEE Communications Surveys & Tutorials,* (99), 1–1. doi:10.1109/COMST.2017.2705720

Taylor, P. D., & Jonker, L. B. (1978). Evolutionary stable strategies and game dynamics. *Mathematical Biosciences, 40*(1–2), 145–156. doi:10.1016/0025-5564(78)90077-9

Tembine, H., Altman, E., El-Azouzi, R., & Hayel, Y. (2010). Evolutionary Games in Wireless Networks. *IEEE Transactions on Systems, Man, and Cybernetics. Part B, Cybernetics, 40*(3), 634–646. doi:10.1109/TSMCB.2009.2034631 PMID:19963703

Tordera, E., Masip-Bruin, X., Garcia-Almiana, J., Jukan, A., Ren, G.-J., Zhu, J., & Farre, J. (2016). *What is a Fog Node? A Tutorial on Current Concepts towards a Common Definition.* Retrieved from http://arxiv.org/abs/1611.09193

Trestian, R., Ormond, O., & Muntean, G.-M. (2011). Reputation-based network selection mechanism using game theory. *Physical Communication, 4*(3), 156–171. doi:10.1016/j.phycom.2011.06.004

Ungureanu, V. (2018). Strategic Form Games on Digraphs. In *Pareto-Nash-Stackelberg Game and Control Theory. Smart Innovation, Systems and Technologies* (pp. 167–197). Cham: Springer; doi:10.1007/978-3-319-75151-1_8

Vassaki, S., Panagopoulos, A. D., Constantinou, P., & Vázquez-Castro, M. A. (2010). Market-Based Bandwidth Allocation for Broadband Satellite Communication Networks. In *2010 Second International Conference on Advances in Satellite and Space Communications* (pp. 110–115). IEEE. 10.1109/SPACOMM.2010.22

Vedhavathy, T. R., & Manikandan, M. S. K. (2018). Triple referee incentive mechanism for secure mobile adhoc networks. *Cluster Computing*, 1–10. doi:10.100710586-017-1631-8

Vermesan, O., Bröring, A., Tragos, E., Serrano, M., Bacciu, D., Chessa, S. … Bahr, R. (2017). Internet of robotic things : converging sensing/actuating, hypoconnectivity, artificial intelligence and IoT Platforms. In Cognitive Hyperconnected Digital Transformation : Internet of Things Intelligence Evolution (pp. 97–155). Academic Press.

Wang, B., Wu, Y., & Liu, K. J. R. (2010). Game Theory for Cognitive Radio Networks: An Overview. *Computer Networks*, *54*(14), 2537–2561. doi:10.1016/j.comnet.2010.04.004

Wang, B., Wu, Y., Liu, K. J. R., & Clancy, T. C. (2011). An anti-jamming stochastic game for cognitive radio networks. *IEEE Journal on Selected Areas in Communications*, *29*(4), 877–889. doi:10.1109/JSAC.2011.110418

Wang, S., Zhang, X., Zhang, Y., Wang, L., Yang, J., & Wang, W. (2017). A Survey on Mobile Edge Networks: Convergence of Computing, Caching and Communications. *IEEE Access: Practical Innovations, Open Solutions*, *5*, 6757–6779. doi:10.1109/ACCESS.2017.2685434

Watanabe, E. H., Menasche, D. S., de Souza e Silva, E., & Leao, R. M. M. (2008). Modeling Resource Sharing Dynamics of VoIP Users over a WLAN Using a Game-Theoretic Approach. In *IEEE INFOCOM 2008 - The 27th Conference on Computer Communications* (pp. 915–923). IEEE. 10.1109/INFOCOM.2008.144

Xiao, Y., Chen, K. C., Yuen, C., Han, Z., & DaSilva, L. A. (2015). A Bayesian Overlapping Coalition Formation Game for Device-to-Device Spectrum Sharing in Cellular Networks. *IEEE Transactions on Wireless Communications*, *14*(7), 4034–4051. doi:10.1109/TWC.2015.2416178

Yan, Y., Huang, J., & Wang, J. (2013). Dynamic Bargaining for Relay-Based Cooperative Spectrum Sharing. *IEEE Journal on Selected Areas in Communications*, *31*(8), 1480–1493. doi:10.1109/JSAC.2013.130812

Yap, K.-K., Huang, T.-Y., Kobayashi, M., Yiakoumis, Y., McKeown, N., Katti, S., & Parulkar, G. (2012). Making use of all the networks around us. In *Proceedings of the 2012 ACM SIGCOMM workshop on Cellular networks: operations, challenges, and future design - CellNet '12* (p. 19). New York: ACM Press. 10.1145/2342468.2342474

Yi, S., Li, C., & Li, Q. (2015). A Survey of Fog Computing: Concepts, Applications and Issues. In *Proceedings of the 2015 Workshop on Mobile Big Data - Mobidata '15* (pp. 37–42). New York: ACM Press. 10.1145/2757384.2757397

Zhang, D., Ileri, O., & Mandayam, N. (2008). Bandwidth exchange as an incentive for relaying. In *2008 42nd Annual Conference on Information Sciences and Systems* (pp. 749–754). IEEE. 10.1109/CISS.2008.4558621

Zhang, L., Afanasyev, A., Burke, J., Jacobson, V., Claffy, K., Crowley, P., ... Zhang, B. (2014). Named Data Networking. *Computer Communication Review*, *44*(3), 66–73. doi:10.1145/2656877.2656887

Zhang, Y., & Guizani, M. (2011). *Game theory for wireless communications and networking*. CRC Press.

Zhang, Y., Lou, W., & Fang, Y. (2004). SIP: a secure incentive protocol against selfishness in mobile ad hoc networks. In 2004 IEEE Wireless Communications and Networking Conference (IEEE Cat. No.04TH8733) (pp. 1679–1684). IEEE. doi:10.1109/WCNC.2004.1311805

Zhang, Z., Long, K., Vasilakos, A. V., & Hanzo, L. (2016). Full-Duplex Wireless Communications: Challenges, Solutions, and Future Research Directions. *Proceedings of the IEEE*, *104*(7), 1369–1409. doi:10.1109/JPROC.2015.2497203

Zheng, G., Liu, S., & Qi, X. (2012). Clustering routing algorithm of wireless sensor networks based on Bayesian game. *Journal of Systems Engineering and Electronics*, *23*(1), 154–159. doi:10.1109/JSEE.2012.00019

Zhong, S., Chen, J., & Yang, Y. R. (2003). Sprite: a simple, cheat-proof, credit-based system for mobile ad-hoc networks. In *IEEE INFOCOM 2003. Twenty-second Annual Joint Conference of the IEEE Computer and Communications Societies (IEEE Cat. No.03CH37428)* (Vol. 3, pp. 1987–1997). IEEE. 10.1109/INFCOM.2003.1209220

Zhu, K., & Hossain, E. (2015). Joint Mode Selection and Spectrum Partitioning for Device-to-Device Communication: A Dynamic Stackelberg Game. *IEEE Transactions on Wireless Communications*, *14*(3), 1406–1420. doi:10.1109/TWC.2014.2366136

Zhu, K., Niyato, D., & Wang, P. (2010). Network Selection in Heterogeneous Wireless Networks: Evolution with Incomplete Information. In 2010 IEEE Wireless Communication and Networking Conference (pp. 1–6). IEEE. doi:10.1109/WCNC.2010.5506371

ADDITIONAL READING

Ackermann, H., Briest, P., Fanghänel, A., & Vöcking, B. (2008). Who Should Pay for Forwarding Packets? *Internet Mathematics*, *5*(4), 459–475. doi:10.1080/15427951.2008.10129168

Akyildiz, I. F., Lee, W.-Y., Vuran, M. C., & Mohanty, S. (2006). NeXt Generation/Dynamic Spectrum Access/Cognitive Radio Wireless Networks: A Survey. *Computer Networks*, *50*(13), 2127–2159. doi:10.1016/j.comnet.2006.05.001

Al-Manthari, B., Nasser, N., & Hassanein, H. (2011). Congestion Pricing in Wireless Cellular Networks. *IEEE Communications Surveys and Tutorials*, *13*(3), 358–371. doi:10.1109/SURV.2011.090710.00042

Alayesh, M. A., & Ghani, N. (2012). Performance of a primary-secondary user power control under rayleigh fast flat fading channel with pricing. In Telecommunications (ICT), 2012 19th International Conference on (pp. 1–5). 10.1109/ICTEL.2012.6221257

Buddhikot, M. M. (2007). Understanding Dynamic Spectrum Access: Models, Taxonomy and Challenges. In 2007 2nd IEEE International Symposium on New Frontiers in Dynamic Spectrum Access Networks (pp. 649–663). 10.1109/DYSPAN.2007.88

Buddhikot, M. M., & Ryan, K. (2005). Spectrum management in coordinated dynamic spectrum access based cellular networks. In First IEEE International Symposium on New Frontiers in Dynamic Spectrum Access Networks, 2005. DySPAN 2005. (pp. 299–307). 10.1109/DYSPAN.2005.1542646

Buttyán, L., & Hubaux, J.-P. (2008). *Security and cooperation in wireless networks : thwarting malicious and selfish behavior in the age of ubiquitous computing*. Cambridge University Press; Retrieved from http://www.cambridge.org/gb/academic/subjects/engineering/wireless-communications/security-and-cooperation-wireless-networks-thwarting-malicious-and-selfish-behavior-age-ubiquitous-computing?format=HB&isbn=9780521873710#xEQOI7PilMoKvAgD.97

Courcoubetis, C., & Weber, R. (2003). *Pricing Communication Networks*. Chichester, UK: John Wiley & Sons, Ltd.; doi:10.1002/0470867175

DaSilva, L. A. (2000). Pricing for QoS-enabled networks: A survey. *IEEE Communications Surveys and Tutorials*, *3*(2), 2–8. doi:10.1109/COMST.2000.5340797

Etkin, R., Parekh, A., & Tse, D. (2007). Spectrum sharing for unlicensed bands. *IEEE Journal on Selected Areas in Communications*, *25*(3), 517–528. doi:10.1109/JSAC.2007.070402

Felegyhazi, M., Cagalj, M., Bidokhti, S. S., & Hubaux, J. P. (2007). Non-Cooperative Multi-Radio Channel Allocation in Wireless Networks. In IEEE INFOCOM 2007 - 26th IEEE International Conference on Computer Communications (pp. 1442–1450). 10.1109/INFCOM.2007.170

Halpern, J. Y. (2007). Computer Science and Game Theory: A Brief Survey. Work, abs/cs/070, 1–17. doi:10.1145/1378704.1378721

Heikkinen, T. (2004). Distributed Scheduling and Dynamic Pricing in a Communication Network. *Wireless Networks*, *10*(3), 233–244. doi:10.1023/B:WINE.0000023858.20849.ab

Heikkinen, T. (2006). A Potential Game Approach to Distributed Power Control and Scheduling. *Computer Networks*, *50*(13), 2295–2311. doi:10.1016/j.comnet.2005.09.010

Heikkinen, T. M. (2002). On Congestion Pricing in a Wireless Network. *Wireless Networks*, *8*(4), 347–354. doi:10.1023/A:1015578321066

Huang, J., Berry, R. A., & Honig, M. L. (2006a). Auction-Based Spectrum Sharing. *Mobile Networks and Applications*, *11*(3), 405–408. doi:10.100711036-006-5192-y

Huang, J., Berry, R. A., & Honig, M. L. (2006b). Distributed interference compensation for wireless networks. *IEEE Journal on Selected Areas in Communications*, *24*(5), 1074–1084. doi:10.1109/JSAC.2006.872889

Ileri, O., Samardzija, D., Sizer, T., & Mandayam, N. B. (2005). Demand responsive pricing and competitive spectrum allocation via a spectrum server. In First IEEE International Symposium on New Frontiers in Dynamic Spectrum Access Networks, 2005. DySPAN 2005. (pp. 194–202). 10.1109/DYSPAN.2005.1542635

Inaltekin, H., Wexler, T., & Wicker, S. B. (2007). A Duopoly Pricing Game for Wireless IP Services. In 2007 4th Annual IEEE Communications Society Conference on Sensor, Mesh and Ad Hoc Communications and Networks (pp. 600–609). 10.1109/SAHCN.2007.4292872

Jia, J., & Zhang, Q. (2008). Competitions and Dynamics of Duopoly Wireless Service Providers in Dynamic Spectrum Market. In *Proceedings of the 9th ACM International Symposium on Mobile Ad Hoc Networking and Computing* (pp. 313–322). New York, NY, USA: ACM. 10.1145/1374618.1374660

Koutsopoulou, M., Kaloxylos, A., Alonistioti, A., Merakos, L., & Kawamura, K. (2004). Charging, accounting and billing management schemes in mobile telecommunication networks and the internet. *IEEE Communications Surveys and Tutorials*, *6*(1), 50–58. doi:10.1109/COMST.2004.5342234

w. Lee, J., Tang, A., Huang, J., Chiang, M., & Calderbank, A. R. (2007). Reverse-Engineering MAC: A Non-Cooperative Game Model. *IEEE Journal on Selected Areas in Communications*, *25*(6), 1135–1147. doi:10.1109/JSAC.2007.070808

MacKenzie, A. B., & Wicker, S. B. (2003). Stability of multipacket slotted Aloha with selfish users and perfect information. In INFOCOM 2003. Twenty-Second Annual Joint Conference of the IEEE Computer and Communications. IEEE Societies (Vol. 3, pp. 1583–1590 vol.3). 10.1109/INFCOM.2003.1209181

Maharjan, S., Zhang, Y., & Gjessing, S. (2011). Economic Approaches for Cognitive Radio Networks: A Survey. *Wireless Personal Communications*, *57*(1), 33–51. doi:10.100711277-010-0005-9

Maillé, P., & Tuffin, B. (2006). Pricing the Internet with Multibid Auctions. *IEEE/ACM Transactions on Networking*, *14*(5), 992–1004. doi:10.1109/TNET.2006.882861

Maskery, M., Krishnamurthy, V., & Zhao, Q. (2009). Decentralized dynamic spectrum access for cognitive radios: Cooperative design of a non-cooperative game. *IEEE Transactions on Communications*, *57*(2), 459–469. doi:10.1109/TCOMM.2009.02.070158

Meshkati, F., Poor, H. V., & Schwartz, S. C. (2007). Energy-Efficient Resource Allocation in Wireless Networks. *IEEE Signal Processing Magazine*, *24*(3), 58–68. doi:10.1109/MSP.2007.361602

Mingbo Xiao, M., Shroff, N. B., & Chong, E. K. P. (2003). A utility-based power-control scheme in wireless cellular systems. *IEEE/ACM Transactions on Networking*, *11*(2), 210–221. doi:10.1109/TNET.2003.810314

Nguyen, D. N., & Krunz, M. (2012). Price-Based Joint Beamforming and Spectrum Management in Multi-Antenna Cognitive Radio Networks. *IEEE Journal on Selected Areas in Communications*, *30*(11), 2295–2305. doi:10.1109/JSAC.2012.121221

Nie, N., & Comaniciu, C. (2005). Adaptive channel allocation spectrum etiquette for cognitive radio networks. In First IEEE International Symposium on New Frontiers in Dynamic Spectrum Access Networks, 2005. DySPAN 2005. (pp. 269–278). 10.1109/DYSPAN.2005.1542643

Niyato, D., & Hossain, E. (2009). Dynamics of Network Selection in Heterogeneous Wireless Networks: An Evolutionary Game Approach. *IEEE Transactions on Vehicular Technology*, *58*(4), 2008–2017. doi:10.1109/TVT.2008.2004588

Niyato, D., Hossain, E., & Han, Z. (2009). Dynamic spectrum access in IEEE 802.22- based cognitive wireless networks: A game theoretic model for competitive spectrum bidding and pricing. *IEEE Wireless Communications*, *16*(2), 16–23. doi:10.1109/MWC.2009.4907555

Novelti, P., Satya, A., & Wardana, A. (2007). Cooperation in Wireless Grids An Energy Efficient MAC Protocol for Cooperative Network with Game Theory Model Cooperation in Wireless Grids – An Energy Efficient MAC Protocol for Cooperative Network with Game Theory Model.

Rasti, M., Sharafat, A. R., & Seyfe, B. (2009). Pareto-efficient and Goal-driven Power Control in Wireless Networks: A Game-theoretic Approach with a Novel Pricing Scheme. *IEEE/ACM Transactions on Networking*, *17*(2), 556–569. doi:10.1109/TNET.2009.2014655

Ren, W., Zhao, Q., & Swami, A. (2008). Power control in spectrum overlay networks: How to cross a multi-lane highway. In *2008 IEEE International Conference on Acoustics, Speech and Signal Processing* (pp. 2773–2776). 10.1109/ICASSP.2008.4518224

Sengupta, S., Anand, S., Chatterjee, M., & Chandramouli, R. (2009). Dynamic pricing for service provisioning and network selection in heterogeneous networks. *Physical Communication*, *2*(1–2), 138–150. doi:10.1016/j.phycom.2009.02.009

Thomas, R. W., DaSilva, L. A., & MacKenzie, A. B. (2005). Cognitive networks. In First IEEE International Symposium on New Frontiers in Dynamic Spectrum Access Networks, 2005. DySPAN 2005. (pp. 352–360). IEEE. 10.1109/DYSPAN.2005.1542652

Vincent, T. L., Vincent, T. L. S., & Cohen, Y. (2011). Darwinian dynamics and evolutionary game theory. *Journal of Biological Dynamics*, *5*(3), 215–226. doi:10.1080/17513758.2010.526306

Wang, D., Comaniciu, C., & Tureli, U. (2006). A Fair and Efficient Pricing Strategy for Slotted Aloha in MPR Models. In *IEEE Vehicular Technology Conference* (pp. 1–5). 10.1109/VTCF.2006.509

Youngmi Jin, Y., & Kesidis, G. (2002). Equilibria of a noncooperative game for heterogeneous users of an ALOHA network. *IEEE Communications Letters*, *6*(7), 282–284. doi:10.1109/LCOMM.2002.801326

Zhang, J., & Zhang, Q. (2009). Stackelberg Game for Utility-based Cooperative Cognitive Radio Networks. In *Proceedings of the Tenth ACM International Symposium on Mobile Ad Hoc Networking and Computing* (pp. 23–32). New York, NY, USA: ACM. 10.1145/1530748.1530753

Zhao, L., Zhang, J., & Zhang, H. (2008). Using Incompletely Cooperative Game Theory in Wireless Mesh Networks. *IEEE Network*, *22*(1), 39–44. doi:10.1109/MNET.2008.4435901

Zhao, Q., & Sadler, B. M. (2007). A Survey of Dynamic Spectrum Access. *IEEE Signal Processing Magazine*, *24*(3), 79–89. doi:10.1109/MSP.2007.361604

Chapter 6

Converged Fi-Wi Passive Optical Networks and Their Designing Using the HPON Network Configurator

Rastislav Róka
FEI STU Bratislava, Slovakia

ABSTRACT

With the emerging mobile applications and needs of ever-increasing bandwidth, it is anticipated that the next-generation passive optical network (NG-PON) with much higher bandwidth is a natural path forward to satisfy these demands and to develop valuable converged fiber-wireless access networks for wireless network operators. NG-PON systems present optical access infrastructures to support various applications of many service providers. Hybrid passive optical networks (HPON) present a necessary phase to future PON networks utilized the optical transmission medium – the optical fiber. For developing hybrid passive optical networks, there exist various architectures and directions. They are specified with emphasis on their basic characteristics. For proposing reliable and survivable architectures, traffic protection schemes must be implemented. For converging Fi-Wi passive optical networks, an integration of optical and wireless technologies into common broadband access network must be considered. Finally, the HPON network configurator as the interactive software tool is introduced.

INTRODUCTION

Emerging applications of advanced end users can be associated with increasing bandwidth demands. Except technological improvements of mobile broadband and broadcast technologies, it is avoidable to consider a reliable support in access networks by appropriate modern and advanced technologies utilizing the fixed transmission medium, above all optical fibers. Optical access networks designed for those applications are called passive optical networks (PON) because only passive optical equipment is located in the remote node (RN). Their main advantages are high reliability, simple maintenance and

DOI: 10.4018/978-1-5225-7570-2.ch006

no need of external power supplies. Moreover, present optical infrastructure is transparent to bit rates and modulation formats of optical information signals. Complete passive optical access networks can be upgraded without substantial changes of this fundamental infrastructure. Also, the wavelength division multiplexing (WDM) technique with separated transmission channels assigned to particular subscribers can be implemented into passive optical networks and, by this way, various specific broadband and multimedia services can be provisioned to each end subscriber (Čuchran & Róka, 2006).

The PON network is in its substance a bi-directional point-to-multipoint system that contains passive (optical fibers, passive optical splitters, couplers, connectors) optical elements in a distribution part of the access network and active optical line and network terminals (OLT, ONT) placed in end terminating points of the access network. The key advantage of this PON approach is locating only passive optical components in the optical distribution network (ODN). Optical transmitters and receivers are located inside buildings; other active optical components are not utilized in the outside plant (Figure 1). By this way, total network costs for installing, operating and maintenance of optical network equipment are markedly decreased.

Current standardized PON networks based on the time-division multiple access (TDMA) have evolved as an access solution to provide simplicity and low operational cost. The most effective utilization of the optical transmission medium can be attained using more wavelengths per one fiber by means of WDM technologies. Moreover, scaling up TDMA-PON networks to several tens of gigabits per second of the aggregate capacity is extremely challenging due to the complexity of optical components and burst mode receivers at such high data rates. Therefore, WDM-PON networks are increasingly considered to deliver ultra-high-speed services by enabling service providers to offer a dedicated wavelength straight to a home or business with limitations in terms of scalability and bandwidth granularity.

From a viewpoint of the architecture, we can classify passive optical networks into following classes with basic characteristics (Ramaswami & Sivarajan, 2001), (Róka, 2014):

- **All Fiber:** Passive optical network (AF-PON),
- **Time Division Multiplexing:** Passive optical network (TDM-PON),

Figure 1. The general PON network architecture

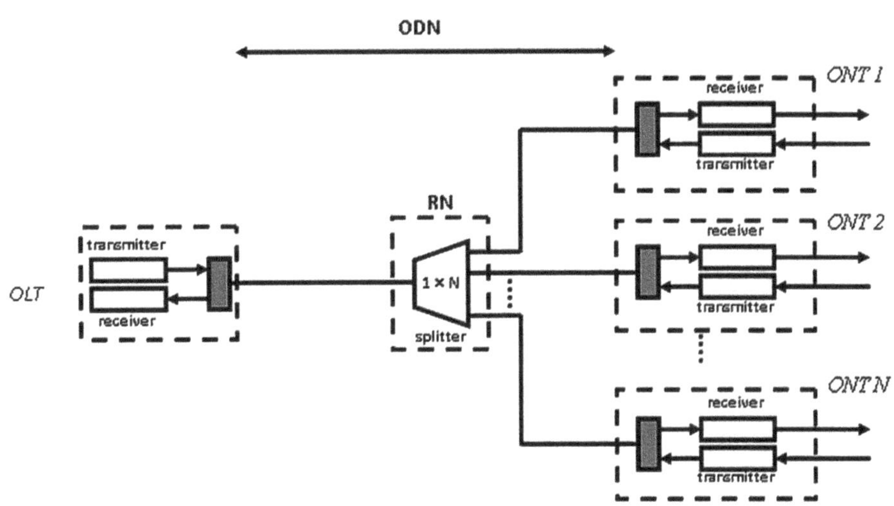

- **Wavelength Division Multiplexing:** Passive optical fiber passive network (WDM-PON),
- **Wavelength Routing:** Passive optical network (WR-PON).

Now, we can shortly present basic steps of a consistent transition from the existing TDM-PON network to the WDM-PON and WR-PON networks (Róka, 2014), (Róka, 2015):

- In the first step, a change of the central office equipment is executed. Specifically, a coupler is adding to combine the original time OLT equipment and the new wavelength OLT equipment into the one-fiber transmission.
- In the second step, appropriate changes in ONT equipment are executed, specifically with adding new wavelength ONT equipment assigned to specific wavelengths for ONT transceivers. These new wavelength ONT terminal can be connected only to the new AWG element, not to a common power splitter. Also, a coupler must be added into the remote node to combine the original 1:N power splitter and the new AWG element with 32 specific wavelength outputs into the one-fiber transmission.
- In the final status, only new wavelength PON system with appropriate OLT and ONT equipment and the AWG element is working. All old TDM-PON system parts in the outside plan are removed, including old OLT TDM-PON equipment, power splitters and old ONT TDM-PON equipment.

There are some basic variations of WDM-PON architectures (Grobe, 2008), (Róka, 2015), they can use a fixed-wavelength laser array or a multi-frequency laser as WDM transmitters. In the OLT equipment, circulators can be used instead of band splitters to enable bidirectional operation over a single fiber.

- The broadcast-and-select (B&S) WDM-PON architecture (Figure 2a) utilizes a passive 1:N splitter in a remote node. It is affected by loss of the passive splitter, as well as the broadcast security issues. No identical ONT units can be used unless both the receiver filters and transmitters are tunable.
- The AWG WR-PON architecture (Figure 2b) replaces the passive splitter by an arrayed waveguide grating (AWG). This scheme offers lower insertion loss and simplified ONT units with no wavelength-selective receivers. However, different (or tunable) ONT wavelengths are still necessary.
- The use of identical ONT units can be facilitated by using a single (shared) wavelength for the upstream (US). This leads to the spectrum slicing (SS) WR-PON architecture (Figure 2c) with a simplified operation. To unify the ONT design, ultra-broadband sources can be used in ONT units as an alternative. In the AWG element, the broadband US signals are spectrally sliced. Spectrum slicing leads to identical ONT units, but it suffers from a poor power budget. Also, the US bandwidth is limited and multiple-access (burst-mode) techniques are required.

For developing future passive optical networks, it is important the investment protection of legacy passive optical networks. Therefore, a strong interest for utilization of old TDM equipment in passive optical networks is present. From this viewpoint, the network sharing for next-generation passive optical network technologies is possible, but it is more effective in a case of hybrid passive optical networks. There are strong differences between costs per home passed for the urban, suburban and rural geotypes. The initial investment cost for operators on shared networks is lower than the cost of one operator deploying the network alone, and this is true regardless of which network architecture or geotype is analyzed

Figure 2. Variations of WDM/TDM-PON network architectures

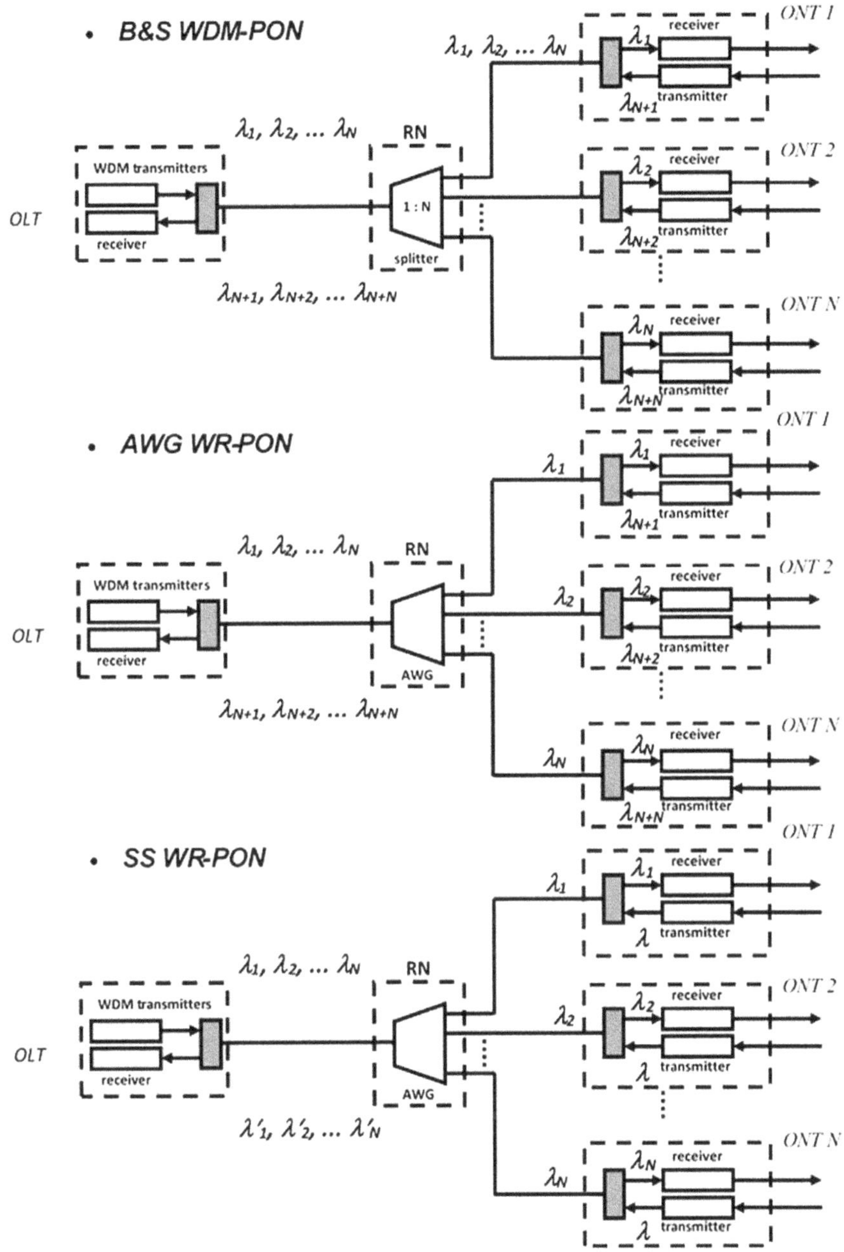

(Schneir, 2014). The cost increase can be explained by additional number of network elements needed to share the network or by the lower number of subscribers achieved by each operator in a network-sharing scheme. To obtain the cost per home connected, the total cost per operator should be divided by the number of subscribers of each operator.

For a reason of comparison, we can present basic steps of a consistent transition from the existing TDM-PON network to the hybrid passive optical network (Róka, 2014), (Róka, 2015):

- In the first step, a coupler is adding to combine the original OLT TDM-PON system and the new OLT TDM/WDM-PON system into the one-fiber transmission. The OLT TDM/WDM-PON equipment is connected to the coupler through dedicated outputs of the common power splitter.
- In the second step, new ONT TDM/WDM-PON equipment is adding and connected to outputs of the original 1:N power splitter.
- In the final status, new hybrid TDM/WDM-PON system with appropriate OLT and ONT equipment and the OLT power splitter is working. Only the old OLT TDM-PON equipment is exchanged, power splitters in the outside plan and old ONT TDM-PON can be re-used without changes.

The long payback periods found in suburban and rural areas explain why operators usually prefer to invest initially in FTTH deployments in urban areas. As the payback period depends on the total cost per operator and a number of subscribers achieved by every operator, there is an increase in a number of years for some PON architectures. For three geotypes, there are differences in the payback period of GPON and XG-PON architectures because both use a multi-fiber deployment and every operator involved in a network sharing has to deploy an additional infrastructure. As costs per home connected are slightly increased when using a network sharing with the TDM/WDM-PON architecture (TWDM-PON), there is only a slight increase in the payback period. For a case of the AWG-based wavelength PON architecture, there is no increase in the payback period.

In the first case, ducts of the feeder segment are already available, which implies that an operator incurs no initial investment for digging or deploying manholes, but this operator needs to pay an annual fee for using the ducts and must deploy the fibers. In the second case, ducts of the feeder and distribution segments are already available and the operator must pay an annual fee and deploy the fiber. For all cases, there are important reductions of costs per home connected achieved when available ducts in the feeder and distribution segments are used. When the passive infrastructure is reused, costs per operator are reduced. However, a network sharing scheme still leads to higher costs for GPON, XG-PON and TWDM-PON architectures than the standalone scenario (Schneir, 2014).

For technological improvements of mobile broadband and broadcast technologies, it is important a reliable support in access networks. Therefore, another challenge is the reliable offer of traffic protection and restoration schemes for different migration strategies in developing future passive optical networks. The migration toward a protected access network can be based on either the access network planning that has overseen its future traffic protection and the installation included dark fibers or on the planning that did not foresee a traffic protection and any additional protection equipment and/or fibers were neither bought nor installed.

There exist various protection deployment approaches to get protected optical access networks based on the proposed fiber layout. The first case is from the greenfield directly to the protected scenario. The second case is represented from the greenfield to the protected scenario through an intermediate unprotected architecture under the first case. The third case is characterized from the greenfield to the protected scenario through an intermediate unprotected architecture under the second case.

The benefit of deploying a traffic protection – a significant reduction of the total costs of ownership compared to the unprotected access in all of considered (rural, urban, dense urban) scenarios – at very low increase of infrastructure expenses a large reduction of operational expenditures can be obtained as a consequence of the reliability performance improvement and the service interruption decrement experienced by users. It can be beneficial to either provide protection functionalities at the time of network deployment or at least install a sufficient amount of fibers in advance. It can be recommended to provide a traffic protection as early as possible (Mas Machuca, 2012).

Future passive optical access networks must be able to provide high sustainable bandwidths on per user basis while keeping capital and operational expenditures as low as possible. Moreover, the large coverage makes possible to reduce the total network costs by merging several optical line terminals into a single one. Furthermore, the growing importance of uninterrupted internet access makes the fault management an important challenge for future optical access networks. Therefore, next-generation passive optical networks (NG-PON) need to provide survivability schemes in a cost-efficient way. The reliability requirements may depend on user profiles. Thus, NG-PON networks should also support the end-to-end protection for some selected users when requested.

For practical implementations of future passive optical networks, a high importance is dedicated to the energy saving in long-reach broadband access networks and energy efficient activities on both optical terminal sides and in converged broadband wireless-optical access networks.

Novel methods of energy efficiency improvements for optical access networks are arisen (Vetter, 2014):

- The sleep mode is a method of the power reduction by turning off parts of the system when the offered traffic is lighter than the total capacity of the system. The goal is to make the average power consumption as much as possible proportional to the traffic load. The idea behind the bit-interleaving protocol is that if a ONT terminal can determine which bits are intended for other units, these bits should not undergo further processing. In the TDM-PON system, only a small part of the total bit rate is destined to a simple ONT terminal, therefore a significant power saving can be realized.
- It is possible to improve the energy efficiency of optical components by exploiting the low subscriber rate and optical budget. By introducing programmable transmitter and receiver circuits, it is also possible to adaptively control launch power and a corresponding receiver signal gain at the transimpedance amplifier in both OLT and ONT terminals to optimize their operating point.
- An additional improvement on the component level in the WDM-PON system is to employ a cooler-less tunable laser or amplifier. It is possible to improve the energy efficiency of the switching fabric by considering the relatively low network utilization and typical aggregation function.

The cyclic sleep mode is the most effective approach to reduce the power consumption when compatibility to existing standards is a given. In the TDM-PON network, the cyclic sleep mode applies to ONT terminals, the OLT terminal with its relatively small contribution is always on. The WDM-PON network consumes more power due to a tunability of the WDM transmitter in ONT terminals, as well as the thermo-electric cooling for stabilizing the wavelengths at the OLT side. The TWDM-PON architecture offers an interesting capability to scale the power consumption of the OLT terminal as a function of the required total capacity. It is done by lighting up an appropriate number of wavelength pairs by all ONT units connected to the shared medium. The OFDMA-PON architecture consumes the highest power due to the need for ADC/DAC and (I)FFT signal processing blocks. By narrowing the spectral band behind the ONT receiver, it is possible to reduce the power consumption incurred by these signal processing blocks (Vetter, 2014).

In the near future, a convergence of optical and wireless technologies will be continued. Mobile base stations can be backhauled through the normal passive optical network and they can be thought of as normal PON ONT elements. Between the central office and the remote base station will be introduced another intermediate node responsible for generation and transmission of control data streams over shorter fiber distances. Moreover, a virtualization of heterogeneous optical-wireless network and IT infrastructures can be mentioned for support of wireless access and backhauling solutions and for cloud and mobile cloud service.

Broadband wireless solutions that are able to support several even mobile end-users in a large areas and the optical technology representing by the PON can bring a large bandwidth, but in certain, fixed and pre-determined optical paths are prepared for convergence. A utilization of advantages from both technologies can provide an access solution by deployment of the hybrid wireless-optical access network, known as the fiber over wireless (FiWi). An infrastructure of the optical part can utilize a termination FTTC or FTTB, whereby wireless access points can widen a broadband access connection up to end-users. The integration of optical and wireless technologies into the FiWi network can ensure a powerful solution in respect of the high broadband potential of optical fibers and the mobility of wireless access equipment. In principle, optical and wireless technologies can mutually complement for building a common network and are able supporting new unavoidable applications and services requesting high bandwidth also in some cases of mobility.

The integrated wireless-optical network can connect several optical and wireless technologies. Elementary factors for selecting of technologies incorporate an available budget, a geographical area (rural or city), a requested QoS guarantee and other requirements related to end-users, e.g. a connection type, a cable interface, a mobile connection, etc. In most cases of converged technologies, a hybrid architecture is selected with the imbedded ONT network element. Optical technologies (including TDM-PON, WDM-PON, …) can be combined with WiMAX, WiFi and LTE wireless technologies. An advantage of optical technologies is above all a possibility for providing high transmission capacities. The most of proposed converged networks utilize optical fibers for supporting the backbone part. Wireless technologies by contrast offer low costs of necessary equipment and mobile broadband access networks. The main aim of the Fi-Wi integration is creating of at once easy available and reliable network.

In this chapter, we focus on a design of a transition stage between TDM-PON and WDM-PON networks and explain possible cooperation between optical fiber and wireless technologies in converged Fi-Wi PON networks. A main reason for this motivation is that at first a number of operating TDM-PON networks is even now high and still rising, at second a utilization of installed optical infrastructures is maximized for transmission capacity's increasing and at third an extreme extension of wireless technologies between subscribers. On the other side, a creating the hybrid HPON network can be used as the upgrade of old networks in many cases with a utilization of relevant parts of the original infrastructure with minimum financial costs and with traffic protection provisioning. Moreover, a consistent transition from only optical HPON network to the passive optical network with converged fiber-wireless technologies is important. Therefore, it is important to identify a right time for creating and building of the converged Fi-Wi PON network. For this purpose, the HPON Network Configurator as an interactive software tool can be used.

REQUIREMENTS FOR NG-PON NETWORKS

The next-generation PON (NG-PON) with much higher bandwidth is a natural path forward to satisfy demands of ever increasing bandwidth. Many network operators are motivated to further develop a valuable optical access network and to leverage such NG-PON systems as a common access infrastructure to support broader market segments. The NG-PON technology must be able to protect the investment of legacy passive optical networks. There are several migration scenarios to meet disparate service provider`s needs. Two of them – service-oriented and service-independent scenarios – reflect recognition that differing service introduction strategies might affect requirements for the NG-PON specifications.

NG-PON technologies can be divided into two categories. The NG-PON1 presents an evolutionary growth with supporting the coexistence with the GPON on the same ODN. The coexistence feature enables seamless upgrade of individual customers on active optical fibers without disrupting services of other customers. The NG-PON2 presents a revolutionary change with no requirement in terms of coexistence with the GPON on the same ODN (Kani et al., 2009; Effenberger et. al., 2009). From another view, there are several possible architectures that could meet the NG-PON requirements. The 10G-GPON system is referred to as the XG-PON, where the Roman numeral X signifies 10 Gbit/s transmission speed, and we can expect its various versions – the XG-PON1, the XG-PON2, the extended reach and wavelength controlled XG-PON, the hybrid DWDM/XG-PON (Effenberger et. al., 2009; Tanaka, 2010).

The largest requirement for NG-PON networks is its coexistence with the operational GPON on the same ODN infrastructure. This presents a challenge due to multiplexing new systems with old ones. For this purpose, we can use both WDM and/or TDM techniques. Each of these methods will have different requirements for the wavelength plan (Effenberger et. al., 2009), (Róka, 2014):

- The WDM technique in both directions,
- The WDM downstream, the TDMA upstream,
- The TDM downstream, the TDMA upstream.

In present days, PON networks with the TDM multiplexing technique are creating in many countries. In the near future, we can expect NG-PON technologies with different motivations for developing of hybrid passive optical networks (Róka, 2014), (Róka, 2015):

1. Creating the new PON network overcoming TDM-PON network possibilities with minimum financial costs. In this case, various optical resources are utilized, but it isn't the full-value WDM-PON network, and the TDM approach is still utilized from various reasons. This HPON network is not very expensive created (in a comparison with the WDM-PON network) and provides a sufficient transmission capacity for customer needs in a long-time horizon.

This variation can be included in the NG-PON1 category, where the WDM filter is installed to combine and separate GPON and XG-PON1 signals into and out of the common ODN infrastructure. The ODN, i.e. optical fibers and the remote node, is not replaced or changed during the migration to the NG-PON1 network.

2. Preparing the transition from the TDM-PON to the WDM-PON network with minimum costs for rebuilding of the existing TDM-PON infrastructure. Such HPON network should satisfy following features:
   a. A backward compatibility with the original TDM-PON architecture and a coexistence of TDM and WDM approaches,
   b. A maximum exploitation of the existing optical infrastructure, optical fibers and optical equipment,
   c. New bonus functions for the network protection and fast traffic restoration in a case of failures.

This variation can be included in the NG-PON2 category, where separate optical fibers and power splitters may be used. Also, a different device replacing the simple power splitter may be used (Kani et. al., 2009). In a case of the extended reach and wavelength controlled NG-PON version, the basic feature is a more controlled ONT wavelength. If we use a wavelength-controlled ONT, then the bandwidth of optical amplifiers can be reduced to about 0,5 nm and this reduction allows a sensitivity to be improved by many dB. By this way, a wavelength-controlled ONT also opens possibility for WDM-based multiplexing upgrades in a future. In a case of the hybrid DWDM/XG-PON network, multiple NG-PON are combined with using a DWDM MUX/DEMUX and a colorless technology for the ONT equipment. Key components for the new system are colorless transmitters (the seed light injected reflective semiconductor optical amplifiers and tunable laser diodes) and WDM filters.

For new deployments (greenfield) (Figure 3a), the impact on existing customers will not be an issue and testing the ODN infrastructure during the construction stage will be very similar to what is being

Figure 3. Greenfield and brownfield deployments for NG-PON networks

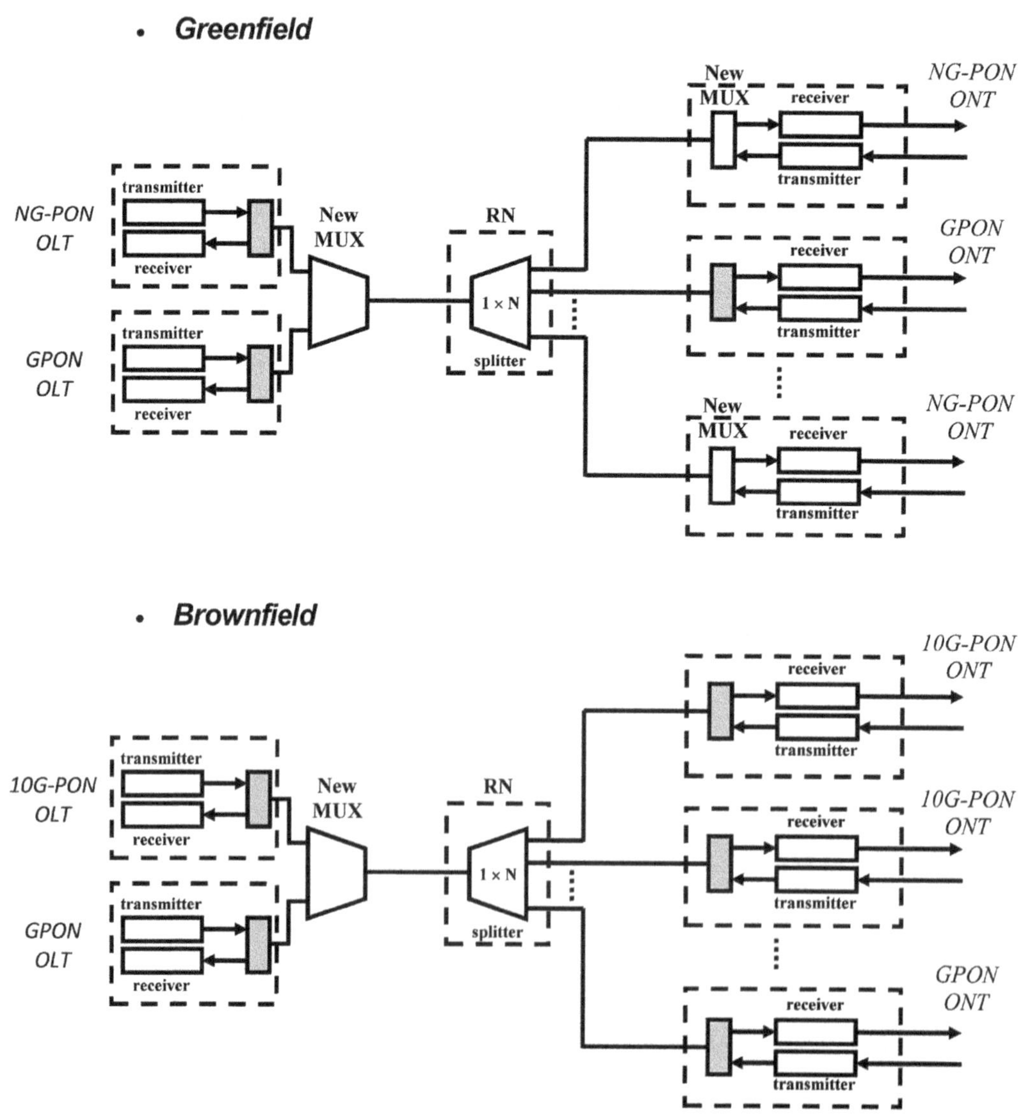

done for current PON networks. However, in order to conduct proper testing during the activation and maintenance phases, new measurement instruments will be required, as current tools must be adapted to new requirements.

In brownfield deployments (Figure 3b) where the next-generation PON technology will coexist with current technology, customer will already be connected to the network, so special care will have to be taken to minimize the effects on subscriber services. This type of deployment (brownfield) will require the installation of new optical components (such as WDM filters) to combine two technologies, and adding these components could temporarily interrupt all services. Furthermore, adding new optical devices in the optical distribution network will add extra loss to the overall loss budget, which could affect existing customers if the previous budget was not sufficient enough to compensate for extra loss in the network. In addition to the filters, next-generation ONT units will also have to be deployed, and power measurements should be taken to ensure that each of these ONT terminals will receive enough power to respect requirements by the different standards (EXFO, 2012).

Bandwidth demands continue to increase as subscribers are constantly adopting new applications and services. Service providers are looking into ways to make their optical networks – long-haul, metropolitan or access – faster. Key drivers and applications for deploying next-generation optical access networks can be easily found anywhere and there can be found many examples in both the business and residential markets. Among all the technologies that could possibly allow service providers to increase the bandwidth per user, two currently stand out to become the technology of choice for next-generation PON networks – NG-PON1 (10G-GPON, 10G-EPON) and NG-PON2 (WDM-PON).

Besides the fact that these technologies can offer higher bandwidth per user, one of the main reasons that these options are ahead of the others is that they are based on PON networks, so service providers who have already deployed FTTx architectures will be able to re-use the same optical distribution network and therefore protect their investment. One interesting characteristic of 10G-GPON and 10G-EPON systems is allowing for concurrent operation with current PON technology (Table 1). Both PON networks are based on the existing PON technologies and are upgraded for 10 Gbit/s transmission rates by WDMA or TDMA technologies. The future NG-PON2 network is expected for coexisting these PON networks (Nakamura, 2013).

The time and wavelength division multiplexing PON technology is a primary system for the NG-PON2 network (Nakamura, 2013). Table 2 describes a few of the main characteristics defining this TWDM-PON technology.

Different TWDM-PON flavors with varying remote node architectures are proposed. Typically, there are three variants of the TWDM-PON architecture according to the RN configuration (Dixit, 2012):

Table 1. Technical specifications for 10 Gbit/s PON systems

Item	10G-EPON (10G Symmetric)	XG-PON
Line rate	Upstream: 10,3125 Gbit/s Downstream: 10,3125 Gbit/s	Upstream: 2,488 Gbit/s Downstream: 9,953 Gbit/s
Transmission bandwidth	10 Gbit/s for upstream and downstream (64B66B coding)	Same as the line rate (scrambled NRZ coding)
Maximum loss	20 / 24 / 29 dB	29 / 31 dB
Services	Ethernet data	Full services (Ethernet, TDM, POTS)

Table 2. Specifications for the TWDM-PON system

Item	
Bandwidth	Upstream: 2,5-10 Gbit/s/channel
	Downstream: 2,5-10 Gbit/s/channel
Wavelength channels	4 - 8
Physical splitting ratio	64
Transmission distance	40 km
Co-existence systems	Legacy PON systems RF-video system

- A wavelength selected TWDM-PON network with power splitters,
- A wavelength split TWDM-PON network with arrayed waveguide gratings,
- A wavelength switched TWDM-PON network with wavelength selective switches.

The wavelength selected variant is a fully flexible solution as the power splitter broadcasts all wavelengths to all users, but has a high insertion loss and poor security due to the use of only optical power splitting. The wavelength split variant has a fixed wavelength allocation at the arrayed waveguide gratings and thus will not be able to serve flexibility advantages. However, the full flexibility is not always required and a partial flexible solution can already give several advantages of flexibility. Overcoming the drawbacks of both previous variants, a partially flexible solution with the wavelength selective switch can be proposed as the wavelength switched variant.

The WDM-overlay application is requiring high capacity and low latency. In this case, the ONT terminal uses a dedicated wavelength in both US and DS directions (Nakamura, 2013). Table 3 describes a few of the main characteristics defining this technology.

Table 3. Specifications for the WDM-overlay system

Item	
Bandwidth	Upstream: 1-10 Gbit/s/channel
	Downstream: 1-10 Gbit/s/channel
Wavelength channels	to be defined
Physical splitting ratio	to be defined
Transmission distance	to be defined
Co-existence systems	Legacy PON systems RF-video system

REQUIREMENTS FOR THE HPON NETWORK

Features of the HPON Network

One of the most promising candidates for the NG-PON architecture is a hybrid WDM/TDM passive optical network. The WDM technique increases the capacity using an additional wavelength layer, while the TDM technique improves the scalability and leads to flexible utilization. A cost-efficient architecture for the HPON network can support different levels of the traffic protection and restoration in order to satisfy availability requirements of both residential and business users (Róka, 2003).

The hybrid passive optical network (HPON) is a hybrid in a way that utilizes both TDM and WDM multiplexing principles together on a physical layer. The HPON network utilizes similar or soft revised topologies as TDM-PON architectures. For downstream and upstream transmissions, TDM and WDM approaches are properly combined, i.e. it is possible to utilize the time-division or wavelength-division multiplexing of transmission channels in the common passive optical architecture.

The first HPON network design is based on principles of the evolutionary architecture from the TDM-PON network utilizing few WDM components (Figure 4). A basic architecture of the optical distribution network that distributes signals to users consists of the one-fiber topology and several topological links connected to the optical network terminals through the remote node. Logically, a connection of the point-to-point type is created between the optical line terminal and the remote node. The remote node consists of either a passive optical power splitter (TDM-PON) or an arrayed waveguide grating (WDM-PON). As resources of optical radiations, two different types of tunable lasers based on dense or coarse wavelength multiplexing for various wavelength areas can be utilized in order to decrease a number of necessary optical sources. A number of OLT tunable lasers is smaller than a number of transmission channels utilized in a network; therefore various subscribers can dynamically share tunable lasers. Of course, there are other possible HPON architectures that can be also included in the HPON Network Simulator.

Figure 4. The evolution of the WDM/TDM-PON network architecture

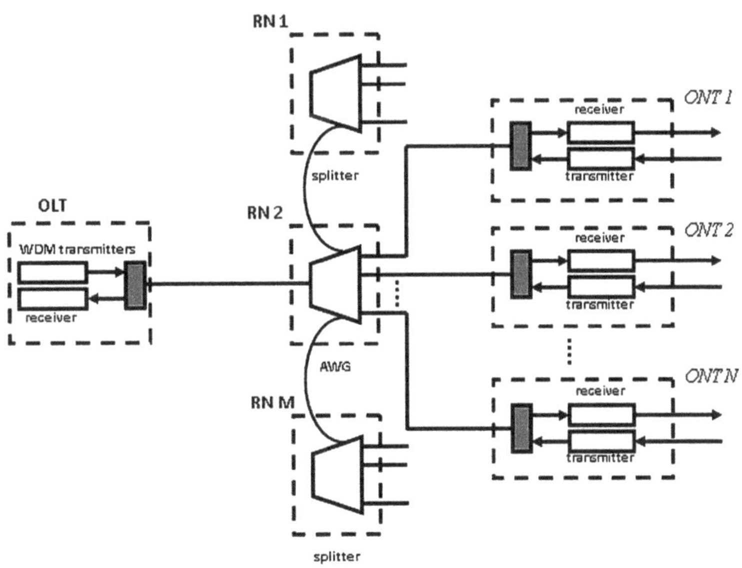

Except the original WDM/TDM-PON network design (Róka, 2015), we can propose other variations of hybrid passive optical networks. In the second design, besides utilizing few WDM components in the original TDM-PON infrastructure, changes of the network topology can be considered with modifications of WDM techniques utilizing in both transmission directions (SUCCESS HPON). In this case, there exists a backward compatibility with previous TDM-PON networks, however, an exchange of TDM ONT equipment is necessary. In the third design, an integration of metropolitan and access networks is included into utilizing WDM technologies (SARDANA HPON). In this case, a combination of the WDM distribution network ring with add/drop nodes and TDM access subnetwork trees seems to be very profitable for the implementation. In the fourth design, active components are utilized in an outside plant and therefore a network reach can be markedly extended (LR-PON).

Although a TDM-PON technology minimizes a number of required optical components using one shared transceiver unit at the central office, it does this at a performance penalty. In a TDM-PON network, the power splitter in a remote node prevents an OTDR from separating and identifying the multiple Rayleigh backscatter signals from each of its distribution fibers. These problems are not present in a WDM-PON network. Since a wavelength splitter is used in place of the power splitter, the splitting loss can be very small. In addition, because a WDM technology provides a point-to-point optical connection, the receiver noise penalty arising in a TDM-PON network does not exist in a WDM-PON network since the bandwidth of each ONT receiver is matched to its data rate. Another relatively important advantage is the ability to completely characterize all the optical fiber paths in a WDM-PON network from the central office. This is possible since at each wavelength a single optical path exists. The GigaWaM project aims to develop essential optical subsystem components required for a future-proof WDM-PON broadband access system providing each end user with up to 1 Gbit/s bidirectional data bandwidth (Prince, 2012).

The WDM-PON technology offers the most straightforward way of capacity increase. It minimizes needed for the TDMA protocol and has a potentially high reach and security. However, the WDM-PON system suffers from some major issues like migration from presently deployed TDM-PON network, low customer fan out and static resource allocation. A new architecture of the wavelength switched TWDM-PON network in which one WSS switch feeds wavelengths to multiple AWG elements is proposed. Hybrid TDMA/WDM-PON (TWDM-PON) network combines the flexibility of the TDM-PON technology with the increased overall capacity of the WDM technology, and it is an important NG-PON candidate. Its advantages are its high fan out, easy migration capabilities and ability to provide higher peak data rates. The most important advantages can be found in network planning, network migration, network extensibility and energy efficiency (Dixit, 2012).

Hybrid Networks for a Smooth Transition

Hybrid networks for a smooth transition from TDM-PON to WDM-PON networks allow a possibility for simultaneous provisioning of both TDM and WDM services. However, new WDM subscribers can be added by exchanging TDM ONT equipment. Configuration changes are controlled from the OLT and no other interventions are necessary on the original TDM subscribers' side. There can be considered various architectures (Róka, 2014):

- The self-renewable HPON with a tree topology,
- The HPON with a video overlay,
- The SUCCESS HPON.

Hybrid Networks With an Integration

Hybrid networks with an integration of the WDM technology with present TDM networks bring an increasing the number of subscribers, transmission rates and network reaches. By this way, a network design can concatenate and combine functionalities of metropolitan and access networks. This integration utilizes the common central office equipment and combines WDM distribution networks with a bidirectional ring topology and OADM nodes that are connecting to TDM access subnetworks. As an example, the SARDANA HPON can be presented (Róka, 2014).

Hybrid Networks for Long Distances

As technologies advance, long-reach access networks have become a reality. Further extended systems enable a number of existing PON networks to be grouped and converted into long-reach systems with DWDM backhaul networks. An attention is focused on the TDM-based long-reach PON network, e.g. SuperPON. Recently, long-reach PON networks with the WDM technology are investigated and hybrid DWDM/TDM long-reach PON systems are proposed (Talli, 2007), (Smolorz, 2009), (Róka, 2014)

- SuperPON,
- LR-PON,
- Hybrid DWDM/TDM Long reach PON,
- Photonic integrated metro and access network PIEMAN.

Traffic Protection in Hybrid HPON Networks

HPON architectures can be proposed with and without protection. They are designed having in mind different possible paths for the network deployment and protection upgrade. The proposed survivable architectures can also be applied to passive optical networks with more than one stage of remote nodes based on power splitters. The hybrid passive optical network can be proposed with different levels of the traffic protection compatible with all the HPON architectures (Róka, 2015), (Róka, 2016):

- **No Protection:** The basic scheme without any protection is modified compared to the conventional deployment. This approach takes into account facilitating an easy and cost-efficient migration toward a reliable next-generation optical access network. For future protection upgrades, a limited number of output power splitter ports is reserved for the end-to-end protection of some selected customers. The main goal is to minimize the amount of new fiber paths required to provide a full protection for some selected users, which could be much more costly than additional costs caused by the increased number of passive components.
- **Protection Up to First Remote Node:** Compared to the unprotected scheme, the extra infrastructure and equipment is needed for protection. The 1:N components in the remote node are replaces by the 2:N ones in order to provide a connection to both the working and protection (backup) feeder fibers. Extra costs of the duplicated resources are shared by all the users connected to each PON network. This approach comes out from previous when duplicated devices are permanently utilized.

- **End-to-End Protection:** Two parallel distribution networks act as a potential backup for each other, which will considerably reduce the need for additional trenching for the traffic protection. To obtain the end-to-end protection, duplicated transceivers are considered at the ONT side to access both working and protection last mile fibers. Furthermore, a new fiber connection for the protected ONT should be added. To decrease the new trenching required for the traffic protection, the available ducts should be utilized.

It can be shown a clear benefit when a network planning is done with possible protection upgraded, which leads to a decrease in investment costs. The longer the protection deployment time, the higher the total capital expenditures. This confirms an importance of the right deployment plan for future passive optical networks (Mahloo, 2014).

Fi-Wi Convergence in Hybrid HPON Networks

The integration of optical and wireless technologies into one broadband access network is marked as the fiber over wireless (Fi-Wi). A lot of wireless-optical scenarios can be divided in two main categories from a viewpoint of the integration:

- **The RoF Architecture:** A radio signal transmission over optical fibers - radio over fiber (RoF) – is focused on integrating the physical layer mainly. Radio frequencies are transmitted through an analog interconnection of optical fibers between the central office and remote antenna units that are used by end-users for data transmitting and receiving. The RoF primary aim is related to modulation and transmission functions. A functionality and a complexity of subscribers units is low because of signal processing functions and resource allocation rules are executed in the central office. Therefore, a costs and power saving is present in subscriber units and by using a distribution MAC protocol IEEE 802.11b in a wireless part, power limitations are arising. However, a centralized RoF function can bring on a troubleshooting point in the hybrid FiWI network because a generated traffic from end-users in a wireless part is directed into an optical part. In a case of the central office failure, this can lead to an enormous impact to total network performance.
- **The R&F Architecture:** Hybrid optical-wireless architectures are focused on integrating two layers – physical and data link. The radio and fiber (R&F) design incorporate self-employed optical and wireless networks that together create one hybrid optical-wireless network and specific features of present networks are inviolate. The R&F structure is able solving limitations of the RoF architecture by using various data link protocols in both areas – optical and wireless - simultaneously. Generally, a realization of the R&F architecture can be executed in two different platforms based on orientation in service provisioning.
    - The first R&F architecture is a hybrid wireless optical access network. In this case, hybrid network architectures are utilizing one wireless base station (BS) directly connected to the optical network by means of optical network terminal (ONT). A typical example is the integration of different PON structures with broadband WiMAX, LTE and/or Wi-Fi wireless technologies.
    - The second R&F architecture is called hybrid wireless optical mesh networks that can broaden a network range by various equivalent wireless nodes of the Ad Hoc network. The most representative method is the wireless-optical broadband access network (WOBAN).

THE HPON NETWORK CONFIGURATOR

The HPON Network Configurator allows comparing possibilities of various passive optical access networks. The created HPON Network Configurator (Róka, 2013), (Róka, 2015) represents real possibilities for a consistent transition from the TDM-PON to the HPON based on various specific parameters – a network capacity from a viewpoint of the physical layer, a number of TDM and WDM network subscribers, a number of exploited wavelength multiplexing types, a growth of the channel capacity by connecting of new subscribers and other feasibilities.

For better understanding, the HPON NetCon manual is inserted into the lower part of the main dialogue window. The manual consists of three simple steps that can help users to direct toward to achieving desired results. In substance, the manual describes three parts of the program:

- Parameters of the optical fiber,
    - A type of the optical fiber,
    - A DWDM multiplexing density,
    - An availability of CWDM and/or DWDM wavelength carriers,
- Parameters of the TDM-PON network infrastructure,
    - A number of TDM networks,
    - A type of the network,
    - A number of subscribers,
    - A OLT-ONT distance,
- Topology selections including the traffic protection,
    - A type of the network topology,
    - A specified HPON network – WDM/TDM-PN, SUCCESS HPON, SARDANA HPON, LR-PON,
- Expected parameters of the selected HPON network:
    - A total network capacity,
    - A total number of subscribers,
    - Power relationships (a transmitting power of light sources, a sensitivity of receivers, an attenuation of the transmission path and particular optical components,
- Traffic protection and restoration schemes.

This network simulator has main dialogue window for designing a transition from original and deployed TDM-PON to new and developing HPON networks. Additional dialogue windows with basic network schemes and short interactive descriptions serve for the specific HPON configuration setup.

Based on parameters of the optical transmission environment (attenuation values, wavelengths), possible CWDM/DWDM carriers are calculated. A number of new DWDM subscribers and a required capacity per one subscriber are inserted. Based on primary input values of the deployed TDM-PON network (a number of TDM nodes, a capacity of the node, a number of subscribers per node), a total capacity of the deployed hybrid network and an average capacity per one subscriber are calculated (only for the downstream direction). Also, a growth of the capacity due to connecting of new subscribers, a number of necessary DWDM transmitters and receivers, a number of utilized CWDM and DWDM wavelengths are evaluated.

The Configuration of the HPON Environment: Its Capabilities and Parameters

The HPON Network Configurator for comparing possibilities of various passive optical access networks is created by using the Microsoft Visual Studio software working with the C++ programming language including the Microsoft foundation class (MFC) library for the graphical interface. The simulation model has one main interactive dialogue window (Figure 5) for inserting and presenting parameters of transitions from TDM-PON to HPON networks. It allows comparing and analyzing four principal approaches for designing and configuring of hybrid passive optical networks. Therefore, additional dialogue windows with configuration and relation of basic network infrastructures can be used for the specific HPON configuration setup. For WDM/TDM-PON and SUCCESS HPON networks, transitions from the original TDM-PON architecture are expressed by GIF animations. For their presentation, a free available CPictureEx class is used. For SARDANA HPON and LR-PON networks, features of networks are presented in a simple picture.

The HPON Network Configurator is working in several steps following algorithm:

1. **Setting Parameters for the Optical Fiber:** A type of the optical fiber (according to the ITU-T standards), the DWDM multiplexing density.
2. **Evaluating Optical Fibers and Wavelengths:** Standard or inserted specific attenuation values in dB/km, a calculation of numbers of utilizable CWDM and DWDM carriers.
3. **Inserting Input Parameters for the Deployed TDM-PON Network:** A number of original TDM networks, a type of the network, a number of subscribers per one network, a distance between the OLT and ONT terminals.
4. **Evaluating TDM-PON Network:** A calculation for the total transmission capacity of the TDM network together with the average capacity per one subscriber, the total number of subscribers and the maximum attenuation of the TDM network; finally, the attenuation class is presented.

Figure 5. The main window of the HPON network configurator

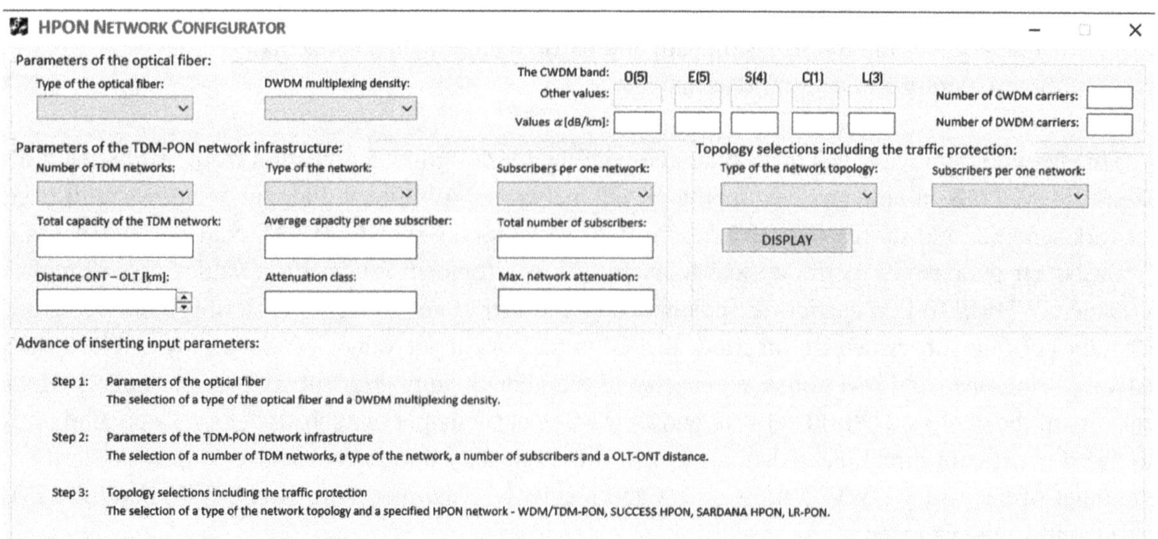

When one of higher splitting ratios of subscribers per network is selected or when a total network attenuation exceeds defined attenuation classes, options for configuration of other hybrid passive optical networks - WDM/TDM-PON, SUCCESS HPON and SARDANA HPON - are automatically deactivated because they don't support this selected splitting ratio. For this case, an option for the Long Reach PON configuration is only active. This step is terminating with the selection of detailed hybrid PON configuration design.

5. **Setting Input Parameters for the Hybrid PON Configuration:** Based on the stored TDM-PON network data and selecting one from possible HPON types.
6. Application input parameters and specific parameters of the HPON network configuration (the total capacity of the hybrid network, the total number of subscribers, the average capacity per one subscriber, the maximum attenuation of the hybrid network between the OLT and ONT terminals, a number and type of used active and passive components) with summing up a type and number of deployed optical components and presenting possibilities for future expanding of hybrid HPON network types is executed.

Thereafter, changes of the topology and a subsequent transition to the selected HPON network are presented to the user. By this way, a user can take a decision to regulate or to exchange primary input values for adapting to traffic demands of the specific HPON network. Thereafter, further possible network extensions are displayed.

Parameters of the Optical Fiber

In the first step, a selection of the optical fiber's type and the DWDM multiplexing density can be executed. A selected type of the optical fiber is presented by the specific attenuation values and by a number of transmission bands. For specifications, various ITU-T recommendations – ITU-T G.652A, G.652B, G.652C, G.652D, G.656, G.657 – together with the "Other values" option can be inserted. Then, specific attenuation coefficients are concretely displayed. Also, a total number of CWDM and DWDM carrier wavelengths for particular bands are presented.

Also, an attenuation of optical fibers for available wavelengths is calculated in specific network configurations. In this case, we prefer attenuation coefficients for common optical fibers according to the ITU-T G.652 (ITU-T, 2009). However, we can incorporate attenuation coefficients for new types of optical fibers according to the ITU-T G.657 (ITU-T, 2009) and the ITU-T G.656 (ITU-T, 2010) and, by this way, evaluate their utilization in the HPON network infrastructure.

The G.652A fiber has a zero value of the chromatic dispersion at the 1310 nm wavelength, so it is suitable for the O band transmission. In the C-band transmission, a calculated value of the chromatic dispersion is around 17 ps/(nm.km).

The G.652B fiber supports higher transmission rates due to decreasing of the polarization mode dispersion.

The G.652C fiber has similar parameters as the G.652A fiber. Moreover, a total attenuation is affected by elimination of the OH- absorption and a transmission in all bands from O up to L. However, this fiber is not optimized to the PMD dispersion.

The G.652D fiber is an alternative with reduced values of the polarization mode dispersion as in a case of the G.652B fiber.

The G.656 fiber is characterized by nonzero values of the chromatic dispersion that allows a suppression of nonlinear effects in C, S and L wavelength bands.

The G.657 fiber has lowered attenuation parameters caused by macrobends.

Parameters of the TDM PON Network Infrastructure

In the second step, a selection and a listing of parameters for the deployed TDM-PON network can be executed. A number and type of networks (EPON, GPON, 10G-EPON, XG-PON), a number of subscribers per one network and a network reach, respectively the OLT-ONT distance (max. 999 km) can be selected. In Table 4, specific parameters of selected PON networks are presented. Then, features of the selected TDM-PON network configuration – a total capacity, an average capacity per one subscriber, a total number of subscribers, the maximum network attenuation and the attenuation class – can be calculated.

In a case that the maximum network attenuation exceeds a specified value depending on parameters of the optical fiber, deployed TDM network types, a number of subscribers and on a distance between OLT and ONT terminals, options of WDM/TDM-PON, SUCCESS HPON, SARDANA HPON network

Table 4. Specific parameters of selected PON networks

	EPON	10G-EPON	GPON	XG-PON
Recommendation	IEEE 802.3ah	IEEE 802.3av	ITU-T G.984	ITU-T G.987
Variations of transmission rates	1G/1G symmetric	10G/10G symmetric	1,25G/1,25G symmetric	10G/2,5G asymmetric
		10G/1G asymmetric	2,5G/1,25G asymmetric	
			2,5G/2,5G symmetric	
Transmission rates of the physical layer	1,25Gbit/s	10,3125Gbit/s	2,48832Gbit/s	9,95328Gbit/s
		1,25Gbit/s	1,24416Gbit/s	2,48832Gbit/s
Attenuation classes	PX10 PX20	PR10, PRX10 PR20, PRX20 PR30, PRX30	A, B, B+, C	Nominal 1 and 2 Extended 1 & 2
Wavelengths [nm]	DS 1480-1500 US 1260-1360	DS 1575-1580 US 1260-1280 or 1260-1360	DS 1480-1500 US 1260-1360 or 1290-1330	DS 1575-1580 US 1260-1280
Reach [km]	< 10, < 20	< 10, < 20	< 20	< 20, < 40
Max. splitting ratio	1:16, 1:32	1:16, 1:32 (also 1:64, 1:128)	1:64, (proprietary 1:128)	1:256

Table 5. Attenuation classes for EPON and GPON technologies

	EPON		GPON		
	PX 10	PX 20	A	B	C
DS Wavelength [nm]	1490		1480-1500		
US Wavelength [nm]	1300		1260-1360		
Max. reach [km]	10	20	up to 20		
Max. achievable attenuation [dB]	20	24	20	25	30
Min. achievable attenuation [dB]	5	10	5	10	15
Splitting ratio	1:16		up to 1:64		

configurations are turned off and a challenge for utilization of the Long Reach PON network is appearing. For selected PON network technologies - EPON and GPON (Table 5), the 10G-EPON (Table 7) and XG-PON (Table 8) technologies, concrete values of the attenuation classes are presented. In the HPON Network Configurator, there are included intervals of values for specific attenuation classes determined in Tables 6 and 9.

Table 6. Attenuation intervals for EPON and GPON technologies

EPON		GPON	
Attenuation Values [dB]	**Attenuation Class**	**Attenuation Values [dB]**	**Attenuation Class**
< 5, 10)	PX 10	< 5, 10)	A
< 10, 21)	PX 10/20	< 10, 15)	A/B
< 21, 25)	PX 20	< 15, 21)	A/B/C
> 25	invalid	< 21, 26)	B/C
		< 26, 30)	C
	-	> 30	invalid

Table 7. Attenuation classes for the 10G-EPON technology

	Low Attenuation Class		Medium Attenuation Class		High Attenuation Class	
	PRX 10	**PR10**	**PRX 20**	**PR 20**	**PRX 30**	**PR 30**
DS Wavelength [nm]	1577 - 2, 1577 + 3					
US Wavelength [nm]	1310 ± 50	1270 ± 10	1310 ± 50	1270 ± 10	1310 ± 50	1270 ± 10
Max. reach [km]	> 10		> 20		> 20	
Max. attenuation [dB]	20		24		29	
Min. attenuation [dB]	5		10		15	
Splitting ratio 10 km	1:16		1:16		-	
Splitting ratio 20 km	-		1:32		1:32	

Table 8. Attenuation classes for the XG-PON technology

	Nominal Attenuation Class			Extended Attenuation Class		
	N1	**N2A**	**N2B**	**E1**	**E2A**	**E2B**
DS Wavelength [nm]	1575 - 1580					
US Wavelength [nm]	1260 - 1280					
Max. reach [km]	> 20					
Max. attenuation [dB]	14-29	16-31	31	18-33	20-35	35
Splitting ratio 20 km	1:256					

Table 9. Attenuation intervals for 10G-EPON and XG-PON technologies

10G-EPON		XG-PON	
Attenuation Values [dB]	**Attenuation Class**	**Attenuation Values [dB]**	**Attenuation Class**
< 5, 10)	PR 10	< 14, 16)	N1
< 10, 15)	PR 10/20	< 16, 18)	N1/N2
< 15, 21)	PR 10/20/30	< 18, 20)	N1/N2/E1
< 21, 25)	PR 20/30	< 20, 29 >	N1/N2/E1/E2
< 25, 29)	PR 30	(29, 31 >	N2/E1/E2
> 29	invalid	(31, 33 >	E1/E2
		(33, 35 >	E2
		> 35	invalid

Attenuation classes for the 10G-EPON technology can be divided into 2 subclasses:

- PRX for the asymmetric transmission with the 1 Gbit/s upstream transmission rate and the 10 Gbit/s downstream transmission rate,
- PR for the symmetric transmission with the 10 Gbit/s transmission rates in both directions.

For the XG-PON technology, there will be two loss budgets denoted Nominal and Extended. The Nominal loss budget is defined with a Class B+ loss budget plus an insertion loss from a WDM1 filter. The link loss will be approximately 28,5 dB to 31 dB at BER = 10-12. The Extended loss budget is defined with a Class C+ loss budget plus an insertion loss from a WDM1 filter (Effenberger, 2009).

The Configuration of Selected Network Topologies and Traffic Protection Schemes

In the third step, hybrid network topologies including the traffic protection are reserved. By using the DISPLAY push button in the main dialogue window, autonomous dialogue windows for the specific hybrid network topology configuration are opened. Then, a configuration of specific network parameters can be proceeding for the WDM/TDM-PON, SUCCESS HPON, SARDANA HPON and LR-PON options. If necessary, RELATION push buttons are prepared for users with short descriptions of basic features for the selected HPON network. Finally, a comprehensive list of basic characteristics – a total capacity of the hybrid network, a total number of subscribers, an average capacity per one subscriber, the maximum network attenuation – calculated for each option are displayed.

As an advanced extension, a new functionality is focused on the traffic protection and restoration schemes. There are two basic topologies considered for of the protection schemes - the tree topology and the ring topology.

Based on selected HPON network configuration, a type and a number of deployed optical components are summarized. For this possibility, allowable optical components are associated according to their location, specifically in OLT and ONT terminals or in the ODN network.

Simultaneously, possibilities for future expansions of the specific hybrid passive optical network are presented with a subsequent transition to the appropriate HPON network architecture. Except equipment enhancement, changes in the network topology are offered to the user.

Applications of the HPON Selection and the Traffic Protection

Providing a certain level of the traffic protection is necessary for next-generation hybrid passive optical networks. However, network providers might prefer to start with the deployment of an unprotected network but considering the future upgrade. Different protection approaches can be proposed, starting from a no-protection scenario towards a proposed architecture with the end-to-end protection:

- First, providers will deploy an unprotected access network. Then, providers offer a traffic protection to the first remote node. This level of protection is necessary in order to prevent a large number of customers being out of service at the same time. Finally, an end-to-end protection is offered on a per-user basis as soon as a business customer requests a reliability performance improvement.
- Second, the transition from a no-protection scheme in the access network directly to a protection to first remote node is provided. The possibility for an end-to-end protection for business users is added in the future if required. This approach is more logical if a provider has the monopoly, and every user in the region has to switch to its network eventually.
- Third, it might be more beneficial to deploy a reliable HPON network in a single step, from a no-protection scheme directly toward an end-to-end protection in the access network. This approach is realizable, if operators know in advance the location and number of all residential and business users in their networks.

In the HPON Network Configurator, a selection of traffic protection types is added as the extension of original configuration windows. In the first case, the WDM/TDM-PON configuration window with extended traffic protections (Figure 6) is opened. After inserting requested parameters of the hybrid WDM/TDM-PON network and also a number of TDM subnetworks, the Apply button can be pushed. Then, a complete set introducing type and number of optical components is introduced together with their basic characteristics.

If NO PROTECTION option is selected, then a reasonable notice in an empty bottom panel is displayed. Also, if no protection scheme is required, then the appropriate box must be confirmed and the single WDM/TDM-PON configuration is no influenced. In any type of the traffic protection is selected, then appropriate calculations and notices in the main dialogue window are changing.

Specially, a condition for the 1:N traffic protection is implemented. This traffic protection type is active only if a number of WDM/TDM-PON networks is more than 2. The considered condition coordinates a mutual relationship between a number of networks and the selected traffic protection type.

The comparison of traffic protection types in WDM/TDM–PON and LR-PON networks is based on following input parameters and particular specifications:

- **In the Main Window:** G.652 D optical fiber, 0,4 nm (50 GHz) DWDM channel spacing, 10 G–EPON 10 Gbit/s technology, 2 TDM networks, 16 subscribers per network, 20 km OLT-ONT distance

Figure 6. The WDM/TDM-PON configuration window with extended traffic protections

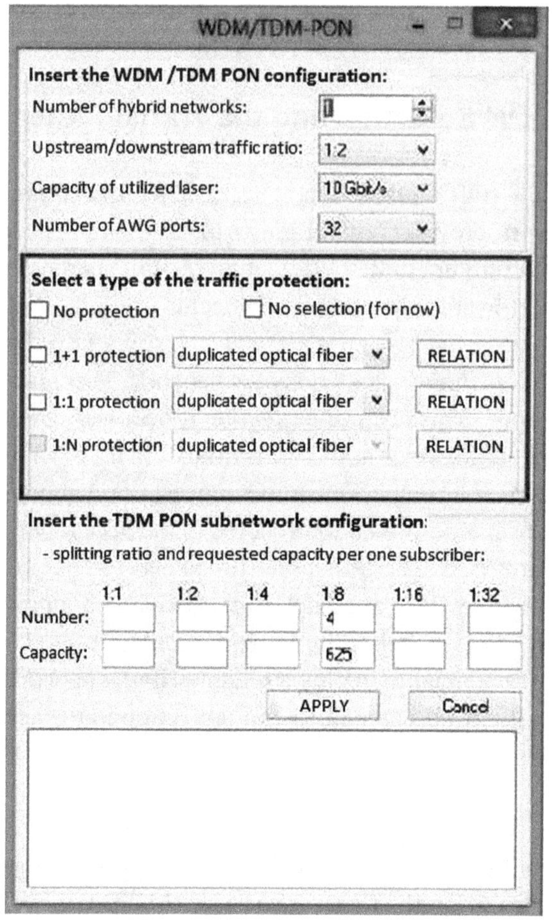

- **In the WDM/TDM-PON Configuration Window:** 2 WDM/TDM-PON networks, 1:2 upstream /downstream traffic ratio, 10 Gbit/s utilized laser rate, 32 AWG ports (a number of subnetworks), 1:8 splitting ratio, 4 TDM subnetworks, 625 Mbit/s channel capacity per subscriber

Consequently, specific parameters and network components can be introduced (Róka, 2016).

For the 1+1 protection type (a duplicated optical fiber), a number of optical components is not increasing excepting a duplication of the optical fiber. However, an increasing of the maximum network attenuation is arising. For other protection types, a number of optical components is increasing and simultaneously network costs are influenced.

For the 1+1 protection type with (duplicated network components), a number of optical components is double increasing when compared to the no-protection scenario. For the 1+1 protection type (the supplementary circuit included in the RN), a number of optical components is double increasing excepting a number of SOA/RSOA amplifiers that can be located in ONT units. For this protection type, 2 power splitters and 2 switches must be located in each RN and therefore network costs are rapidly enhanced.

For the 1:1 protection type (a duplicated optical fiber), a number of optical components is not increasing as it is present for other protection types. Moreover, a maximum network attenuation is not increased. Therefore, a utilization of this protection type has no negative impact on network costs.

For other protection types (with duplicated network components and the supplementary circuit), a number of optical components is increasing and network costs are expressively enhanced. But, more complex traffic protection and restoration is provided.

For the LR-PON network, the same conclusion is valid. A difference is in accessing to optical amplifiers utilization. Each amplifier type has a different maximum amplification level: Raman max. 25 dB, EDFA max. 35 dB and SOA max. 30 dB. Using of these active components, a maximum network attenuation can be accommodated.

For the 1:N protection type (a duplicated optical fiber), optical components are the same. However, it is unavoidable to realize a mutual network interconnection for the 1:N protection provisioning.

In the second case, the SUCCESS HPON configuration window with extended traffic protections (Figure 7) is opened. After inserting requested input parameters for TDM nodes and WDM nodes, the

Figure 7. The SUCCESS HPON configuration window with extended traffic protections

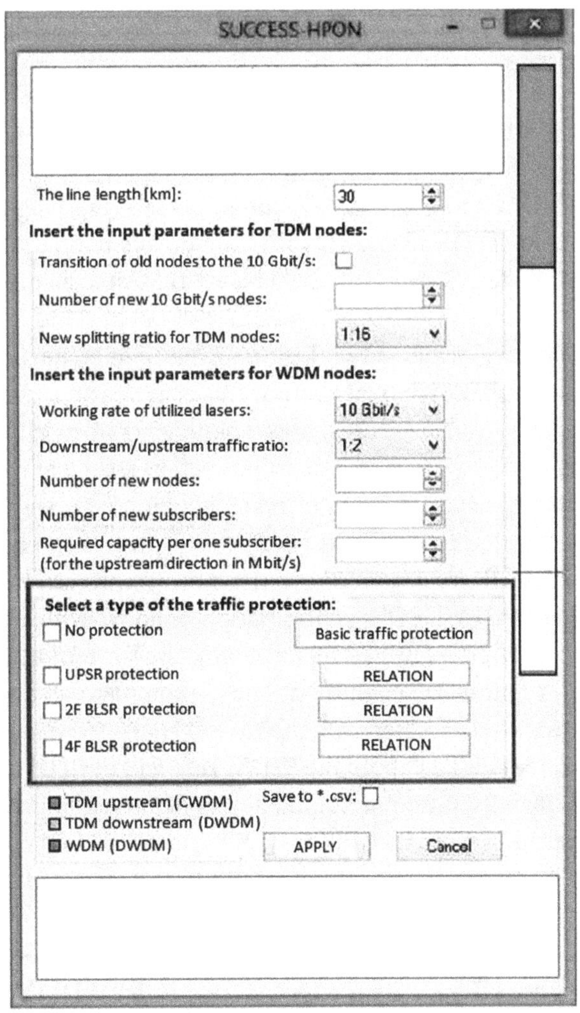

APPLY button can be pushed. Then, a complete set introducing type and number of optical components is introduced together with their basic characteristics. Moreover, a graphical presentation of the bandwidth allocation between TDM downstream, TDM upstream and WDM wavelengths is presented.

If NO PROTECTION option is selected, then a reasonable notice in an empty bottom panel is displayed. Also, if no protection scheme is required, then the appropriate box must be confirmed and the single SUCCESS HPON configuration is no influenced. In any type of the traffic protection is selected, then appropriate calculations and notices in the main dialogue window are changing.

At particular different levels of the traffic protection, there are implemented following possibilities of the selection specifications:

- A number of optical fibers,
- A utilization of the transmission capacity,
- An activity at the simple and multiple fiber interruptions or at the node failure,
- Installation and deployment costs,
- A complexity.

As in the previous case, the comparison of traffic protection types in the SUCCESS HPON network is based on following input parameters and particular specifications:

- **In the Main Window:** G.652 D optical fiber, 0,4 nm (50 GHz) DWDM channel spacing, 10 G–EPON 10 Gbit/s technology, 2 TDM networks, 16 subscribers per network, 20 km OLT-ONT distance
- **In the SUCCESS HPON Configuration Window:** 60 km ring length, 5 new 10 Gbit/s nodes, 1:16 new splitting ratio for TDM nodes, 1:2 upstream/downstream traffic ratio, 10 Gbit/s utilized laser rate, 8 new WDM nodes, 50 new subscribers, 1000 Mbit/s channel capacity per subscriber

Consequently, specific parameters and network components can be introduced (Róka, 2016).

For all protection types, a number of optical components is increasing. This leads to a more complex protection and restoration against traffic failures; however network costs are significantly influenced.

The 4F-BLSR protection type provides the highest protection level from a viewpoint of the physical layer in the ring topology. In this case, a number of optical components is quadruple comparing to no-protection network traffic scenario. For 2F-BLSR and UPSR protection types, a number of components is the same; however the 2F-BLSR provides more comprehensive protection provisioning.

In the third case, the SARDANA HPON configuration window with extended traffic protections (Figure 8) is opened. After inserting requested input parameters for remote nodes, TDM trees and the WDM ring length, the APPLY button can be pushed. Then, a complete set introducing type and number of optical components is introduced together with their basic characteristics.

If the complex solution is required, i.e. both the WDM ring and the TDM tree are requested for traffic protection provisioning, than 24 different combinations of protection specifications are available. For each the WDM ring protection type (UPSR, 2F-BLSR and 4F-BLSR) considered, a selection from TDM tree protection types (1+1 and 1:1) with four possibilities of selected specifications varying by a number of utilized optical components can be executed.

If only the WDM ring part will be protected and the TDM tree part remains without protection, than three protection possibilities are offered. In an opposite case if the WDM ring part remains without a protection and only the TDM tree part will be protected, than eight protection possibilities are offered. Of course, a possibility for unprotected WDM ring and TDM tree parts is also included. Ultimately, 36 various combinations of traffic protection provisioning for the SARDANA HPON network are prepared.

As in previous cases, the comparison of traffic protection types in the SARDANA HPON network is based on following input parameters and particular specifications:

- **In the Main Window:** G.652 D optical fiber, 0,4 nm (50 GHz) DWDM channel spacing, 10 G–EPON 10 Gbit/s technology, 2 TDM networks, 16 subscribers per network, 20 km OLT-ONT distance
- **In the SARDANA HPON Configuration Window:** 16 connected RN, max. number of 32 TDM trees, 4 lasers with 1 Gbit/s (EPON/GPON) rate, 28 lasers with 10 Gbit/s rate, 1:64 splitting ratio, 3 km access fiber length, 17 km WDM ring length

Consequently, specific parameters and network components can be introduced (Róka, 2016).

Other Considered Parameters of HPON Networks

A Total Network Capacity

In the TDM network, a capacity is based on selected architectures and on the OLT transmitter capacity. In the WDM network, a capacity is depending on the transmitter capacity, because each subscriber has a dedicated own wavelength. Therefore, the total WDM capacity is a number of subscribers (or wavelengths) multiplied by the transmission capacity of transmitters. In the HPON network, TDM parts are unchanged and WDM tunable lasers can be shared by several subscribers, then a total network capacity is depending also on a number of utilized tunable lasers.

Figure 8. The SARDANA HPON configuration window with extended traffic protections

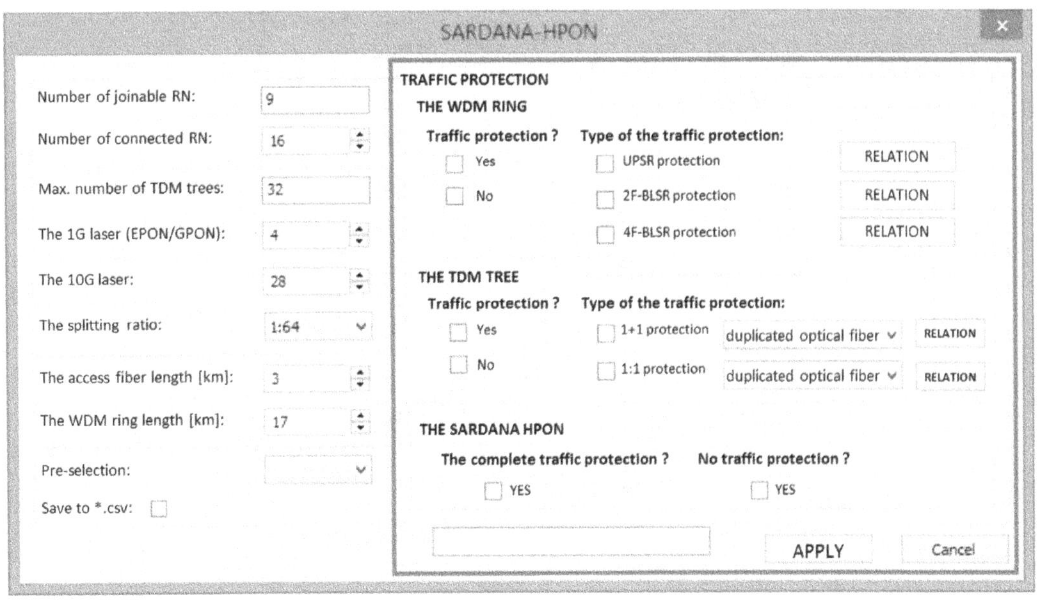

An Availability of Wavelengths

In the WDM/TDM-PON network, 2 sets of wavelengths can be utilized - for CWDM and for DWDM systems. Because these bands are overlapped, a number of available wavelengths is depending on utilized bandwidths and on a density of the wavelength allocation (100 GHz, 50 GHz or 25 GHz channel spacing). In the simulation program, a dependency between wavelengths is exactly scheduled (Róka, 2010), (Róka, 2011).

Now, it is appropriate to clarify a relationship between a number of available CWDM and DWDM wavelengths in the network (Figure 9). The less CWDM wavelengths are used, the more DWDM wavelengths can be utilized in a relevant spectrum. Simulation calculations are based on a following relationship (1):

$$\lambda_D = \frac{20 \cdot (18 - \lambda_C) \cdot 125}{\Delta\lambda} - \lambda_C \qquad (1)$$

where λ_C assigns a number of used TDM nodes, respectively CWDM wavelengths, λ_D is a maximum number of available DWDM wavelengths for WDM subscribers, $\Delta\lambda$ is the interval between DWDM channels (25, 50 or 100 GHz). This relationship describes a maximum possible number of available DWDM wavelengths that can be utilized by WDM subscribers in a dependency on a number of utilized CWDM wavelengths.

In the SUCCESS HPON network, a situation is different. In the first step of a transition to the HPON, a network topology is changing from various point-to-multipoint infrastructures to the one ring by means of the one optical fiber. Therefore, DWDM wavelengths (one wavelength per one TDM node)

Figure 9. The relationship between numbers of available wavelengths at various channel allocations

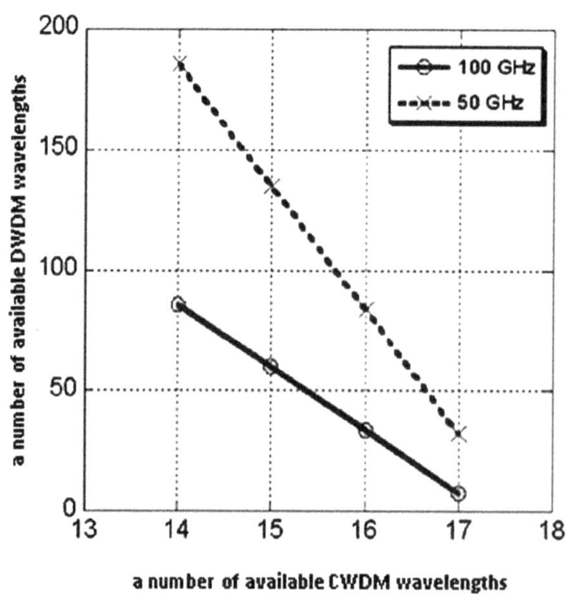

are utilizing in the downstream direction. In the upstream direction, each TDM node utilizes a different wavelength from the CWDM grid.

In the SARDANA HPON network, transmissions in both directions are limited by using of wavelengths in the C-band due to the EDFA operational range. So, the utilization of DWDM systems can be varied with different densities of the channel spacing.

In the LR-PON network, a selection from three types of optical amplifiers –EDFA, RAMAN and SOA – is possible. These optical amplifiers have different characteristics from a viewpoint of wavelengths. The EDFA amplifier is working in the C band only (and special dopped optical fibers are necessary), Raman and SOA amplifiers are working in O, E, S, C and L bands of the wavelength spectrum.

A number of CWDM and DWDM (Róka, 2003) possible wavelengths in particular transmission bands is presented in Table 10. By reducing of the DWDM channel spacing, negative influences of nonlinear effects (e.g. FWM) can be practically increased in the real optical transmission media.

Power Relationships

In the HPON Network Configurator, total power relationships are depending on specific network characteristics and applied optical component parameters. We prefer real values of optical components utilized in passive optical networks (Table 11).

A calculation of the total network attenuation is coming out from attenuation of optical components used in the remote node, a number of remote nodes and a type of coupling elements and parameters (specific attenuation coefficient and length) of the specified optical fiber. For the total network attenuation, a maximum achievable value from the network viewpoint is under consideration. The final relationship consists of two parts – the fiber attenuation together with insertion loss of the remote node and the attenuation in a remote node alone.

The Configuration of the WiMAX Access Part

At the designing of wireless access parts in the specified HPON network topology, it is possible to determine an architecture the network convergence. By using of check boxes, a selection of the architecture type is realized.

Possibilities for network convergence architectures are:

- Without utilizing microwave connections,
- An independent architecture,

Table 10. A number of DWDM and CWDM wavelengths for particular transmission bands

Bands	O 1260-1360 nm	E 1360-1460 nm	S 1460-1530 nm	C 1530-1565 nm	L 1565-1625 nm	Total Number of Possible Wavelengths
DWDM (0,8 nm, 100 GHz)	-	-	50	25	75	150
DWDM (0,4 nm, 50 GHz)	-	-	100	50	150	300
DWDM (0,2 nm, 25 GHz)	-	-	200	100	300	600
CWDM (20 nm)	5	5	4	1	3	18

Table 11. Attenuation specifications of HPON optical components

Symbol	Description	Value
a_{FIBER}	the maximum fiber attenuation in wavelength bands	ITU-T standards
a_{DWDM}	the maximum fiber attenuation in the DWDM wavelength bands	0,25 dB/km
L_{ACCESS}	the access fiber length	optional
L_{RING}	the ring length	optional
a_{FILTER}	the attenuation of the WDM filter	0,4 dB
a_{AWG}	the AWG attenuation	5 dB
$a_{50:50}$	the 50:50 power splitter attenuation	4,4 dB
$a_{90:10}$	the 90:10 power splitter attenuation	0,8:12 dB
	the 1:4 splitter attenuation	7,5 dB
	the 1:8 splitter attenuation	11 dB
	the 1:16 splitter attenuation	14,1 dB
$a_{SPLIT1:N}$	the 1:32 splitter attenuation	17,4 dB
	the 1:64 splitter attenuation	21,0 dB
	the 1:128 splitter attenuation	22,5 dB
	the 1:256 splitter attenuation	26,4 dB
a_{TDM-RN}	the TDM node attenuation (including connectors)	1,5 dB
a_{WDM-RN}	the WDM node attenuation (including connectors)	1 dB
$a_{ADD/DROP}$	the attenuation of added/dropped wavelengths	1,2 dB
$a_{ISOLATOR}$	the attenuation of the isolator	0,3 dB
a_{CON}	the connector attenuation	0,2 dB
	loss of a splice	0,15 dB
	loss of the fiber span	0,25 dB/km

- A hybrid/combined architecture,
- The MoF architecture.

If necessary, RELATION push buttons are prepared for users with short descriptions of basic features for the selected architecture of the Fi-Wi network.

In the HPON Network Configurator, a configuration of the WiMAX access part can be realized inserting main input parameters for the base station (BS):

- A channel bandwidth,
- A modulation technique,
- A SU-BS,
- A visibility.

For the channel bandwidth, possibilities 5 MHz and 10MHz are allowed corresponding to the 802.16d standard.

Figure 10. A possible configuration of the base station in the WDM/TDM-PON network

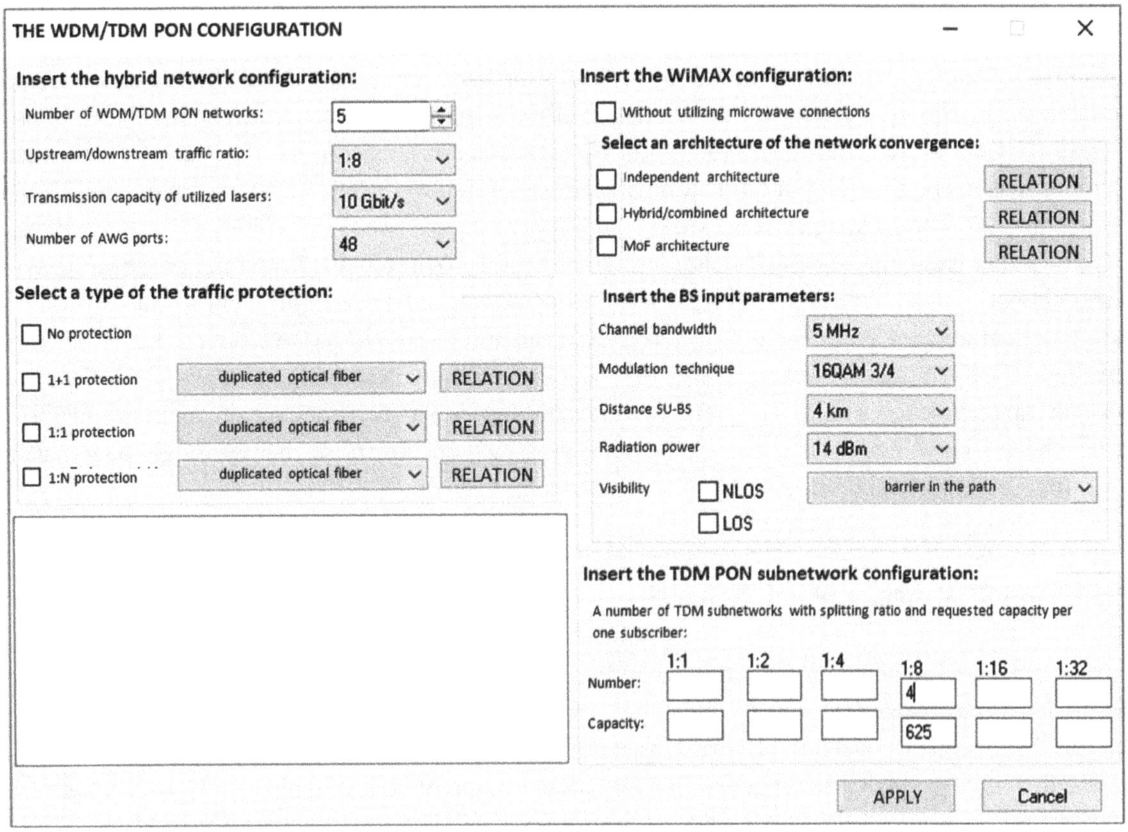

For the modulation technique, a selection from eight combinations using four modulations 2x BPSK 1/2, 2x QPSK 1/2, 2x QPSK 3/4, 2x 16QAM 1/2, 2x 16QAM 3/4, 2x 64QAM 2/3, 2x 64QAM 3/4, 2x 64QAM 5/6 is prepared, where „2x" label means an utilization of the dual antenna on the base station. This dual polarization allows using both types - vertical and horizontal - polarizations. At the configuration, the same modulation for both types of polarization in the frame of one broadcasting external unit is possible. We can suppose an ideal angle level of the antenna when a signal level is stable and the same for each polarization. For this case, we don`t suppose the automatic modulation regulation (AMR) function because ideal conditions provide a stability and a modulation is not oscillating during the transmission depending on external influences, a line saturation, …

For the distance between subscriber units (SU) and the base station, a scale of 1, 2, 4, 6, 8 and 10 km is included.

For the radiation power, differentiated values from 10 to 20 dBm are prepared. The radiation power allows boosting a signal level for the purpose of achieving better parameters in given microwave connection. Presented values are given based on physical hardware possibilities in BS.

For the visibility, one from possibilities the line of sight (LoS) and the non-line of sight (NLOS) can be selected. At the NLOS selection, following specifications can be used – a barrier in the path and a building. Both NLOS examples worsen a signal level between SU and BS significantly comparing to the LOS possibility.

FUTURE RESEARCH DIRECTIONS

NG-PON systems present optical access infrastructures to support various applications of many service providers. Technological improvements of wireless broadband and broadcast technologies will require a reliable support in optical access networks. Therefore, appropriate advanced converged technologies utilizing the optical transmission medium and wireless transmission can be expected in a near future and more reasons begin effect on utilization and implementation of the WDM multiplexing technique in converged Fi-Wi passive optical networks.

On the other hand, the NG-PON technology must be able to protect the investment of legacy passive optical networks. Therefore, a strong interest for utilization of old TDM equipment in converged Fi-Wi passive optical networks is present. Naturally, an expansion of hybrid passive optical networks seems to be a practical solution. However, an area of research directions in converged HPON networks is very wide and can include e.g. a mutual convergence of technologies between optical metropolitan and optical/ wireless access networks, an utilization of power saving methods for subscriber terminals, an expansion of dynamic bandwidth and wavelength allocation mechanism, etc.

CHALLENGES AND OPEN ISSUES

In the near future, a convergence of optical and wireless technologies will be continued. Mobile base stations can be backhauled through the hybrid passive optical network and they can be thought of as common ONT elements. Between the central office and the remote base station will be introduced another intermediate node responsible for generation and transmission of control data streams over longer fiber distances. Broadband NG-PON networks can be deployed by three main scenarios to address different cases with diverse degrees of consolidation – the urban, the passive extended reach and the active extended reach. With these, also various converged Fi-Wi PON deployments could be included into possible future extensions. Moreover, a variety of users - residential, corporate/business and telco – must be considered. And, therefore, more flexible HPON architectures should be considered.

For practical implementation, there are some open issues. A high importance is dedicated to the power saving in long-reach broadband optical metropolitan and access networks and energy efficient efforts on both terminal sides in converged broadband converged optical fiber-wireless access networks. From this viewpoint, the HPON possibilities with reliable traffic protection provisioning can also play an important role. For enhancing broadband NG-PON networks, DSP-based approaches can be utilized for flexible per-wavelength rate upgrades and for extending reach of optical metropolitan and access systems. The virtualization of converged heterogeneous wireless-optical network with reconfigurable OLT elements and ICT infrastructures can be mentioned for supporting wireless and metallic access and backhauling solutions and for cloud services, the HPON Network Configurator must be ready for incorporation also these future deployments. Therefore, possible HPON deployments can be extended with these prospective variations of future broadband NG-PON networks.

CONCLUSION

With the emerging mobile applications and needs of ever increasing bandwidth, it is anticipated that the next-generation passive optical network (NG-PON) with much higher bandwidth is a natural path forward to satisfy these demands and for wireless network operators to develop valuable converged fiber-wireless access networks. NG-PON systems present optical access infrastructures to support various applications of many service providers. thus also for emerging applications of advanced mobile broadband and broadcast technologies associated with increasing bandwidth demands.

Hybrid passive optical networks (HPON) present a necessary phase of the future transition between PON classes with TDM and/or WDM multiplexing techniques utilized on the optical transmission medium – the optical fiber. For developing hybrid passive optical networks, there exist various architectures and directions. For proposing reliable and survivable architectures, traffic protection schemes must be implemented for various HPON networks. For converging Fi-Wi passive optical networks, an integration of optical and wireless technologies into common broadband access network must be considered. We focused on a design of converged Fi-Wi PON networks based on the HPON network integrated optical fiber and wireless technologies. A main reason for this motivation is that at first a number of operating TDM-PON networks is even now high and still rising, at second a utilization of installed optical infrastructures is maximized for transmission capacity's increasing and at third an extreme extension of wireless technologies between subscribers. a creating the hybrid HPON network can be used as the upgrade of old networks in many cases with a utilization of relevant parts of the original infrastructure with minimum financial costs and with traffic protection provisioning. Moreover, a consistent transition from only optical HPON network to the passive optical network with converged fiber-wireless technologies is important. Therefore, it is important to identify a right time for creating and building of the converged Fi-Wi PON network.

The HPON Network Configurator as an interactive software tool can be used for identifying a right time for creating and building of the full-value WDM-PON network. Its main aim is helping users, professional workers, network operators and system analysts to design, configure, analyze and compare various possible hybrid passive optical networks. This software program can be simply, fast and effectively upgraded for further variations of future broadband NG-PON networks.

ACKNOWLEDGMENT

This work is a part of research activities conducted at Slovak University of Technology Bratislava, Institute of MICT, within the KEGA agency project - 007STU-4/2016 “Progressive educational methods in the field of telecommunications multiservice networks” and the VEGA agency project - 1/0462/17 “Modeling of qualitative parameters in IMS networks”.

REFERENCES

Abbas, H. S., & Gregory, M. A. (2016). The next generation of passive optical networks: A review. *Journal of Network and Computer Applications*, *67*(5), 53–74. doi:10.1016/j.jnca.2016.02.015

Arevalo, G. V., Hincapié, R. C., & Gaudino, R. (2017). Optimization of multiple PON deployment costs and comparison between GPON, XGPON, NGPON2 and UDWDM PON. *Optical Switching and Networking*, *25*(7), 80–90. doi:10.1016/j.osn.2017.03.003

Čuchran, J. & Róka, R. (2006). *Optocommunication systems and networks*. Bratislava, Slovakia: Slovak University of Technology Publishing house.

Dixit, A. (2012). Wavelength switched hybrid TDMA/WDM-PON: A flexible next-generation optical access solution. *International Conference on Transparent Optical Network*, 14, 1-5.

Effenberger, F., Mukai, H., Kani, J., & Rasztovits-Wiech, M. (2009). Next-generation PON - Part III: System specification for XG-PON. *IEEE Communications Magazine*, *47*(11), 58–64. doi:10.1109/MCOM.2009.5307467

Effenberger, F., Mukai, H., Soojin Park, & Pfeiffer, T. (2009). Next-generation PON - Part II: Candidate systems for next-generation PON. *IEEE Communications Magazine*, *47*(11), 50–57. doi:10.1109/MCOM.2009.5307466

EXFO. (2012). *FTTH PON guide - Testing passive optical networks*. Quebec City, Canada: EXFO Electro-Optical Engineering Inc.

Grobe, K., & Elbers, J. P. (2008). PON in adolescence: From TDMA to WDM-PON. *IEEE Communications Magazine*, *46*(1), 26–34. doi:10.1109/MCOM.2008.4427227

Gutierrez, L. (2010). Next-generation optical access networks: from TDM to WDM. In *Trends in Telecommunication Technologies*. London, UK: Intechopen Limited. doi:10.5772/8473

Kani, J., Bourgart, F., Cui, A., Rafel, A., Campbell, M., Davey, R., & Rodrigues, S. (2009). Next-generation PON - Part I: Technology roadmap and general requirements. *IEEE Communications Magazine*, *47*(11), 43–49. doi:10.1109/MCOM.2009.5307465

Kazovsky, L. G. (2011). *Broadband optical access networks*. Wiley Interscience, John Wiley & Sons. doi:10.1002/9780470910931

Mahloo, M., Jiajia Chen, Wosinska, L., Dixit, A., Lannoo, B., Colle, D., & Mas Machuca, C. (2014). Toward reliable hybrid WDM/TDM passive optical networks. *IEEE Communications Magazine*, *52*(2), S14–S23. doi:10.1109/MCOM.2014.6736740

Mas Machuca, C., Jiajia Chen, & Wosinska, L. (2012). Cost-efficient protection in TDM PONs. *IEEE Communications Magazine*, *50*(8), 110–117. doi:10.1109/MCOM.2012.6257535

Nakamura, H. (2013). *NG-PON2 technologies*. NTT Access Network Service Systems Laboratories, NTT Corporation. doi:10.1364/NFOEC.2013.NTh4F.5

Prat, J. (2008). Next-generation FTTH passive optical networks: Research towards unlimited bandwidth access. New York: Springer Science+Business Media. doi:10.1007/978-1-4020-8470-6

Prat, J. (2016). Roadmap on optical communications: Access networks. *Journal of Optics*, *18*(6), 27–28.

Prince, K., Gibbon, T. B., Rodes, R., Hviid, E., Mikkelsen, C. I., Neumeyr, C., ... Monroy, I. T. (2012). GigaWaM - Next-generation WDM-PON enabling gigabit per-user data bandwidth. *Journal of Lightwave Technology*, *30*(10), 1444–1454. doi:10.1109/JLT.2012.2185039

Ragheb, A. M., & Fathallah, H. (2011). Performance analysis of next generation PON architectures. *International Conference on High-capacity Optical Networks and Emerging Technologies*, 8, 339-345. 10.1109/HONET.2011.6149766

Ramaswami, R., & Sivarajan, K. N. (2001). *Optical networks - A practical perspective*. San Francisco, CA: Morgan Kaufmann Publishers.

Róka, R. (2003). The utilization of the DWDM/CWDM combination in the metro/access networks. *Joint 1st Workshop on Mobile Future & Symposium on Trends in Communications, 1*, 160-162.

Róka, R. (2010). The utilization of the HPON Network Configurator at designing of passive optical networks, *International Conference on Telecommunication and Signal Processing*, 33, 444-448.

Róka, R. (2011). The extension of the HPON Network Configurator at designing of NG-PON networks, *International Conference on Telecommunication and Signal Processing*, 34, 79-84.

Róka, R. (2012). The designing of NG-PON networks using the HPON Network Configurator. *Journal of Communication and Computer*, *9*(6), 669–678.

Róka, R. (2013a). The analysis of SUCCESS HPON networks using the HPON Network Configurator. *Advances in Electrical and Electronic Engineering*, *1*(5-SI), 420–425.

Róka, R. (2013b). The analysis of SARDANA HPON networks using the HPON Network Configurator. *Advances in Electrical and Electronic Engineering*, *11*(6), 522–527. doi:10.15598/aeee.v11i6.866

Róka, R. (2014). Broadband NG-PON networks and their designing using the HPON Network Configurator. In *Convergence of Broadband, Broadcast, and Cellular Network Technologies*. Hershey, PA: IGI Global. doi:10.4018/978-1-4666-5978-0.ch012

Róka, R. (2015). *Hybrid PON networks - features, architectures and configuration*. Saarbrücken, Germany: LAP Lambert Academic Publishing.

Róka, R. (2016). Optimization of traffic protection schemes for utilization in hybrid passive optical networks. *International Journal of Application or Innovation in Engineering & Management*, *5*(9), 107–116.

Schneir, J. R., & Xiong, Y. (2014). Cost analysis of network sharing in FTTH/PONs. *IEEE Communications Magazine*, *52*(8), 126–134. doi:10.1109/MCOM.2014.6871680

Smolorz, S. (2009). Next generation access networks: PIEMAN and beyond. *International Conference on Photonics in Switching*, 1-4.

Talli, G. (2007). Integrated metro and access network: PIEMAN. *European Conference on Networks and Optical Communications*, 12, 493-500.

Tanaka, K., Agata, A., & Horiuchi, Y. (2010). IEEE 802.3av 10G-EPON standardization and its research and development status. *Journal of Lightwave Technology*, *28*(4), 651–661. doi:10.1109/JLT.2009.2038722

ITU-T Telecommunication Standardization Sector. (2009). *Characteristics of a single-mode optical fiber and cable.* Recommendation G.652.

ITU-T Telecommunication Standardization Sector. (2009). *Characteristics of a bending-loss insensitive single-mode optical fiber and cable for the access network.* Recommendation G.657.

ITU-T Telecommunication Standardization Sector. (2010). *Characteristics of a fiber and cable with non-zero dispersion for wideband optical transport.* Recommendation G.656.

Vetter, P., Suvakovic, D., Chow, H., Anthapadmanabhan, P., Kanonakis, K., Lee, K., ... Lannoo, B. (2014). Energy-efficiency improvements for optical access. *IEEE Communications Magazine*, *52*(4), 136–144. doi:10.1109/MCOM.2014.6807958

KEY TERMS AND DEFINITIONS

Fi-Wi PON: The converged access network based on the cooperation between optical fiber and wireless technologies for connecting more subscriber terminals.

HPON: The hybrid passive optical network with utilization of both TDM and WDM multiplexing principles together on a physical layer with assigning different time slots and/or wavelengths for particular subnetworks or subscriber terminals.

HPON Network Configurator: The interactive software tool for designing, configuring, analyzing, and comparing various possible hybrid passive optical networks.

NG-PON: The next-generation PON as a natural path forward to satisfy demands of ever-increasing bandwidth and to protect the investment of legacy passive optical networks.

PON: The access network based on the optical transmission medium with only passive optical components locating in the outside environment.

TDM Multiplexing: Sharing a common optical path by various transmission channels in the upstream direction assigning different time slots for particular subscriber terminals.

WDM Multiplexing: Sharing a common optical path by various transmission channels in downstream or upstream directions assigning different wavelengths for particular subnetworks or subscriber terminals.

Section 2

QoS Provisioning Solutions

Chapter 7
QoS-Aware Green Communication Strategies for Optimal Utilization of Resources in 5G Networks

Ganesh Prasad
National Institute of Technology Silchar, India

Deepak Mishra
Linköping University, Sweden

Ashraf Hossain
National Institute of Technology Silchar, India

ABSTRACT

With increase in demand of data traffic with no compromise on the underlying quality of service (QoS), the coexistence problem arises due to high electricity consumption by the network architecture which results in a huge CO2 emission and thereby causing various health hazards. Efficient utilization of the resources can reduce the cost of power consumption which will increase the economy-characteristics of the network. The resource consumption can be reduced under an intelligent technology-neutral policies which optimizes the deployment of the network architecture along with their transmit power paving the way for fifth generation (5G) in green wireless communications. On another front, the ultra-dense deployment of the small cells can increase the frequency reuse factor as well as help in reducing the energy consumption. This chapter designs the energy efficient networks while satisfying the underlying QoS by joint optimization of available resources depending on the interoperability challenges in terrestrial, underwater acoustic, and free space optical (FSO) communications.

INTRODUCTION

This chapter describes a quality-of-service (QoS)-aware energy-efficient network in terrestrial, underwater acoustic and free space optical (FSO) communications according to the interoperability challenges in their transmission links as shown in Figure 1.

DOI: 10.4018/978-1-5225-7570-2.ch007

Figure 1. Optimal resource allocation in cooperative underwater acoustic, FSO, and terrestrial networks while tackling the underlying interoperability challenges

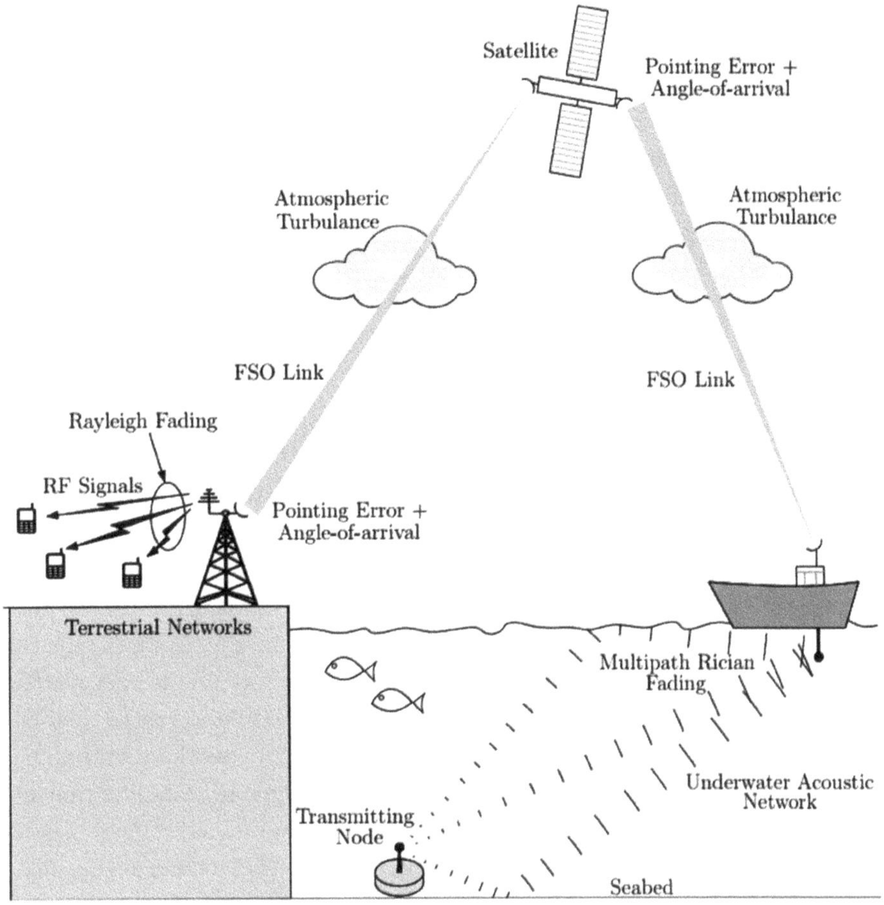

The demand of data is increasing day by day without any compromise with QoS. For improving the coverage of the user equipments (UEs), various works have been done on the deployment of base stations (BSs) in cellular networks. Deployment in grid-based network where the shape of the cells is hexagonal is less tractable (ElSawy, Hossain & Haenggi, 2013). Placement of BSs based on homogeneous Poisson point process (HPPP) and binomial point process (BPP) is realistic (Andrews, Baccelli & Ganti, 2011; Srinivasa & Haenggi, 2010), but deterministic deployment of BSs using the distribution of UEs gives significant performance enhancement (Su et al, 2014). The power consumption in the cellular network is nearly 1% of the world wide total energy consumption (Fettweis, & Zimmermann, 2008). Therefore, it leads to a need for an energy-efficient network design for minimizing the operational cost of the network while satisfying the quality-of-service (QoS) of the network.

Also, the energy-efficient network design in underwater acoustic networks (UANs) has gained significant research interest in recent years (Darehshoorzadeh & Boukerche, 2015). As the underwater acoustic channel has large delay and restricted bandwidth in long-range communications. Therefore, for satisfying a minimum data rate, it is required to transmit the signal through a relay in a cooperative communication by reducing the hop length which also gives an energy-efficient design (Stojanovic,

2007). Thus, this chapter investigates the green network by jointly optimizing the power allocation (PA) and relay placement (RP) in a dual-hop UAN where the direct link is absent.

As the optical communication in wireless channel is an alternative of fulfilment of demand of high data rate with less power consumption in congested radio frequency (RF) spectrum, this has enforced a research interest in free space optical (FSO) communication. The usage of FSO communication ranges from home network to satellite communication (Son, & Mao, 2017). However, the range of FSO communication is limited by lots channel impairments due to atmospheric turbulence, pointing error, and angle-of-arrival (AOA) fluctuations (Huang, & Safari, 2017). Again, using relay-assisted cooperative communication, the link length can be reduced which is also helpful in reducing the channel impairments in FSO communications (Saidi, Tourki, & Hamdi, 2016).

BACKGROUND

From the state-of-the-art, most of the works have deployed the BSs and UEs according to HPPP. Andrews et al. (2011) have evaluated that HPPP is more tractable than conventional grid method for the deployment. But it has been shown that BPP is more tractable than HPPP with respect to distribution of points and size of the network (Srinivasa & Haenggi, 2010). Placement of BSs based on homogeneous Poisson point process (HPPP) and binomial point process (BPP) is realistic (Andrews, Baccelli & Ganti, 2011; Srinivasa & Haenggi, 2010), but deterministic deployment of BSs using the distribution of UEs gives significant performance enhancement (Su et al, 2014). Therefore, in the energy-efficient network design, the BSs are deterministically deployed according to the distribution of UEs by BPP.

Mugume et al. (2015) have minimized the operation cost of the network by optimizing the transmit power, BS densities, and the deployment factor whereas multiple parameters like transmit power, BSs densities, number of BS antenna, and number of UEs per cell are optimized by Verenzuela et al. (2016). Peng et al. (2015) have shown that the joint optimization of transmit power and BS density while satisfying the coverage constraint is converged to a fixed value. Gonzalez-Brevis et al. (2011) have separately optimized location, PA and then number of BSs and location. From the state-of-the-art, the joint optimization of placement of BSs, their number and transmit power has not been evaluated.

Liu et al. (2017) and Kam et al. (2014) have designed an energy-efficient UAN by optimizing the location of the relays, whereas Babu et al. (2012) have investigated optimal PA. Nouri et al. (2016), Cao et al. (2010), and Doosti-Aref et al. (2017) have shown that significant performance enhancement is achieved by optimizing the PA and RP in cooperative UAN. Therefore, this chapter designed an energy-efficient cooperative UAN by jointly optimizing the PA and RP while minimizing the outage probability for satisfying the minimum data rate.

In case of cooperative FSO communication, Abou-Rjeily et al. (2014), Sun et al. (2017), and Hassan et al. (2017) have optimized the power consumption of the network by considering only channel fading due to atmospheric turbulence. Abou-Rjeily et al. (2014) and Sun et al. (2017) have analysed the energy-efficient network by considering intensity modulation and direct detection (IM/DD), but Hassan et al. (2017) have considered coherent detection because of its lots of advantages in the performance enhancement. From the state-of-the-art, the joint optimization of power over source and relay nodes in a realistic channel is required to be investigated.

The aim of this chapter is to explore the opportunities of design for QoS-aware energy-efficient networks in both terrestrial and underwater communications which paves the way for next generation green wireless communications. The interoperability challenges and optimal solution in the design over the two networks have been described. In case of terrestrial cellular networks, it has been shown that the power consumption can be significantly reduced by jointly optimizing the location of BSs, their density and transmit power. Since their joint optimization is nonconvex and combinatorial, the optimal solution can be obtained by optimizing the three parameters individually. Whereas, in case of UANs, the joint optimization of PA and RP is more challengeable than terrestrial wireless networks due to completely frequency-selective nature of the underwater channel in fading, path loss, and channel noise. From the recent development, it is discoursed that although the problem is generalized convex, it is complicated and not tractable for the optimal solution in continuous frequency domain. Therefore, the chapter describes a recently proposed solution methodology where the optimal design has been realized by discretising the operating bandwidth into sub-bands because in each sub-band the frequency-selective channel behaves like a flat channel. However for large number of sub-bands, the optimal solution faces high computational complexity. It is presented using the state-of-the-art that the complexity can be reduced by computing the optimal solution using tight near optimal approximation algorithm where the optimal solution is realized using one equation irrespective of the number of sub-bands. Finally, the challenges in designing energy-efficient cooperative FSO network with its anticipated solution methodology are discussed briefly as a future direction.

The rest of the chapter is organized as follows. In first section, the optimal deployment strategy of the BSs in a circular field is described. Then it is presented that the optimal location, density and transmit power of BSs can be individually optimized for their joint solution. Via numerical investigations, various insights on the variation of different parameters on the optimization are realized along with comparative study for the performance enhancement. In next section, the design of QoS-aware relay-assisted DF UAN is described using the recent developments. Then it is presented that the optimization problem is generalized convex and has global optimal solution which is tractable in discrete frequency domain. In following discussion, it is discoursed that the computational complexity can be significantly reduced by a tight near optimal approximation algorithm. Via numerical investigation, various insights on optimal PA and RP are realized with a comparative investigation of the joint optimization against a benchmark scheme. Lastly, the chapter is concluded by summarizing the key take-away points and presenting some future research directions.

EFFICIENT ULTRA-DENSE BASE STATION DEPLOYMENT FOR QOS-AWARE COMMUNICATIONS

Here the strategies for optimal deployment of BSs have been discoursed in both the scenarios when the number of UEs is high and when it is moderate.

Network Topology in the Terrestrial Communication

In the network as described by Prasad et al. (2018), N_U UEs are placed by BPP where the BSs are deployed deterministically for the efficient coverage. Intracell and intercell interference are assumed to be mitigated by orthogonal multi-access techniques (Andrews, J. G., 2005). It gives the advantage in obtaining the optimal solution with less complexity. The cells over the circular field are generated efficiently without coverage hole using Voronoi-tessellation (Liu, C. H. & Wang, L. C., 2015). Based on that Prasad et al. (2018) have generated the cells using the sectoring $mk+t$ as shown in Figure 2.

Sectoring mk for $t=0$ and $mk+1$ for $t=1$ is differ with respect to absence or presence of a BS at the center of the circular field. Initially, the circular field is divided into k equal sectors each of angle θ and along each sector, m BSs are placed in their associated cells along the symmetric line of each sector. All generated sectors in the circular field are having same properties. Location of the BSs is measured from the center which is given as $\boldsymbol{d}=\{d_i; i\in\mathcal{I}\}$, where $\mathcal{I}=\{1-t,2-t,\ldots,m\}$ and d_i is the location of the ith BS.

The transmit signal over the channel is distorted by a path loss and frequency selective Rayleigh fading. The channel power gain is given by $h_{n,i}r_{n,i}^{-\alpha}$, where α is the path loss exponent, $h_{n,i}$ is power of the channel coefficient, $r_{n,i}$ is distance of nth UE from the ith BS. The distribution of $h_{n,i}$ is exponential with mean value 1. The signal-to-noise ratio (SNR) at the UE is expressed as $\gamma_{n,i}=\frac{P_t h_{n,i}}{\sigma^2 r_{n,i}^{\alpha}}$, where P_t is the transmit power, σ^2 is the variance of the additive which Gaussian noise, and T is the threshold SNR for detecting the signal. The coverage probability can be expressed as (Mekikis et al., 2015, eq. (8)):

$$P_{\text{cov}}^{n,i}=\mathbb{E}_{r_{n,i}}\left[e^{-\frac{T\sigma^2 r_{n,i}^{\alpha}}{P_t}}\right]=\int_0^{r_{u,i}} e^{-\frac{T\sigma^2 r_{n,i}^{\alpha}}{P_t}} f_{n,i}(r_{n,i},\boldsymbol{d_i})dr_{n,i} \tag{1}$$

where $f_{n,i}(r_{n,i},\boldsymbol{d_i})$ is the probability density function of $r_{n,i}$.

Figure 2. Sectoring (a) $\mathcal{M}: mk=N_B$ for $t=0$, (b) $\mathcal{M}: mk+1=N_B$ for $t=1$

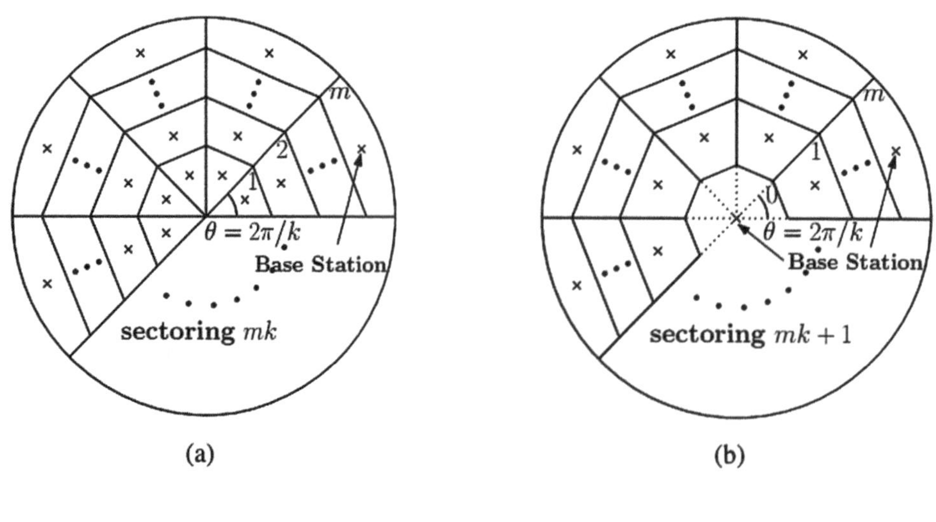

Optimal Deployment Strategy for High Density of UEs

As discoursed by Prasad et al. (2018), when the number of UEs N_U in a finite field is asymptotically high then farthest UE associated with a cell lies at the farthest point Euclidean distance from the BS. Therefore in this case the optimal location of BSs depends on the minimization of farthest point Euclidean distance over all cells which can be expressed as $\min_{\boldsymbol{d}} \max_i r_{u,i}$. As shown in Figure 3, it can be further written as $\min_{\boldsymbol{d}} \max\{d_1, l_1, l_2, \ldots, l_m\}$. Here l_i is the distance of the vertex of the cell from the BS, $r_{u,i} = \max\{l_{i-1}, l_i\}\ \forall i \in \mathcal{I} \setminus 1$, and $r_{u,i} = \max\{d_i, l_i\}$ for $i = 1$. The problem achieves the minimum value $l_{\min}$ at $\boldsymbol{d}^*$, when farthest point Euclidean distance of all cells becomes equal, i.e., $r_{u,i} = l_{\min} = l_i = d_1\ \forall i \in \mathcal{I}$. Using the trigonometric relationship with some algebraic simplification, $l_{\min}$ with corresponding optimal location $\boldsymbol{d}^*$ has been listed in Table 1 for different sectorings. The optimal sectoring for a given N_B is determined by choosing a sectoring from a given set of sectorings which gives the least value of $l_{\min}$ as depicted in Figure 4. The range of N_B for which a sectoring is optimal has been listed in fourth column of the table.

Optimal Deployment Strategy for Moderate Density of UEs

For moderate density of UEs over the circular field, the optimal location of BSs depends on both the farthest point Euclidean distance as well as the distribution of area around the boundaries as shown in Figure 5 (Prasad, G., Mishra, D. & Hossain, A., 2018). Here for calculation of actual location of the BS, first the location is computed for minimizing the farthest point Euclidean distance. Thereafter the probable region around the optimal location is evaluated where the actual location lies. In case of single BS deployed at the center of the circular field (cf. Figure 5(a)), the distribution of area around the boundaries is same. Therefore, the probability of lying of farthest UE around the boundaries is almost same. So, the actual optimal location $d^*_{0,\text{act}}$ lies exactly at the optimal location d^*_0 which is computed by minimiz-

Figure 3. Deployment of BSs in a sector for t = 0

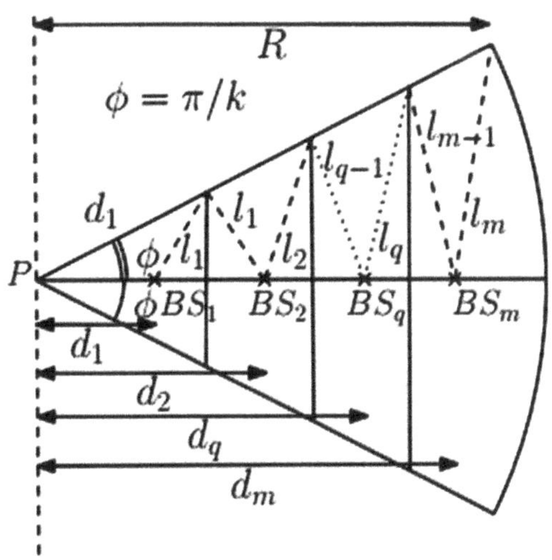

Figure 4. Variation of $l_{\min}$ with N_B in each cell for different type of sectoring

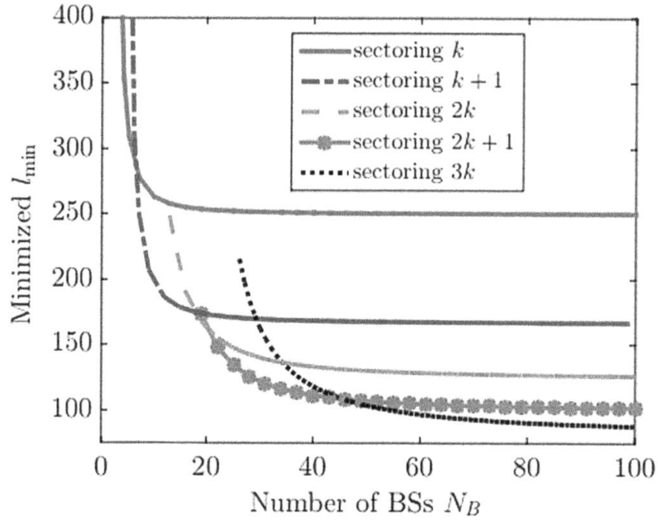

Figure 5. (a) Actual location $d^*_{0,\text{act}}$ in a circular field, (b) $d^*_{i,\text{act}}$ in i^{th} cell for a given N_B

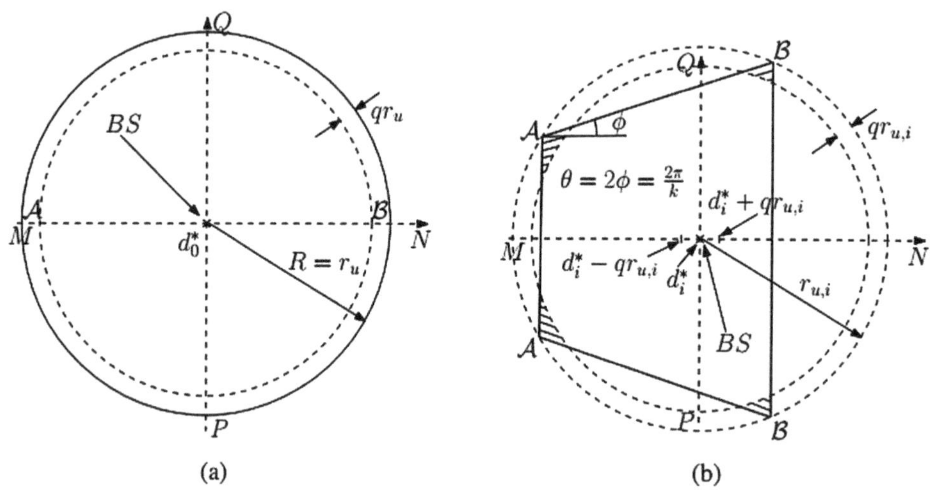

ing the farthest point Euclidean distance. For a BS associated with a cell as shown in Figure 5(b), the area around the boundaries about the MN axis is symmetric, whereas it is asymmetric about the axis PQ. Therefore, if the area $2\mathcal{A}$ around the boundaries is greater than the area $2\mathcal{B}$ then there is higher probability that the farthest UE lies on left side. So, for minimizing the farthest UE's distance from the BS, the actual location $d^*_{i,\text{act}}$ lies on left side to the optimal location d^*_i and vice versa. Here d^*_i is realized from minimizing the farthest point Euclidean distance of the cell from the associated BS. From extensive simulations, it has been interpreted that performance enhancement achieved by deploying the BSs based on minimizing the farthest point Euclidean distance is almost same as the deployment based on minimizing the farthest UE's distance.

Table 1. Optimal sectoring $\mathcal{M}^*$ and the corresponding minimized value $l_{\min}$ of $\max_i r_{u,i}$ for a given number of BSs N_B at their optimal location $\mathbf{d}^*$

Optimal sectoring $\mathcal{M}^*$	Minimized value $l_{\min}$ of $\max_i r_{u,i}$	Optimum location $\mathbf{d}^*$ of the BSs	Range of N_B
k	$R\sin\left(\frac{\pi}{N_B}\right)$	$d_1^* = R\cos\left(\frac{\pi}{N_B}\right)$	$N_B = 3$
	$\frac{R}{2\cos\left(\frac{\pi}{N_B}\right)}$	$d_1^* = \frac{R}{2\cos\left(\frac{\pi}{N_B}\right)}$	$N_B \in \{4,5,6\}$
$k+1$	$\frac{R}{4\cos^2\left(\frac{\pi}{N_B-1}\right)-1}$	$d_0^* = 0, d_1^* = \frac{2R\cos\left(\frac{\pi}{N_B-1}\right)}{4\cos^2\left(\frac{\pi}{N_B-1}\right)-1}$	$N_B \in \{7,8,\ldots,17\}$ $\bigcup\{19\}$
$2k$	$\frac{R}{4\cos\left(\frac{2\pi}{N_B}\right)\cos\left(\frac{4\pi}{N_B}\right)}$	$d_1^* = \frac{R}{4\cos\left(\frac{2\pi}{N_B}\right)\cos\left(\frac{4\pi}{N_B}\right)}$, $d_2^* = \frac{R\left(1+\cos\left(\frac{4\pi}{N_B}\right)\right)}{4\cos\left(\frac{2\pi}{N_B}\right)\cos\left(\frac{4\pi}{N_B}\right)}$	$N_B \in \{18,20,\ldots,44\}$
$2k+1$	$\frac{R\left(1+2\cos\left(\frac{4\pi}{N_B-1}\right)\right)}{16\cos^2\left(\frac{2\pi}{N_B-1}\right)\cos^2\left(\frac{4\pi}{N_B-1}\right)-1}$	$d_0^* = 0,\ d_1^* = \frac{2R\left(1+2\cos\left(\frac{4\pi}{N_B-1}\right)\right)\cos\left(\frac{2\pi}{N_B-1}\right)}{16\cos^2\left(\frac{2\pi}{N_B-1}\right)\cos^2\left(\frac{4\pi}{N_B-1}\right)-1}$, $d_2^* = \frac{4R\left(1+2\cos\left(\frac{4\pi}{N_B-1}\right)\right)\cos\left(\frac{4\pi}{N_B-1}\right)\cos\left(\frac{2\pi}{N_B-1}\right)}{16\cos^2\left(\frac{2\pi}{N_B-1}\right)\cos^2\left(\frac{4\pi}{N_B-1}\right)-1}$	$N_B \in \{21,23,\ldots,45\}$
$3k$	$\frac{R\cos\left(\frac{3\pi}{N_B}\right)}{\left(2\cos\left(\frac{6\pi}{N_B}\right)+1\right)\left(\cos\left(\frac{12\pi}{N_B}\right)+\cos\left(\frac{6\pi}{N_B}\right)\right)}$	$d_1^* = \frac{R\cos\left(\frac{3\pi}{N_B}\right)}{\left(2\cos\left(\frac{6\pi}{N_B}\right)+1\right)\left(\cos\left(\frac{12\pi}{N_B}\right)+\cos\left(\frac{6\pi}{N_B}\right)\right)}$, $d_2^* = \frac{R\cos\left(\frac{3\pi}{N_B}\right)\left(1+2\cos\left(\frac{6\pi}{N_B}\right)\right)}{\left(2\cos\left(\frac{6\pi}{N_B}\right)+1\right)\left(\cos\left(\frac{12\pi}{N_B}\right)+\cos\left(\frac{6\pi}{N_B}\right)\right)}$, $d_3^* = \frac{R\cos\left(\frac{3\pi}{N_B}\right)\left(1+2\cos\left(\frac{6\pi}{N_B}\right)+2\cos\left(\frac{9\pi}{N_B}\right)\right)}{\left(2\cos\left(\frac{6\pi}{N_B}\right)+1\right)\left(\cos\left(\frac{12\pi}{N_B}\right)+\cos\left(\frac{6\pi}{N_B}\right)\right)}$	$N_B \in \{48,51,\ldots\}$

SMART POWER CONTROL TECHNIQUES FOR OPERATIONAL COST MINIMIZATION

In this section, the number of BSs and PA are jointly optimized for realizing the minimized operational cost over the network.

Operational Cost Model

The operational power cost model over a BS is expressed as below (Peng, J., Hong, P. & Xue, K., 2015; Liu et al., 2017):

$$P_{BS} = a_B P_t + b_B \tag{2}$$

In the above linear cost model, a_B is a constant that scales the transmit power which depends on the number of power amplifiers, their number, b_B represents the power cost due to processing, cooling and battery backup. Using equation (2), the total operational power cost is $N_B P_{BS}$ which has to be minimized for the design of QoS-aware green communication while satisfying the minimum coverage demand of the UEs. The numerical results for optimal deployment strategy with minimization of operational power cost have been discoursed in the next section.

Problem Formulation

As discussed earlier the coverage over the region is determined by the average coverage probability over the farthest UE which is expressed as (Prasad et al. 2018, eq. (4)):

$$P_{\text{cov}}^{far,i} = \int_0^{r_{u,i}} e^{-\frac{T\sigma^2 r_{far,i}^{\alpha}}{P_t}} f_{far,i}(r_{far,i}, \boldsymbol{d_i}) dr_{far,i} \tag{3}$$

Here $f_{far,i}(r_{far,i}, \boldsymbol{d_i})$ denotes the PDF of farthest UE's distance $r_{n,i}$. Using the coverage constraint as given in equation (3), the joint problem for minimizing the operation power cost by optimizing transmit power P_t, location $\boldsymbol{d}$, and corresponding number of BSs N_B can be expressed as [Prasad et al. 2018, eq. (5)]:

$$(\text{P0}): \underset{N_B, P_t, \mathbf{d}}{\text{minimize}} \quad \text{N}_B \left[a_B P_t + b_B \right],$$

subject to

$$C1: P_{\text{cov}}^{\text{far},i} \geq 1 - \epsilon, \forall i \in \mathcal{I},$$

$$C2: N_B = \{1, 2, \ldots, N_{B,\max}\},$$

$$C3: 0 \leq P_t \leq P_{t,\max},$$

$$C4: 0 \leq d_i \leq d_{i,\max}, \forall i \in \mathcal{I},$$

Here constraint C1 makes sure that coverage probability over the farthest UE greater than the acceptable threshold $1 - \epsilon$, where $0 < \epsilon \ll 1$. The box constraints C2 and C3 gives the boundary conditions on the transmit power P_t and d_i, respectively. As the problem (P0) is nonconvex combinatorial, it is discretized into three individual problems to find the optimal solution.

Optimal Power Allocation

For high coverage demand, the coverage probability given by constraint C1 of problem (P0) should be high or the corresponding threshold ϵ must be quite less. To ensure at least 90% coverage or $\epsilon \leq 0.1$, the coverage probability can be approximated with error less than 0.05% as (Prasad et al. 2018, eq. (7)):

$$P_{\text{cov}}^{far,i} \approxeq 1 - \frac{T\sigma^2}{P_t} \int_0^{r_{u,i}} r_{far,i}^{\alpha} f_{far,i}(r_{far,i}, \boldsymbol{d_i}) dr_{far,i} \geq 1 - \epsilon, \forall i \in \mathcal{I} \tag{4}$$

The percentage error further decreases with the coverage demand. After some algebraic simplification, the optimal PA over whole BSs deployed in the homogeneous network can be expressed as (Prasad et al. 2018, eq. (9)):

$$P_t^* \approxeq \max_i \left\{ \frac{T\sigma^2}{\epsilon} \int_0^{r_{u,i}} r_{far,i}^{\alpha} f_{far,i}(r_{far,i}, \boldsymbol{d}_i^*) dr_{far,i} \right\} \tag{5}$$

Here note that the optimal PA given by equation (5) is a function of acceptable threshold ϵ and the optimal location $\boldsymbol{d}_i^* = \{d_{i-1}^*, d_i^*, d_{i+1}^*\}$ which is the optimal placement of the BSs in $(i-1)^{th}$, i^{th}, and $(i+1)^{th}$ cells, respectively. Also, the optimal PA is evaluated for a given N_B for which the optimal sectoring has been taken from Table 1.

Iterative Scheme for Evaluation of Optimal Number of BSs

As the problem (P0) is unimodal in N_B (Prasad, G., Mishra, D. & Hossain, A., 2018), using the computed optimal location and optimal PA in equation (5), the optimization of number of BSs N_B can be evaluated using Algorithm 1 and Algorithm 2 (Prasad et al. 2018, pages 7, 8).

Using Algorithm 1, the total operational cost is calculated for a given number of BSs N_B. Here, first optimal sectoring $\mathcal{M}^*$ is evaluated using Table 1, then optimal PA at each BS can be evaluated at their optimal locations. Finally using equation (5) the optimal PA is computed for the homogeneous network by which the total operational power cost can be computed. In Algorithm 2, golden search has been used for the iterative solution of optimal number of BSs N_B^*. Thus, the optimized total operational cost is given as $N_B^* P_t^*$. The algorithms converges to the optimal solution as the golden search reduces the search range logarithmically.

Numerical Results

The default system parameters used in evaluating the numerical results has been listed in Table II. The optimization techniques have been compared against a fixed benchmark scheme with setting given as $N_B = 35$, $P_t = 4$ W, $\mathcal{M}: k$, $d_1 = 250$ m for $N_B \geq 2$; $d_1 = 0$ for $N_B = 1$.

Algorithm 1. Calculating operational cost function $f(N_B)$ for a given N_B and predefined set of system parameters

Input: Number of BSs $N_B \leq N_{B,max}$ and all other system parameters as defined in Section II

Output: Operational power cost

1: Using Table I, find the optimal location $\mathbf{d}^*$ and sectoring $\mathcal{M}^*$ for a given value of N_B

2: **for** $i = 1 - t$ to m **do,**

3: Calculate the minimum value of power allocation at d_i^* in i^{th} cell using eq. (4)

4: Using eq. (5), find the optimal power allocation P_t^* for all BSs by taking the maximum transmit power over $m - t$ values calculated in step 3

5: Calculate the operational cost as $N_B(a_B P_t^* + b_B)$.

Algorithm 2. Iterative scheme to obtain optimal operational cost

Input: Bounds N_B^l, N_B^u, and acceptable tolerance ς

Output: Optimal operational cost along with joint optimal solution $(N_B^*, \mathbf{d}^*, \mathcal{M}^*, P_t^*)$

1: Calculate $N_B^p = \lceil N_B^u - 0.618 \times (N_B^u - N_B^l) \rceil$
2: Calculate $N_B^q = \lfloor N_B^l + 0.618 \times (N_B^u - N_B^l) \rfloor$
3: Calculate $f(N_B^p)$ and $f(N_B^q)$ using Algorithm 1
4: Set $\Delta_N = N_B^u - N_B^l$
5: **while** $\Delta_N > \varsigma$ **do**
6:     **if** $f(N_B^p) \leq f(N_B^q)$ **then**
7:         Set $N_B^u = N_B^q$, $N_B^q = N_B^p$, and $N_B^p = \lceil N_B^u - 0.618 \times (N_B^u - N_B^l) \rceil$
8:     **else**
9:         Set $N_B^l = N_B^p$, $N_B^p = N_B^q$, and $N_B^q = \lfloor N_B^l + 0.618 \times (N_B^u - N_B^l) \rfloor$
10:     Calculate $f(N_B^p)$ and $f(N_B^q)$ using Algorithm 1
11:     Set $\Delta_N = N_B^u - N_B^l$
12: Calculate $N_B^* = \left\lceil \frac{N_B^u + N_B^l}{2} \right\rceil$
13: Calculate optimal operational cost $f(N_B^*)$
14: Using Table I, find the optimal location $\mathbf{d}^*$ and sectoring $\mathcal{M}^*$ by substituting $N_B = N_B^*$
15: By substituting optimal deployment of BSs as obtained in steps 12 and 14 in (4), the optimal power allocation P_t^* for all BSs is obtained

Table 2. Default system parameters and their values

S. No.	System parameter	Symbol	Value
1	Radius of the circular field	R	500 m
2	Side length of the square field	a	$500\sqrt{\pi}$ m
3	Number of UEs	N_U	120
4	Maximum number of deployed BSs	$N_{B,\max}$	35
5	Coefficient of power consumption	a_B	5.5
6	Additive power consumption	b_B	32 W
7	Maximum transmit power	$P_{t,\max}$	5 W
8	Acceptable threshold	ϵ	10^{-2}
9	SNR threshold	T	−10 dB
10	Path loss exponent	α	4
11	Noise power	σ^2	−70 dBm
12	Maximum density of BSs	$\lambda_{\max}$	5×10^{-5} m^{-2}

In Figure 6, the variation of average coverage probability with P_t has been shown which entails that the improvement in coverage probability is not much more when the deployed number of BSs $N_B = 2$ instead of 1. But, for $N_B = 3$ and 4, significant improvement take place because of large reeducation in farthest point Euclidean distance.

The operational power cost minimization using the two deployment strategies has been compared in Figure 7. It approves the earlier discussion that deployment based on minimization of farthest point

Figure 6. Validation of coverage probability variation with P_t

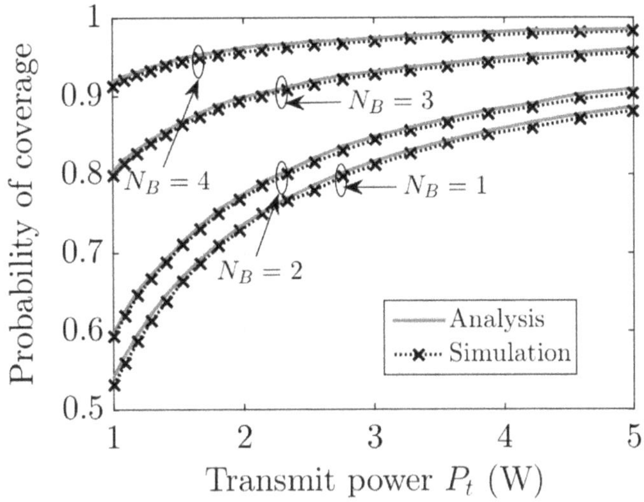

Figure 7. Comparison of the two deployment which are based on actual and fixed optimal location

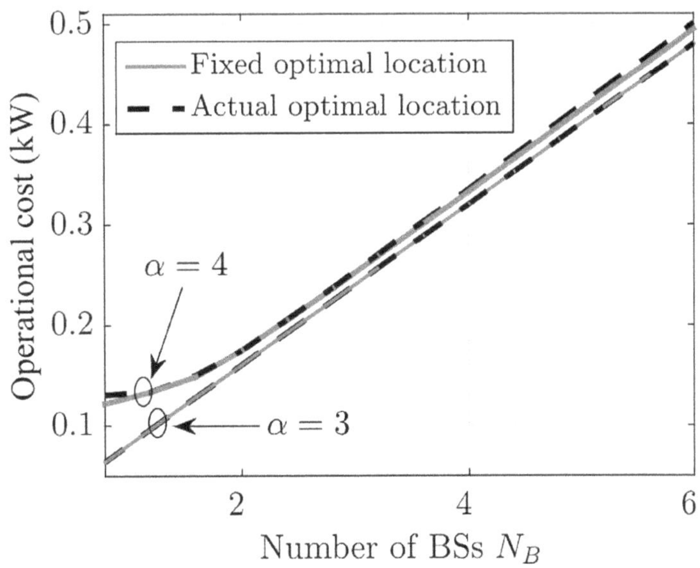

Euclidean distance almost gives the same performance as it can be achieved by minimization of farthest UEs distance.

It can be observed in Figure 8 that when the transmit power by the BSs is low then higher number of BSs is required for meeting the quality of coverage demand, whereas for high transmit power the N_B^* is low. But, the optimized operational power cost is almost same due to the trade-off between the transmit power and number of BSs requirement where the operational cost is the product of the two.

The performance degradation in minimization of operational power cost due to non availability of $\aleph_c$ number of cells has been depicted in Figure 9. As for low coverage demand, the requirement of

Figure 8. Variation of optimized entities with quality of coverage for a given transmit power

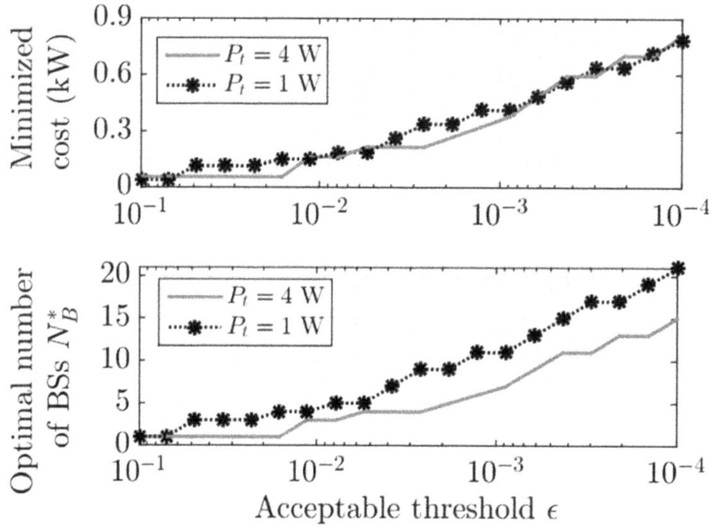

Figure 9. Performance degradation due to non-availability of few cells for deployment of the BSs

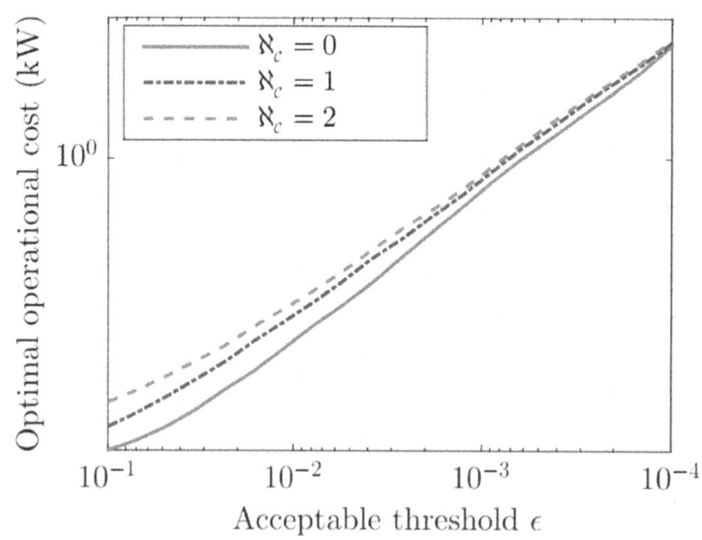

number of BSs is less, therefore higher degradation take place even when 1 or 2 randomly picked cells are not available for the deployment of BSs. Whereas, for very high coverage demand, high number of BSs causes negligible degradation in the performance.

In Figure 10, the three optimization schemes, ONB: optimal number of BSs with $P_t = 4$ W, OPA: optimal PA with $N_B = 35$, and the joint optimization scheme are compared against the fixed benchmark scheme. The joint optimization always performs better than ONB and OPA. If we compare ONB and OPA, initially ONB performs better than OPA, but for very high coverage demand, there is always need of high number of BSs where the optimization of PA dominates over optimization of number of BSs. In next section, the optimization of PA along with relay location has been studied in a cooperative UAN.

Figure 10. Comparison of various optimization techniques against a benchmark scheme

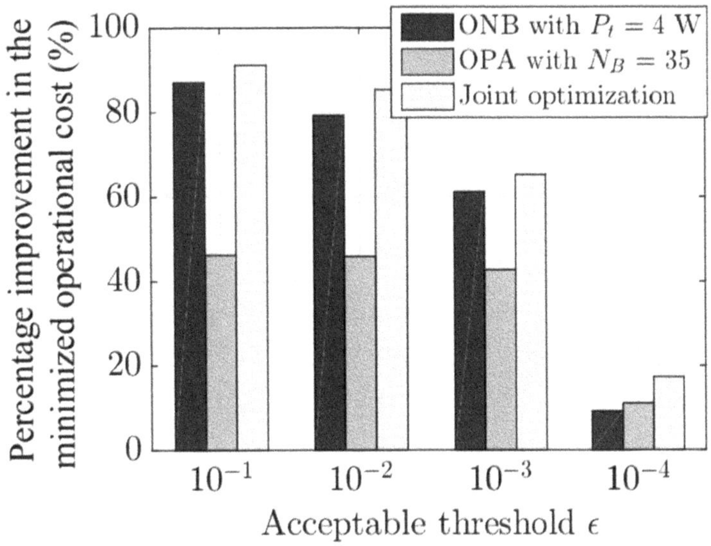

Optimal Resource Allocation in Cooperative UAN

Network Topology

In cooperative UAN, the information transfer from the source $\mathcal{S}$ to destination $\mathcal{D}$ takes place through relay $\mathcal{R}$ in two slots: first from $\mathcal{S}$ to $\mathcal{R}$, then from $\mathcal{R}$ to $\mathcal{D}$. Using the channel model given by Stojanovic (2007) and Cao et al. (2010), the SNR received at node j from the transmit node i is expressed as (Prasad, G., Mishra, D. & Hossain, A., 2018, eq. (1)):

$$\gamma_{ij}(f) = S_i(f)G_{ij}(f)\left[a(f)\right]^{-d_{ij}} d_{ij}^{-\alpha}\left[N(f)\right]^{-1} \tag{6}$$

where $S_i(f)$ is power spectral density (PSD) of the transmit signal by the node i, $G_{ij}(f)$ is the Rician fading power gain, α is the spreading factor, $N(f)$ is PSD of noise, $a(f)$ is absorption coefficient. For Rice factor $K \leq 8000$ of the frequency-selective fading channel, the complementary CDF of $\gamma_{ij}(f)$ is given as (Mishra, D., & De, S., 2017, eq. 10):

$$\Pr[\gamma_{ij}(f) > x] = e^{-\mathcal{A}\left(\frac{2(1+K)\beta x}{\bar{\gamma}_{ij}(f)}\right)^{\mathcal{B}}} \tag{7}$$

Here $\mathcal{A} = e^{\phi(\sqrt{2K})}$ and $\mathcal{B} = \dfrac{\varphi(\sqrt{2K})}{2}$ and the functions $\phi(v)$ and $\varphi(v)$ is described in (Mishra, D., & De, S., 2017, eqs. 8(a), 8(b)).

Problem Formulation

In a DF realy-assisted UAN, the end-to-end outage probability in the absence of direct link is expressed as (Mishra, D. & De, S., 2017):

$$p_{out} = \Pr\left(\int_0^{B_W} \frac{1}{2} \log_2(1 + \min\{\gamma_{SR}(f), \gamma_{RD}(f)\}) df \le r \right) \tag{8}$$

For minimizing the outage probability given in equation (8) through jointly optimizing the PA and RP, the joint problem can be formulated as (Prasad, G., Mishra, D. & Hossain, A., 2018, problem (P0)):

$$(\text{P1}): \underset{S_S(f), S_R(f), d_{SR}}{\text{minimize}} \; p_{out},.$$

subject to

$$C1: d_{SR} \ge \delta,$$

$$C2: d_{SR} \le D - \delta,$$

$$C3: \int_0^{B_W} (S_S(f) + S_R(f)) df \le P_B,$$

Here the constraint C1 and C2 restrict the location of the relay with minimum separation between the two nodes is δ. C3 it the constraint on the total power budget P_B for efficiently utilizing the available energy resources. It is shown by Prasad et al. (2018) that p_{out} is pseudoconvex with respect to its underlying variables. Therefore, the problem (P1) is generalized-convex which has a global solution.

Equivalent Optimization

As the problem (P1) is intractable in continuous frequency domain (Babu, A. V. & Joshy, S., 2012), it is discretized to compute its optimal solution. In discrete domain, the number of discrete frequency subbands n is large enough such that difference between p_{out} as defined in equation (8) is nearly same as $\widehat{p_{out}}$ given in equation (9) which is expressed as (Prasad, G., Mishra, D. & Hossain, A., 2018, eq. (5)):

$$\widehat{p_{out}} \triangleq \Pr\left(\sum_{q=1}^{n} \frac{\Delta f}{2} \log_2(1 + \min\{\gamma_{SR_q}, \gamma_{RD_q}\}) \le r \right) \tag{9}$$

In the dual-hop communication, the qth sub-band of $\mathcal{SR}$ and $\mathcal{RD}$ links are coupled with each other. Note that larger the value of n, $\widehat{p_{out}}$ matches more with p_{out}. The received SNR at node j of ij link is given as: $\gamma_{ij_q} = P_{i_q} G_{ij_q} a_q^{-d_{ij}} d_{ij}^{-\alpha} [N_q \Delta f]^{-1}$ in which P_{i_q} is the power allocated, S_{i_q} is the power spectral density (PSD) over the qth sub-band of the ith transmitting node, a_q is absorption coefficient, N_q is additive noise, G_{ij_q} and $c_{ij_q} = \mathbb{E}[G_{ij_q}]$ are the channel power gain and its expectation in q^{th} sub-band of the ij link. In each sub-band, each parameter is flat and depends on its center frequency.

As the CDF of a random variable monotonically decreases with its expectation (Mishra, D., De, S. & Krishnaswamy, D., 2017, Theorem 1), minimization of $\widehat{p_{out}}$ in (9) is equivalently can be realized by maximization of the mean of the underlying variable

$$\frac{\Delta f}{2} \mathbb{E}[\log_2 \prod_{q=1}^{n} (1 + \min\{\gamma_{\mathcal{SR}_q}, \gamma_{\mathcal{RD}_q}\})].$$

$\mathbb{E}[\prod_{q=1}^{n} (1 + \min\{\gamma_{\mathcal{SR}_q}, \gamma_{\mathcal{RD}_q}\})]$ is a jointly pseudoconcave function as the logarithmic transformation is monotonically increasing. Assuming that all the sub-bands are independent to each other, equivalently problem (P1) can be written as

$$(\text{P2}): \underset{\{P_{\mathcal{S}_q}, P_{\mathcal{R}_q}\}_{q=1}^{n}, d_{\mathcal{SR}}}{\text{maximize}} \prod_{q=1}^{n} (1 + \mathbb{E}[\min\{\gamma_{\mathcal{SR}_q}, \gamma_{\mathcal{RD}_q}\}])$$

subject to

$$C1, C2, \widehat{C3}: \sum_{q=1}^{n} (P_{\mathcal{S}_q} + P_{\mathcal{R}_q}) \leq P_{\mathcal{B}}.$$

Here the constraint $\widehat{C3}$ gives the condition on the utilization of power budget. From (Prasad, G., Mishra, D. & Hossain, A., 2018), the expectation

$$\mathbb{E}[\min\{\gamma_{\mathcal{SR}_q}, \gamma_{\mathcal{RD}_q}\}] = \overline{\gamma}_q = \frac{\beta}{N_q} \left[\left(\frac{a_q^{d_{\mathcal{SR}}} d_{\mathcal{SR}}^{\alpha}}{c_{\mathcal{SR}_q} P_{\mathcal{S}_q}} \right)^{\mathcal{B}} + \left(\frac{a_q^{D-\delta-d_{\mathcal{SR}}} (D - \delta - d_{\mathcal{SR}})^{\alpha}}{c_{\mathcal{RD}_q} P_{\mathcal{R}_q}} \right)^{\mathcal{B}} \right]^{\frac{-1}{\mathcal{B}}}.$$

As the problem (P2) is convex due to pseudoconcave objective function and convex constraint, Lagrangian function $\mathcal{L}_1$ can be expressed as:

$$\mathcal{L}_1 = \prod_{q=1}^{n} \left(1 + \mathbb{E}\left[\min\left\{ \gamma_{\mathcal{SR}_q}, \gamma_{\mathcal{RD}_q} \right\} \right] \right) - \lambda \mathcal{J} \tag{10}$$

Here λ, $\mathcal{J} \triangleq \left(\sum_{q=1}^{n} (P_{\mathcal{S}_q} + P_{\mathcal{R}_q}) - P_B \right)$. Using the KKT conditions by taking the derivative with respect to underlying random variables, along with $\mathcal{J}$, $(2n+2)$ system of equations have been expressed as:

$$P_{\mathcal{S}_q} = P_{\mathcal{R}_q} c_{\mathcal{RD}_q} \left[c_{\mathcal{SR}_q} a_q^{D-\delta-2d_{\mathcal{SR}}} \left(\left(D-\delta\right) d_{\mathcal{SR}}^{-1} - 1 \right)^{\alpha} \right]^{-1} \left[\left(Q_q \beta c_{\mathcal{SR}_q} \left[N_q \lambda \Delta f a_q^{d_{\mathcal{SR}}} d_{\mathcal{SR}}^{\alpha} \right]^{-1} \right)^{\frac{\mathcal{B}}{\mathcal{B}+1}} - 1 \right]^{\frac{1}{\mathcal{B}}} \tag{11a}$$

$$P_{\mathcal{R}_q} = P_{\mathcal{S}_q} c_{\mathcal{SR}_q} a_q^{D-\delta-2d_{\mathcal{SR}}} \left(\left(D-\delta\right) d_{\mathcal{SR}}^{-1} - 1 \right)^{\alpha} c_{\mathcal{RD}_q}^{-1} \left[\left(Q_q \beta c_{\mathcal{RD}} \left[N_q \lambda \Delta f a_q^{D-d_{\mathcal{SR}}-\delta} \left(D - d_{\mathcal{SR}} - \delta \right)^{\alpha} \right]^{-1} \right)^{\frac{\mathcal{B}}{\mathcal{B}+1}} - 1 \right]^{\frac{1}{\mathcal{B}}} \tag{11b}$$

$$\sum_{q=1}^{n} \beta c_{\mathcal{SR}_q} c_{\mathcal{RD}_q} P_{\mathcal{S}_q} P_{\mathcal{R}_q} \left[N_q \Delta f \right]^{-1} V_q^{\frac{-\mathcal{B}}{\mathcal{B}+1}} \left(c_{\mathcal{RD}_q} P_{\mathcal{R}_q} a_q^{d_{\mathcal{SR}}} d_{\mathcal{SR}}^{\alpha} \right)^{\mathcal{B}} \left[\begin{array}{l} T_q \left(\ln a_q + \alpha \left(D - \delta - d_{\mathcal{SR}} \right)^{-1} \right) \\ - \left(\ln a_q + \alpha d_{\mathcal{SR}}^{-1} \right) \end{array} \right] = 0 \tag{11c}$$

where

$$Q_q = \frac{\beta}{\lambda N_q \Delta f} \left[\left(\frac{c_{\mathcal{SR}_q}}{a_q^{d_{\mathcal{SR}}} d_{\mathcal{SR}}^{\alpha}} \right)^{\frac{\mathcal{B}}{\mathcal{B}+1}} + \left(\frac{c_{\mathcal{RD}_q} a_q^{-(D-\delta-d_{\mathcal{SR}})}}{(D-\delta-d_{\mathcal{SR}})^{\alpha}} \right)^{\frac{\mathcal{B}}{\mathcal{B}+1}} \right]^{\frac{\mathcal{B}+1}{\mathcal{B}}} \tag{12a}$$

$$T_q = (\beta c_{\mathcal{SR}_q} [\lambda N_q \Delta f a_q^{d_{\mathcal{SR}}} d_{\mathcal{SR}}^{\alpha}]^{-1})^{\frac{\mathcal{B}}{\mathcal{B}+1}} - 1 \tag{12b}$$

$$V_q = (c_{\mathcal{SR}_q} P_{\mathcal{S}_q} a_q^{D-\delta-d_{\mathcal{SR}}} (D-\delta-d_{\mathcal{SR}})^{\alpha})^{\mathcal{B}} + (c_{\mathcal{RD}_q} P_{\mathcal{R}_q} a_q^{d_{\mathcal{SR}}} d_{\mathcal{SR}}^{\alpha})^{\mathcal{B}} \tag{12c}$$

For large value of n, it is quite difficult to solve $(2n+2)$ equations, therefore next a low-complexity approximation algorithm has been proposed.

Approximation Algorithm

In the proposed algorithm, first the optimization is done within a sub-band of the two links for a given RP, then the optimization across the sub-bands are evaluated. Finally, the relay location is optimized after solving a single equation.

Optimal PA Within a Sub-Band

For a given power budget along a sub-band $P_{t_q} = P_B / n$, $\bar{\gamma}_q$ is maximized by evaluating $\partial\bar{\gamma}_q / \partial P_{\mathcal{S}_q} = 0$ after substituting $P_{\mathcal{R}_q} = P_{t_q} - P_{\mathcal{S}_q}$ which gives the optimal power allocations $P^*_{\mathcal{S}_q}$ and $P^*_{\mathcal{R}_q} = \mathcal{Z}_q P^*_{\mathcal{S}_q}$ across the sub-bands, where

$$\mathcal{Z}_q \triangleq \left(c_{\mathcal{SR}_q} a_q^{D-\delta-d_{\mathcal{SR}}} \left(D-\delta-d_{\mathcal{SR}}\right)^{\alpha} \left[c_{\mathcal{RD}_q} a_q^{d_{\mathcal{SR}}} d_{\mathcal{SR}}^{\alpha} \right]^{-1} \right)^{\frac{\mathcal{B}}{\mathcal{B}+1}} \tag{13}$$

Optimal PA Across the Sub-Bands

As $P_{\mathcal{R}_q} = \mathcal{Z}_q P_{\mathcal{S}_q}$, $\{P_{\mathcal{R}_q}\}_{q=1}^{n}$ can be eliminated from $\mathcal{L}_1$ which can be expressed as:

$$\mathcal{L}_2 = \prod_{q=1}^{n} \left(1 + P_{\mathcal{S}_q} [\mathcal{K}_q]^{-1}\right) - \lambda\left(\sum_{q=1}^{n} P_{\mathcal{S}_q} (1+\mathcal{Z}_q) - P_B\right) \tag{14}$$

where

$$\mathcal{K}_q = N_q \Delta f a_q^{d_{\mathcal{SR}}} d_{\mathcal{SR}}^{\alpha} \left(1 + \mathcal{Z}_q^{\frac{1}{\mathcal{B}}}\right)^{\frac{1}{\mathcal{B}}} \left[\beta c_{\mathcal{SR}_q}\right]^{-1}.$$

Now the KKT conditions for evaluating optimal PA across the sub-bands are given as:

$$\frac{\partial \mathcal{L}_2}{\partial P_{\mathcal{S}_q}} = \frac{1}{\mathcal{K}_q} \prod_{j=1, j\neq q}^{n} \left(1 + \frac{P_{\mathcal{S}_j}}{\mathcal{K}_j}\right) - \lambda(1+\mathcal{Z}_q) = 0 \tag{15a}$$

$$\lambda\left(\sum_{q=1}^{n} (1+\mathcal{Z}_q) P_{\mathcal{S}_q} - P_B\right) = 0 \tag{15b}$$

It is evident that (15a) is not satisfied by $\lambda^* = 0$. So for $\lambda^* > 0$, $\{P^*_{\mathcal{S}_q}\}_{q=1}^{n}$ and λ^* are computed as (Prasad, G., Mishra, D. & Hossain, A., 2018):

$$P^*_{\mathcal{S}_q} \triangleq \frac{P_B + \sum_{j=1}^{n} (1+\mathcal{Z}_j)\mathcal{K}_j - n(1+\mathcal{Z}_q)\mathcal{K}_q}{n(1+\mathcal{Z}_q)} \tag{16a}$$

$$\lambda^* \triangleq (1+\mathcal{Z}_1)(\mathcal{K}_1 + P_{\mathcal{S}_1})^{n-1} \prod_{j=1}^{n} \frac{1}{\mathcal{K}_j (1+\mathcal{Z}_j)} \tag{16b}$$

Optimization of RP

Now after substituting (16a) and (16b) into $\mathcal{L}_2$, the new Lagrangian function $\mathcal{L}_3$ is a function of single variable d_{SR}. By evaluating $\frac{\partial \mathcal{L}_3}{\partial d_{SR}} = 0$, optimal relay location d_{SR}^* which is after substitution in (16a) gives $P_{S_q}^*$ and $P_{R_q}^*$ can be computed using the relationship $P_{R_q}^* = \mathcal{Z}_q P_{S_q}^*$. Here note that irrespective of the value of n, the optimal point can be evaluated by solving only one equation.

Numerical Results

The default system parameters for the numerical results are: the operating frequency range is 5 to 15 kHz, direct link length $D = 10$ km, location of relay $d_{SR} = 5$ km, number of sub-bands $n = 260$, threshold data rate $r = 1$ kbps, Rice factor $K = 3.01$ dB, spreading factor $\alpha = 1.5$, and power budget $P_B = 100$ dB re μ Pascal. In the benchmark scheme, the uniform PA (UPA), i.e., $P_{S_q} = P_{R_q} = P_B / (2n)$ is allocated over all sub-bands and relay location is fixed at $d_{SR} = 0.5D$.

In Figure 11, the optimal PA (OPA) is compared against fixed PA (FPA) scheme where uniform power is allocated at the sub-bands of $\mathcal{S}$ or $\mathcal{R}$. OPA always performs better in outage minimization, also it can be observed that with lower PA over the sub-bands at the source nodes through FPA, the relay location d_{SR} is shifted leftward and vice versa to minimize the outage probability. Also, the performance of actual and approximation algorithm in the outage performance is almost same.

The three optimization schemes, ORP: optimal relay placement with uniform PA across the all sub-bands, OPA: optimal PA with a fixed relay location $d_{SR} = 0.5D$, and the joint optimization always perform better than the benchmark scheme as shown in Figure 12. Also, joint optimization outperforms compared to the ORP and OPA. Significant performance enhancement is achieved through joint and ORP when the assymetricity in the two links increases.

Figure 11. Variation of outage probability with location of relay in fixed power allocation scheme

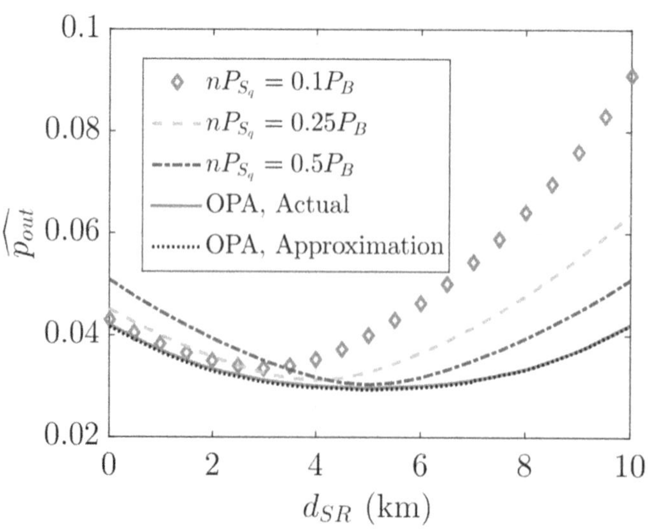

Figure 12. Performance comparison (in %) of different optimization schemes against a benchmark scheme

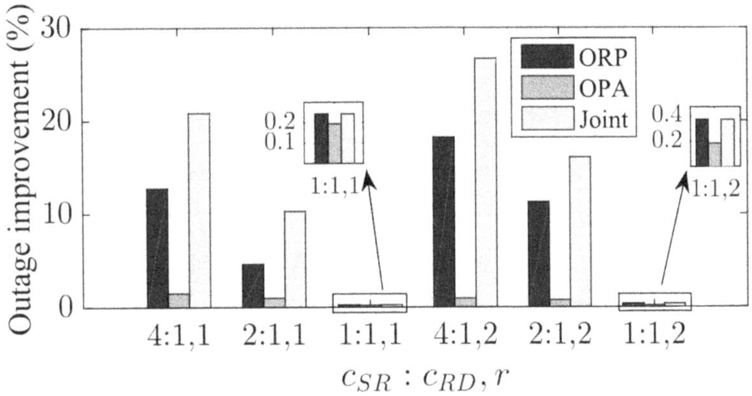

CONCLUSION AND FUTURE DIRECTIONS

The design of QoS-aware energy-efficient network in terrestrial and UAN according to the interoperability challenges in their communication links has been described in this chapter. In the terrestrial cellular network, the total operational power cost can be minimized by jointly optimizing the deployment, PA and density of the BSs over the field of interest. In case of UAN, the channel is entirely frequency-selective compared to the terrestrial channels. Also, the optimal solution for the joint optimization of PA and RP is intractable in continuous frequency domain. Therefore, the solution can be realized by discretizing the frequency domain, but for large number of sub-bands in the discrete frequency domain, it is cumbersome to solve large number of system of equations. So, a low-complexity approximation algorithm has been presented using the recent developments to find the global points.

As mentioned in the introduction section of the chapter that apart from the terrestrial and underwater networks, FSO networks is also gaining interest in recent developments. However challenges in FSO networks for joint optimization of PA over source and relay nodes arises due to fading caused by atmospheric turbulence, pointing error, and AOA which will be investigated in future. Also, due to high data rate in FSO, the QoS based on limiting the bit-error is suitable for the QoS-aware energy-efficient network design. Harvesting energy from the dedicated radio frequency (RF) source(s) has also recently grown as a promising technology for practically realizing the goal of perpetually operating IoT networks (Mishra, D. & De, S., 2016). Consequently, the smart communication strategies and resource allocation protocols as presented in this chapter can be merged together with these ones, along with the usage of green signal processing techniques (Mishra, D. & Alexandropoulos, G. C., 2018) for multi-antenna wireless powered communications, to move towards the goal of sustainable 5G networks.

REFERENCES

Abou-Rjeily, C., & Haddad, S. (2014). Inter-relay cooperation: A new paradigm for enhanced relay-assisted FSO communications. *IEEE Transactions on Communications*, *62*(6), 1970–1982. doi:10.1109/TCOMM.2014.2316262

Andrews, J. G. (2005). Interference cancellation for cellular systems: A contemporary overview. *IEEE Wireless Communications*, *12*(2), 19–29. doi:10.1109/MWC.2005.1421925

Andrews, J. G., Baccelli, F., & Ganti, R. K. (2011). A tractable approach to coverage and rate in cellular networks. *IEEE Transactions on Communications*, *59*(11), 3122–3134. doi:10.1109/TCOMM.2011.100411.100541

Arnold, O., Richter, F., Fettweis, G., & Blume, O. (2010, June). Power consumption modeling of different base station types in heterogeneous cellular networks. In *Future Network and Mobile Summit, 2010* (pp. 1–8). IEEE.

Babu, A. V., & Joshy, S. (2012). Maximizing the data transmission rate of a cooperative relay system in an underwater acoustic channel. *International Journal of Communication Systems*, *25*(2), 231–253. doi:10.1002/dac.1266

Cao, R., Yang, L., & Qu, F. (2010, April). On the capacity and system design of relay-aided underwater acoustic communications. In *Wireless Communications and Networking Conference (WCNC)*, 2010 *IEEE* (pp. 1-6). IEEE.

Darehshoorzadeh, A., & Boukerche, A. (2015). Underwater sensor networks: A new challenge for opportunistic routing protocols. *IEEE Communications Magazine*, *53*(11), 98–107. doi:10.1109/MCOM.2015.7321977

Doosti-Aref, A., & Ebrahimzadeh, A. (2017). Adaptive Relay Selection and Power Allocation for OFDM Cooperative Underwater Acoustic Systems. *IEEE Transactions on Mobile Computing*.

ElSawy, H., Hossain, E., & Haenggi, M. (2013). Stochastic geometry for modeling, analysis, and design of multi-tier and cognitive cellular wireless networks: A survey. *IEEE Communications Surveys and Tutorials*, *15*(3), 996–1019. doi:10.1109/SURV.2013.052213.00000

Fettweis, G., & Zimmermann, E. (2008, September). ICT energy consumption-trends and challenges. *Proceedings of the 11th international symposium on wireless personal multimedia communications,* 2, 6.

Gonzalez-Brevis, P., Gondzio, J., Fan, Y., Poor, H. V., Thompson, J., Krikidis, I., & Chung, P. J. (2011, May). Base station location optimization for minimal energy consumption in wireless networks. In *Vehicular Technology Conference (VTC Spring), 2011 IEEE 73rd* (pp. 1-5). IEEE.

Hassan, M. Z., Leung, V. C., Hossain, M. J., & Cheng, J. (2017, December). Delay-QoS Aware Adaptive Resource Allocations for Free Space Optical Fronthaul Networks. In *GLOBECOM 2017-2017 IEEE Global Communications Conference* (pp. 1-6). IEEE.

Huang, S., & Safari, M. (2017). Free-Space Optical Communication Impaired by Angular Fluctuations. *IEEE Transactions on Wireless Communications*, *16*(11), 7475–7487. doi:10.1109/TWC.2017.2749219

Kam, C., Kompella, S., Nguyen, G. D., Ephremides, A., & Jiang, Z. (2014). Frequency selection and relay placement for energy efficiency in underwater acoustic networks. *IEEE Journal of Oceanic Engineering*, *39*(2), 331–342. doi:10.1109/JOE.2013.2245057

Liu, C. H., & Wang, L. C. (2015, June). Random cell association and void probability in poisson-distributed cellular networks. In *Communications (ICC), 2015 IEEE International Conference on* (pp. 2816-2821). IEEE.

Liu, L., Ma, M., Liu, C., & Shu, Y. (2017). Optimal Relay Node Placement and Flow Allocation in Underwater Acoustic Sensor Networks. *IEEE Transactions on Communications*, *65*(5), 2141–2152. doi:10.1109/TCOMM.2017.2677448

Mekikis, P. V., Kartsakli, E., Lalos, A. S., Antonopoulos, A., Alonso, L., & Verikoukis, C. (2015, June). Connectivity of large-scale WSNs in fading environments under different routing mechanisms. In *Communications (ICC), 2015 IEEE International Conference on* (pp. 6553-6558). IEEE.

Mishra, D., & Alexandropoulos, G. C. (2018). Transmit Precoding and Receive Power Splitting for Harvested Power Maximization in MIMO SWIPT Systems. *IEEE Transactions on Green Communications and Networking*, *2*(3), 774–786. doi:10.1109/TGCN.2018.2835409

Mishra, D. & De, S. (2016, April). Energy harvesting and sustainable M2M communication in 5G mobile technologies. *Internet of Things (IoT) in 5G Mobile Technologies, Springer-Verlag Handbook*, 99–125.

Mishra, D., & De, S. (2017). i^2RES: Integrated Information Relay and Energy Supply Assisted RF Harvesting Communication. *IEEE Transactions on Communications*, *65*(3), 1274–1288. doi:10.1109/TCOMM.2016.2637915

Mishra, D., De, S., & Krishnaswamy, D. (2017). Dilemma at RF Energy Harvesting Relay: Downlink Energy Relaying or Uplink Information Transfer? *IEEE Transactions on Wireless Communications*, *16*(8), 4939–4955. doi:10.1109/TWC.2017.2704084

Mugume, E., So, D. K., & Alsusa, E. (2015, December). Energy efficient deployment of dense heterogeneous cellular networks. In *Global Communications Conference (GLOBECOM)*, 2015 *IEEE* (pp. 1-6). IEEE.

Nouri, H., Uysal, M., & Panayirci, E. (2016). Information theoretical performance analysis and optimisation of cooperative underwater acoustic communication systems. *IET Communications*, *10*(7), 812–823. doi:10.1049/iet-com.2015.0640

Peng, J., Hong, P., & Xue, K. (2015). Energy-aware cellular deployment strategy under coverage performance constraints. *IEEE Transactions on Wireless Communications*, *14*(1), 69–80. doi:10.1109/TWC.2014.2329858

Prasad, G., Mishra, D., & Hossain, A. (2018). Joint Optimization Framework for Operational Cost Minimization in Green Coverage-Constrained Wireless Networks. *IEEE Transactions on Green Communications and Networking*, *2*(3), 693–706. doi:10.1109/TGCN.2018.2828092

Prasad, G., Mishra, D., & Hossain, A. (2018). Joint Optimal Design for Outage Minimization in DF Relay-assisted Underwater Acoustic Networks. *IEEE Communications Letters*, *12*(8), 1724–1727. doi:10.1109/LCOMM.2018.2837019

Saidi, H., Tourki, K., & Hamdi, N. (2016, November). Performance analysis of PSK modulation in DF dual-hop hybrid RF/FSO system over gamma gamma channel. In *Signal, Image, Video and Communications (ISIVC), International Symposium on* (pp. 213-216). IEEE.

Son, I. K., & Mao, S. (2017). A survey of free space optical networks. *Digital Communications and Networks, 3*(2), 67-77.

Srinivasa, S., & Haenggi, M. (2010). Distance distributions in finite uniformly random networks: Theory and applications. *IEEE Transactions on Vehicular Technology*, *59*(2), 940–949. doi:10.1109/TVT.2009.2035044

Stojanovic, M. (2007). On the relationship between capacity and distance in an underwater acoustic communication channel. *Mobile Computing and Communications Review*, *11*(4), 34–43. doi:10.1145/1347364.1347373

Su, G., Li, L., Lin, X., & Wang, H. (2014, July). On the optimal small cell deployment for energy-efficient heterogeneous cellular networks. In *Proceedings Ubiquitous and Future Networks (ICUFN),* 2014 Sixth International Conf (pp. 172-175). IEEE.

Sun, Z. G., Yu, H., & Zhu, Y. J. (2017). A Superimposed Relaying Strategy and Power Allocation for Outdoor Visible Light Communications. *IEEE Access: Practical Innovations, Open Solutions*, *5*, 9555–9561. doi:10.1109/ACCESS.2017.2700880

Verenzuela, D., Björnson, E., & Sanguinetti, L. (2016, May). Optimal design of wireless networks for broadband access with minimum power consumption. In *Communications (ICC), 2016 IEEE International Conference on* (pp. 1-6). IEEE.

Chapter 8
Managing 5G Converged Core With Access Traffic Steering, Switching, and Splitting:
From Hybrid Access to Converged Core

Toktam Mahmoodi
King's College London, UK

Stephen H. Johnson
British Telecommunications plc, UK

Massimo Condoluci
King's College London, UK

Vicknesan Ayadurai
Ericsson Research, Sweden

Maria A. Cuevas
British Telecommunications plc, UK

Mischa Dohler
King's College London, UK

ABSTRACT

This chapter discusses the ongoing work around hybrid access and network convergence, with particular emphasis on recent works on ATSSS in 3GPP. Three main aspects are analyzed: policy enforcement, integration with 5G QoS framework, interaction with underlying multi-path transport protocol. The chapter also provides some preliminary testbed results showing the benefits of ATSSS in the management of multiple accesses analyzing some primary performance indicators such as achievable data rates, link utilization for aggregated traffic, and session setup latency. The chapter also provides some results by considering two examples of realization of ATSSS policies to avoid inefficiency in link utilization and to allow the fulfillment of data rate requirements.

DOI: 10.4018/978-1-5225-7570-2.ch008

INTRODUCTION

The increasing availability of heterogeneous accesses offers an opportunity for exploitation in order to boost network capacity and present new business opportunities as for instance discussed by Raj et al. (2010). The exploitation of fixed and mobile broadband (FBB and MBB, respectively) introduces several benefits, among those: (i) boosting end-user performance; (ii) load balancing; (iii) network optimization; (iv) "always best connected"; (v) session continuity when leaving coverage of one access (i.e., failover).

The exploitation of multiple accesses can be accomplished via different approaches. A first option is to exploit the aggregation of multiple available links towards one destination to boost the data rate of a session. Available links can be managed at the end-points (i.e., the end-device and the remote server) thanks to utilization of transport protocols such as multipath TCP (MPTCP), which is in charge for advertising the availability of multiple links and managing traffic aggregation. In this case, the end-points are unaware of the status of the two networks, and the FBB and MBB links are unaware of the aggregation that is performed. Therefore, this limits the opportunities for network operators to optimize the traffic within their networks.

In order to exploit in a more effective way the availability of multiple links, another approach is the so-called fixed-mobile convergence (FMC), where a service provider is aware of the availability of the different links and exploits this capability directly in a coordinated way. FMC enables the possibility to deliver any service, anywhere and via any access technology. Given the fact that, historically, FBB and MBB networks have been built separately and operated independently, FMC can be realized by exploiting a hybrid access gateway (HAG) considering the terminology used by the Broadband Forum (BBF) in BBF TR-348. The HAG is an aggregation point for the traffic from/to the Border Network Gateway (BNG) and the Packet Gateway (PGW) of the fixed and mobile networks, respectively, as seen in Figure 1(a). In this case, an operator managing the two networks is aware of the multiple accesses and, thus, can properly manage the traffic across them. In addition to improved user performance, the HAG allows to: (i) improve reliability, as traffic can be switched to another access if the performance degrades due to either congestion or mobility issues; (ii) enable seamless experience, as a service can now be accessed via either FBB or MBB. The latter aspects might also enable rapid service deployment by providing services such as high-speed broadband usually delivered via, e.g., fixed access through mobile networks in regions where fixed access is difficult and costly (or vice versa). More details on the improvements introduced through the exploitation of the HAG are given by Samdanis et al. (2017).

Nevertheless, some limitations are due to the fact that the BNG/PGW nodes as well as the HAG have a limited visibility of the network (e.g., layer 1/2 and mobility status is not available) and this limits the possibilities for the overall optimization of the end-to-end paths. Recently, the 3rd Generation Partnership Project (3GPP) has been working towards the concept of having a converged core network on the road to the standardization of the 5G system architecture, which can be found in 3GPP TS 23.501. This solution, shown in Figure 1(b), allows network operators to manage multiple accesses in a coordinated way by considering information related to service (e.g., QoS), network (e.g., statistical performance) and user (e.g., subscription type). To take full advantage of this feature, 3GPP TR 23.793 introduces the access traffic steering, switching and splitting (ATSSS) capabilities within the 5G CN in order to optimize and effectively drive the behavior of the network. The goal is twofold: (i) defining ATSSS policies to map specific services to the access(es) that better support their QoS requirements (i.e., steering) and this is of particular importance to effectively deliver services with strict QoS constraints such as ultra-reliable low latency communications (URLLC); (ii) guaranteeing reliability for ongoing sessions in case of performance degradation (e.g., due to mobility or congestion) by moving to another access or simultaneously using multiple accesses (switching and splitting, respectively). Hence, ATSSS allows service providers to apply the best connectivity for optimal network utilization and end-user experience. In addition, network operators can guarantee consistent performance across the user equipment (UE) as all UEs equipped with the same protocol stack will be able to support the same set of ATSSS policies. The 5G network architecture has to be enhanced with ATSSS capabilities on both user- and control-plane (UP and CP, respectively) in order to support the exploitation of multiple accesses. In the CP, new capabilities are needed to generate and enforce ATSSS policies as well as to trigger commands (switching or splitting) to drive the utilization of the available accesses. To this aim, new policies are needed, containing information about the parameters to monitor in order to trigger, switch or split. The UP has to be enhanced with multi-path capabilities that can guarantee session continuity for switching or splitting and, in order to support this, the UP has to provide additional performance monitoring on both session and access levels. The ATSSS framework needs to be integrated within the ongoing 5G design, by means of having an ATSSS policy delivery integrated within network procedures currently under definition in 5G. Finally, ATSSS needs to be integrated within the 5G protocol stack and, in particular, with the 5G QoS framework to guarantee a proper treatment of the traffic via exploited accesses. The aim of this paper is to provide a technical overview of activities on ATSSS in 5G and above mentioned aspects will be presented in detail in the remainder of this chapter.

The aim of this chapter is to present the status of FMC activities, focusing on both hybrid access and converged CN approaches in order to discuss their capabilities in terms of network configuration. The chapter summarizes the state of the art for hybrid access as well as discussing the recent advances by the 3GPP in supporting the enforcement of ATSSS policies in converged networks. The chapter also analyzes the role played by MPTCP in hybrid access and converged core scenarios as well as the improvements needed in order to better cope with these approaches. A testbed where a HAG aggregates the traffic of a FBB and an in-house commercial-grade LTE network has been developed, the chapter presents some results of experiments conducted with such testbed to understand how capacity and latency imbalances between FBB and MBB links affect the overall performance when using MPTCP to aggregate the traffic. The results also analyze the benefits of FMC in reducing the session setup time, as this aspect has some benefits in improving the network utilization and the radio link efficiency. Finally, results will show some examples of policies to be used for ATSSS in order to improve the utilization of FBB and MBB links.

Figure 1. FMC with a HAG inter-connecting FBB and MBB (a), FMC with converged core network (b)

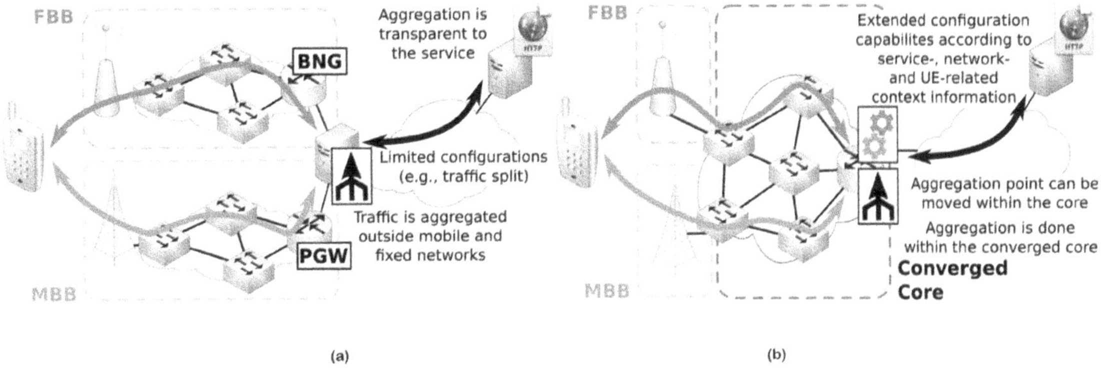

BACKGROUND

Hybrid Access

The BBF is investigating different aspects related to hybrid access with the aim to define and examine scenarios, evaluate possible business models, and drive the technical roadmap to align the industry and to interwork with other standardization bodies.

The reference architecture to move from non-converged networks towards a fully hybrid access network is presented in Figure 2, where a hybrid customer premises equipment (HCPE) offers a coordinated and simultaneous access over different networks via a hybrid access gateway (HAG), which can be placed either at the BNG or PGW. To carry the traffic between the HAG and the HCPE, BBF mainly considers different transport options: layer 3 (L3) overlay tunneling, L3 network-based tunneling and L4 multipath, like MPTCP.

MPTCP is designed with the twofold purpose of improved throughput and resiliency. By building upon widely used TCP and having stringent networking compatibility goals, MPTCP integrates well into many networks and coexists easily with middleboxes like firewall and Network Address Translation (NAT) already in place, which helped to expedite its uptake as for instance described in IETF RFC 8041 where several "real world" use cases of MPTCP are presented.

Alternative means of hybrid access are being investigated in literature. Samdanis et al. (2017) describe several hybrid access use cases and provides a comprehensive overview of the different transport options discussed in both 3GPP and BBF standardization bodies, focusing on architectural concepts and review the implications in control- and user-plane (CP and UP, respectively) and on the role of programmable networks enabled by Network Functions Virtualization (NFV) in supporting enhanced traffic distribution schemes. Leitao et al. (2016) focus on the way towards a unified network for policy and QoS convergence, however authors do not mention the hybrid access use case of network convergence. Coninck et al. (2016) discuss the benefits of hybrid access in terms of bandwidth aggregation, increased reliability and service continuity, with details on the implementation of a carrier-grade MPTCP proxy which allows service providers to implement traffic steering policies using the available access networks.

Figure 2. Hybrid Access reference architecture from BBF TR-348

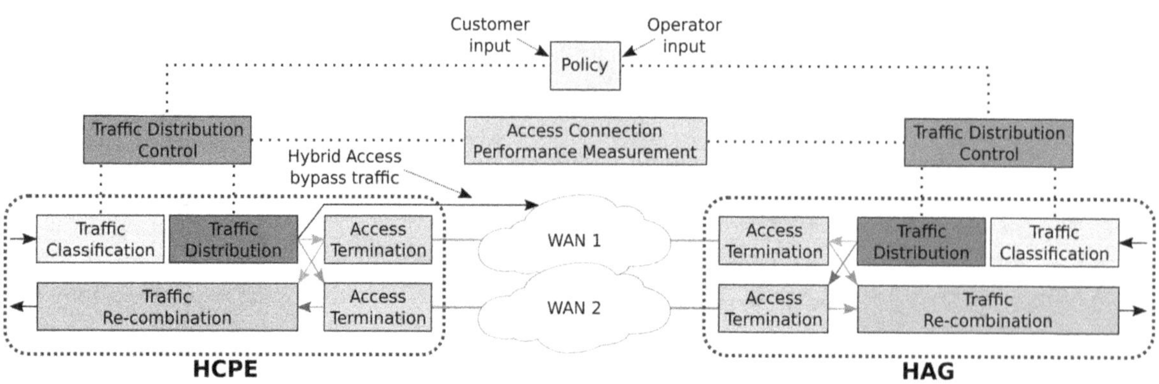

5G Converged Core

The 5G system architecture is defined in 3GPP TS 23.501, and the main components are depicted in Figure 3 which are:

- **Access and Mobility Management Function (AMF):** It handles registration, connection, reachability, and mobility management, access authentication and authorization.
- **Session Management Function (SMF):** It handles session establishment, modification and release, selection of CP and UP functions and configuration of traffic steering of UP entities.
- **Policy Function (PCF):** It provides policy rules to CP entities to govern the network behavior.
- **Unified Data Management (UDM):** It includes the User Data Repository (UDR) containing user subscription data.
- **User Plane Function (UPF):** It interconnects the 5G network to external networks and acts as anchor point for intra-access mobility. The UPF is in charge for packet forwarding/routing, QoS handling and traffic usage reporting.
- **Application Function (AF):** It interacts with the PCF to implement application-tailored policy control.
- **Network Data Analytics (NWDA):** It generates statistical information about network performance.

The support of untrusted non-3GPP is achieved by means of the Non-3GPP InterWorking Function (N3IWF), which extends CP/UP functionalities towards the non-3GPP access with a N2 interface to the AMF and a N3 interface to the UPF. The N3IWF allows devices accessing the 5G core through non-3GPP access to support non-access stratum (NAS) signaling (i.e., N1 interface). The availability of a CP connection via the N3IWF allows the CN to manage a UE connected via a non-3GPP access in a similar way as it were connected via 3GPP access thanks to the availability of functionalities such as NAS signaling and policy enforcement.

Figure 3. 5G architecture as in 3GPP TS 23.501 with ATSSS capabilities

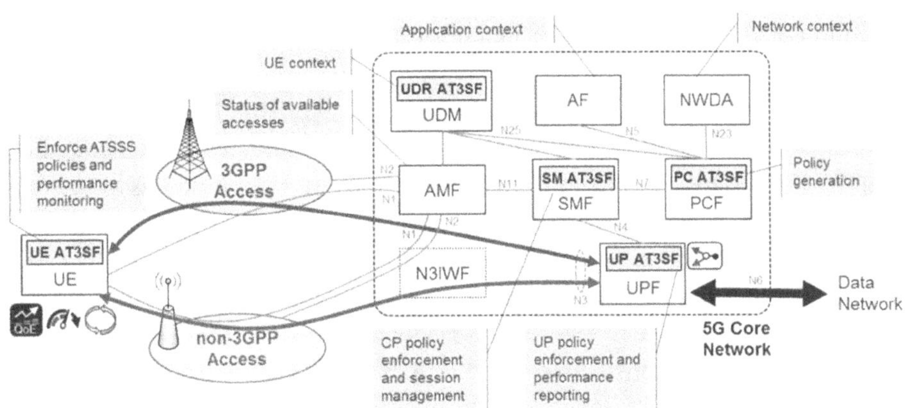

ACCESS TRAFFIC STEERING, SWITCHING AND SPLITTING (ATSSS)

Use Cases and Related Benefits

They key benefits of ATSSS from an operator's perspective are largely around increasing the quality of experience for the user. One important use case is improving the experience of using a multi-connected mobile device; the main issues are associated with poor throughput and edge of coverage experience, and lack of seamless mobility. The typical user behavior to resolve the first two of these issues will be to switch off Wi-Fi on the device which will only lead to poor experience later when the WiFi needs to be switched back on again. ATSSS offers the potential to improve this experience by combining networks to boost the throughput potentially by splitting traffic flows across both accesses. However, not all traffic can be given this treatment, either due to latency requirements, where the experience will be better if the traffic flow is steered or switched to the best performing network, or due to access restrictions where the service is only available on, for example, the local private network, such as a printer.

Another important operator use case is the residential gateway that provides fixed broadband access to either a home or business. In some locations, the fixed line speed available to the residential gateway is limited. Offering a multi-connected residential gateway with 5G ATSSS capabilities would enable the operator to provide a boosted throughput experience similar to that described above, in addition to an increase resilience to network faults and fast provisioning.

One key aspect of providing ATSSS in a 5G converged core that underlies both these use cases, is that the identity of the mobile device or residential gateway is the same across all access networks, and so the same consistent treatment can be applied to traffic such as policy control regardless of access.

ATSSS Architecture

Figure 3 shows one of the approaches being considered in the ATSSS study 3GPP TR 23.793. The 5G system architecture is enriched by means of:

- **UDR - Access Traffic Steering Switching and Splitting Function (UDR-AT3SF):** It holds UE ATSSS subscription data for operator service and user profiles, e.g., subscription profiles (primary, etc.), list of allowed accesses and related priority, access-specific restrictions (e.g., maximum allowed data rate).
- **Policy Control Access Traffic Steering Switching and Splitting Function (PC-AT3SF):** It defines ATSSS policies to drive the behavior of the network. Policies are generated on a UE- and application-basis, with the twofold aim of selecting the most suitable access to steer an application according to its QoS needs and of effectively exploiting the availability of multiple accesses to guarantee adequate QoS for an ongoing session.
- **Session Management Access Traffic Steering Switching and Splitting Function (SM-AT3SF):** It receives policies from the PC-AT3SF and is responsible for ATSSS policy enforcement and session management of all sessions between the 5G CN and UE. It can receive traffic usage reports from either the UE and the CN and, based on such traffic usage reports, it may send commands (e.g., change access or access forbidden) to either UE or CN to optimize ATSSS behavior.

- **User Plane Access Traffic Steering Switching and Splitting Function (UP-AT3SF):** It is responsible for enforcement in the UP of the CN of ATSSS policy rules for downlink traffic. It generates traffic usage reports to be sent to SM-AT3SF.
- **UE - Access Traffic Steering Switching and Splitting Function (UE-AT3SF):** It is the point for ATSSS policy rule enforcement for UE-initiated traffic (i.e., uplink direction) and it might also generate traffic reports sent via the UP-AT3SF or as CP signaling via N1 (i.e., NAS).

ATSSS Policies

The flow chart related to ATSSS policies is depicted in Figure 4. ATSSS policies are generated by the PC-AT3SF. ATSSS steering policies extend the existing UE Route Selection Policy (URSP) framework defined in 3GPP TS 23.502 and 3GPP TS 23.503 and contain information such as:

- **"App" Identifier:** It is used by the packet filter to identify the packets to apply the policy to.
- **Default Access:** It might also contain information about a specific RAT (e.g., 4G or 5G).
- **ATSSS Mode:** It may be one of steered, switched or split.
- **Criteria:** It indicates the criteria to be monitored for the App under consideration. Examples are throughput, packet delay budget, and packet error rate. Such criteria are used as thresholds to trigger changes of the access(es) to be used according to the ATSSS mode of the policy.
- **Measurement Window:** It is the time window used to measure the criteria indicated in the policy.

Figure 4. ATSSS policy generation and enforcement

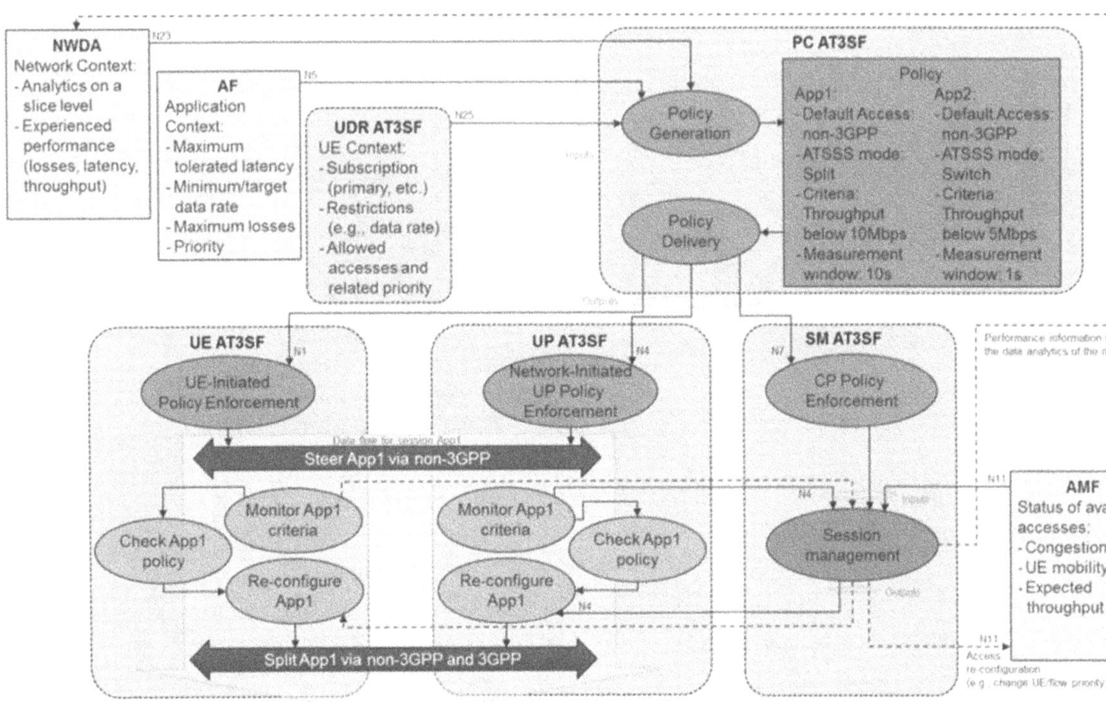

Policies are generated with input from the UDR-AT3SF to consider the capabilities and restrictions across the available accesses according to UE subscription type. The AF provides the PC-AT3SF with application-related input, with focus on QoS parameters.

ATSSS policies are delivered to the UE-AT3SF during UE network registration as an extension of the set of URSP policies. In case of UE- or network-initiated ATSSS traffic, related policies are enforced at UE-AT3SF, SM-AT3SF and UP-AT3SF during the session establishment. UE-AT3SF and UP-AT3SF monitor the criteria defined in the policy to check whether an ATSSS threshold is reached and thus to split/switch the traffic as defined by the ATSSS policy.

The SM-AT3SF receives traffic reports by the UP-AT3SF. Additional reports might also be gathered from the UE-AT3SF. The SM-AT3SF exploits above discussed reports for session management purposes, where ATSSS decisions can be taken to optimize network behavior such as improved balancing mechanisms. Information available at the SM-AT3SF might be enhanced with the availability of access-related information from the AMF, to guarantee that ATSSS decisions will consider the current status of available accesses to avoid, e.g., switching the traffic to a congested access that cannot guarantee the needed QoS.

ATSSS and QoS Framework and Multi-Path Protocol

The ATSSS protocol stack is depicted in Figure 5. ATSSS works as a middleware in between the application and the access-related layers. At the UE, ATSSS policies generated by the PC-AT3SF extend the set of URSP rules delivered during the registration phase and are used to configure the packet filter. In

Figure 5. ATSSS protocol stack

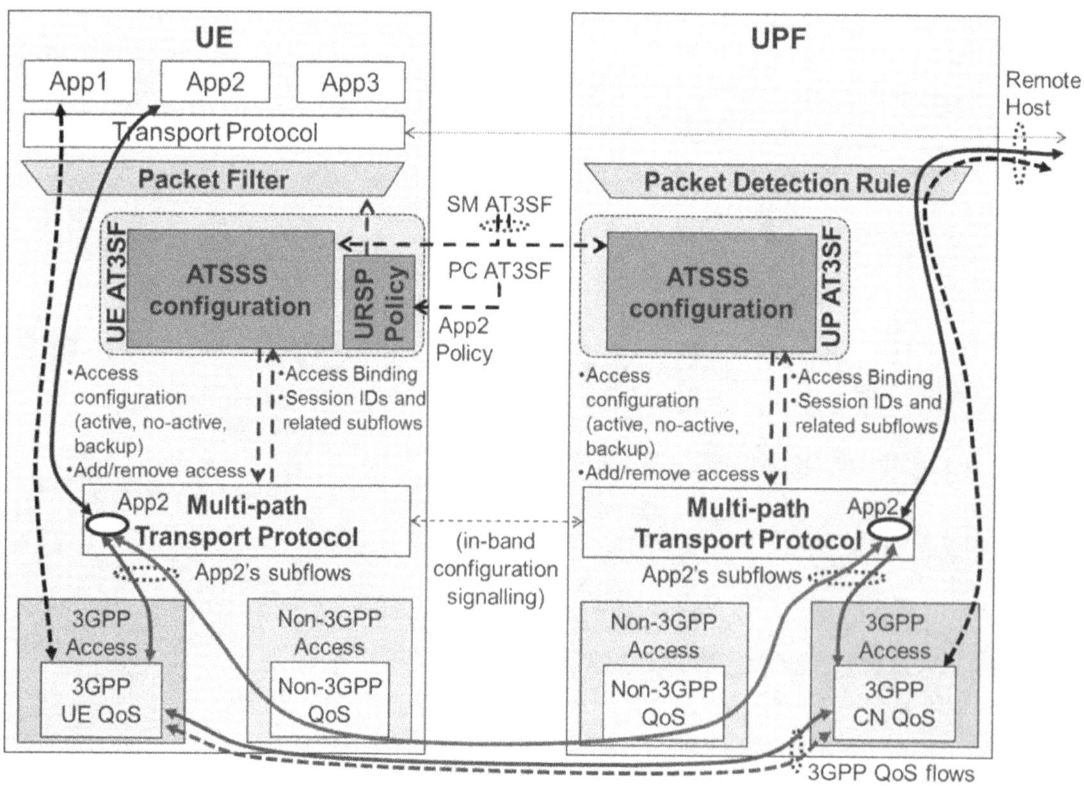

this way, traffic from specific applications can be mapped as ATSSS-traffic (e.g., App2 in the example in Figure 5) and redirected to a layer able to manage the availability of multiple paths (referred to as multi-path transport protocol in Figure 5). Filtering can be implemented by considering the 5-tuple IP/port source and destination and application ID (or a subset of). When a session is established, the SM-AT3SF will be in charge of delivering the related ATSSS rules to the ATSSS configuration entities at the UE-AT3SF and UP-AT3SF, which are responsible for properly configuring the management of available accesses. A multi-path protocol such as MPTCP can be used to guarantee session continuity when switching access or in-sequence delivery in case of splitting.

The ATSSS needs to configure and to inform the multi-path protocol with the following parameters:

- ID (or filtering information) of the connection the configuration refers to.
- Definition of the accesses to be used and related priorities for resilience purposes.
- Capabilities to add/remove access in case an access becomes available/unavailable for an ongoing session.

Additional configuration options might also include: selection of the transport protocol type (e.g., TCP or UDP), configuration of buffer sizes for each access, configuration of transmission window strategies for TCP flows. To implement above listed features, the following information is required by ATSSS from the multi-path protocol: access binding, to guarantee the identification of available accesses; identification of current sessions at both session and subflow levels.

For ongoing sessions, if needed, the SM-AT3SF can update the ATSSS configuration (e.g., switch access). In this case, the update might happen at both UE-AT3SF (via NAS) and UP-AT3SF (via N4) and the update is reflected with an update configuration of the parameters of the multi-path protocol. To reduce the CP signaling via NAS, the update can be triggered at the UP-AT3SF and the novel configuration might be enforced at the multi-path protocol in the CN and delivered at the UE-side via in-band signaling of the multi-path protocol, if this feature is available. This solution might also guarantee a quicker re-configuration compared to NAS signaling, as delivered directly with data transfer.

TESTBEDS

The aim of the testbeds presented in this chapter is to create a platform enabling hybrid access over FBB and MBB. The testbed depicted in Figure 6 comprises of physical LTE, 4G core network and Ethernet links, that provide a realistic setup. In order to conduct an analysis considering performance as in real deployments, the testbeds have been configured by taking into account statistical analysis FBB and 4G MBB performance in UK provided by Ofcom (2015, 2017).

The testbed consists of an MPTCP-capable HCPE which communicates with an MPTCP-capable HAG via an operator which runs FBB and MBB. HCPE and HAG are Linux-based computers with an experimental in-house developed implementation of MPTCP. A FreeBSD-based router acts as a common anchor point for FBB and MBB and connects both networks to the HAG via a high-speed Gigabit Ethernet to accommodate traffic from/to both networks.

The HCPE is multi-homed with MBB and FBB interfaces. The MBB link is made via an LTE dongle which dials up to an in-house commercial-grade LTE eNB. Traffic from/to the eNB is managed by a PGW for the UP traffic and by a Mobility Management Entity (MME) for the control-plane traffic. MBB's RTT

Figure 6. General setup of the testbed

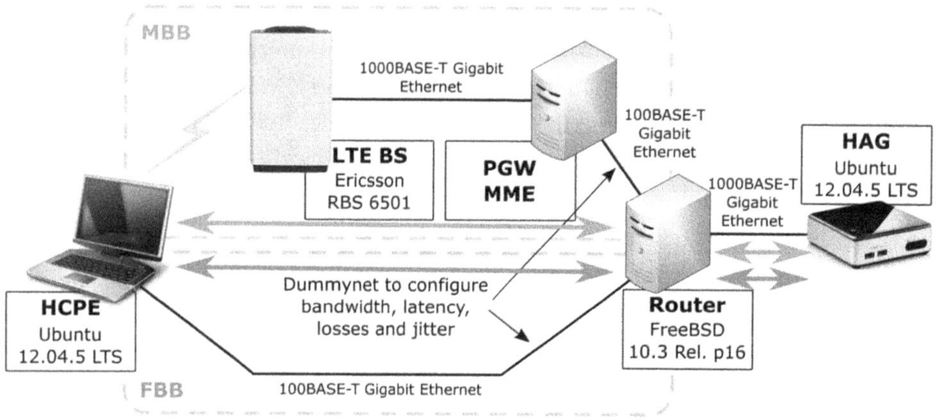

is ~53ms, with throughput performance ~20Mbps and ~5Mbps in downlink and uplink, respectively. The second interface of the HCPE is connected via Ethernet directly to the FreeBSD router. Employing the dummynet tool on this interface, we emulate the FBB environment denoted as "up to 76Mbps FTTC" by Ofcom (2017), with a RTT equal to ~13ms and average rates of ~70Mbps and ~20Mbps in downlink and uplink, respectively.

The HCPE and the HAG act as the end-points in our experiments, hosting the applications used for testing. TCP-based traffic is managed by running several applications targeting the transmission of a 100MB file: scp has been used in both downlink and uplink directions to consider an application with a complex session setup mechanism; iperf has been used for HCPE-HAG traffic; wget for HAG-HCPE traffic. Network analysis tools such as tcpdump and ifstat are used to carry out measurements over the system.

Fixed-Mobile Converge

In this analysis, the aim is to analyze the behavior of FBB and MBB links when using FMC (without ATSSS). Figure 7 focuses on the link rate for the cases without FMC (i.e., MBB only and FBB only) and with FMC (i.e., FBB and MBB used simultaneously). This analysis shows that, when having FMC, the downlink and uplink link rates are ~80Mbps and ~18Mpbs, respectively, meaning that the aggregated link rate is not equal to the sum of the link rates of aggregated accesses. Comparing the performance of FBB and MBB when aggregated with FMC, the utilization of the FBB and MBB links is reduced down to 85% and 82%, respectively, compared to the case when FBB and MBB are used without FMC. On average, the impact is more marked in the downlink, which exhibits the highest imbalance between the capacity and latency between the two accesses. This is an important aspect to consider when using protocols such as MPTCP. This behavior is due to fact that MPTCP has to manage the traffic of two links with different capacities and RTTs, and this might have the side effect of reducing the utilization of one link if an acknowledgment from another path is missing due to losses or delays. Summarizing above results, our testbed shows that FMC effectively improves UE's performance in terms of data rate, but also the imbalances in capacity/RTT of the available links affect the gains in terms of aggregated performance and this depends on the transport protocol managing the traffic aggregation.

Figure 7. Impact of fixed-mobile convergence on link utilization

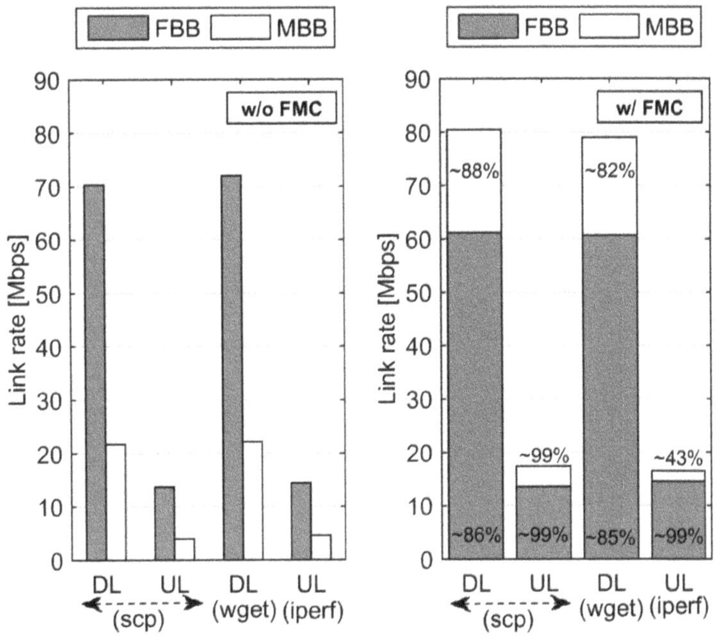

One aspect analyzed in our testbed is the session setup time, measured as the time interval from the moment the application is triggered to the moment the first data packet is transmitted over the radio interface. This analysis allows to understand two aspects: (i) the "waiting" time for the end-user before the effective session start; (ii) the utilization of the available links. The second aspect is interesting from a network point of view, because session setup is composed of small packets for, e.g., key exchange, mutual authentication, and this might impact the overall utilization of the links and especially the spectral efficiency of MBB.

From the analysis for scp in the downlink case, the top plot in Figure 8 shows that the session setup time for FBB and MBB only cases are ~400ms and ~900ms, respectively. The bottom plot in Figure 8 shows that in case of FMC the application traffic starts after ~300ms, i.e., both FBB and MBB links can be used ~100ms and ~600ms earlier compared to the case without FMC. This behavior is given by the fact that also the session setup can happen in parallel via the different links with FMC, i.e., FMC further improves the user performance such that the user will be able to start a session earlier than if having a single FBB or MBB link. In addition, from a network point of view, FMC allows resources to be used more efficiently for the transmission of (possibly huge) data packets instead of being underutilized for a long-time interval.

ATSSS

Apart from the in-house built MPTCP implementation, the testbed realises ATSSS-like capabilities via numerous newly created software modules, which communicate with each other via control messages. A Policy Module to drive the traffic behavior runs on the HAG and enforces policy at both HAG and UE, through changing the MPTCP configuration (i.e., priority of accesses). A Measurement Module

Figure 8. Impact of fixed-mobile convergence on session setup time

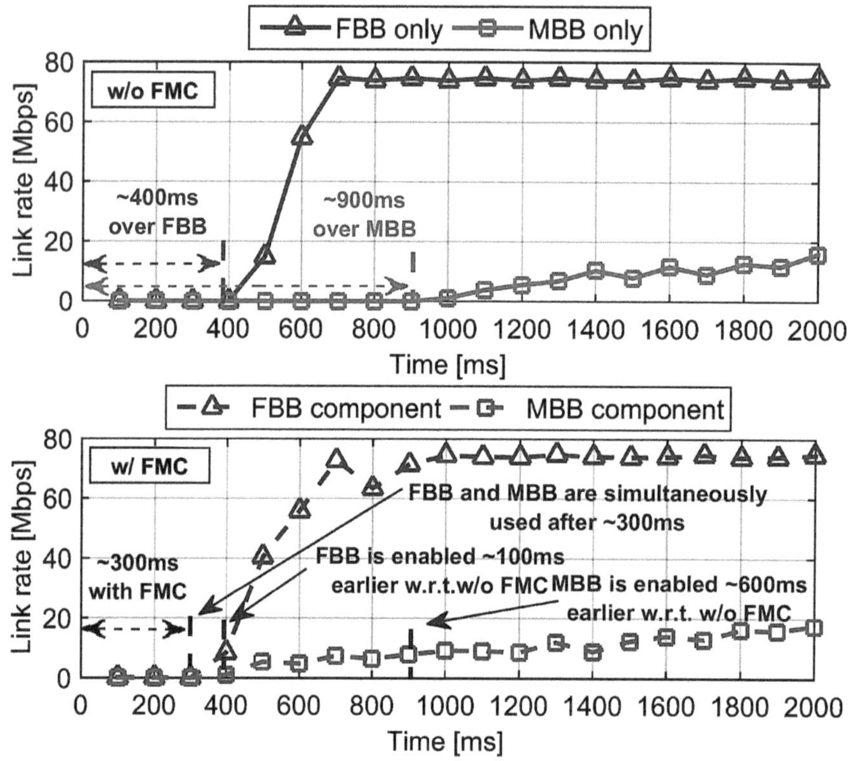

implemented in software on the router performs real-time session-based measurements of the traffic from/to each access. These measurements are sent to the Policy Module on the HAG, which monitors policy-defined criteria and provides the ATSSS control for ongoing sessions. Such decisions cause the Policy Module to reconfigure the behavior of MPTCP at the HAG (i.e., modify the list of access priorities), and the updated configuration is delivered at the UE via MPTCP in-band signaling for use in uplink control. In the following analysis, the focus on downlink direction, and iperf TCP traffic has been used for traffic generation.

For the first analysis, the following ATSSS policy based on traffic duration has been developed: traffic is transmitted initially via FBB, if the network detects that the traffic has been continuously transmitted for 5 seconds, the MBB is enabled and the traffic is split by simultaneously using FBB and MBB. The reason behind this policy is the following: different links may be associated to different costs, and from an operator perspective it could be preferable to avoid using expensive connections to transmit a small amount of data. In our testbed, we considered only one application, in real implementations the measurements on traffic duration can be performed with different granularities (also for application or application class or on a QoS flow level) and different activation thresholds can be defined.

The results of this analysis are depicted in Figure 10. To analyze the impact of the policy described above, the first scenario considered is Case A, where the traffic involves a file transfer of 20MB. In case of FMC without ATSSS, both links are used simultaneously since the application is triggered, with the remark that MBB link requires ~100ms w.r.t. FBB before being fully utilized for data transfer. As a result, the overall file transfer lasts ~2.4s, where both FBB and MBB links are utilized simultaneously

Figure 9. Fixed-mobile convergence testbed enhanced with ATSSS capabilities

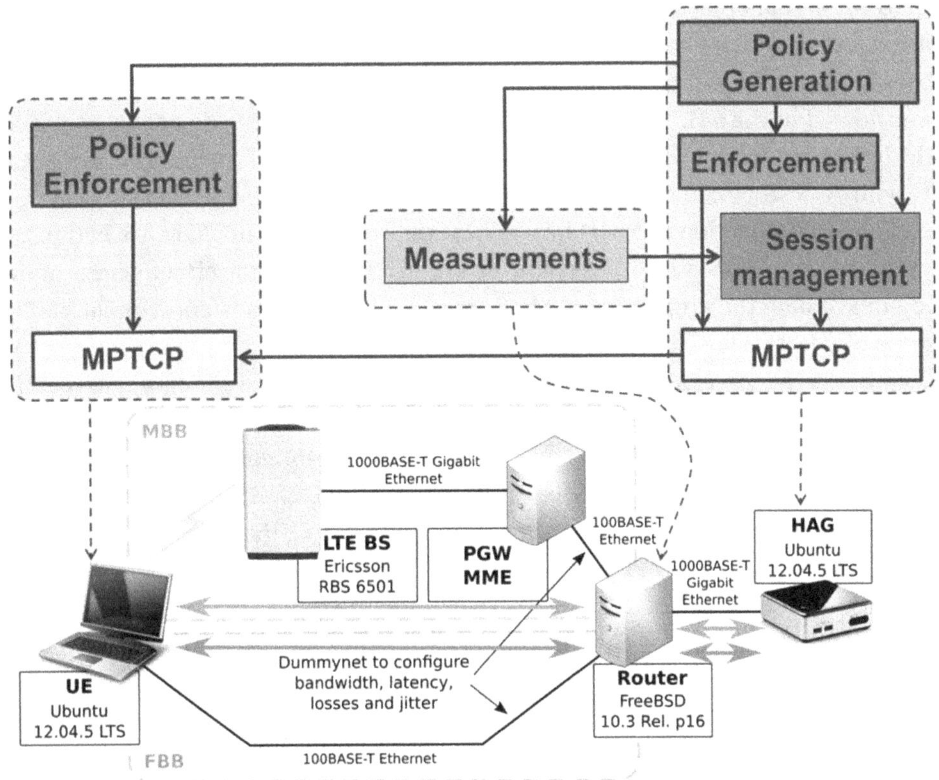

Figure 10. ATSSS policy based on traffic duration

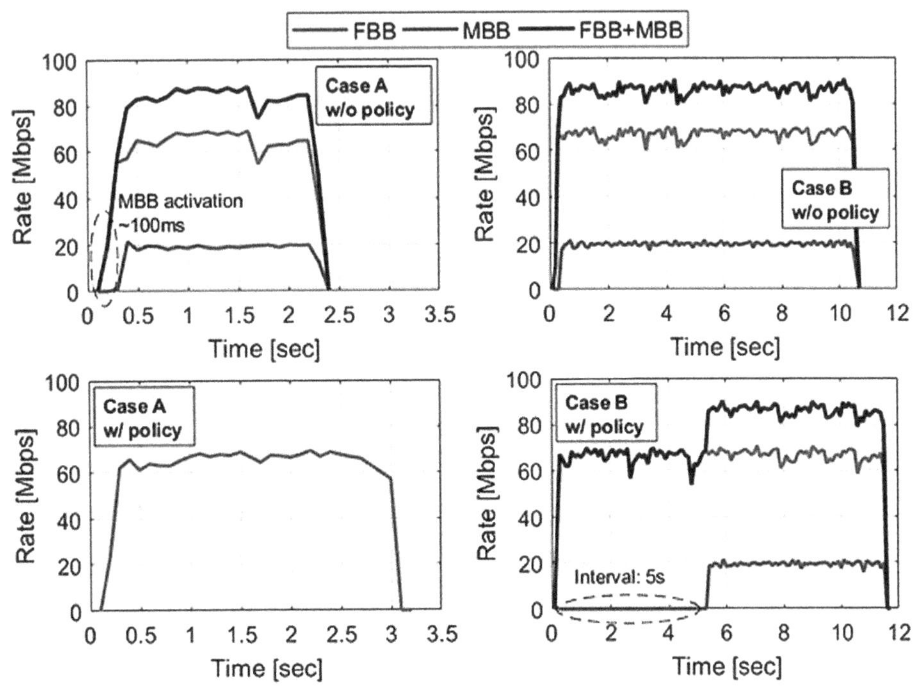

for the whole duration of the transfer. When considering Case A with ATSSS, the small file size allows to use only FBB for the data transfer, which in this case lasts ~3.1s (i.e., MBB has not been activated as the traffic lasted less than 5s). Combining the two results, the overall file transfer when using ATSSS policy has been extended of ~700ms with the advantage of saving ~2.4s of MBB utilization. Another scenario is considered in Case B, where now the file size to be transferred is equal to 100MB. In this case, when FMC is used without ATSSS, the file transfers lasts ~10.7s while it is equal to ~11.7s in case of FMC with ATSSS policy based on traffic duration. The overall file transfer duration has been increased of ~1s, with a reduction of MBB link utilization from ~10.7s down to ~6.7s (i.e., a savings of ~38% in terms of time the MBB is up). The results in Figure 10 testifies the effectiveness of the designed ATSSS policy in avoiding the utilization of MBB link for short file transfers without hardly affecting the performance of file transfer.

The following analysis considers the following ATSSS policy based on data rate requirement: the network aims at offering a data rate of at least 55Mbps to its users, ATSSS is applied by monitoring the performance in terms of link rate of user traffic via FBB and by enabling the MBB link to boost the data rate up to the data rate requirement of 55Mbps. For this analysis, iperf has been used with a file size transfer of 100MB. In Figure 11, three cases are considered: Case C, the user is able to reach ~70Mbps via FBB; Case D, where the user is able to reach ~50Mbps via FBB; Case E, where the user reaches ~40Mbps via FBB. Above listed cases can be associated to different scenarios: (i) the data rate of FBB is affected by the distance of the user from the cabinet (the longer the distance, the lower the data rate of FBB due to longer cabling); (ii) FBB performance varies during the day or during the week due to increased/decreased utilization; (iii) FBB is affected by a sudden reduction of data rate due to anomalies of hardware fault. The results of this analysis are depicted in Figure 11, which shows how ATSSS is

Figure 11. ATSSS policy based on data rate requirement

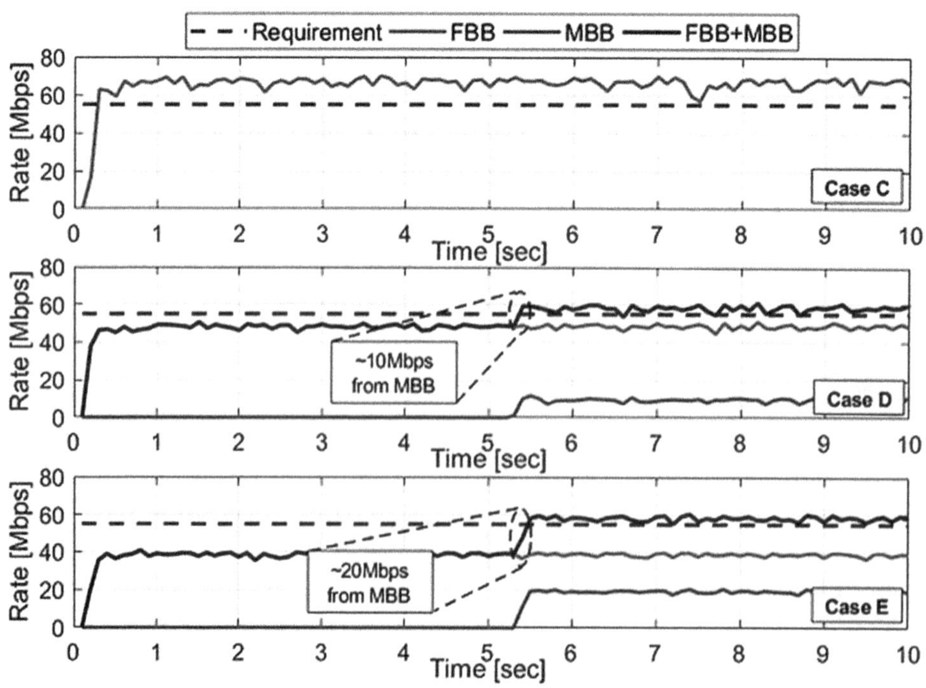

able to boost the data rate for Cases D and thus allowing the traffic to reach the target requirement of 55Mbps. It is worth to underline that the ATSSS policy is able to understand how much traffic from MBB is needed to cope with FBB limitations, thus boosting the traffic with ~10Mbps and ~20Mbps via MBB for Cases D and E, respectively. This policy allows a network operator to offer to users high data rates even in cases where the capabilities of FBB are limited, without incurring in inefficiencies of using MBB for short data transfers (in this analysis, policies on data rate requirement and traffic duration are jointly used).

FUTURE RESEARCH DIRECTIONS

From the results presented in the previous Section, the following outputs can be summarized:

- FMC allows boosting of UE performance, but the multi-path protocol strategy exploited to simultaneously utilize multiple links might limit the gains in terms of aggregated data rate if unbalanced capacity/RTT figures are not managed properly.
- FMC can exploit the availability of different accesses to shorten the session setup time, with consequent benefits from an end-user and network efficiency points of view.
- ATSSS policies to drive the access selection for a new session or to change the access for an ongoing session are beneficial for operators to properly optimize the traffic within their networks to avoid inefficiencies without negatively affecting the users or to guarantee the agreed requirements.

Supporting policies for ATSSS requires several enhancements. Generalizing the policies presented in this chapter, to generate subscriber-oriented policies, the policy framework needs to take into consideration information such as subscriber priority, to guarantee that a given subscriber will be served according to their profile. In addition, to effectively satisfy session requirements, the policy framework needs to retrieve application-oriented policy (e.g., minimum requested data rate, maximum tolerated latency) to avoid steering or switching a session to use an access that is unable to satisfy the service needs. This aspect is of particular interest in order to effectively support services with strict QoS requirements such as URLLC. This also means that information about the ability to satisfy some traffic requirements needs to be available at the network side. Once policies are generated and enforced, switching and splitting can be used to react to changes such as mobility, access availability and congestion. To enable this capability, two components are needed: (i) monitoring of ongoing sessions to allow the network to understand if service requirements are currently satisfied by the selected access; (ii) link with UP policy enforcement points (i.e., UE and UPF) to enable switching/splitting of existing traffic. These two aspects need to be properly investigated, especially regarding the delivery of policies and commands to the UE from the network.

Another aspect to consider is the management of multiple links. As shown the experiments of this chapter, the presence of links with unbalanced capacity/RTT might result in underutilization of available accesses compared to the cases when such accesses are managed without aggregation (i.e., no FMC). Nevertheless, even when the links are balanced in terms of average capacity and delays, issues such as multi-user multi-link scheduling becomes very relevant in large scale deployments. Although FMC should be independent of the transport protocol used to aggregate the available links, effective solutions to avoid the reduction of the link utilization when aggregating multiple links are thus beneficial to further boost the performance of FMC. An example on this direction is provided by Wu et al. (2015), who investigate the problem of load distribution for real-time traffic over multipath networks and propose a goodput-aware load distribution model that takes into consideration path status estimation to accurately sense the quality of each transport link, flow rate assignment to optimize the aggregate goodput of input traffic, and deadline-constrained packet interleaving to mitigate consecutive losses. Another aspect requiring further investigation is related to the size of data packets, as MPTCP-like solutions might be better designed for large packets while other approaches with ideally less overhead should be considered for traffic dealing with small-packets, as for instance analyzed by Grinnemo et al. (2015).

A final remark is that FMC should work regardless of the transport protocol type. Currently, MPTCP is considered as a possible solution for TCP connections, while further improvements are currently under investigation for UDP. Other approaches, such as QUIC, might offer alternative solutions to manage multiple links for UDP connections with less overhead compared to MPTCP.

CONCLUSION

This chapter discussed the ongoing activities on convergence between fixed and mobile networks. Different solutions have been presented considering the standardization efforts by BBF and 3GPP, with focus on hybrid access and aggregated core network with ATSSS capabilities. On the latter aspect, the paper described the further benefits introduced with ATSSS in terms of enforcing policies for traffic management in a unified way in both FBB and MBB. Future directions have been identified and discussed, which can be used to drive the research on this topic.

A testbed has been developed to analyze some preliminary results achieved with FMC, focusing on aspects such as link utilization and session setup time when using MPTCP as a transport protocol to aggregate FBB and MBB. In addition to boosting end-user performance in terms of throughput, results underlined the benefits introduced with FMC in reducing the session setup time, with consequent gains in enhancing network utilization and radio link efficiency. The testbed has been extended with ATSSS-like capabilities, and two examples of policies have been presented: one policy aiming at avoiding the exploitation of MBB link for short file transfers, thus avoiding inefficiencies in using expensive links for small data; one policy on data rate requirement aiming at boosting the data rate to reach a pre-defined rate requirement in case it is not fulfilled via FBB.

ACKNOWLEDGMENT

This work has been supported by the industry grant to King's College London "Low latency Research: Beyond Convergence" funded by British Telecommunications plc and by the industry grant to King's College London "5G and Tactile Internet" funded by Ericsson.

REFERENCES

3GPP. System Architecture for the 5G System (Release 15). TS 23.501.

3GPP. Procedures for the 5G System (Release 15). TS 23.502.

3GPP. Policy and Charging Control Framework for the 5G System (Release 15). TR 23.503.

3GPP. Study on Access Traffic Steering, Switch and Splitting support in the 5G system architecture (Release 15). TR 23.793.

BBF. Hybrid Access Broadband Network Architecture. TR-348.

Coninck, Q. D., Baerts, M., Hesmans, B., & Bonaventure, O. (2016). Observing real smartphone applications over multipath TCP. *IEEE Communications Magazine*, *54*(3), 88–93. doi:10.1109/MCOM.2016.7432153

Grinnemo, K. J., & Brunstrom, A. (2015). A first study on using MPTCP to reduce latency for cloud based mobile applications. *IEEE Symposium on Computers and Communication (ISCC)*.

IETF. Use Cases and Operational Experience with Multipath TCP. RFC 8041.

Leitao, F., Ros, R. D. C., & Riu, J. R. (2016). Fixed-mobile convergence towards the 5G era: Convergence 2.0: The past, present and future of FMC standardization. *IEEE Conference on Standards for Communications and Networking (CSCN)*.

Ofcom. (2015). *Measuring mobile broadband performance in the UK. 4G and 3G network performance*. White paper.

Ofcom. (2017). *UK Home broadband performance: The performance of fixed line broadband delivered to UK residential consumers*. White paper.

Raj, M., Narayan, A., Datta, S., Das, S. K., & Pathak, J. K. (2010). Fixed mobile convergence: Challenges and solutions. *IEEE Communications Magazine*, *48*(12), 26–34. doi:10.1109/MCOM.2010.5673069

Samdanis, K., Leitao, F., Oechsner, S., Riu, J. R. I., Ros, R. D. C., & Fabregas, G. (2017). From Interworking to Hybrid Access Systems and the Road toward the Next-Generation of Fixed-Mobile Convergence. *IEEE Communications Standards Magazine*, *1*(1), 36–43. doi:10.1109/MCOMSTD.2017.1600536ST

Wu, J., Yuen, C., Cheng, B., Shang, Y., & Chen, J. (2015). Goodput-Aware Load Distribution for Real-Time Traffic over Multipath Networks. *IEEE Transactions on Parallel and Distributed Systems*, *26*(8), 2286–2299. doi:10.1109/TPDS.2014.2347031

KEY TERMS AND DEFINITIONS

Access Traffic Steering Switching and Splitting (ATSSS): A network capability to configure policies to statically or dynamically manage multiple accesses, either switching from one access to another or simultaneously using multiple accesses.

Hybrid Customer Premises Equipment (HCPE): A device able to simultaneously exploit multiple accesses.

Fixed Broadband (FBB): Wired access (e.g., ADSL).

Fixed Mobile Convergence (FMC): A network which is composed of both fixed (e.g., ADSL) and mobile (e.g., 3G, 4G, 5G) accesses.

Hybrid Access Gateway (HAG): A network component that inter-connects two networks at the level of the border router.

Mobile Broadband (MBB): Wireless access (e.g., 3G, 4G, 5G).

Quality of Service (QoS): A set of characteristics that identify the type of treatment a certain traffic should receive in the network.

Chapter 9
Handoff Management in Macro-Femto Cellular Networks

Muhammad Faheem Mustafa
COMSATS University Islamabad – Wah, Pakistan

Ayaz Ahmad
COMSATS University Islamabad – Wah, Pakistan

Raheel Ahmed
COMSATS University Islamabad – Wah, Pakistan

ABSTRACT

With the rapid increase in data traffic and high data rate demands from cellular users, conventional cellular networks are becoming insufficient to fulfill these requirements. Femto cells are integrated in macro cellular network to increase the capacity, coverage, and to fulfill the increasing demands of the users. Time required for handoff process between the cells became more sensitive and complex with the introduction of femto cells in the network. Public internet which connect the femto base station with the mobile core network induces higher latency if conventional handoff procedures are also employed in macro-femto cell network. So, handoff process will become slower and network operation will become insufficient. Some standards, procedures, and protocols should be defined for macro-femto cell network rather than using existing protocols. This chapter presents a comprehensive survey of handoff process, types of handoff in macro-femto cell network, and proposed methods and schemes for frequent and unnecessary handoff reduction for efficient network operation.

INTRODUCTION

Over the last few years, there is a tremendous increase in the ratio of smart phones and tablets etc. As a result, the number of users in the network are increasingly aggressively and their requirements of bandwidth and throughput are increasing exponentially in recent times. Cisco research group has shown a 39 fold increase in the data traffic and throughput demand from year 2009 to 2014 (Reardon 2010). Due to this rapid increase in the data traffic and users demand of higher throughput for their multime-

DOI: 10.4018/978-1-5225-7570-2.ch009

dia application like video streaming, video conferencing and online gaming etc. Should we keep small cells or femto cells? Conventional wireless cellular network was becoming short of resources which was severely degrading the QoS of the users. To cope up with these limitations of the network, concept of heterogeneous network was evolved which integrates the small cells like femto, micro and pico cell in the conventional network and forming multi-tier communication. The next generation wireless communication networks are promising in terms of providing the larger bandwidth and higher throughput while assuring the quality of service (QoS) to the users for both voice and different data and multimedia services. Femtocell networks provides good quality of services with enhanced throughput to the users in the indoor environment with lower cost. Transmitter and receiver deployed closer to the user increase the capacity of wireless link as well as it creates dual benefits of better quality of links and more spatial reuse of spectrum resources (Liu et al. 2010). So, femtocell is one of the best approaches for the heterogeneous convergence networks for coverage area extension and capacity maximization.

A femtocell is a small base station with shorter communication range and low transmit power also referred to as home-enhanced NB (H(e)NB). This base station supports fewer users as compared to macro base station and employs the frequency reuse concept to enhance the system capacity. Femto cells are normally installed indoor to provide communication coverage to the users affected penetration losses and low SINR of macro cell signals (Chandrasekhar, Andrews, and Gatherer 2008a). One of the most important feature of a cellular communication system is the freedom of mobility but in this system, this service will be provided by supporting handoff process also known as handover from one cell to another without any interreption. Handoff is the process of changing the frequency, time slot, channel, spreading code, or combination of them from one base station to another while communication session is in progress. It is often initiated when any user crosses the cellular region or degradation of QoS in current channel, Different types of handoff algorithms are investigated by research community which are based on received signal strength (RSS), user's velocity, dwell time, SINR value and users association with predefined threshold have been studied in (Halgamuge et al. 2005). The threshold sets a minimum handoff criteria for handoff initiation but hysteresis model adds a margin H(hysteresis threshold) to the threshold of RSS or dwell time from the serving BS.

The integration of femtocell with conventional macrocell networks is not very simple and easy because it leads several types of problems i-e, radius size of femtocell is very small which means, we required a large number of femto cells up to several hundred within a macrocell. If a UE attempts handoff between macrocell and femtocell then it will experience more severe SINR degradation than UE moving between macrocell at the speed. In otherwords, we can say that interference also increases by deploying the femotocells within a macrocell. Interferrence also increase as we increase the femtocells within a macrocell.

An effective and comprehensive solution for these problems may lead towards efficient femtocell network operation (Bae, Ryu, and Park 2011). The key advantage of femto cell technology from user's perspective is that, users required no any other equipment for communication setup. Due to this advantage, provision of high data rate and the deployment cost of femto cell is very low increasing the capacity of the system. This makes the femto cells feasible for deployment at a large scale which is the ultimate objective of this technology (Nasser, Hasswa, and Hassanein 2006) (Akiyama et al. 2003). Femto cells are connected with core network through internet broadband backhaul connection whereas macro cell are connected with core network through dedicated lines. Handoff time between macro-macro cell is maximum up to 100ms whereas handoff between macro-femto cell or femto-femto cell takes more than 200ms due to longer transmit time required for packet transfer in broadband internet (Dimou et al. 2009).

Hence, current handoff procedures defined for macro cellular network will not work well for femto cell integrated heterogeneous environment. In conventional cellular networks the standardized legacy handoff processes are controlled and executed by the mobile core network (MCN). User equipment cache (UE) associated by which macro base station is also a part of this core network. In standardized handoff process from one base station to another which is managed by the MCN, it is assumed that the latency for handoff process is minimum between the macro base stations connected with the same core network (Leung 1992).

Femtocells are densely deployed and integrated with the macro cells cellular network which will impact on the architecture of current cellular systems (Zhang and De la Roche 2011). Integration of femto cells with the conventional macro cells will make the network structure multi-tier i.e. two different communication tiers for macro cell and femto cells would be formed. In these communication tiers, macro cell will be served to provide cellular coverage to the mobile users and users at longer distance whereas, femto cell is served to fulfill the higher throughput demand of the indoor and comparatively static users with service provision of different multimedia applications with short range base stations. Although two-tier communication network increases the sum rate of the network and also extends the coverage area but topology of two-tier networks also induces some new design challenges, like cross-tier interference at downlink or uplink of the communication due to the spectrum sharing and reuse at both communication tiers.

Handoff between femto-femto cell and macro-femto cell is challenging as it is difficult to find4 the most appropriate femto cell among several candidate target cells for optimal handoff execution (Habeeb and Qadeer 2009). In femto-femto cell handoff scenario, it is necessary to maintain a minimum neighbor femtocell list because it will take much more power for scanning the multiple neighbor femtocells for handoff and will cause a significant MAC overhead as femtocell are deployed with high density. Seamless handoff is executed if neighboring femto cells coordinates with each other to maintain the effective neighborhood list. Large neighbor cell list causes unnecessary scanning for candidate handoff cell and takes much longer time for handoff initiation process with selected optimal candidate cell. Neighbor cell list should be precise and as short as possible and it also must considers the hidden femto cells so that signaling message to large number of neighbors is avoided and network resources are not overburdened and utilized efficiently.

In this book chapter, we have thoroughly elaborated the concept of small cell network and its emergence, deployment of femto cells in two-tier heterogeneous network and handoff process in two-tier communication network. Handoff decision making criteria and different scenarios of handoff are also elaborated in detail. This paper is categorized in four main section. Section-I provides the introduction and overview of macro-femto network and enlightens the handoff process and its importance in two-tier communication network. Section-II shows the overall management of handoff and its initiation process. In section-III different handoff decision making criterias are discussed due to which handoff function is induced in the network. In section-IV different handoff scenarios like macro-femto, femto-femto and femto-macro cell handoffs are elaborated in detail and different proposed schemes for these handoff scenarios are presented. Section-V shows the issues and bottlenecks for handoff process in the network and identifies some future research directions to be explored. Section-VI provides an overview and concludes the paper in an effective manner.

HANDOFF MANAGEMENT

In wireless multi cell network, each cell has some communication range which depends upon its cell size, communication area and transmission power (Racz, Temesvary, and Reider 2007). If a user wants to continue its communication session outside the host cell communication area it must have to handoff to the other cell or base station. Handoff is also known as handover and it is a process of changing the radio frequency band, time slot, code words or combination of these parameters depending upon the employed access technique while session is not terminated (Kim and Lee 2010). The procedure of handoff mostly occurs at the cell edge or if the strength of the received signal is low from a specific threshold. In femto-cell networks, both inbound and outbound handoffs are executed. Inbound handoff is executed from upper communication tier to the lower (i.e. from macro cell to femto cell) and outbound handoff is a process of handing-off from lower communication tier to the upper (i.e. from femto cell to macro cell). Issues in the femto cell network becomes more critical due to multi-tier communication network with large number of base stations having reduced coverage area. In femto cell scenario, handoff can occur between femto-femto cell, femto-macro cell and macro-femto cells. Handoff between macro-femto cells is more complicated as the user will experience more interference than users handing off between macro cells. Femto cell network deployment with overlaid macro cell and their handoff scenarios are shown in Figure 1.

A macro cell optimization mechanism is proposed in (Lee et al. 2010) which is based on macro-pico cells configuration but it focuses on in-bound mobility of user, i.e., from macro cell to pico cell. Another handoff mechanism based on femto cell is proposed in (Kim and Lee 2010) which focuses on out-bound mobility, i.e., user requests for handoff from femto cell to macro cell. To avoid frequent handoffs in multi-tier small cell network concept of umbrella cells is also introduced (Furukawa and Akaiwa 1994) . This is the solution for users with high mobility in the small cell network and it covers several small cells under its coverage area. Transmission power inside the umbrella cell area is increased as compared to the small cells. When user with high mobility or frequent handoff is detected it is handed off to the umbrella cell with whom it can stay connected for longer distance and time. Hence, reducing the number of handoffs and network load (Furukawa and Akaiwa 1994) (Chowdhury, Jang, and Haas 2011) (Chowdhury, Jang, and Haas 2010a).

Handoffs process is classified in two types i.e. *hard handoff* and *soft handoff*. Hard handoff is also referred to as break before make connection in which radio link of the user is terminated from current BS and then established with the target BS (either macro or small cell). This process is fast enough that call does not drop and user does not observe any disconnectivity. Soft handoff is also referred to as make before break connection in which first radio link with the target base station is established and handed over and then previous connection is disconnected. In femto cell network, with overlaid macro cell, users are prone to uplink interference due to large asymmetry between downlink and uplink signal strengths. Soft handoff in femto cell network mitigates this problem as currently deployed symmetric coverage is not beneficial in heterogeneous deployment (Balachandran et al. 2012). A cooperative uplink reception scheme is also proposed in (Balachandran et al. 2012) which mitigates the cross cell interference by employing the virtual soft handoff scheme. If handoff occurs between two small cells or two sectors of a macro cell then it is referred to as *softer handoff* (Chowdhury, Jang, and Haas 2010b). In small cell scenario with multiple micro, femto and pico cells in open access mode, handoff occurs very often which is hard handoff due to its easy implementation and support in OFDMA networks. These handoffs reduces the network performance and induces the signaling overhead. Several handoff reduction techniques like

Figure 1. Dense femto cell deployment with overlaid macro base station and their handoff scenarios

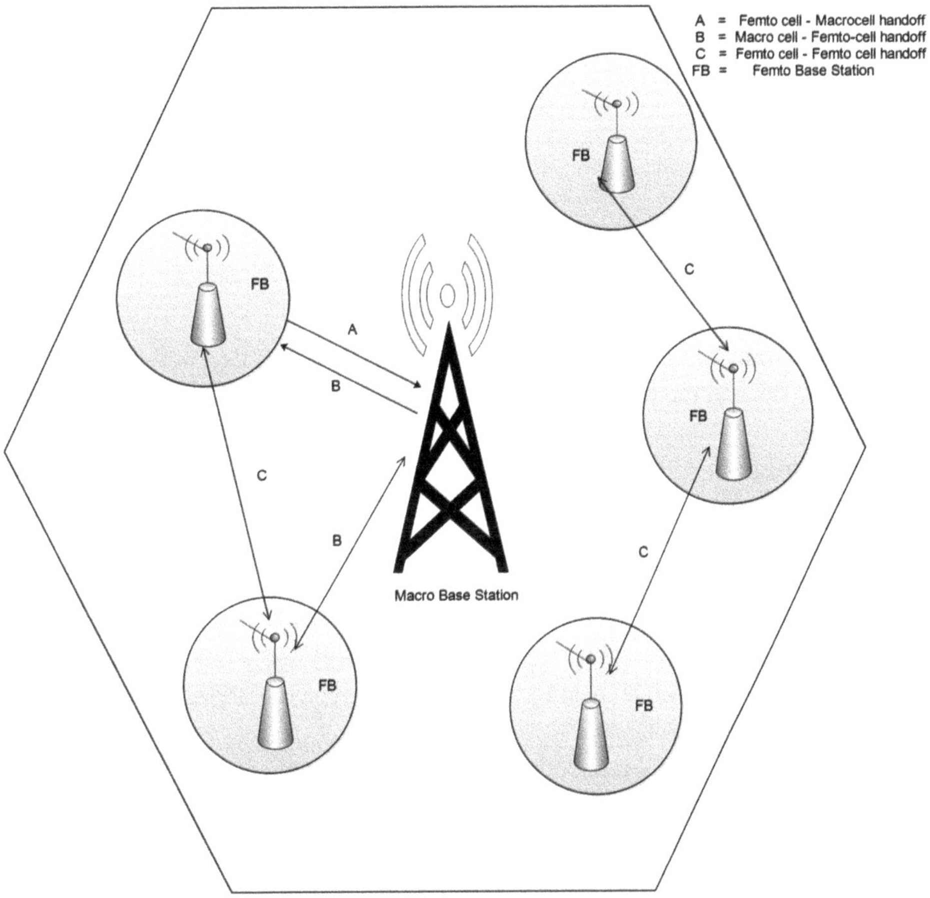

HeNB aided virtual handoff (HaVHO), self-organizing network (SON), HeNB policy function (HeNB PF) and mufti-threshold dependency algorithm (MTDA) are employed in small cell network but some of these techniques reduces the data rate of the network (Andrews et al. 2012) (Chandrasekhar, Andrews, and Gatherer 2008b) (Chowdhury, Jang, and Haas 2010a) (Pantisano et al. 2014). In (Lee et al. 2013) author has investigated the fast cell selection in the two-tier small cell network and has compared it with the hard handoff. Author shows the significant improvement in throughput of the network depending on the backhaul connection. If backhaul connection has large capacity then throughput is high and if backhaul capacity is limited then it shows improvement for low bit rate.

Handoff Initiation Schemes

As discussed earlier, there are several handoff reduction techniques like HaVHO, SON, HeNB PF and MTDA are employed in small cell network but considering all these schemes, handoff decision making process can be divided into three main categories which are centralized, semi distributed and distributed handoff scheme. Comparisons of these categories are given in Table 1. Centralized Handoff method is also known as network controlled handoff. In this method, network makes complete handoff decision (i.e.

Table 1. Handoff decision making categories

Handoff Control	Advantages	Disadvantages	Delay	Application Area	References
Centralized	Fully Network control handoff	Not suitable for femto cells	1 sec or 2 sec	Macro Base Station	(Rappaport 1996)
Distributed	Autonomous, Suitable for femto cells	Not suitable for Macro cells	1 sec	Inter Femto Base Station	(Dharmaraja, Trivedi, and Logothetis 2003) (Tawil, Pujolle, and Demerjian 2008)
Semi-distributed	Partially Network Controll handoff	Not suitable for Inter femto cells	10msec	Macro to femto Base Station	(Vrzic and Fong 2004)

channel switching, data transmission and network switching). Handoff decision is based on the number of users at a particular BS and the received signal strength. This handoff decision at the network level have overburdened the network and brought complexity in it. Centralized handoff method is not suitable for femto cell network overlaid with macro cell due to the fact that there are many femto cells in a region and each cell has to maintain a long neighbor list and handoff signaling is also increased (Hsieh et al. 2014). Delay associated with this type of handoff process is approximately 1s or 2s (Rappaport 1996).

Semi distributed Handoff method is also known as MS-assisted Handoff. In this handoff method as the name MS-assisted indicates, MS makes the handoff measurements and network (MSC) makes the decision of handoff based on the received signal strength (RSS) and signal to noise and interference ratio (SINR) measurements of MS. Delay associated with this type of handoff process is approximately 1s (Vrzic and Fong 2004).

Handoff method mostly used in small cells and multi-tier network is distributed Handoff. It is also known as MS controlled handoff. In this method, MS measures the RSS and interference from all nearby cells and initiates the handoff process, if RSS from a particular cell is higher than the current one by a pre-defined threshold value then the MS requests to target cell for allotment of lowest interference radio channel. In distributed handoff scheme, each MS is fully authorized for the handoff process at its own. This type of handoff schemes are most appropriate for higher mobility and multi cell scenarios. Main advantage of this scheme is its short handoff time i.e. 100ms (Dharmaraja, Trivedi, and Logothetis 2003) (Tawil, Pujolle, and Demerjian 2008). In multi cell scenario, where many femto cells are operating within the macrocell coverage area, and macro cell may not contain information of all neighboring femto cells. Due to which macro cell cannot transfer all the control information to the femto cell. In this scenario, MS controlled handoff schemes would be most suitable to measure and decide the handoff to a particular femto or macro base station. This scheme is not much efficient in single cell scenario because of reliable and efficient centralized handoff scheme, emerging device to device communication and sufficient BS resources for handoff process (Yeh et al. 2008). A distributed handoff mechanism in small cell network is proposed in (Wu, Chu, and Fang 2013a). In this mechanism, a stochastic election process is incorporated for selection of candidate femto cells in handoff of UE. This stochastic election process is evaluated by Monte Carlos simulation which efficiently selects the small number of candidate femto cells for handoff. Performance analysis shows that distributed handoff mechanism performs well as compared to centralized mechanism in dense small cell scenario and induces less uncoordinated interference in the small cell communication tier.

Handoff process with different scenarios in two-tier communication network and different handoff decision making criterias in the network are given in Figure2 which are also elaborated comprehensively in the following sections.

HANDOFF DECISION MAKING CRITERIA

Handoff is a feature of mobile communication especially in macro cell femto cell network where users moves more randomly and frequently to and from the cell communication area. So handoff is more often in this type of network. There are some particular limits and thresholds at which handoff is initiated by either network or mobile entity. These handoff criteria are set very carefully and precisely as they significantly contributes towards network efficiency. The major handoff criteria in macro cell femto cell network are surveyed below.

Figure 2. Handoff scenarios and decision-making criteria

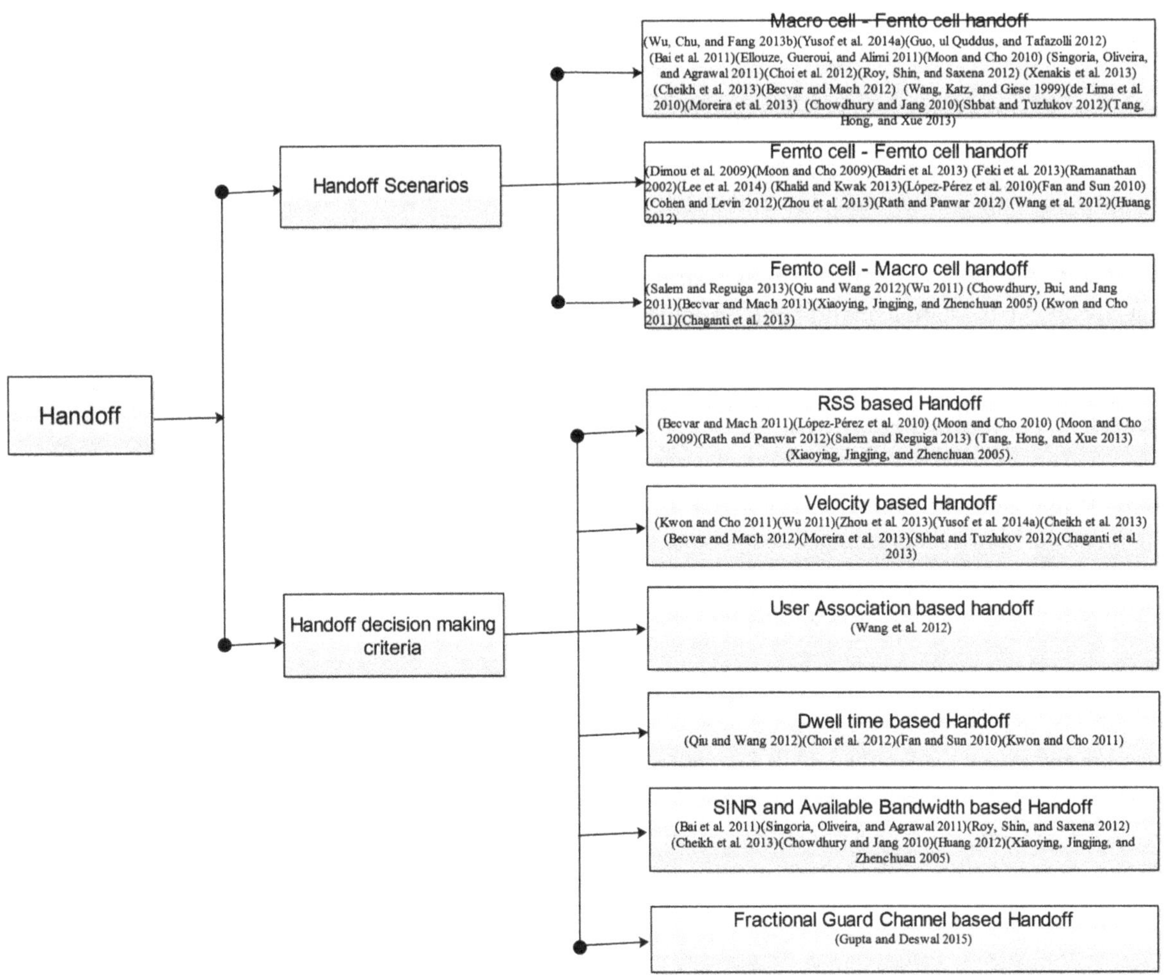

RSS Based Handoff

This handoff scheme is based on received signal strength by mobile user from different base stations with in the area. Each base station keeps the minimum but effective neighborhood list for efficient handoff operation. Mobile user receives the signals from multiple base stations in macro femto co-deployment scenario and during mobility when user has to handoff to some other base station, it makes the handoff decision based on RSS value, i.e. user gets connected to that base station from which it receives the strongest signal. It is an effective criterion for handoff operation which does not includes complex calculations and reduces the signaling overhead of the network.

Velocity Based Handoff

Another criterion for handoff operation is based on the speed of the user equipment (UE). This is an efficient criteria to reduce the frequent handoff load of the network in small cell deployment scenario. In this handoff criteria, decision is based on the velocity of the user, i.e, if UE is moving with higher velocity then femto cell would not be considered as a potential target for handoff and if velocity is lower than both femto and macro cells would be considered for handoff. It improves the QoS of the user as well as the network efficiency.

User Association Based Handoff

Handoff decision is also made based on the association of the user with a particular base station. In this criterion, UE maintains the list of neighborhood base stations and when handoff is required it checks the RSS and SINR values of neighbor base stations and handoffs to the pre-decided base station neglecting that any other base station has higher RSS value in the vicinity ensuring that RSS value of the pre-decided base station must be higher than threshold value. This criteria induces the fairness in the network.

Dwell Time Based Handoff

Handoff decision function is also taken on the basis of dwell time which is the time spent by a UE in a particular cell for its communication without a handoff. If a dwell time of UE is less for a particular cell then handoff will be more frequent. Dwell time of UE is also helpful in measuring the speed of the UE which can be further used for handoff decision.

SINR and Available Bandwidth Based Handoff

Noise in the signal received by a UE is another factor which induces the handoff. If the ratio of noise in received signal is high and available bandwidth for UE is also low then it will reduce the throughput of the user and leads to the inefficient operation of network resources which further reduce the efficiency of the system. UE always look for higher bandwidth and higher SINR value to maximize the throughput. So we may conclude from this, greater the SINR value, lower will be the handoff ratio.

HANDOFF SCENARIOS IN FEMTO CELL NETWORK

In femto cell network with integration of macro base station there are three main scenarios in which handoff process is executed i.e. macro-femto, femto-femto and femto-macro cell handoffs. These scenarios with the existing handoff strategies are thoroughly reviewed in the following sections and also illustrated in Table 2.

Macro Cell: Femto Cell Handoff

In this scenario handoff occurs from macro cell to femto cell due to mobility of UE. Handoff from macrocell to femto cell due to user's mobility is difficult as compared to the two later scenarios due to the large number of candidate femto cells in the user's vicinity and some characteristics of femto cells, e.g., access mode, RSS and radio channel availability. In this handoff scenario, communicaion with multiple femto cells for handover function and a list of large number of neighboring femto cells and macro cell are involved, which may induce a large delay in handoff process and signaling overhead in the network. Different schemes and algorithms proposed for handoff in (Wu, Chu, and Fang 2013a) – (Tang, Hong, and Xue 2013) are illustrated. A macro-femto handoff scheme is proposed in (Wu, Chu, and Fang 2013a). This scheme proposes a handoff mechanism based on "Dutch auction" in which MS works as an auctioneer and candidate femto cell works as a bidder. Furthermore, a stochastic process is also employed for electing the bidders so that the number of candidate bidders can be reduced. Link quality is improved significantly by employing this scheme but there is no coordination among the neighbor femto cells and neighbor discovery is repeated for each handoff process in this scheme. In (Yusof et al. 2014a), authors have proposed a mechanism for macro-femto cell handoff for outdoor users which uses context information from both network and user for femto cell discovery and better handoff decision. Context

Table 2. Handoff scenarios in multi-tier femto cell network

Sr. No	Handoff Scenario	Handoff Scheme	Related Reference
1	Macro cell-Femto cell handoff.	Dutch auction, reactive data bicasting HeNB policy finction(HeNB-PF),RSS value, Dwell time, multi armed bandit, Context prefetching (CoPS), User's velocity,MMNN scheme .	(Wu, Chu, and Fang 2013b)(Yusof et al. 2014a)(Guo, ul Quddus, and Tafazolli 2012) (Bai et al. 2011)(Ellouze, Gueroui, and Alimi 2011)(Moon and Cho 2010)(Singoria, Oliveira, and Agrawal 2011)(Choi et al. 2012)(Roy, Shin, and Saxena 2012) (Xenakis et al. 2013)(Cheikh et al. 2013)(Becvar and Mach 2012) (Wang, Katz, and Giese 1999)(de Lima et al. 2010)(Moreira et al. 2013) (Chowdhury and Jang 2010)(Shbat and Tuzlukov 2012)(Tang, Hong, and Xue 2013)
2	Femto cell-Femto cell handoff.	Feedback technique, HeNB aided virtual handoff (HaVHO), SON, queuing theory, fractional guard channel, backbone capacity, dwell time, Ant colony optimization(ACO), Bayseian estimation.	(Dimou et al. 2009)(Moon and Cho 2009)(Badri et al. 2013) (Feki et al. 2013)(Ramanathan 2002)(Lee et al. 2014) (Khalid and Kwak 2013)(López-Pérez et al. 2010)(Fan and Sun 2010) (Cohen and Levin 2012)(Zhou et al. 2013)(Rath and Panwar 2012) (Wang et al. 2012)(Huang 2012)
3	Femto cell-Macro cell handoff.	Multi threshold dependency alforithm, Velocity and QoS, Temporary area partners (TAP), Multi egress network mobility(MEN-NEMO), dwell time and velocity, RSSI.	(Salem and Reguiga 2013)(Qiu and Wang 2012)(Wu 2011) (Chowdhury, Bui, and Jang 2011)(Becvar and Mach 2011)(Xiaoying, Jingjing, and Zhenchuan 2005) (Kwon and Cho 2011)(Chaganti et al. 2013)

information includes the candidate femto cells list and their access modes from network side and location and speed of the user from user side. Results shows reduced power consumption for outdoor users without compromising the QoS parameters. Another macro – femto based handoff scheme is proposed in (Guo, ul Quddus, and Tafazolli 2012) which is based on reactive data bicasting. As compared to the conventional handoff schemes based on data forwarding which induces a large downlink interruption time, data bicasting scheme uses very limited extra network resource in terms of buffer memory and reduces the downlink interruption time and avoids the packet loss. In this scheme, when the network mobility management entity (NMME) receives a handoff message from the current cell, it determines the target candidate cell and asks the source gateway for data bicasting to both source and destination end. When handoff process is completed, this buffered data is deleted. Using this data bicasting scheme, packet loss is avoided and seamless handoff is executed.

A policy based unnecessary handoff reduction strategy between macro – femto cell is proposed in (Bai et al. 2011). In this scheme authors have proposed a HeNB policy function (HeNB PF) for handoff which takes into account the user type (authorized or unauthorized in case of CSG), access mode of the candidate small cell and current load of the small cell for handoff decision. This scheme makes an intelligent decision for selecting the candidate small cell for handoff i.e. if user type and access mode conditions are satisfied but radio resource is not available or having low SINR value, it will not select that particular cell for handoff and vice versa. Advantages of this scheme are minimization of unnecessary handoff, reduction in handoff failure probability, load reduction of shared resources in open access femto cell and QoS assurance. Additional computations increase signaling overhead of the system. Another macro – femto cell efficient handoff scheme is proposed in (Ellouze, Gueroui, and Alimi 2011) . This scheme is based on two factors, first is minimization of connection break time by minimizing the number of scanned target femto cells and second is handoff selection decision i.e., weather to select the femto cell or macro cell for handoff based on user QoS profile. In this scheme, handoff is decided according to the user's QoS requirements. This scheme reduces the handoff delay as well as balances the load among neighboring small cells and macro cell. Another handoff algorithm is proposed in (Moon and Cho 2010) for inbound mobility of the user. This algorithm combines the RSS value of serving cell (i.e. macro base station) and target femto cell by considering the deviation in their transmit power for efficient handoff. Derivation criteria for handoff from these combined RSS values is based on combination process. This combination process is described by the RSS from macro BS and a combination factor which is performed when RSS value of target femto cell is greater than pre-defined threshold. This scheme selects the optimal candidate femto cell and reduces the frequency of frequent handoff. Another RSS based in-bound handoff strategy for macro/femto hierarchal network is proposed in (Tang, Hong, and Xue 2013). In this scheme large asymmetry in the transmit powers of macro cell and femto cell is reflected. Handoff decision is made based on this large asymmetry in the transmit power. Simulation results shows the improvement in the handoff process for hierarchal network.

A distributed handoff strategy to reduce the unnecessary handoff between WIMAX macro cell and small cells is proposed in (Singoria, Oliveira, and Agrawal 2011). Small cells are integrated to minimize the shadowed area for WIMAX macro cell. In this strategy, a new CAC protocol is defined which takes into account the users mobility, signal to interference and noise ratio (SINR) and available bandwidth so that MS executes the handoff process and there is no load on FAP for handoff process. By taking all these factor into account, unnecessary handoffs are reduced. Another SINR based multi objective handoff solution for macro cell-small cell handoff scenario is proposed in (Roy, Shin, and Saxena 2012). Most of the handoff schemes for LTE network are based on SINR. Proposed solution in (Roy, Shin, and

Saxena 2012) considers the SINR as well as the available bandwidth for optimal target cell selection. Consideration of these factors reduces the failure probability of successful handoff and handoff delay time as well as lead to the best cell selection among all available cells. A handoff process is proposed in (Choi et al. 2012) which is based on available data volume. For this purpose, a "*dwell time*" is defined which is the duration for which a user stays in a particular cell. When a handoff is desired, firstly a *dwell time* is estimated by MS then it calculates the available data volume for both macro and femto cell. Handoff between macro cell and femto cell is performed only if available data volume of femto cell is greater than that of macro cell. This scheme increases the total throughput of the system and reduces the number of handoffs as compared to the RSS based schemes but it doesn't works well for high mobility users and fairness among the users is compromised. Enhanced ICIC mechanism for interference reduction is proposed in (Xenakis et al. 2013) which enhances the handoff performance and also maximizes the throughput of the network for users of higher and medium level mobility. This mechanism uses the multi armed bandit (MAB) method and results of these parameters are checked and verified by 3GPP mobility robustness optimization indicators. Result shows improved throughput of the network and reduction in the handoff process failure.

An efficient call admission control based handoff algorithm for macro/femto environment is proposed in (Cheikh et al. 2013) . Motivation of this algorithm is to provide good quality communication during handoff process and reduction of unnecessary handoffs. This algorithm takes into account the UE's direction, its velocity, SINR value and resource availability in selecting the target FAPs list for possible handoff. Comparison with conventional SINR based handoff schemes has been done to validate the simulation results which shows the improvement in signal quality and unnecessary handoff reduction. Another handoff algorithm for macro/femto environment is proposed in (Becvar and Mach 2012) which can be implemented for all three macro-femto cell, femto-femto cell and femto-macro cell environment. This algorithm triggers the handoff decision based on received signal strength indicator (RSSI), velocity of the UE and periodic triggering. Periodic triggering parameter with combination of UE's velocity, enforces the MS to handoff to femto cell which offloads the traffic towards femtocell for its maximum utilization. Results validates the reduced unnecessary handoff and efficient FAP utilization ratio. Another macro to femto handoff scheme is proposed in (Shbat and Tuzlukov 2012) . In this scheme handoff is based on policy function of femto cell which specially includes the velocity information of UE and CSG access mode information of target femto cells and other parameters like available bandwidth and SINR value for optimal cell selection. All candidate femto cells for handoff are evaluated on above mentioned parameters and best among all is selected according to the objectives.

A context pre-fetching (CoPS) based handoff scheme for WiMAX small cell environment is proposed in (Wang, Katz, and Giese 1999) . It performs the context retrieval and pre-registration of data path before handoff happens in actual to reduce the lengthy handoff preparation phase. Two triggering mechanisms (periodic triggering and signal based threshold triggering) in accordance with mobility scenarios are also implemented to initiate the CoPS. Approximately 17% improvement in handoff time is achieved by employing this scheme but some additional computational overhead is also induced in the network. A handoff mechanism based on reducing handoff cost (RHC) is proposed in (de Lima et al. 2010) . Objective of this mechanism is to minimize the frequent handoffs and enhance the traffic offloading probability of femto cells. RHC uses handoff delay timer to minimize the frequent handoffs. Due to this delay timer handoff reduction and traffic offloading ratios are improved and results shows the effective achievement of objective. A modified handoff procedure is proposed in (Moreira et al. 2013) with integration of femto cell in macro cell network. Main objective of this procedure is load balancing among different cells and different communication tiers. This modified procedure makes the handoff decision based on number of users in a cell and their velocity. If velocity of the user is lower than 30km/h then handoff will be done with femto cell otherwise hand off will be done with macro cell to minimize the frequent handoff load. By employing this scheme, better load balancing and QoS is achieved as compared to the RSS and dwell time based handoff methods. Another handoff mechanism for small and medium scale WCDMA femto/macro cell network is proposed in (Chowdhury and Jang 2010). In this mechanism, M/M/N/N queuing scheme based CAC strategy is proposed for minimizing the handoff call dropping probability and new call blocking probability whereas utilizing the available bandwidth efficiently. Proposed scheme takes into account the signal to interference level and user's authentication to reduce the candidate femto cells for handoff. Due to this, signaling overhead of the network and size of neighbor femto cells list is reduced. Results shows better performance in terms of bandwidth utilization and unnecessary handoff reduction. Comparison among different handoff decision making criteria are presented in Table 3.

Table 3. Comparison between different handoff decision making criteria

Sr. No	Handoff decision criteria	Advantages	Disadvantages	Suitable Application Scenario	Related References
1	Received Signal Strength based handoff	Optimal target cell selection and avoidance of unnecessary handoff	Inefficient for users with high mobility	Hetrogeneous and Microcellular Network	(Moon and Cho 2010)(Tang, Hong, and Xue 2013) (Badri et al. 2013)(López-Pérez et al. 2010) (Rath and Panwar 2012)(Salem and Reguiga 2013) (Becvar and Mach 2011) (Xiaoying, Jingjing, and Zhenchuan 2005)
2	Velocity based handoff	Reduced handoff execution time, efficient FAP utilization and load balancing ratio	Enhanced signaling and overhead and may induces interference MUEs	-	(Yusof et al. 2014a)(Cheikh et al. 2013) (Becvar and Mach 2012)(Moreira et al. 2013) (Shbat and Tuzlukov 2012)(Zhou et al. 2013) (Wu 2011)(Kwon and Cho 2011) (Chaganti et al. 2013)
3	Handoff	Network capacity enhancement, better utilization of femto cell resources and frequent handoff reduction	Fairness among the users is compromised may arise cross-tier interference	-	(Wang et al. 2012)
4	Dwell time based handoff	Enhancement in total throughput of the system and minimization of unecessary handoffs	Inefficient for high mobility users introduces complexity in the system and unfair for users accessibility	GPRS/ EDGE Networks	(Choi et al. 2012) (Fan and Sun 2010) (Qiu and Wang 2012) (Kwon and Cho 2011)
5	SINR and available bandwidth available handof	Reduced co-tier and cross-tier interference and best target cell selection	Frequent handoffs for high mobility users and inefficient utilization of small cell resources.	Hetrogeneous Networks	(Bai et al. 2011)(Singoria, Oliveira, and Agrawal 2011) (Roy, Shin, and Saxena 2012)(Cheikh et al. 2013) (Chowdhury and Jang 2010)(Shbat and Tuzlukov 2012) (Huang 2012)(Xiaoying, Jingjing, and Zhenchuan 2005)
6	Fractional guard channel based handoff	Reduced call drop probability for handoff calls and improved system capacity	Long queue time for the new built calls.	Two tier Networks such as macro-femto and interfemto	(Khalid and Kwak 2013)

Femto Cell: Femto Cell Handoff

In this scenario, handoff occurs from one femto cell to another femto cell due to mobility of UE. Handoff decision making process among different femto cells is also difficult due to more than one candidate cells. This makes it difficult to select the target cell. Handoff among the femto cells is a slow process due to the fact that a message takes about 200ms to transmit via public internet network as compared to the macro cell handoff message which takes about 100ms (Dimou et al. 2009). Different handoff schemes to minimize this delay and to optimize the handoff process among the small cells are proposed in (Moon and Cho 2009)–(Huang 2012) to improve the network efficiency. In (Moon and Cho 2009) author has proposed the femto – femto cell handoff scheme based on a feedback technique and air interface existence as signaling system. In the feedback technique, target femto cell receives an effective SINR value from the UE which helps to enhance the speed of handoff process and air interface allows to exchange the priority list of UE made initially to help in handling the handoff process faster and easily. In (Badri et al. 2013) authors have proposed a HeNB aided virtual handoff (HaVHO) mechanism in which neighbor femto cells exchange the cooperation messages when they receive interference or low signal quality. In this mechanism, when a femto cell detects strong interference it requests the neighbor HeNB to relay the data traffic for its user without performing a handoff. This relay of the data is referred to as virtual handoff and it leads towards handoff minimization.

Another handoff scheme for femto cell network is proposed in (Feki et al. 2013). The proposed scheme considers the interference, energy efficiency and user mobility into account for handoff decision process. This scheme selects that femto cell as a target for handoff which maintains the minimum channel gain for continuity of the service and minimizes the average transmit power of MS by utilizing the standard signal measurements. Another femto – femto cell handoff scheme is proposed in (Ramanathan 2002) to minimize the unnecessary handoff. This scheme is based on self-organizing network (SON). SON estimate the position of UE in a building or particular area by using the position reference signal (PRS). Handoff triggering is executed on the basis of above said information which avoids the unnecessary handoff intelligently. In (Lee et al. 2014), authors have proposed a femto – femto cell handoff scheme based on queuing theory and cell selection algorithm. This scheme maintains fairness among the users and minimizes the handover failure if any fault occurs in the femto cell. In cell selection algorithm, UE selects the target femto cell on the basis of information of those UEs which have initiated the handoff previously. This previously transition flow is described by the cell selection matrix "α". Queuing model is employed to represent the dynamics of femto cell. Drawback of the proposed scheme is that it induces signaling overhead by broadcasting the cell selection information by the previously handed off UEs. A fractional guard channel based handoff scheme is presented in (Gupta and Deswal 2015). In this scheme authors have presented a new call admission control policy which reserves a fractional part of available bandwidth for handoff calls and also allows the new built calls to utilize the fractional bandwidth in addition by a factor *"decimal fraction"* after a limited queue time. This bandwidth allocation to the new built calls reduces the call blocking probability of the network but induces the queue time which prolongs the session initiation process for new built calls.

A received signal strength (RSS) based femto-femto cell efficient handoff scheme is proposed in (López-Pérez et al. 2010). This scheme is based on the knowledge of neighbor cells. In this scenario, each cell has to maintain a minimum but effective neighbor cells list for efficient femto-femto cell handoff. This schemes takes the RSS into account and detects the sub channel from both open and hidden neighbor cell. Neighbor list maintained for small number of neighbors reduces the scanning power consumption and inclusion of hidden FAPs decreases the handoff failure probability. A handover strategy between femto cells is also proposed in (Fan and Sun 2010). In this strategy, proposed algorithm takes into account the femto backbone capacity and time spent in femto cell by a particular user "*tc*". If backbone capacity of the FAP in terms of delay and bandwidth are not satisfied and value of "*tc*" is lesser than "*tc threshold*" which is defined according to FAP load than handover is initiated. Increase in "*tc*" will decrease the handoff frequency. Another handoff scheme based on optimal call admission control (CAC) for small cell network is proposed in (Cohen and Levin 2012). This handoff scheme maximizes the network performance for fast growing network applications like adaptive TCP video streaming, video conferencing and online gaming. Frequent handoff in small cell environment will effect these application due to packet loss in handoff. Earlier complete forward or complete block algorithms were used for data packets which were inefficient methods. In this scheme, selective forward algorithm is used which employs the elastic forwarding buffer to make the runtime decision i.e. which packet has to be forwarded and which has to be blocked to the handed off small cell. This scheme maximizes the throughput and minimizes the throughput variance during the handoff process but complexity of the network is increased and additional resources are occupied in terms of bandwidth during handoff process.

A fast handoff scheme for femto-femto cell handoff is proposed in (Zhou et al. 2013). Femto cells are connected with the core network through public internet and packet transfer through this mode is slower as compared to the conventional cellular network. A prefetch-based fast handoff scheme is proposed in (Zhou et al. 2013) which employs the speed of UE as a handoff decision criterion and modifies the process of message flow of conventional handoff scheme used in LTE system for prefetching the higher layer data to the target femto cell. This makes the handoff process faster without making any modification to the system architecture. An automated handoff optimization mechanism is proposed in (Rath and Panwar 2012). In femto cell coverage area, there are some special areas where frequent users mobility leads to frequent handoff process and some areas are general where mobility does not leads towards handoff. The proposed mechanism is based on ant colony optimization (ACO) algorithm which differentiates between special area and general area and assigns different hysteresis margins for these areas within a femto coverage area. If RSRP of neighbor cell is equal to HM in terms of dBs in special area then handoff is initiated. It solves the problem of different handoff parameter calculations for UEs of different areas with different mobility and reduces the frequent handoff rate.

A cell association based handoff scheme is proposed in (Wang et al. 2012). In this scheme authors have proposed an algorithm for user association by making tradeoff between network capacity and fairness. Based on this algorithm, another algorithm is developed for handoff between the small cells using Bayesian estimation. Posterior probability is computed upon this estimation to find the available BSs and best one is selected on the basis of coverage for the user. Authors in (Huang 2012) have proposed a handoff algorithm for femto cell network in which a new handoff decision parameter is defined which combines the femto/energy capacity, SINR, channel nature and user priority. This algorithm balances the user load among macro and femto cells while consuming the minimum possible energy and reduces the unnecessary handoff.

Femto Cell: Macro Cell Handoff

In this scenario handoff occurs from small cell to macro cell due to mobility of UE. Handoff decision making process is simpler as compared to the earlier described two scenarios. There is only one candidate cell with the higher power for handoff so comparison and best cell selection overheads are reduced. In mobility management and handoff process call admission control (CAC) is very important, i.e, how to execute the handoff process while signaling overhead is kept minimum. In this regard, different handoff schemes proposed for femto – macro cell handoff scenario are reported in (Salem and Reguiga 2013) –(Chaganti et al. 2013)and also shown in Table 3. In (Salem and Reguiga 2013) authors have presented a handoff scheme for TD-SCDMA femto cell network. Based on this scheme author has proposed mufti-threshold dependency algorithm (MTDA) for call admission control. In MTDA, handoff is triggered by interdependent thresholds like RSS, bandwidth availability and access permission instead of RSS threshold only which leads to efficient handoff execution and minimizes the unnecessary handoff. A scheme based on RSS threshold value has been proposed in (Becvar and Mach 2011). In this scheme a procedure for femto hybrid access mode is proposed. In conventional access system, if a non-CSG MS comes in femto cell coverage area with handoff request, macro cell avoids the unnecessary handoff signaling by the knowledge of femto access list. Proposed procedure depends upon adaptive threshold like RSS based on QoS for femto cell access and handoff process which is initiated by femto cell. This minimizes the handoff signaling overhead in the network. Likewise, another RSS based intra cell handoff scheme to cope up the cross-tier interference is proposed in (Xiaoying, Jingjing, and Zhenchuan 2005). In this scheme if a base station either macro or femto detects the cross-tier interference and signal quality degradation it can re-assign the power and sub channel resources to the user. This scheme is applicable for both closed and open access modes as cross-tier interference is more often in close access mode and handoff is more frequent in open access mode. In (Qiu and Wang 2012), authors have proposed a hybrid access mode based novel handoff scheme between femto – macro cells. In this scheme feature of the femto – macro hierarchal network are utilized with specialized knowledge of dwell time threshold, adaptive hysteresis and differentiated treatment and better system performance is achieved. It performs load balancing for hybrid access femtocell but fairness is not maintained among the users by this scheme.

In (Wu 2011), the authors have presented a handoff procedure between femto – macro cell in which handoff decision is based on the speed of UE and QoS. Speed of UE is categorized in three different threshold levels which are below 20km/h, from 20km/h to 30km/h and above 30km/h. For the later two cases, handoff is performed with the macrocell without considering the femto cell due to higher user mobility. But in the first case (below 20km/h), neighboring femto cells are also considered for handoff but users may face the QoS problem due to attenuation in the outdoor signal propagation which can cause the failure in the handoff process. Another handoff scheme for femto – macro cell is proposed in (Chowdhury, Bui, and Jang 2011) in which energy efficient handoff is performed with low latency. In this scheme, concept of temporary area partners (TAP) is introduced by the authors which consists of mobile UE connected through device to device communication. This TAP network is used to establish the connection with the macro cell by using the S1interface instead of X2 interface which uses the public internet as a backhaul connection and it is slow in nature. By using this technique, latency of handoff process is reduced up to some extent

An urban environment handoff scheme based on reactive and proactive concept is proposed in (Kwon and Cho 2011). Here UE's velocity and dwell time (time spent by a user in a particular cell) are taken into account for handoff decision. Upper and lower thresholds are set for UE's velocity which decides that whether user has enough time to handoff to a femto cell or not. If velocity is high than all femto cells are removed from target handoff cell list and handoff is performed with macro cell only. Performance analysis shows the improvement in call dropping probability. A handoff scheme for high speed user in femto network environment is proposed in (Chaganti et al. 2013). This scheme is based on multiple egress network mobility (MEN-NEMO) method which tends to reduce frequent handoff, handoff latency and poor QoS experience to the user. This method is usually used in high speed trains for provision of internet. In this scheme two antennas are deployed on the train, i.e, one on the head (engine) of the train and second on the tail of the train. They performs their functions alternatively for carrying user's traffic and performing handoff process execution. Multiple egress network interface prevents the users from bad QoS service and transient transmission during the handoff process. Simulation results shows improvement in frequent handoff reduction by employing this scheme in high mobility environment but large computational complexity and continuous neighbor list updating induces signaling overhead.

FUTURE RESEARCH DIRECTIONS

Small cells are deployed to enhance the capacity of macro cell based cellular network. Integration of these small cells in the core network with fair resource allocation is a challenging task. For mass deployment of femto cells in the hot spot areas, it is necessary to develop some distributed interference. Management/ mitigation and efficient handoff schemes are also used to minimize the frequent and unnecessary handoffs while ensuring the QoS parameters of both macro and femto cell UEs. This enhances the coverage area and system capacity as a secondary objective. Some already proposed handoff schemes for all three scenarios of handoff between macro-femto cells are already elaborated thoroughly in this survey paper. Performance or some parameters of these schemes are still needed to be investigated further for optimal results. In this regard, some future research areas are highlighted in this section which are given as under.

Handoff Validation and Comparative Performance Analysis

Design and development of simple small cell network layouts doesn't often replicate their optimized deployment in real life implementation. Although some of the existing algorithms for handoff decision making are supplemented with built in performance analysis parameters, but validating the performance of these HO decision algorithms in real life systems or simulation setups is still a critical issue for femto cell network. Beside the independent performance analysis in real environment, comparative performance analysis among different HO algorithm should also be carried out. Evaluation of other important parameters like data rate, power consumption and interference measurement of the proposed algorithms, should be an integral part of femtocell-specific HO decision algorithms.

CONCLUSION

With the emergence of new technologies and increase in the users and their data demand, network structure has been shifting towards multi-tier and heterogeneous network and small cells have been introduced. In this survey paper, handoff process with its types, proposed methods and schemes for minimization of frequent and unnecessary handoffs are discussed thoroughly as handoff process becomes more frequent and complex in the two-tier macro-femto cell networks. All three scenarios for handoff, i.e, macro-femto, femto-femto and femto-macro are elaborated in detail with handoff decision making criteria. In the later part of this survey paper, future research directions and challenges are also highlighted for motivation for future work to be done in this area of research.

REFERENCES

Akiyama, Oodaira, Ishikawa, Hisamoto, & Watanabe. (2003). The Umbrella Cell: A Logic-Process-Compatible 2T Cell for SOC Applications. In *VLSI Circuits, 2003. Digest of Technical Papers. 2003 Symposium On* (pp. 89–92). IEEE.

Andrews, J. G., Claussen, H., Dohler, M., Rangan, S., & Reed, M. C. (2012). Femtocells: Past, Present, and Future. *IEEE Journal on Selected Areas in Communications*, *30*(3), 497–508. doi:10.1109/JSAC.2012.120401

Badri, T. F. Z., Saadane, R., Wahbi, M., & Samir, M. (2013). Handover Management Scheme in LTE Femtocell Networks. *International Journal of Computer Science & Information Technology*, *5*(3), 89–100. doi:10.5121/ijcsit.2013.5307

Bae, Ryu, & Park. (2011). *Analysis* of Handover Failures in LTE Femtocell Systems. In *Telecommunication Networks and Applications Conference (ATNAC), 2011 Australasian* (pp. 1–5). IEEE.

Bai, Wang, Liu, & Zhang. (2011). A Policy-Based Handover Mechanism between Femtocell and Macrocell for LTE Based Networks. In *Communication Technology (ICCT), 2011 IEEE 13th International Conference On* (pp. 916–920). IEEE.

Balachandran, Kang, Karakayali, & Rege (2012). Virtual Soft Handoff Enabled Dominant Interference Cancellation for Enhanced Uplink Performance in Heterogeneous Cellular Networks. In *Wireless Communications and Networking Conference Workshops (WCNCW), 2012 IEEE* (pp. 109–114). IEEE.

Becvar & Mach. (2011). On Enhancement of Handover Decision in Femtocells. *In Wireless Days (WD), 2011 IFIP* (pp. 1–3). IEEE.

Chaganti, Sathya, Ahammed, Rex, & Tamma. (2013). Efficient Son Handover Scheme for Enterprise Femtocell Networks. In *Advanced Networks and Telecommuncations Systems (ANTS), 2013 IEEE International Conference On* (pp. 1–6). IEEE.

Chandrasekhar, V., Andrews, J. G., & Gatherer, A. (2008a). Femtocell Networks: A Survey. *IEEE Communications Magazine*, *46*(9).

Cheikh, Ayari, Langar, Pujolle, & Saidane. (2013). Optimized Handover Algorithm for Two-Tier Macro-Femto Cellular LTE Networks. In *Wireless and Mobile Computing, Networking and Communications (WiMob), 2013 IEEE 9th International Conference On* (pp. 608–613). IEEE.

Choi, Seo, Nam, Park, & Cho. (2012). Handover Decision Algorithm Based on Available *Data* Volume in Hierarchical Macro/Femto-Cell Networks. In *Communications and Electronics (ICCE),* 2012 *Fourth International Conference On* (pp. 145–149). IEEE.

Chowdhury, Bui, & Jang. (2011). Neighbor Cell List Optimization for Femtocell-to-Femtocell Handover in Dense Femtocellular Networks. In *Ubiquitous and Future Networks (ICUFN), 2011 Third International Conference On* (pp. 241–245). IEEE.

Chowdhury & Jang. (2010). *Handover Control for WCDMA Femtocell Networks.* ArXiv Preprint ArXiv:1009.3779.

Chowdhury, Jang, & Haas. (2010). *Network Evolution and QoS Provisioning for Integrated Femtocell/Macrocell Networks*. ArXiv Preprint ArXiv:1009.2368.

Chowdhury, Jang, & Haas. (2011). Cost-Effective Frequency Planning for Capacity Enhancement of Femtocellular Networks. *Wireless Personal Communications*, *60*(1), 83–104.

Cohen & Levin. (2012). Handovers with Forward Admission Control for Adaptive TCP Streaming in LTE-Advanced with Small Cells. In *Computer Communications and Networks (ICCCN), 2012 21st International Conference On* (pp. 1–7). IEEE.

Dharmaraja, S., Trivedi, K. S., & Logothetis, D. (2003). Performance Modeling of Wireless Networks with Generally Distributed Handoff Interarrival Times. *Computer Communications*, *26*(15), 1747–1755. doi:10.1016/S0140-3664(03)00044-6

Dimou, K., Wang, M., & Yang, Y. (2009). Handover within 3GPP LTE: Design Principles and Performance. In Vehicular Technology Conference Fall (VTC 2009-Fall), 2009 IEEE 70th (pp. 1–5). IEEE.

Ellouze, R., Gueroui, M., & Alimi, A. M. (2011). MacroFemto Cell Handover with Enhanced QoS in Mobile WiMAX. In *Wireless Telecommunications Symposium (WTS)*, 2011 (pp. 1–6). IEEE.

Fan & Sun. (2010). Access and Handover Management for Femtocell Systems. In *Vehicular Technology Conference (VTC 2010-Spring), 2010 IEEE 71st* (pp. 1–5). IEEE.

Feki, Capdevielle, Roullet, & Sanchez. (2013). Handover Aware Interference Management in LTE Small Cells Networks. In *Modeling & Optimization in Mobile, Ad Hoc & Wireless Networks (WiOpt), 2013 11th International Symposium On* (pp. 49–53). IEEE.

Furukawa & Akaiwa. (1994). A Microcell Overlaid with Umbrella Cell System. In *Vehicular Technology Conference, 1994 IEEE 44th* (pp. 1455–1459). IEEE.

Guo, Quddus, & Tafazolli. (2012). Seamless Handover for LTE Macro-Femto Networks Based on Reactive Data Bicasting. *IEEE Communications Letters*, *16*(11).

Gupta & Deswal. (2015). Survey of Handover Schemes in Femtocellular Networks. *International Journal of Science, Engineering and Technology Research*, *4*(4).

Habeeb & Qadeer. (2009). Interference Evaluation and MS Controlled Handoff Technique for Femtocell. In *Internet, 2009. AH-ICI 2009. First Asian Himalayas International Conference On* (pp. 1–5). IEEE.

Halgamuge, Vu, Rarnamohanarao, & Zukerman. (2005). Signal-Based Evaluation of Handoff Algorithms. *IEEE Communications Letters*, *9*(9), 790–792.

Hsieh, P.-J., Lin, P.-H., & Lee, Y.-C. (2014). CoPS: Context Prefetching Handover Scheme on 4G Outdoor Small Cell Testbed. In *Modeling and Optimization in Mobile, Ad Hoc, and Wireless Networks (WiOpt), 2014 12th International Symposium On* (pp. 50–56). IEEE.

Huang. (2012). A Handover Scheme for TD-SCDMA Femtocell Networks. In *Computer and Information Technology (CIT), 2012 IEEE 12th International Conference On* (pp. 948–952). IEEE.

Khalid & Kwak. (2013). Handover Optimization in Femtocell Networks. In *ICT Convergence (ICTC), 2013 International Conference On* (pp. 122–127). IEEE.

Kim & Lee. (2010). Handover in UMTS Networks with Hybrid Access Femtocells. In *Advanced Communication Technology (ICACT), 2010 The 12th International Conference On* (pp. 904–908). IEEE.

Kwon & Cho. (2011). Load Based Cell Selection Algorithm for Faulted Handover in Indoor Femtocell Network. In *Vehicular Technology Conference (VTC Spring), 2011 IEEE 73rd* (pp. 1–5). IEEE.

Lee, C.-W., Chuang, M.-C., Meng, C. C., & Sun, Y. S. (2014). Seamless Handover for High-Speed Trains Using Femtocell-Based Multiple Egress Network Interfaces. *IEEE Transactions on Wireless Communications*, *13*(12), 6619–6628. doi:10.1109/TWC.2014.2364179

Lee, Huang, Lin, Su, & Wang. (2013). Reducing Handover Cost for LTE Femtocell/Macrocell Network. In *Globecom Workshops (GC Wkshps), 2013 IEEE* (pp. 4988–4993). IEEE.

Lee, Shin, Lim, & Hong. (2010). Effects of Time-to-Trigger Parameter on Handover Performance in SON-Based LTE Systems. In *Communications (APCC), 2010 16th Asia-Pacific Conference On* (pp. 492–496). IEEE.

Leung, V. C. M. (1992). Internetworking Wireless Terminals to Local Area Networks via Radio Bridges. In *Wireless Communications, 1992. Conference Proceedings., 1992 IEEE International Conference on Selected Topics In* (pp. 126–129). IEEE.

de Lima, Ghaboosi, Bennis, MacKenzie, &Latva-aho. (2010). A Stochastic Association Mechanism for Macro-to-Femtocell Handover. In *Signals, Systems and Computers (ASILOMAR), 2010 Conference Record of the Forty Fourth ASILOMAR Conference On* (pp. 1570–1574). IEEE.

Liu, Hong, Xue, & Peng. (2010). Conflict Avoidance between Mobility Robustness Optimization and Mobility Load Balancing. In *Global Telecommunications Conference (GLOBECOM 2010), 2010 IEEE* (pp. 1–5). IEEE.

López-Pérez, Valcarce, Ladányi, de la Roche, & Zhang. (2010). Intracell Handover for Interference and Handover Mitigation in OFDMA Two-Tier Macrocell-Femtocell Networks. *EURASIP Journal on Wireless Communications and Networking*, (1).

Moon, J.-M., & Cho, D.-H. (2009). Efficient Handoff Algorithm for Inbound Mobility in Hierarchical Macro/Femto Cell Networks. *IEEE Communications Letters*, *13*(10).

Moon & Cho. (2010). Novel Handoff Decision Algorithm in Hierarchical Macro/Femto-Cell Networks. In *Wireless Communications and Networking Conference (WCNC), 2010 IEEE* (pp. 1–6). IEEE.

Moreira, Lage, Gomes, Radwan, & Rodriguez. (2013). Context-Aware Macro-Femto Handover for Energy Efficiency of Outdoor Users. In *Future Network and Mobile Summit (FutureNetworkSummit), 2013* (pp. 1–10). IEEE.

Nasser, N., Hasswa, A., & Hassanein, H. (2006). Handoffs in Fourth Generation Heterogeneous Networks. *IEEE Communications Magazine*, *44*(10), 96–103. doi:10.1109/MCOM.2006.1710420

Pantisano, F., Bennis, M., Saad, W., Debbah, M., & Latva-Aho, M. (2014). Improving Macrocell-Small Cell Coexistence through Adaptive Interference Draining. *IEEE Transactions on Wireless Communications*, *13*(2), 942–955. doi:10.1109/TWC.2013.120613.130617

Qiu & Wang. (2012). A Novel Handover Strategy in LTE Networks with Hybrid Access Femtocells. In *Wireless Communications, Networking and Mobile Computing (WiCOM), 2012 8th International Conference On* (pp. 1–5). IEEE.

Racz, Temesvary, & Reider. (2007). Handover Performance in 3GPP Long Term Evolution (LTE) Systems. In *Mobile and Wireless Communications Summit, 2007. 16th IST* (pp. 1–5). IEEE.

Ramanathan, Subramanian (2002). Self-Organizing Mobile Wireless Station Network.

Rappaport, T. S. (1996). Wireless Communications: Principles and Practice (vol. 2). Prentice Hall.

Rath & Panwar. (2012). Fast Handover in Cellular Networks with Femtocells. In *Communications (ICC), 2012 IEEE International Conference On* (pp. 2752–2757). IEEE.

Reardon, M. (2010, February 9). *Cisco Predicts Wireless Data Explosion* [Press Release]. Cisco.

Roy, A., Shin, J., & Saxena, N. (2012). Multi-Objective Handover in LTE Macro/Femto-Cell Networks. *Journal of Communications and Networks (Seoul)*, *14*(5), 578–587. doi:10.1109/JCN.2012.00016

Salem & Reguiga. (2013). Context-Aware Handoff Selection Algorithm in Femtocell Networks. In *Wireless Communications and Mobile Computing Conference (IWCMC), 2013 9th International* (pp. 243–247). IEEE.

Shbat & Tuzlukov. (2012). Handover Technique between Femtocells in LTE Network Using Collaborative Approach. In *Communications (APCC), 2012 18th Asia-Pacific Conference On* (pp. 61–66). IEEE.

Singoria, Oliveira, & Agrawal. (2011). Reducing Unnecessary Handovers: Call Admission Control Mechanism between WiMAX and Femtocells. In *Global Telecommunications Conference (GLOBECOM 2011), 2011 IEEE* (pp. 1–5). IEEE.

Tang, H., Hong, P., & Xue, K. (2013). HeNB-Aided Virtual-Handover for Range Expansion in LTE Femtocell Networks. *Journal of Communications and Networks (Seoul)*, *15*(3), 312–320. doi:10.1109/JCN.2013.000054

Tawil, Pujolle, & Demerjian. (2008). Distributed Handoff Decision Scheme Using MIH Function for the Fourth Generation Wireless Networks. In *Information and Communication Technologies: From Theory to Applications, 2008. ICTTA 2008. 3rd International Conference On* (pp. 1–6). IEEE.

Vrzic & Fong. (2004). *Semi-Distributed Scheduling Scheme for the Reverse Link of Wireless Systems*. Academic Press.

Wang, Katz, & Giese. (1999). Policy-Enabled Handoffs across Heterogeneous Wireless Networks. In *Mobile Computing Systems and Applications, 1999. Proceedings. WMCSA'99. Second IEEE Workshop On* (pp. 51–60). IEEE.

Wang, Li, Yu, & Qiu. (2012). Automated Handover Optimization Mechanism for LTE Femtocells. In *Communications and Networking in China (CHINACOM), 2012 7th International ICST Conference On* (pp. 601–605). IEEE.

Wu, Chu, & Fang. (2013). The Periodic Scan and Velocity Decision Handover Scheme for next Generation Femtocell/Macrocell Overlay Networks. In *ICT Convergence (ICTC), 2013 International Conference On* (pp. 201–206). IEEE.

Wu. (2011). A New Handover Strategy between Femtocell and Macrocell for LTE-Based Network. In *Ubi-Media Computing (U-Media), 2011 4th International Conference On* (pp. 203–208). IEEE.

Xenakis, Passas, Merakos, & Verikoukis. (2013). Energy-Efficient and Interference-Aware Handover Decision for the LTE-Advanced Femtocell Network. In *Communications (ICC), 2013 IEEE International Conference On* (pp. 2464–2468). IEEE.

Zhang, Wu, & Zhang. (2005). Optimization and Performance Analysis of Handover Process in Micro-Cell Systems. In *Wireless Communications, Networking and Mobile Computing, 2005. Proceedings. 2005 International Conference On* (pp. 1047–1051). IEEE.

Yeh, Talwar, Lee, & Kim. (2008). WiMAX Femtocells: A Perspective on Network Architecture, Capacity, and Coverage. *IEEE Communications Magazine, 46*(10).

Yusof, Zainali, Nasir, & Ya'Acob. (2014). Handover Adaptation for Load Balancing Scheme in Femtocell Long Term Evolution (LTE) Network. In *Control and System Graduate Research Colloquium (ICSGRC), 2014 IEEE 5th* (pp. 242–246). IEEE.

Zhang, J., & De la Roche, G. (2011). *Femtocells: Technologies and Deployment.* John Wiley & Sons.

Zhou, Hu, Mao, Agrawal, & Reddy. (2013). Cell Association and Handover Management in Femtocell Networks. In *Wireless Communications and Networking Conference (WCNC), 2013 IEEE* (pp. 661–666). IEEE.

ADDITIONAL READING

Arshad, R., ElSawy, H., Sorour, S., Al-Naffouri, T. Y., & Alouini, M.-S. (2017). Velocity-aware handover management in two-tier cellular networks. *IEEE Transactions on Wireless Communications, 16*(3), 1851–1867. doi:10.1109/TWC.2017.2655517

Bhoite, K. S., & Gengaje, S. (2018). Handover Management in Two-Tier Femtocell–Macrocell Network. *Wireless Personal Communications, 98*(3), 2849–2866. doi:10.100711277-017-5004-7

Chandrasekhar, V., Goldsmith, A., Krishna, S., & Madan, R. K. (2018). Interference management and network performance optimization in small cells.

Chih-Lin, I., Greenstein, L. J., & Gitlin, R. D. (1993). A microcell/macrocell cellular architecture for low-and high-mobility wireless users. *IEEE Journal on Selected Areas in Communications, 11*(6), 885–891. doi:10.1109/49.232297

Claussen, H. (2007). Performance of macro-and co-channel femtocells in a hierarchical cell structure. In *Personal, Indoor and Mobile Radio Communications, 2007. PIMRC 2007. IEEE 18th International Symposium on* (pp. 1–5). IEEE.

Claussen, H., Ho, L. T., & Samuel, L. G. (2008a). An overview of the femtocell concept. *Bell Labs Technical Journal, 13*(1), 221–245. doi:10.1002/bltj.20292

Claussen, H., Ho, L. T., & Samuel, L. G. (2008b). Self-optimization of coverage for femtocell deployments. In *Wireless Telecommunications Symposium, 2008. WTS 2008* (pp. 278–285). IEEE. 10.1109/WTS.2008.4547576

Huber, K. D., Flynn, J. J., & Mansfield, W. G. (2018). Interface for access management of femto cell coverage.

Huber, K. D., Mansfield, W. G., & Flynn, J. J. (2018). Location-based services in a femtocell network.

Jo, H.-S., Mun, C., Moon, J., & Yook, J.-G. (2009). Interference mitigation using uplink power control for two-tier femtocell networks. *IEEE Transactions on Wireless Communications, 8*(10).

Lee, K.-Y., Kim, S., Park, Y.-S., Bae, B.-J., & Lee, B.-Y. (2018). Method and apparatus for identifying microcells in macrocells in wireless communication systems, and handover method and system using same.

Liang, Y., Goldsmith, A., Foschini, G., Valenzuela, R. A., & Chizhik, D. (2008). Evolution of base stations in cellular networks: Denser deployment versus coordination. In *Communications, 2008. ICC'08. IEEE International Conference on* (pp. 4128–4132). IEEE.

Mishra, A., Shin, M., & Arbaugh, W. (2003). An empirical analysis of the IEEE 802.11 MAC layer handoff process. *Computer Communication Review, 33*(2), 93–102. doi:10.1145/956981.956990

Patini, J. (2018). Targeting communications in a femtocell network.

Radeluscu, A. D., Chande, V., Chen, J. M., Nanda, S., Singh, D., & Yavuz, M. (2018b). Apparatus and methods of hand-in to a femto node.

Tinnakornsrisuphap, P., Singh, D., Radulescu, A. D., & Patel, C. S. (2018). Method and system for determining location of mobile devices in a femtocell.

Yao, G., Cao, J., Liu, X., & Siebert, J. (2018). Enabling fast handoff in dense 802.11 WMNs via temporal-spatial knowledge and opportunistic probing. *Journal of Reliable Intelligent Environments, 4*(1), 57–70. doi:10.100740860-018-0057-2

Zhao, X., Ma, H., Jin, Y., & Yao, J. (2018). Measuring Instability of Mobility Management in Cellular Networks. *IEEE Network, 32*(5), 138–144. doi:10.1109/MNET.2018.1700204

KEY TERMS AND DEFINITIONS

Femto Cell: A femtocell is a small base station with shorter communication range and low transmit power also referred to as home-enhanced NB (H(e)NB). Femto cells are normally installed indoor to provide communication coverage to the users affected penetration losses and low SINR of macro cell signals.

Handoff Process: Handoff is the process of changing the frequency, time slot, channel, spreading code, or combination of them from one base station to another while communication session is in progress.

Hard Handoff Process: A hard handoff is a handoff technique used with cellular networks that requires the user's connection to be entirely broken with an existing base station before being switched to another base station.

HeNB: Femto cells which are normally installed indoor to provide communication coverage to the users affected penetration losses and low SINR of macro cell signals.

Macro Cells: A macro cell or macro site is a cell in a mobile phone network that provides radio coverage served by a high-power cell site (tower, antenna or mast). Generally, macro cells provide coverage larger than microcell and femto cell.

Small Cells: A broader term of femto cell which is more widespread in the industry is small cell, with femto cell as a subset. Micro cell is also included in small cells.

Soft Handoff Process: A soft handoff is a technique in which a cellular user is switched to another base station before broken the connection with an existing base station. In this technique, call of user is not disconnected when a user switches from one base station to another.

Chapter 10
Enhancing Mobile Data Offloading With In-Network Caching

Xu Zhang
Xi'an University of Technology, China

Yue Cao
University of Northumbria, UK

Linyu Peng
Waseda University, Japan

Junhuai Li
Xian University of Technology, China

ABSTRACT

This chapter discusses the most recent solutions for accelerating mobile data delivery, focusing on the different issues for achieving this goal. These issues include the efficiency of mobile data offloading and in-network caching. By characterizing different offloading options from the aspect offloading efficiency, dynamically offloading cellular traffic onto cost-effective options is able to achieve great performance gains. On the other hand, with concerns on tightly coupled dependency on original data sources, mobile users could easily suffer from the scarcity of content object resources. Researchers further account for the possibility of decoupling content availability dependency from original content sources. And a key enabling factor could be the recently emerged technology (i.e., in-network caching) whereby the emerging information-centric networking (ICN) performs its critical functionalities. Coupling ICN with mobile cellular network could be an attractive solution in order to resolve the tension between the tremendous growth of mobile content traffic and backhaul bottleneck.

DOI: 10.4018/978-1-5225-7570-2.ch010

1. INTRODUCTION

This chapter discusses the most recent solutions for accelerating mobile data delivery, focusing on the different issues for achieving this goal. The objective of this chapter is to review in detail these issues, which include the efficiency of mobile data offloading, the impact of in-network caching on the alleviation of ever-growing mobile content demands, incentives of how these techniques could contribute to 5G research and finally the challenges in bringing the concept into reality.

Mobile data offloading enables to dynamically deliver much of the cellular traffic through complementary wireless technologies. These techniques utilize cost-effective options to provide ubiquitous connectivity for mobile Internet access. Various open research areas where mobile data offloading will show great efficiency are explained. This includes the eNodeB or base station (BS) offloading that makes use of existing cellular infrastructures (i.e., BSes), the Wi-Fi and Femtocell offloading that deploy cost-effective facilities (i.e., access points), and device-level offloading (i.e., device-to-device (D2D) offloading) that utilizes end users' equipment (e.g., smartphones and tablets, etc.). Offloading efficiency are discussed to show the effectiveness specific operation regimes can achieve.

There are great concerns on tightly coupled dependency on original data sources, and mobile users could easily suffer from the scarcity of content object resources. Typical examples are P2P-like content delivery networks, where end users (or user devices) are both consumers and contributors. By additionally accounting for such issues, researchers further account for the possibility of decoupling content availability dependency from original content sources. And a key enabling factor to achieve such goal could be the recently emerged technology, i.e., in-network caching, whereby the emerging Information-Centric Networking (ICN) performs its critical functionalities.

Coupling ICN with mobile cellular network could be an attractive solution especially for the 5G network (Cao, 2015; Zhang, 2017), in order to resolve the tension between the tremendous growth of mobile content traffic and backhaul bottleneck. The implications of in-network caching on network efficiency are discussed, highlighting various approaches focuses on the different issues for enhancing the mobile data offloading efficiency. These approaches include independent in-network caching scheme and cooperative caching approach. Some of the main challenges and incentives are elucidated. The chapter finally summarizes there research efforts and provides an overview on the mobile content delivery enhanced with offloading coupled with ICN techniques.

Service providers aim to provide *pervasive connectivity* nowadays, whereby *service velocity* could be substantially improved by enabling services at the network edge. Researchers as well as service providers have commonly agreed that offloading along with in-network caching enables great performance gains for the above two goals. And the following discussions in this chapter are based on the two aspects.

The rest of the chapter is organized as follows. Open research issues with mobile data offloading are discussed in Section 2. Section 3 elucidates the most representative approaches of in-network caching for mobile content delivery, followed by Section 4 wherein incentives and challenges are concerned. And the chapter is summarized in Section 5, wherein conclusions are also made.

2. MOBILE DATA OFFLOADING OPEN RESEARCH ISSUES

Cellular traffic data has been observed a dramatic increase globally. The proliferation of mobile device (e.g., smartphones, tablets) plays a key role in such massive growth, largely driven by the popularity of online social networking services. In fact, end users are more than ever in huger for *data-intensive* applications, e.g., video streaming, etc. However, it is commonly believed that the cellular capacity is being overloaded already and advanced cellular technologies (4G, LTE) cannot scale up enough to satisfy such appetite (Cisco, 2017).

To ease the pain of carriers' concerns on constrained mobile networks, a cost-effective approach is to automatically *offload* cellular data traffic onto small-cells (e.g., Femtocells) and WiFi networks via complementary network technologies (Andrews, 2012; Suh, 2016; Zeydan, 2016). With offload, an optimal network available is selected (e.g., examined and chosen by the mobile device), and the subscriber is automatically connected and authenticated. Mobile offload enables to free up cellular resources especially in high-traffic venues, e.g., city centers and airports. And the ubiquitous connectivity demanded by customers can be supported by offloading techniques with fast downloading speed. The focus of this section is on mobile data offloading open research issues. The most representative approaches are discussed, including pros and cons.

2.1 Wi-Fi Offloading

According to Cisco reports, more than half of total mobile data traffic was offloaded through WiFi or Femtocells in 2016 and will reach 63% by 2021 (Cisco, 2017). WiFi offers high data rates with low cost that normally works on unlicensed spectrum. And hence, there is no interference with existing cellular networks. WiFi offload has been recognized by the wireless industry as an effective as well as economical approach to satisfy growing customer demand for mobile data.

Actually, mobile network operators have deployed a number of WiFi access points (AP) (Suh, 2016; Zeydan, 2016), such as AT&T, Vodafone and Orange. The integration of WiFi into the mobile network is becoming essential so as to maximize the benefits of both. As illustrated in Figure 1, by offloading cellular data onto WiFi networks, such as content (or video) data and even voice services, more network capacity is able to be added in an affordable and flexible way. Customers are enjoying good service

Figure 1. A simple illustration of WiFi offload paradigm

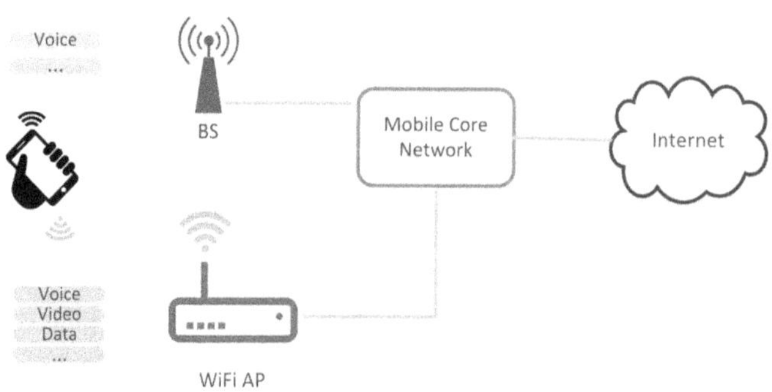

experiences as well, such as desirable downloading speed and better voice services especially indoors. In fact, most WiFi-enabled mobile devices are designed to favor WiFi connections whenever available.

Integration of WiFi networks with the mobile core will become the mass-market business opportunity for upcoming 5G network operators. Clearly, it is a strong trend that 5G and Wi-Fi will be integrated into one service. Incentives for WiFi offload are mainly driven by the following well-known factors.

- **Cost-Effectiveness:** To accommodate adequate data service, many more base stations (BSs) or access points are required. Compared to costly BSs deployment (e.g., expensive real-estate cost), carrier-class WiFi deployments allow to save massive capital cost for mobile operators, up to 40% on network CAPEX according to Ref. (Aptilo, 2013).
- **Smart-Driven Devices:** The widespread WiFi-capable devices favor WiFi networks when available. And these smart devices are capable of choosing the best network available intelligently and dynamically. The intelligence in the devices enables to constantly measure the connection quality, and to actively interact with mobile core to make connection shift decisions.

By enabling Wi-Fi access with mobile core integration, WiFi network can be treated as an alternative Radio Access Network (RAN), with full mobility. Offloaded data traffic is backhauled to the mobile core through GPRS Tunneling (GTP) protocol. With this approach, WiFi traffic is treated as mobile broadband and service operators can have complete control over subscribers' connection to WiFi networks. This means that the existing cellular policy control or charging mechanisms can all be applied. Therefore, personalized services over WiFi networks are able to be offered with enhanced customer experiences. This may require to integrate policy and charging with the Wi-Fi service management system as well.

2.2 Small Cell Offloading

Unlike WiFi, small cell (e.g., Femtocell) operates on licensed spectrum via a low-power cellular base station with low-range coverage (radius up to 10 m). In comparison, the macro-cell is controlled by the base station with wider range (radius up to 30 km) and higher power, as illustrated in Figure 2. Since the same licensed frequency band is shared by both small cell and macro-cell, no additional physical sup-

Figure 2. Small cell offloading illustration

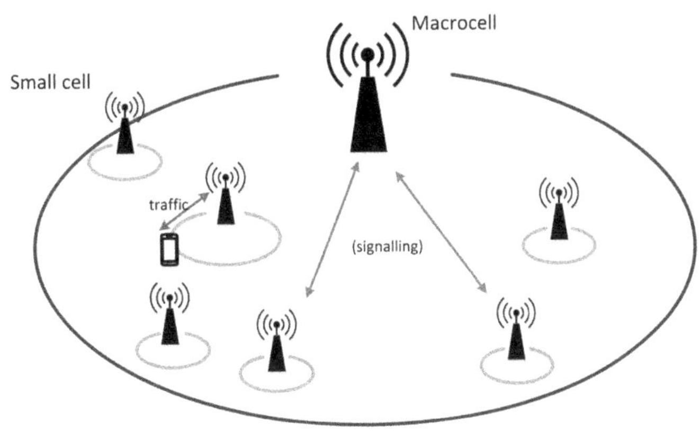

port is required at the side of user equipment, such as a WiFi radio unit. However, there is interference between Femtocell and macro network, and between nearby Femtocells, as well.

As a typical example, Femtocell is designed as a solution to offer better voice and data services (Andrews, 2012) for poor indoor performance limited by macro-cell. With backhaul, Femtocell connects to the mobile network operator usually through broadband Internet connection (e.g., cable modem or DSL). Femtocells are usually deployed by customers, and macro cell BS does not have to actively oversee Femtocell BS so this is practically suitable for large deployment. Therefore, femtocell is considered to be energy-efficient and of high security.

With small cells offloading, cellular data traffic is diverted onto small cell BS via wireless access points, which is similar to WiFi offload. However, due to different broadband frequencies adopted, they differ in the following aspects.

- **Operators Guaranteed Service Quality:** Since small cells operates on licensed spectrum same as cellular networks, operators can have full control over the traffic and offer premium services.
- **Interference:** Since the same spectrum channels are shared between macro and small cells, coordination between them is needed to mitigate such conflict.

Often, small cells and WiFi work together especially in locations with high traffic demands, such as dense metro areas (e.g., airports and city centers), with small cells more frequently deployed outdoors and Wi-Fi indoors. The marginal cost of deploy a WiFi access point or a small cell is very low, so it is feasible to have Wi-Fi built-in modules for small cells.

2.3 BS Offloading

With locally cached popular contents, the potential of utilizing existing cache facilities at RAN is expected to satisfy users' demands more efficiently. Specifically, cache-enabled BSs (or evolved NodeBs, eNBs) play key role in data offloading at RAN (Golrezaei, 2012). Many research works have exploited RAN caching at eNBs that have shown great efficiency on traffic load reduction at mobile core (Zhang, 2017).

Figure 3. A simple example of BS offloading

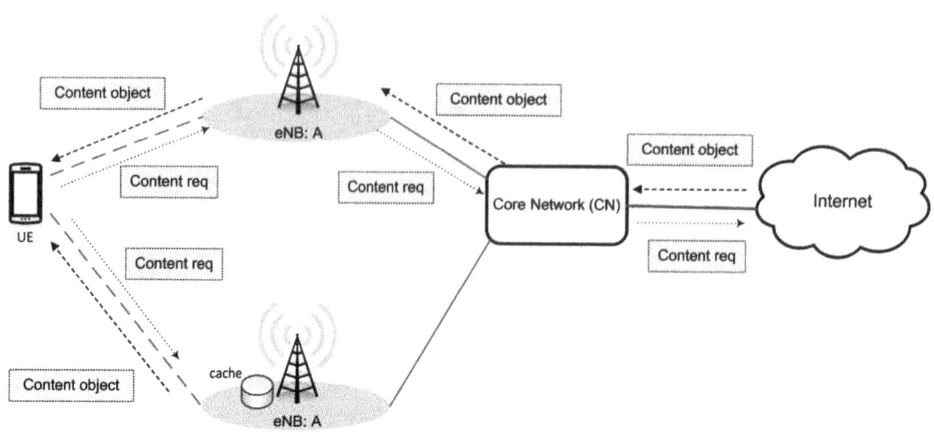

BSs offloading provides desirable service experiences for customers, and enables to effectively alleviate backhaul congestion issues with intelligent content placement planning. Content objects are often split into data chunks that can be delivered via different paths to the requestor. As will be discussed in Section 3, eNBs are required to be equipped with content-aware functionality if equipped with the in-network caching capability (Zhang, 2017; Koponen, 2007; Jacobson, 2009). As such, eNBs or BSs are capable of intercepting and caching passing-through data items.

Figure 3 illustrates a simple example of eNB (or BS) offloading paradigm, wherein cache-enabled eNBs are incorporated at RAN. Without caching capability at eNB, content requests generated from user equipment (UE) have to reach far into the Internet through mobile core network (CN), where the content provider (CP) located. As compared, UEs can be satisfied locally at eNBs with great feasibility if the demanded data items are fully (or partially) cached. And hence the backhaul traffic is considerably alleviated given ever-growing UE demands on mobile data.

For example in Figure 3, assuming that the content object is requested at UE side, which is originally delivered from CP by splitting into several data chunks. Suppose eNB *A* has cached several chunks of the content object. Mobile users subscribing to eNB *B* request for the content object. Instead of fetching all content chunks from the original source, content demand could be partially served locally by nearby eNB *A* alternatively, via coordination between eNB *A* and eNB *B*.

BS offloading can be highly effective if the local eNB has knowledge of caching states at nearby eNBs (Zhang, 2017). To facilitate such capability, coordination between adjacent eNBs needs to be enabled, which can be realized through the X2 interface (Rangan, 2012). With cooperation, service providers are able to achieve the following performance goals as compared to independent BS offloading.

- Lower access delay for customers without resorting to further remote source CP
- Elimination of potential bottlenecks on backhaul links
- Reduction on cache redundancy among cache-enabled eNBs

It is commonly believed that a transformed mobile base station (e.g., BS offload-enabled) can greatly relive the hunger for computational capacity and meet the needs for robust networking.

2.4 D2D Offloading

Cellular offloading via device-to-device (D2D) communication can be a particular enhancement technique for data distribution in 5G networks. Different from other offloading options dependent more or less on cellular infrastructure, direct communication between UEs in proximity is enabled without use of the mobile network facilities. By forming into a device-to-device (D2D) networking group, UEs operate as ad hoc relays in the following two ways: a) either to connect to non-congested BSs (Andreev, 2015; Lin, 2014) or b) to cooperatively download the content. By concept, the BS unicasts original data items of the content to a selected group of UEs, which then multicast them to each other over local ad hoc networks using multi-hop cooperation.

Due to the opportunistic communication nature with D2D, non-real time content delivery is often offloaded for delay-tolerant networks. And D2D in 5G may even allow for better service experiences, in terms of decreased communication latency and higher downloading rate with reduced energy consumption. Possible applications are thus driven by user proximity, such as social networking and local exchange of information (e.g., advertisements and vehicular communication between smart cars). In particular, D2D networking could be a key enabling factor in the support for national security/public safety in case of not-spot network coverage area, where local connectivity is provided by devices at least.

D2D links can operate either on licensed bands or unlicensed bands. Unlicensed band D2D protocols operates on WiFi-direct or bluetooth are without network control, while protocols on licensed bands are with the help of cellular networks. With network-assistance, a centralized infrastructure (e.g., orchestrated by BSs) assists and controls the operation of D2D communication (Andreev, 2015; Lin, 2014).

For example in Figure 4, network control is applied when D2D shares with cellular the same licensed spectrum bands. With network assistance, device discovery is managed by the controller. The D2D link quality effect can be detected by the smart device locally, which is then reported to the serving controller (e.g., BS) when an offload mode shift need is detected. In comparison, with unlicensed spectrum adopted, D2D links can be established without centralized control. However, the discovery process can be very time and energy consuming in this way.

In fact, mobile devices are becoming smart capable of decision-making on switching between different access types (e.g., Wi-Fi or cellular). However, with network-assistance, customers could experience benefits especially on minimizing energy consumption and accelerating device discovery (Fodor, 2012). And operators can also gain from such functionality for concerns on traffic management, policy, charging and security, etc.

With network control, D2D communication could achieve potential gains as the following.

- **Energy and Time Consumption Gain:** D2D paring and device discovery could be significantly improved with the assistance of network infrastructure, e.g., BSs. For example, the scan for other wireless technologies can be avoided, the transmission and reception of discovery signals can be synchronized to save UEs frequent information exchange.
- **Capacity Gain:** By enabling spectrum resources sharing with cellular networks, the network capacity can be greatly improved with spectrum efficiently shared among multi- concurrent D2D links. By confining radio transmissions to the D2D connection, even higher spectrum reuse gain can be achieved as compared to LTE small cells.
- **Data Rate and Latency Gain:** High peak rates can be achieved when devices in proximity and actively engaged in data propagation. Since devices communicate over a direct link, the end-to-end latency may be reduced as compared to cellular communication, while a busy BS can be the bottleneck.

Figure 4. An example of cellular offloading via D2D communication

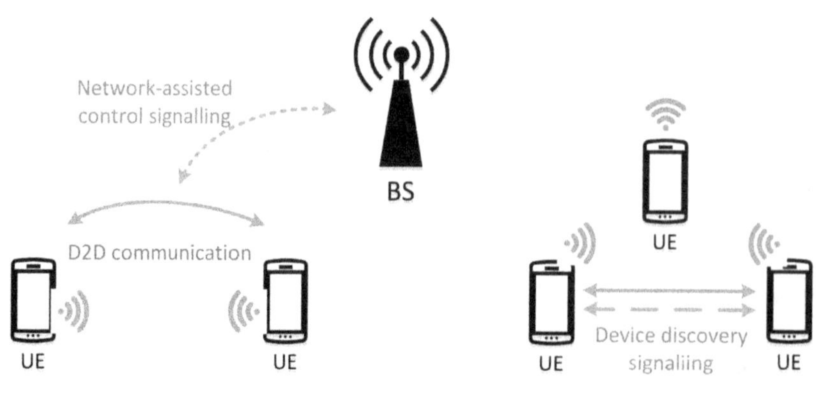

With D2D offloading, cooperation among mobile devices can be made within a single hop range or to multi-hops extent. However, one hop cooperation proves to be computationally efficient. And a tradeoff between delay and offloading extent exists, which proves to be NP-complete. To boost performance gains, D2D offloading bundled with BS offloading can significantly alleviate traffic pressure over cellular networks.

2.5 Mobile Data Offloading Efficiency

An efficient offloading approach is able to maximize the offloaded cellular traffic volume with improvement on performance gains, in terms of access delay, user data rate, network capacity and energy consumption, etc. The criteria is tightly coupled with each other and tradeoffs exist between one or two metrics as discussed previously. And thus a discussion on specific performance metric is meaningless without account for careful designs on the following operation process.

- **When to Offload:** Such issue has been discussed previously with respect to specific offload technique. To efficiently offload traffic from cellular networks, the key issue is to manage seamless *shift control between different offload options*, especially in a heterogeneous network scenario with various options co-existing (e.g., WiFi, small cell, BS and D2D offload). Such shift control can be performed by smart devices locally, or assisted by cellular infrastructure to save more energy on devices. When the delay users can tolerant is beyond a threshold, an offload shift will be triggered. In particular, the interworking of two options, such as WiFi offload integrated into cellular networks, will offer great performance gains. Meanwhile, when there are multiple offload options available, some level of strategy is need to hierarchically shift to a favored offload mode. Such strategized operation can be assisted by cellular network, so as to achieve energy and computation efficiency.
- **Where to Offload:** This could be a concern particularly with D2D offload. The rational is that a group of D2D candidates needs to be selected prior to commencing D2D communication sessions, wherein the original data is delivered to the group via cellular communication. However, how to choose such target UE group can be a challenge with concerns on tradeoffs between desirable service experiences and network performance gains. With network assistance, the BS can help to choose a set of user devices identified as target-UEs. The optimal UE-set selection proves to be an NP-hard problem (Han, 2012). Typical target-set selection algorithms are well-known as Greedy, Heuristic and Random algorithm. If mobile users actively participate in delivering content data, the performance among different set selection algorithms is not very significant. As such, a simple Random algorithm is effective enough to achieve desirable performance benefits.
- **What to Offload:** While alleviating traffic load from cellular networks, BSs and D2D offload need to consider *what to offer* to customers from their local cache. To intelligently provide such functionality, BSs and target-set UEs can be equipped with *in-network caching functionality*, which will be discussed in detail in Section 3. With such in-network replica intelligence, BSs (or target-UEs) are capable of intercepting passing through popular contents and selectively making cache decisions. When content items are pushed to UE level, the BS would assist D2D systems to select content items (Andreev, 2015). For instance, top popular content data items could be pushed by the BS to the target-UEs, a) by off-line operation with pre-selected content catalog, or b) by on-

line procedure managed in real-time the push-decision, which could rely on the demand history recorded at the BS. As such, data items cached at a BS is able to serve subscribed customers, and users would be benefited from nearby UEs with local caches for direct data delivery via D2D communication. Details regarding this part will be elaborated in Section 3.

3. IN-NETWORK CACHING FOR MOBILE CONTENT DELIVERY

The incentives of mobile edge offloading (e.g., BS offloading, D2D offloading, etc.) could be enhanced greatly by the capability of ICN-based caching technologies, whereby caching at the core or edge network in 5G-based network environments can be supported by sophisticated caching mechanisms at the device-level.

Offloading along with in-network caching technologies are among the most popular solutions for enabling efficient multimedia content delivery services. Such a feature will substantially increase the network efficiency when more and more end users become actually mobile users with smartphones, and thus will substantially improve the service in future 5G environments, given that mobile data will substantially dominate in future content distribution applications.

3.1 A Brief Survey on Information-Centric Networking

As the rapid growth towards the popularity of content distribution applications, end users are more caring about the content itself that can be obtained from the Internet. Consequently, tremendous overlay traffic can be incurred over the network due to users' appetite for data-intensive applications. In fact, the cellular networks is already overloaded especially by the mobile data traffic (Cisco, 2017). On the other hand, substantial access burden at the content provider side can be increased considerably due to their limited uploading bandwidths capacity, compared to the massive data demanded.

In order to provide more efficient content delivery, ICN (Koponen, 2007; Jacobson, 2009) has therefore emerged as an optimized method to relieve the above issues through caching inside network elements, e.g., at cache-enabled eNodeBs. A key feature of ICN is to deploy *in-network caching* that content objects can be cached within the network for local access by future interested clients, so as to alleviate the content access burden from the content provider's side. Namely, caching is deployed in ICN as an inherent network capability rather than an overlay service as conventional caching promoted, e.g., the Web caching or the CDN caching.

Generally, in-network caching policy can be categorized as two main types (Alimi, 2013; 2011; Rossini, 2012): *independent* and *coordinated caching*. With independent caching, individual network elements make their own caching decisions based on local information. Within this scope, a caching everything everywhere policy is generally utilized (Jacobson, 2009), whereby the concept is to cache every data item retrieved at all network devices along the delivery path.

However, given the limited capacity of network devices in comparison to the large number of content overlay traffic, the management of content distribution among the device side, or the in-network caching is crucial. Some researches thus promote content distribution among intermediate nodes in a selective way (Psaras, 2012), by taking into consideration some local information, such as cache capacity, in the caching decisions.

Coordinated caching (Guo, 2012; Cho, 2012; Ming, 2014), on the other hand, requires network elements to make caching decisions in a coordinated manner by sharing their state of local caching information with each other, in order to provision the cache capacity more efficiently and thus improve the network efficiency as well as users' perceived service quality.

In the following subsections, the main ICN architecture is presented to give an overview of how in-network caching operates. Different caching policies will be elaborated and compared in the next section, followed by a more detailed discussion on the concerns of caching efficiency.

3.1.1 Overview of the ICN Architecture

Next, the most representative ICN architecture is introduced by describing its main components to give an overview of the operation of an ICN-deployed network before bringing out in the next section its main idea of in-network caching.

Several architectures regarding the ICN concept have been proposed (Koponen, 2007; Jacobson, 2009; Ain, 2009; Ahlgren, 2010), among which Ref. (Jacobson, 2009) is the most representative paradigm that introduced content-centric networking (CCN) paradigm. The main components of a CCN node include a content store entity, a pending interest table (PIT) and a forwarding information base (FIB), as shown in Figure 5. The Request Packet Process and the Data Packet Process are key functions wherein in-network caching is involved.

Figure 5. ICN architecture overview

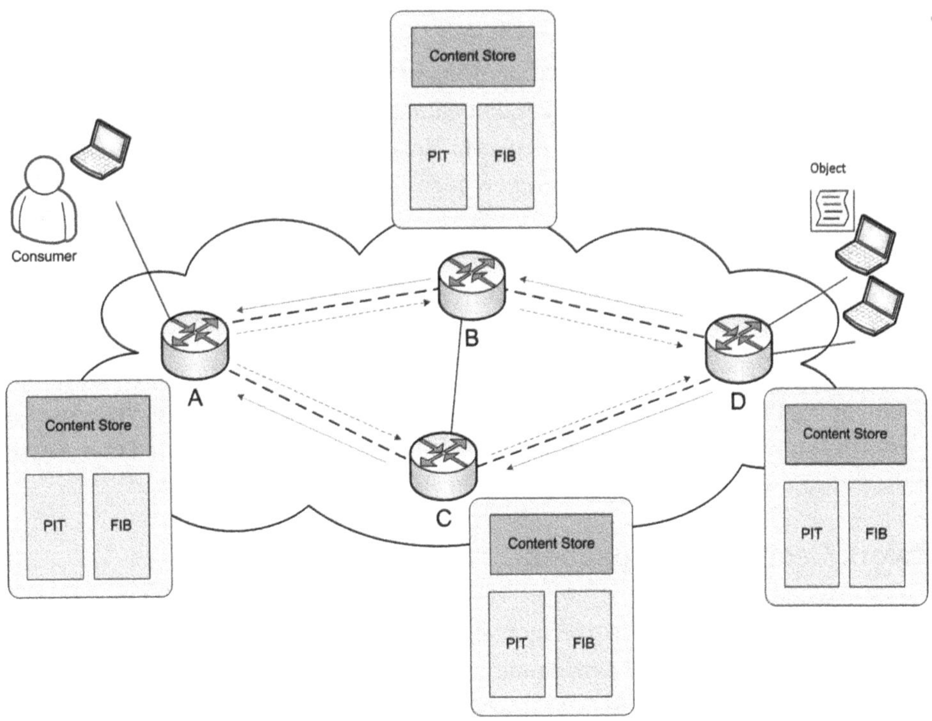

- **Request Packet Processing**

A data consumer, having access to the Internet via an access point (or network element) *A* in Figure 5, requests for a content object located next to *D* by broadcasting its interest over all available connectivity, e.g., via *A-B-D* and *A-C-D* in Figure 5 (the interest can be decomposed as several interests for small data chunks that constitutes the content object). Any intermediate network element along the delivery path can intercept the interest and locally serve it if it has a replica cached at the Content Store entity. A longest-match lookup on the Content Name in the Interest packet will perform the cache check. Otherwise, if the interest matches one of the PIT entries, simply the Interest's arrival face will be recorded and the interest will not be forwarded upstream. Such operation is referred to as the interest aggregation so as to avoid repeated forwarding and to reduce traffic redundancy. If no matches can be found at PIT entries, the Interest packet would be forwarded upstream towards the source data if there is a matching FIB entry, and a new PIT entry is created for this Interest and also its arrival face. Otherwise, the Interest will be simply discarded if there is no match at all, since such interest could arise from malicious behaviours.

- **Data Packet Processing**

The Data packet retrieving is relatively simple as compared to the Interest packet processing. The Data packet simply follows the symmetric reverse path of the PIT entries back to the requester as illustrated in Figure 5, e.g., via *D-B-A* and *D-C-A*. Only a PIT match along the delivery path can enable a caching decision upon an arrival of a Data packet. The reason is straightforward, since a Content Store match only means a duplicated data and thus the Data packet is discarded. A FIB match occurs when there is no PIT match, and it can arise from a malicious behaviour, and thus the Data packet is discarded as well. In particular, based on the CCN paradigm, each content chunk is cached at every cache along the delivery path, referred to as the *caching everything everywhere policy*. Such universal caching can certainly achieve a high caching performance in terms of desirable cache hits at individual caches due to its aggressive caching for each incoming content chunk. However, huge *cache redundancy* can exist since duplicated caching exist in the whole network. Besides, cache operational cost can as well arise due to great cache evictions incurred for each new incoming chunk. These issues will be discussed in detail in the next subsection.

It should be noted that in-network caching is a key feature in the context of ICN, which differentiates it from traditional caching schemes, such as the Web caching or CDN caching that deploy the caching as an overlay service. Rather, the ICN deploys caching as an inherent network capability in order to enable more efficient content distribution and achieve network efficiency. Henceforth, the understanding of the in-network caching is important to help to have an in-depth understanding of device-level mobile data offloading. Next different in-network caching policies are elaborated and detailed comparisons are also dicussed, in the purpose of providing a hint to design an optimal caching policy.

3.2 In-Network Caching Approaches

The motivation of in-network replica is to effectively alleviate the problem of increasingly constrained mobile networks through device-level data chunk caching. This technology has been introduced to optimise core network resources, and also to achieve better Quality of Services (QoS) including reduced content access time as well as end-to-end transmission delay. Researchers have explored intelligent algorithms and mechanisms for content chunk caching according to specific content delivery application characteristics such as content availability/popularity, as well as end user behaviours including group join/leaving and mobility patterns (Zhang, 2015). Optimised chunk caching and replacing algorithms are designed at the device level in order to best utilise the content caching resources.

Next, the in-network caching polices are elaborated to provide a general understanding. Note the caching policy is orthogonal to the cache replacement policy and thus the two can operate separately. Generally, in-network caching policy can be categorized as two main types: *independent* and *coordinated caching*.

3.2.1 Independent In-Network Caching Scheme

Within the scope of independent caching, individual network elements independently make caching decisions based on local information, without awareness of caching state of others. The main representative approaches in this scope can be categorized into two groups: the *universal caching* approach (Jacobson, 2009; Rosensweig, 2009; Arianfar, 2010), and the *selective caching* approach (Zhang, 2015; Psaras, 2012).

- **Universal Caching:** As shown in Figure 6, the caching decision is made solely based on local information. Network elements randomly (or aggressively) caches incoming chunks if none of the incoming chunk has been stored at the cache currently (Jacobson, 2009; Rosensweig, 2009; Arianfar, 2010). And if the cache is fully occupied, cache replacement policy will be adopted to make room for the new one, which will be discussed later. In this way, each data item of the content object can be cached everywhere within the delivery scope as shown in Figure 6. It should be noted that such caching policy makes caching decisions locally without awareness of duplicated caching elsewhere or content diversity property, e.g., chunk popularity degrees associated with different content objects.

Figure 6. An independent caching overview

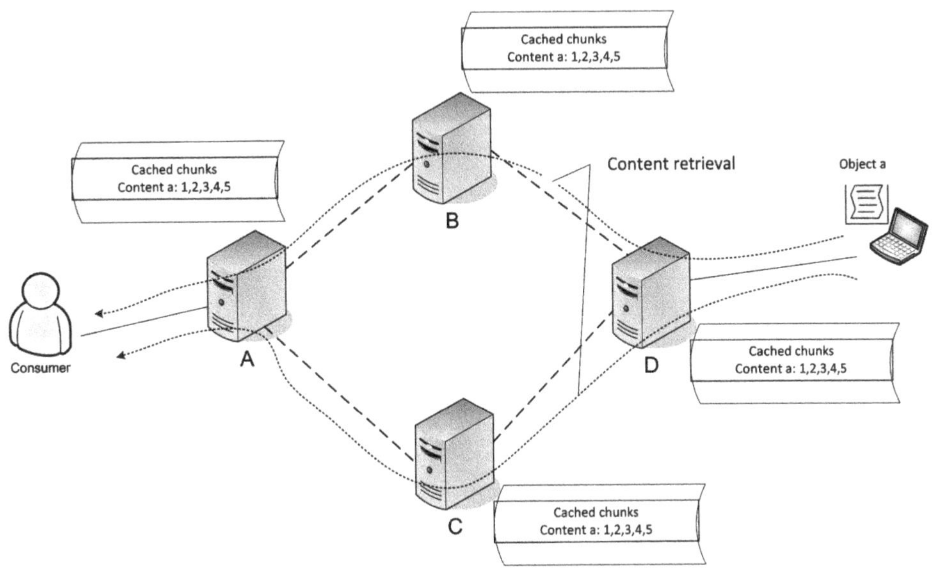

- **Selective Caching:** Based on the universal caching approach, the network cache resources are not efficiently utilized, especially given the limited cache capacity at individual network devices. As a result, an aggressive caching can easily result in cache redundancy. Towards this end, efficient caching approaches are proposed (Zhang, 2015; Psaras, 2012).

A simple example of selective caching is a *fixed rate caching decision* approach (Arianfar, 2010) And its caching logic is performed uniformly over the network element with a fixed probability for every incoming content chunk. For example, if the probability is set at 0.5 for each incoming chunk to be cached, each cache will store approximately half amount of the demanded content object at some point when the network is stable. As such, a certain cache space would be released for upcoming content chunks and the cache redundancy could be reduced to a certain level. Intuitively, higher set of the fixed probability value imply higher cache hits due to more chunks to be cached. And the probability equals to one based on the universal caching, wherein severe cache redundancy will be triggered.

Some research works are more concerning the issue of *where to cache*. They suggest to deploy in-network caches at some selected localities, such as inside access networks rather than at every cacheable network device (Katsaros, 2011), in order to localize overlay traffic to reduce mobile core network traffic and also to achieve the performance improvement of end users as well.

However, *what to cache* is not a big concern with in-network caching approaches as discussed above. And a simply universal caching policy is usually adopted. However, more strategized caching approaches (Zhang, 2015; Psaras, 2012) emerged aiming to achieve more efficient content distribution across the network. These policies concerns more about the content object features. Specifically, content popularity/availability (Zhang, 2015), and context information at local (e.g., the cache capacity and cache locality) are taken into account to make caching decisions. With the consideration of cache capacity, content chunks tend to be stored at relatively larger caches along the delivery path. In case of multiple same size caches as candidates, the one located in proximity to requesters will be preferred. As such, the content access latency will be reduced and the network resources can be more efficiently managed, especially in the heterogeneous caches network scenario. For instance, if caches in Figure 6 all have identical cache capacity, more popular content chunks will be cached at network devices closer to the consumer side based on this strategized caching policy, e.g., more chunks would be cached towards to cache *A* rather than at cache *D*.

3.2.2 Cooperative In-Network Caching Scheme

However, a certain level of duplicated caching can still exist even if a carefully selective caching policy is adopted. The main rational is that there is no context information exchanged between individual caches regarding their cached contents. And the lack of coordination among caches can also result in the case that a request unable to be served at a cache would be forwarded towards the original source, while the demanded data is cached somewhere near to the cache already.

In the purpose of eliminating the cache redundancy so as to achieve more efficient in-network caching, a few research works have been engaged in the context of coordinated in-network caching efforts (Zhang, 2017; Cho, 2012).

A key enabling factor of coordinated caching is cooperation among neighboring caches. By concept, network elements within the same community (e.g., e.g., caches along the same delivery path or network devices nearby) are able to share with each other caching status. This includes context information, such as the popularity degree of incoming content chunks and the list of cached contents. As such, content demands could be served in a whole neighborhood range rather than at a single cache. With awareness of such context information available, caches can leverage it when making caching decisions, such that cache redundancy can be further reduced.

Generally, coordinated caching enables significant improvement on caching performance in mainly two ways. Firstly, cache hits could be greatly increased by forwarding requests to nearby caches. And thus the access burden from content providers can be relieved considerably. And secondly, the coordination operation allows caches to be aware of cached states within the overall neighbours, such that a duplicated caching can be avoided.

Take an example in Figure 7 for the illustration of a simple coordination scenario with two objects traversing, content object *a* and *b*, respectively. Each content object is divided into five data chunks of a unified size. And the cache size is denoted as the number of chunks, assumed to be a five-chunk capacity. Assuming a case wherein neighbouring caches of one-hop distance coordinate with each other via caching status sharing and requests serving. For individual network elements, either a universal caching or a selective policy can be adopted. As shown in the figure, neighbouring caches are able to hold diverse content chunks due to coordination.

However, network devices beyond one-hop distance can still hold the same chunks, such as cache *A* and *D*, or cache *B* and *C*, since it is beyond the communication distance. If the cache redundancy in a neighbourhood is calculated as $\frac{\#\ of\ chunks\ duplicated\ cached}{\#\ of\ chunks\ cached}$. Then with awareness of caching status in the neighbourhood of *A*-*B*-*C*, the cache redundancy intensity can be obtained as $\frac{2\times 2}{5\times 3} = 4\,/\,15$ for content *a*, and $\frac{2\times 3}{5\times 3} = 6\,/\,15$ for content *b*, and the overall cache redundancy can be deduced with

Figure 7. A simple coordinated in-network caching illustration

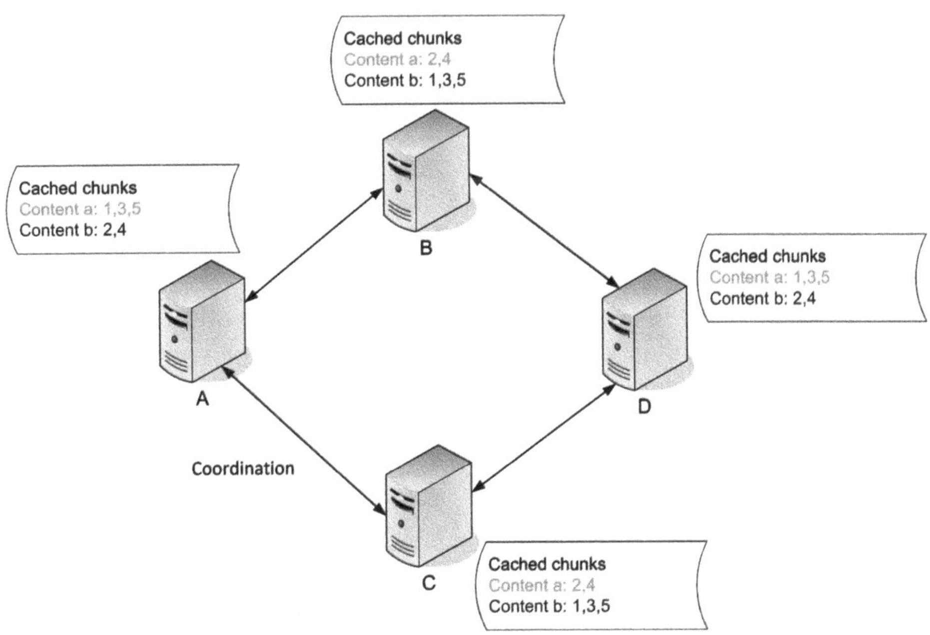

all contents as $\frac{\frac{4}{15}+\frac{6}{15}}{2}=\frac{1}{3}=33\%$. As compared with non-coordination caching approaches, wherein the cache redundancy intensity can be up to 100% based on a universal caching policy, while a reduction of 67% at least can be achieved under coordination in this case.

Theoretically, the ideal performance gains could be achieved when neighbouring caches exchange the whole list of cached contents, so as to completely avoid duplicated caching. However, the signalling overhead can be another concern. Thus a trade-off exists between the goal of cooperation (maximising the diversity of cached items) and the performance goal of minimizing the access latency.

3.2.3 Cache Replacement Policies

When the cache is fully occupied, a cache replacement policy needs to be adopted to evict cached data chunks. In comparison, cache replacement is orthogonal to caching policy, and thus the two can operate independently without affecting each other's decisions. Generally, there are two main eviction policies, known as the LFU (Least Frequently Used) and LRU (Least Recently Used). Based on these two eviction policies a combination of a Least Frequently and Recently Used policy is also implemented (Lee, 2001). A more simplified eviction policy known as the RND (random replacement policy) is considered as an efficient replacement policy especially for ICN in (Arianfar, 2010), due to the line-speed consideration in ICN caching operations and a RND is easy to implement in practice to make ICN scalable.

LFU is the eviction policy to remove the item in the cache with the smallest number of accesses. However, such eviction policy is hard to be implemented in practice in constant time. In Comparison, LRU is a relatively simpler policy easily implemented in practice as compared to the LFU. It replaces items that are so far have not been accessed in the longest time. LRFU is a combination of LRU and LFU, which leverages between these two eviction policies to promote the most benefits of both. A weight value is associated to each chunk to weigh its preference degree for staying in the cache, wherein a constant parameter allows for a trade-off between recency and frequency. Chunks with the lowest weight value will be evicted. A simple random eviction policy (Rossini. 2012), however, is proved to be able to perform fairly well to achieve desirable performance gains.

3.3 In-Network Caching Efficiency

The caching efficiency is essentially associated with the following performance aspects: *cache hits*, *cache evictions*, and *cache redundancy*.

- **Cache Hits:** Are important to evaluate a caching policy, which is often expressed as a ratio that refers to the fraction of requests served at a cache for a given amount of time. Intuitively, a caching-everything-everywhere policy enables high cache hits due to its aggressive caching. However, it can suffer from low caching efficiency due to its data-insensitive caching nature. In comparison, a more strategized caching policy takes into account important context information so as to perform intelligent caching decisions' making. Typical examples are (Zhang, 2015; Cho, 2014; Ming, 2014) based on content popularity and routing information, such as distance-aware caching. As such, popular content items tend to be more frequently served at caches, and latency can be reduced due to demanded data stored at closer caches. If coordination is further enabled between adjacent caches, cache hits can be further improved due to the possibilities for serving content requests are increased to larger scale. Namely, requests unable to be served locally re forwarded to nearby caches that may store the demanded data (Zhang, 2017; Cho, 2012).

- **Cache Evictions:** Are also important to expresses the efficiency of a caching policy. The operational cost for a cache is mainly attributed to cache evictions. And thus high cache hit ratio along with low number of cache evictions usually defines the efficiency for a caching policy. With the universal caching, high cache hits do exist based on evaluations of Ref. (Zhang, 2015). Nevertheless, severe cache evictions are incurred at the same time due to its content-blind caching strategy. As compared, a more selective caching can reduce huge cache evictions owing to its intelligent content-aware caching decision. It promotes more popular data items to be cached while less unpopular caches are discarded or evicted. With coordinated caching policy, cache evictions can be further reduced as compared to the independent selective caching. The rational is the caching knowledge shared via coordination cooperation among adjacent caches. As such, unnecessary evictions can be avoided if the to-be-cached popular data items are already cached nearby.
- **Cache Redundancy:** Is the metric that can measure the efficiency of the management of network resources. As shown previously, the cache redundancy refers to the intensity of data replicas distributed over the network. From the network's perspective, a low cache redundancy is expected to more efficiently utilize cache resources. However, huge cache redundancy can be incurred under a universal caching as dicussed previously as shown in Figure 6. Although a selective caching can reduce a certain level of the cache redundancy by taking into account some context information (e.g., the locality of the cache, or the popularity degrees of content items), cache redundancy still exists under such caching strategy among network elements due to their independency on caching operations. In comparison, coordinated caching can greatly reduce the cache redundancy. With awareness of neighbouring caches' caching states, duplicated caching can be avoided. However, concerning the signalling scalability, an exchange of a full list of content items among adjacent caches can be costly in terms of communication overhead. On the other hand, the coordination scope among caches can be enabled in various scales that may achieve different performance results. However, by enabling neighbouring caches with one-hop distance coordinating with each other, the performance gains are sufficient enough (Zhang, 2017; 2015). A more strategized consideration promotes selecting caches with the most connection degrees in the network, which are engaged in coordination with their nearby caches (Wang, 2013). The rational behind this is that caches with high connection degree can have great potential to serve more requests from nearby nodes.

4. INCENTIVES AND CHALLENGES

Largely driven by the alleviation of cellular traffic load on networks, mobile data offloading can offer improved network capacity, lower cost and desirable user data rate. These benefits provide incentives for customers to remain connected on the cellular network, most of the time. With offload, the potential incentives include:

- Minimal changes to existing network protocols and interface, by avoiding costly deployment, i.e., dependence on expensive real-estate cost.

- Assured service quality in a simple way with regard to deployment.
- However, new challenges are implied mainly in the following aspects:
    - **Interference Management:** It is one of the challenges for D2D offload, when devices operates in the same band as cellular communications. For example, interference will happen when two adjacent devices share the same band but access to different networks (e.g., one with D2D mode and the other with the cellular network). Due to the coexistence of D2D and conventional cellular, harmful interference can be caused. Two types of interference can be invoked: a) cellular-D2D user interference and b) D2D-D2D user interference. By adopting some power control strategies performed by BS in a centralized way, the cellular-D2D can be relieved. And operations managed by devices in a distributed manner can alleviate the D2D-D2D interference. However, how to design an effective and efficient strategy has always been a major challenge (Lin, 2014; Han, 2012).
    - **Mobility Management:** Since mobile data is offloaded from mobile networks that naturally supports seamless handovers for users in mobility, it is important to maintain service continuity with offload cases. However, the challenge is how to efficiently handle the mobility management in a heterogeneous access network scenario, i.e., in case of multiple offload options available (e.g., WiFi/small cell, BS, D2D offload). Some strategized offload shift policies need to be designed to make efficient decisions, such as to identify the right time to initiate the shift and to trigger optimal selections on the most appropriate interface. Meanwhile, due to D2D natures, high latency and signaling overhead are some of the major issues (Lin, 2014), especially with energy concerns that devices can be out of battery quickly when served as relay. With account for offload efficiency, desirable service continuity becomes challenging for users in mobility.
    - **Security and Privacy Issues:** As in D2D communication, since the user data is transmitted through other users' devices, the security and privacy become crucial which must be maintained. The BS can work as a trusted party that is responsible for authentication of the relaying devices and proper encryption on data, so as to maintain the privacy of the users. Without intervention of BS, devices have to establish trust based on local knowledge that can be vulnerable to malicious attacks, which is an open research problem in this case (Tehrani, 2014).

By taking advantage of network-assistance, concerns discussed above could be greatly relieved, e.g., by adopting the key generation and distribution mechanisms already available in mobile networks. However, the centralized fashion with the network-control incur concerns with issues inherently with such operation, such as communication overhead and network scalability.

Key techniques of in-network caching are discussed previously, in the hope of providing hints on the design of efficient offload paradigm. However, remaining challenges still need to be addressed in order to design an efficient in-network caching policy in a cellular network.

- **Trade-Off:** Between users' perceived service quality and the network efficiency. From a consumer's perspective, it is ideal that every content demand can be served. While from the network's perspective, a desirable caching policy should be more selective rather than aggressive, so as to efficiently distribute diverse contents among caches with limited capacities. As such, more popular content items tend to be cached and served by intermediate nodes while less popular content items are discarded and may only be reached from original sources. With content items suffering from availabilities at original sources, access latency could be increased for end users. Hence, the trade-off between users and the underlying network is one concern worth noticing.

- **Communication Overhead:** Is one major concern in such coordination deployment. In order to keep caches to be updated, network elements need to periodically exchange information on their locally cached data items, in order to avoid unnecessarily duplicated caching between nearby caches. As a result, potentially huge information will be incurred for exchange. Bloom Filters (BFs) (Liu, 2012) are one technique effective for aggregating information on cached data items in signalling between coordinated caches in order to achieve scalability.
- **Cooperation Scope:** Another research issue concerning how far in the neighbourhood caches should search. Most researches keep cooperation at one-hop scale. The main purpose is to ensure the access latency at a demanded level with an accepted cache redundancy. Thus a *trade-off* exists between the goal of cooperation (maximising the diversity of cached items) and the performance goal of minimizing the access latency. An optimal degree of redundancy is also seen as a challenging problem (Wang, 2013). To avoid such a dilemma, the scope of cooperation is usually kept at one-hop scale, which frees researchers from routing concerns and also benefits the reduction of signalling overhead.
- **Legal Issues:** The content items cached may incur copy right issues for some content providers, if the economic benefits are on the main concern of illegally wide distribution of the contents. Such issue needs to be taken into account when making caching decisions, to make sure the cached data has already acquired the certificate from the content provider. Nevertheless, such certificate needs the cooperation with the content owners, and the strategy regarding the distribution of the economic utilities between the network operator and the content provider is still unknown.

5. SUMMARY AND CONCLUSION

Offloading along with in-network caching is among the most popular solutions for enabling efficient multimedia content delivery services in upcoming 5G environments. Researchers as well as service providers have commonly agreed that the two techniques enable great performance gains by enabling services at the network edge, with regard to enhanced pervasive connectivity as well as service velocity. Given that mobile data will substantially dominate in future content distribution applications, such a feature will substantially increase the network efficiency when more end users become actually mobile users with smart user devices.

There is no doubt that mobile data offloading is a major opportunity for mobile network operators, which enables feasible data distributions with cost-effective solutions. By pushing the communication mode towards the network edge, it has the potential to achieve efficient network operation for the concerns of future cellular services. Meanwhile, offload brings forward many challenging issues that remain as open research problems, which requires further research standardization efforts.

By promoting ICN-based caching technologies at the core or edge network in 5G-based network environments, the incentives of mobile edge offloading could be enhanced greatly. In-network caching can be supported by sophisticated caching logics at the network-element or user device-level. Several standardization activities also advocate efficient in-network caching strategy. The Internet Engineering Task Force (IETF) have proposed two options to cache content chunks within the network, either implicitly or explicitly. With implicit caching, network nodes passively cache passing-through content chunks, while the explicit policy enables caches to explicitly request for content chunks to be cached locally. Most research works are based on implicit caching concept, while the explicit caching is not gaining much research interests due to its complexity for deployment in reality.

ACKNOWLEDGMENT

The work is supported by Waseda University, grant program for promotion of international joint research FY2018 "Towards Contact-Aware Proximity (CAP) Service via Opportunistic D2D Relay".

REFERENCES

Ahlgren, B., D'ambrosio, M., Dannewitz, C., Eriksson, A., & Golic, M. J. (2010). *Second netinf architecture description.* 4WARD EU FP7 Project, Deliverable D-6.2 v2. 0. Retrieved from http://www.4ward-project.eu/

Ain, M., Trossen, D., Nikander, P., Tarkoma, S., Visala & Kjällman, J. (2009). *D2. 3–architecture definition, component descriptions, and requirements.* Deliverable, PSIRP 7th FP EU-funded project, 11.

Alimi, R., Rahman, A., Kutscher, D., Yang, Y., Song, H., & Pentikousis, K. (2013). *DECoupled Application Data Enroute (DECADE)* (No. RFC 7069).

Alimi, R., Rahman, A., & Yang, Y. (2011). *A survey of in-network storage systems* (No. RFC 6392).

Andreev, S., Galinina, O., Pyattaev, A., Johnsson, K., & Koucheryavy, Y. (2015). Analyzing assisted offloading of cellular user sessions onto D2D links in unlicensed bands. *IEEE Journal on Selected Areas in Communications, 33*(1), 67–80. doi:10.1109/JSAC.2014.2369616

Andrews, J. G., Claussen, H., Dohler, M., Rangan, S., & Reed, M. C. (2012). Femtocells: Past, present, and future. *IEEE Journal on Selected Areas in Communications, 30*(3), 497–508. doi:10.1109/JSAC.2012.120401

Aptilo Networks. (2013). *Seamless Wi-Fi Offload: A Huge Business Opportunity for Mobile Carriers.* White paper.

Arianfar, S., Nikander, P., & Ott, J. (2010, November). On content-centric router design and implications. In *Proceedings of the Re-Architecting the Internet Workshop* (p. 5). ACM.

Arianfar, S., Nikander, P., & Ott, J. (2010, June). Packet-level caching for information-centric networking. In *ACM SIGCOMM*. ReArch Workshop.

Cao, Y., Wang, N., Kamel, G., & Kim, Y. J. (2015). An electric vehicle charging management scheme based on publish/subscribe communication framework. *IEEE Systems Journal.*

Cho, K., Lee, M., Park, K., Kwon, T. T., Choi, Y., & Pack, S. (2012, March). Wave: Popularity-based and collaborative in-network caching for content-oriented networks. In *Computer Communications Workshops (INFOCOM WKSHPS), 2012 IEEE Conference on* (pp. 316-321). IEEE.

Cisco Visual Networking Index. (2017, March). *Global Mobile Data Traffic Forecast Update, 2016-2021.* White Paper.

Fodor, G., Dahlman, E., Mildh, G., Parkvall, S., Reider, N., Miklós, G., & Turányi, Z. (2012). Design aspects of network assisted device-to-device communications. *IEEE Communications Magazine*, *50*(3), 170–177. doi:10.1109/MCOM.2012.6163598

Golrezaei, N., Shanmugam, K., Dimakis, A. G., Molisch, A. F., & Caire, G. (2012, March). Femtocaching: Wireless video content delivery through distributed caching helpers. In INFOCOM, 2012 Proceedings IEEE (pp. 1107-1115). IEEE.

Guo, S., Xie, H., & Shi, G. (2012, May). Collaborative forwarding and caching in content centric networks. In *International Conference on Research in Networking* (pp. 41-55). Springer.

Han, B., Hui, P., Kumar, V. A., Marathe, M. V., Shao, J., & Srinivasan, A. (2012). Mobile data offloading through opportunistic communications and social participation. *IEEE Transactions on Mobile Computing*, *11*(5), 821–834. doi:10.1109/TMC.2011.101

Jacobson, V., Smetters, D. K., Thornton, J. D., Plass, M. F., Briggs, N. H., & Braynard, R. L. (2009, December). Networking named content. In *Proceedings of the 5th international conference on Emerging networking experiments and technologies* (pp. 1-12). ACM.

Katsaros, K., Xylomenos, G., & Polyzos, G. C. (2011). MultiCache: An overlay architecture for information-centric networking. *Computer Networks*, *55*(4), 936–947. doi:10.1016/j.comnet.2010.12.012

Koponen, T., Chawla, M., Chun, B. G., Ermolinskiy, A., Kim, K. H., Shenker, S., & Stoica, I. (2007, August). A data-oriented (and beyond) network architecture. *Computer Communication Review*, *37*(4), 181–192. doi:10.1145/1282427.1282402

Lee, D., Choi, J., Kim, J. H., Noh, S. H., Min, S. L., Cho, Y., & Kim, C. S. (2001). LRFU: A spectrum of policies that subsumes the least recently used and least frequently used policies. *IEEE Transactions on Computers*, *50*(12), 1352–1361. doi:10.1109/TC.2001.970573

Lin, X., Andrews, J., Ghosh, A., & Ratasuk, R. (2014). An overview of 3GPP device-to-device proximity services. *IEEE Communications Magazine*, *52*(4), 40–48. doi:10.1109/MCOM.2014.6807945

Liu, H., De Foy, X., & Zhang, D. (2012, August). A multi-level DHT routing framework with aggregation. In *Proceedings of the second edition of the ICN workshop on Information-centric networking* (pp. 43-48). ACM. 10.1145/2342488.2342498

Ming, Z., Xu, M., & Wang, D. (2014, August). Age-based cooperative caching in information-centric networking. In *Computer Communication and Networks (ICCCN), 2014 23rd International Conference on* (pp. 1-8). IEEE. 10.1109/ICCCN.2014.6911725

Psaras, I., Chai, W. K., & Pavlou, G. (2012, August). Probabilistic in-network caching for information-centric networks. In *Proceedings of the second edition of the ICN workshop on Information-centric networking* (pp. 55-60). ACM. 10.1145/2342488.2342501

Rangan, S., & Madan, R. (2012). Belief propagation methods for intercell interference coordination in femtocell networks. *IEEE Journal on Selected Areas in Communications*, *30*(3), 631–640. doi:10.1109/JSAC.2012.120412

Rosensweig, E. J., & Kurose, J. (2009, April). Breadcrumbs: Efficient, best-effort content location in cache networks. In INFOCOM 2009, IEEE (pp. 2631-2635). IEEE.

Rossini, G., & Rossi, D. (2012, September). A dive into the caching performance of content centric networking. In *Computer Aided Modeling and Design of Communication Links and Networks (CAMAD), 2012 IEEE 17th International Workshop on* (pp. 105-109). IEEE.

Suh, D., Ko, H., & Pack, S. (2016). Efficiency analysis of WiFi offloading techniques. *IEEE Transactions on Vehicular Technology*, *65*(5), 3813–3817. doi:10.1109/TVT.2015.2437325

Tehrani, M. N., Uysal, M., & Yanikomeroglu, H. (2014). Device-to-device communication in 5G cellular networks: Challenges, solutions, and future directions. *IEEE Communications Magazine*, *52*(5), 86–92. doi:10.1109/MCOM.2014.6815897

Wang, J. M., Zhang, J., & Bensaou, B. (2013, August). Intra-AS cooperative caching for content-centric networks. In *Proceedings of the 3rd ACM SIGCOMM workshop on Information-centric networking* (pp. 61-66). ACM. 10.1145/2491224.2491234

Zeydan, E., Tan, A. S., Mester, Y., & Özcan, G. (2016, April). Integration and management of Wi-Fi offloading in service provider infrastructures. In *Network Operations and Management Symposium (NOMS), 2016 IEEE/IFIP* (pp. 1153-1156). IEEE. 10.1109/NOMS.2016.7502979

Zhang, X., & Cao, Y. (2017). A Cooperation-Driven ICN-based Caching Scheme for Mobile Content chunk Delivery at RAN. *Wireless Communications and Mobile Computing Conference (IWCMC), 13th International*.

Zhang, X., Wang, N., Vassilakis, V. G., & Howarth, M. P. (2015). A distributed in-network caching scheme for P2P-like content chunk delivery. *Computer Networks*, *91*, 577–592. doi:10.1016/j.comnet.2015.08.035

ADDITIONAL READING

Moustafa, H., & Zeadally, S. (Eds.). (2012). *Media networks: Architectures, applications, and standards*. CRC Press. doi:10.1201/b12049

Chapter 11
Principles and Enabling Technologies of 5G Network Slicing

Zoran Bojkovic
University of Belgrade, Serbia

Bojan Bakmaz
University of Belgrade, Serbia

Miodrag Bakmaz
University of Belgrade, Serbia

ABSTRACT

5G mobile systems can be comprehended as highly flexible and programmable E2E networking infrastructures that provide increased performance in terms of capacity, latency, reliability, and energy efficiency while meeting a plethora of diverse requirements from multiple services. Network slicing is emerging as a prospective paradigm to meet these requirements with reduced operating cost and improved time and space functionality. A network slice is the way to provide better resource isolation and increased statistical multiplexing. With dynamic slicing, 5G will operate on flexible zone of the network, permitting varying, adaptable levels or bandwidth and reliability. In this chapter, a comprehensive survey of network slicing is presented from an E2E perspective, detailing its origination and current standardization efforts, principal concepts, enabling technologies, as well as applicable solutions. In particular, it provides specific slicing solutions for each part of the 5G systems, encompassing orchestration and management in the radio access and the core network domains.

INTRODUCTION

Fifth generation (5G) mobile networks are expected to create multitenant ecosystem with extremely increased performance for dedicated use-cases and specific types of services in order to simultaneously satisfy various users' demands. In such an environment network slicing emerged as one of the key pro-

DOI: 10.4018/978-1-5225-7570-2.ch011

spective technologies which support different requirements through a common network infrastructure. In this sense, service necessity, isolation and support on a common level of physical infrastructure are assured. It means that on-demand services can be fully supported, while network resources are efficiently allocated according to the users' requirements. With network slicing, 5G applications will operate on the flexible zone, allowing them to draw adaptable levels of bandwidth and reliability.

Although the network slicing concept is still immature, the potential enabling technologies, such as software defined networking (SDN) and network functions virtualization (NFV), have many feasible researches with practical solutions (Ordonez-Lucena, Ameigeiras, Lopez, Ramos-Munoz, Lorca & Folgueira, 2017). Based on SDN and NFV many user-centric service slicing strategies were proposed. By means of SDN and NFV, operators can provide high degree of flexibility and programmability, allowing legacy functions to be partitioned or migrated in data centers, advancing virtual architectures (Zhang, Liu, Chu, Long, Aghvami, & Leung, 2017).

Currently, network slicing is in the main focus of both, academia and industry, as well as different standardization bodies, e.g., International Telecommunication Union (ITU), 3rd Generation Partnership Project (3GPP), European Telecommunications Standards Institute (ETSI), etc. At the moment, efforts are toward developing 5G mobile systems which are in a position to deploy slicing of different structures and sizes (Katsalis, Nikaein, Schiller, Ksentini, & Braun, 2017).

This chapter aims to provide an updated state of the art in research activities when dealing with the 5G network slicing, together with identified and analyzed opportunities and challenges. To this end, a common framework for bringing together and evaluating various approaches from open literature in concise and holistic manner is presented. The main objective is to analyze current maturity of prospective solutions according architectural segments they target and to identify remaining gaps. This chapter can provide reliable backbone for future standardization and research activities.

The remainder of this chapter is organized as follows. The next section presents service requirements in 5G mobile systems. Following that, a brief overview of network slicing origination and main principles are provided. As enablers of network slicing, different virtualization technologies are introduced. Then, specific slicing approaches for radio access domain and core domain are analyzed. Finally, some open research challenges are elaborated and relevant conclusions are provided.

SERVICE REQUIREMENTS IN 5G SYSTEMS

5G mobile systems can drastically change the architecture and nature of communications. Many use cases are emerging with diverse requirements in terms of data rate, latency, connection density, mobility, reliability, spectrum and energy efficiency. These use cases may be broadly categorized in the three generic services (ITU-R, 2015), i.e., enhanced mobile broadband communications (eMBBC), massive machine-type communications (mMTC), sometimes referred as massive Internet of Things (mIoT), and ultra-reliable low-latency communications (uRLLC), a.k.a. mission-critical services, based on corresponding key performance indicators (KPIs), as presented in Figure 1. In this case KPIs can be treated as technical requirements for 5G services. 5G is anticipated to make it possible to efficiently enable diverse services, connecting a pool of varied devices while accessing diverse networks.

At the same time, the 5G Infrastructure Public Private Partnership (5G-PPP), as a joint initiative between the European Commission and industry, brought interest into the requirements of 5G through the

Figure 1. Technical requirements for 5G services

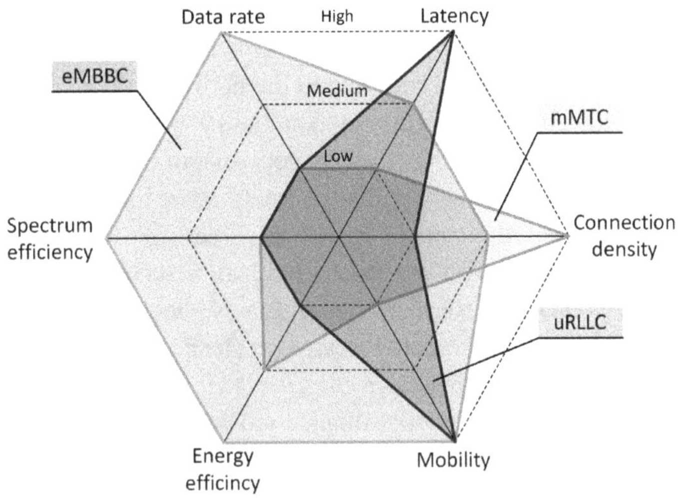

Mobile and wireless communications Enablers for Twenty-twenty (2020) Information Society (METIS) project. The network capabilities such as extremely fast (data rates of 10 Gb/s), great service in a crowd (traffic volume of 9 Gb/h), super real-time and reliable connection (E2E latency less than 5ms and reliability in order of 10^{-5}), as well as ubiquitous thing communicating ($3x10^5$ devices/cell) are supporting rapid time-to-market for brand new services and reducing management costs.

Recently, 3GPP SA WG1 has completed its first technical specification on *Service requirements for the 5G system* (3GPP TS 22.261, 2018). This specification contains both requirements on underlying capabilities and performance targets. 5G capabilities can be characterized by:

- Scalability and ubiquitous support for plethora heterogeneous services and vertical markets.
- Resource efficiency for existing and new services ranging from low data rate servicing (e.g., mMTC) to high bitrate multimedia services (e.g., virtual and augmented reality).
- Energy efficiency and power supply optimization.
- Capability exposure to allow third party Internet service/content providers to manage network slices and deploy applications in the operator's hosting environment.
- Connectivity from remote devices via relay and seamless service connectivity between indirect and direct connections.

It should be noted that this specification defines performance targets for different scenarios when considering service areas (urban and rural wide-area, indoor hotspots, dense urban area, etc.) and applications (broadband access in a crowd, tactile interaction, transport systems, etc.). For example, user experienced data rates vary from 1 Gb/s in downlink for indoor hotspots to 15 Mb/s in uplink for airplanes connectivity. Also, latency targets are as low as 0,5 ms for tactile interaction, while capacity targets can be as high as 15 Tb/s/km^2 with 250 000 users/km^2 for indoor hotspot such as office environments. These service requirements are used as starting guidelines for the normative work in other 3GPP WGs.

NETWORK SLICING ORIGINATION AND PRINCIPLES

The period 1960s, when the IBM operating system CP-40 with implemented resource virtualization was developed, can be considered as pioneering step to the network slicing concept (Goldberg, 1974). Virtualization was widely adopted through data centers in the 1970s, while a decade later it was applied into the computer networks. In the late 1980s, overlay networks of nodes connected over logical links and forming a virtual environment over a physical infrastructure, are introduced. They can be seen as an early form of network slicing, combining heterogeneous resources over various administrative domains. During the last decade, several testbeds on network virtualization technologies for promoting research on a clean slate network are conducted (Berman, Chase, Landweber, Nakao, Ott, Raychaudhuri, Ricci, & Seskar, 2014). The main idea was to promote research on clean state network, while addressing consolidate resources and mobile environment.

Recently, Next Generation Mobile Network Alliance introduced network slicing defined as a concept for running multiple logical networks as independent business operations on a common physical infrastructure (NGMN Alliance, 2016). Each slice represents an independent virtualized E2E network and allows operators to run deployments based on different architectures in parallel. A network slice as a logical E2E construct is self-constrained, having customized functions also including phase in the user equipment (UE) and using network function (NF) chains for delivering services to a given group of users (Nikaein, Schiller, Favraud, Katsalis, Stavropoulos, Alyafawi, Zhao, Braun, & Korakis, 2015). Key aspects necessary to realize the network slicing are: resources, virtualization, orchestration, and isolation.

Resource is manageable unit defined by a set of attributes or capabilities that can be used for service distribution. In this sense, there are two types of resources to be considered, i.e., NFs and infrastructure resources. NFs as functional blocks provide network capabilities in order to support and realize the particular service to use case demands. NFs can be physical and/or virtualized. As for infrastructure resources, they include computing hardware, storage capacity, and radio interface. Virtualization enables resource sharing among slices. Network virtualization enables new business models with novel actors, such as, for example, infrastructure providers, end users, etc. The allocation of resources to each slice can be either statically or dynamically. With static slicing, each slice is assigned a fixed portion of the physical resources during entire service time. This amount of resources usually corresponds to the peak service requirement. Despite its simplicity, static slicing does not result in an efficient utilization of the physical resources. In the case of time-varying service requirements, a static slicing approach may result in the overprovisioning of physical resources. Therefore, a more promising solution is dynamic slicing (Raza, Fiorani, Rostami, Öhlen, Wosinska, & Monti, 2018).

Orchestration is required to separate processes for creating, managing, and distribution of services. In network slicing, orchestration cannot be performed by a single centralized entity, not only because of the complexity and broad scope of orchestration tasks, but also because of its necessity to preserve management independence. Each slice must be managed independently as a separated network.

This is one of a major requirement for isolation, which need to be satisfied in order to operate parallel slices simultaneously on a common shared underlying substrate. Isolation is an E2E issue, regardless of the congestion and performance levels of offered slices. Security and reliability in terms of isolation, means that attacks or failures occurring in one slice must not have an impact on other slices. Each slice must have independent security functions (Arfaoui, Bisson, Blom, Borgaonkar, Englund, Félix, Klaedtke, Nakarmi, Näslund, O'hanlon, Papay, Suomalainen, Surridge, Wary, & Zahariev, 2018).

In order to enable on demand mobility management in 5G systems, a mobility driven network slicing (MDNS) was proposed (Hucheng, Shanzhi, Ming, & Yan, 2017). MDNS takes individual mobility support requirements into account while customizing networks for different services. Within this framework, the actual levels of required mobility support are determined by a mobility description system, and network slice profiles with the corresponding mobility management schemes are defined by a network slice description function. By instantiating the network slices, each mobile terminal can select the network slice with the most appropriate mobility management scheme.

Network Slicing-Based 5G System Presentation

As shown in Figure 2, there are three main layers in network slicing-based 5G system (Foukas, Patounas, Elmokashfi, & Marina, 2017): infrastructure, NF layer and service layer as well as management and orchestration (MANO) entity. This framework is favorable for presentation and analysis of fundamental slicing concepts and benefits.

Infrastructure layer refers to the physical network infrastructure spanning both radio access network (RAN) and core network (CN). It also includes: deployment, control and management of the infrastructure. The allocation of resources to slices and the way they can be managed to higher layers is included, too. NF layer encapsulates all the operators that are related to the configuration and life cycle management of NFs offering E2E services. When considering service layer, it should be noted that the most important property that distinguishes network slicing in 5G from other forms of slicing, is its E2E nature and the possibility to express a service through a high-level description and to map it to appropriate infrastructural elements and NFs. Finally, MANO entity translates user cases and service models into slices by changing NFs, mapping them to infrastructure resources, while configuring and monitoring each slice during its life cycle.

Figure 2. Network slicing-based 5G system representation

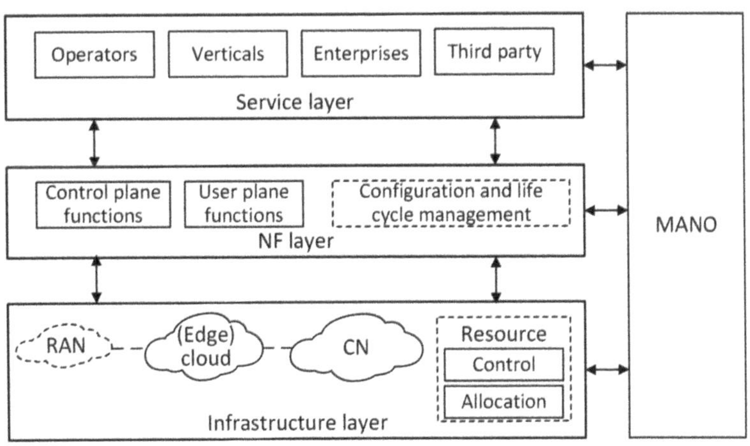

Slicing Benefits

5G slice is a bundle of network services functions, applications, resources and equipment. The perceived advantages are significant, particularly useful in the design of next generation wireless networks, such as slice isolation, simplified service chains, flexible VNF settings, and transparent management. The entirely isolation allows simpler and efficient design of each slice and as a consequence overall system, too. The goal is to meet requirements offered by the slice tenant. In addition, network failures, overload, or security attacks, in one slice will not affects the operation of other slices in the system. Simplified service chains means that in network slicing, each service can rely on a different subset of functions. NFV introduces an additional degree of freedom regarding the placement, of these functions in the network. Intelligent placement may improve network performance and reduce the operating costs. Finally, network slicing provides an abstraction of the physical resources and makes slice management transparent to the slice tenant.

MAIN NETWORK SLICING ENABLING TECHNOLOGIES

Network slicing is found to the realization of 5G making feasible to invoke new applications within new large additional investments. The modularity of the NF chains leads directly to the process of creating new services. As for the operators they feel that the 5G family of technologies come from the possibility to provide value-added solutions for requirements (Samdanis, Costa-Perez & Sciancalepore, 2016). 5G will guarantee levels of network performance with respect to quality of service (QoS) metrics such as data rate, latency, availability, reliability, etc. For realizing this vision, mobile systems need to evolve into more virtualized, cloud like environment, using prospective technologies such as SDN and NFV. The growing interest in these technologies is motivated by the novelty of the overall context, specifically their sustainability and high performance. Together with cloud, edge and fog computing, these technologies can be seen as facets of a broad innovation wave, called softwarization (Manzalini & Crespi, 2016).

Software Defined Networking

SDN simplifies network management by decoupling control and data plane and by utilizing logically centralized intelligence in form of programmability and open access. SDN was originally designed for flexible and dynamic control of data path functions like traffic routing and management. However, characteristics such as flexibility, scalability, and robustness, become significant stimulus for SDN implementation in the slicing concept. A SDN controller alleviates third parties over the virtualizer, and through the means of an agent, it provides multi-tenancy support. Each tenant is assigned a policy that determines its programming capabilities for the underlying data layer using the data control plane function.

The SDN-based network architecture provided by Open Networking Foundation (ONF TR-526, 2016) features a unified plane, where hierarchical controllers are used to achieve differentiated services in user access layers. The major SDN architectural components are resources and controllers (Figure 3).

By applying the SDN architecture, the client context provides the complete abstract set of resources (resource group) and supporting control logic that constitutes a slice, including the complete collection of related client service attributes. Resource group includes infrastructure resources and NFs, as well as network services. SDN controller mediates between clients and resources, acting simultaneously as server

Figure 3. SDN network slicing architecture provided by Open Networking Foundation

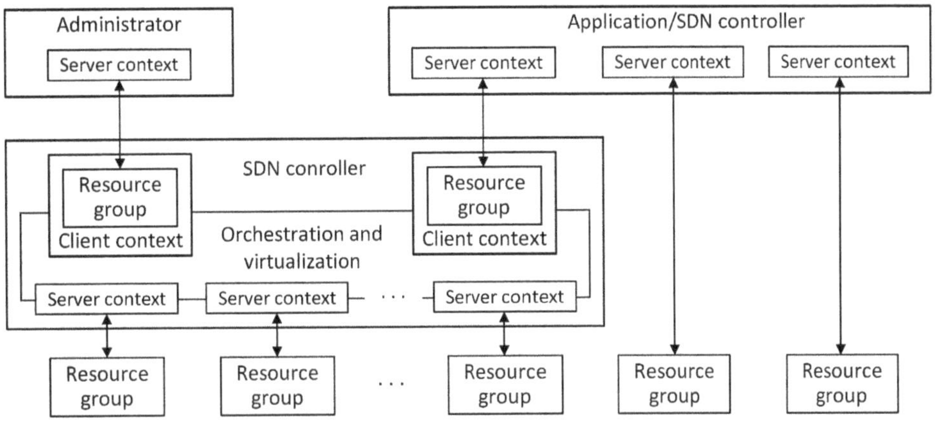

and client via client and server contexts, respectively. This is in accordance with the key principles of network slicing, having in mind a wide range of service demands, which need to be satisfied in a flexible and cost-effective way (Ordonez-Lucena, Ameigeiras, Lopez, Ramos-Munoz, Lorca & Folgueira, 2017).

Client context provides all the necessary information to the controller in order to support and communicate with a given client. It comprises a resource group and a client support function. On the other hand, server context represents all the information the controller needs to interact with a set of resources, included in a resource group.

SDN controller perform virtualization and orchestration functions in the process of transforming the resource groups accessed through server contexts to those defined in separate client contexts. Through virtualization, the SDN controller carries out abstraction and sharing of the underlying resources. In that way, each client context provides a specific resource group that can be used by the client associated with that context to realize desired service. When performing orchestration, the SDN controller optimally distributes the selected resources to such separate resource groups. The interplay of both controller functions enables the fulfillment of the diverse demands from all clients while providing the isolation among them. The SDN architecture also includes an administrator, with instantiating and configuring functions, including the creation of both server and client contexts, as well as their joint policies.

Network Function Virtualization

It is obvious that SDN architecture lacks capabilities that are crucial for efficient management of network slices life cycle and corresponding structural resources. Regarding this fact, the NFV approach is ideal to play this role, because it can manage the infrastructure resources and orchestrates the allocation of such resources. With NFV, services are described as a forwarding graph (Abdelwahab, Hamdaoui, Guizani, & Znati, 2016) of connected NFs. Here, forwarding graph defines the sequence of NFs that process different E2E traffic flows. Generally, virtualization can be conducted on mobility management entities, serving gateways, baseband processing units (e.g., medium access control (MAC), radio link control (RLC), radio resource control (RRC), etc.), switching function, traffic balancing, service centers, etc.

To achieve full potential from the management and orchestration functionalities, appropriate cooperation between SDN and NFV is required. It is important to note that SDN and NFV frameworks are not dependent on each other, although these two concepts are mutually beneficial. It means that the NFs can be deployed without SDN and vice-versa. However, integration of SDN and NFV architectures into a common framework is a complex issue (ETSI, 2015). The NFV framework defines:

1. Virtual network functions (VNFs) which represent software components of NFs deployed in virtual environments.
2. NFV infrastructure (NFVI) which comprises the logical building blocks, i.e., storage, computing, network and corresponding hardware.
3. MANO which is responsible for full automation, management and coordination of VNFs and NFVI.

In the context of network slicing, this framework provides service chaining, resource-oriented VNF embedding, as well as management of VNFs.

Cloud, Edge and Fog Computing

Cloud computing can be seen as a solution for serious challenges such as high availability, load balancing, and high performances. The long term solution for growing traffic demands on the current cellular architecture is the possibility of excluding computing beyond data centers toward mobile end-user, providing E2E connectivity as a cloud service. A set of techniques such as mobile node function virtualization and caching are also introduced for the on-demand provision of a decentralized and elastic network as a cloud service over a distributed network of cloud computing data servers. Operators can enter the cloud computing market and create new value-added services and experiences by integrating industry content and applications.

Cloud computing make user to obtain much more real-time applications to utilize 5G network efficiently. This can be realized through application, platform, as well as infrastructure segment. Application is based on-demand software services. They vary in their pricing and distribution schemes. The platform segment refers to products that are used to develop Internet. Finally, the infrastructure can be seen as the backbone of the entire concept.

For enabling a network slicing process, edge computing together with cloud computing offers a single or multiple platforms. In addition, edge computing enables diverse applications, data management and service acquisition in close proximity to end users. In that way a kind of edge-centric networking will be allowed, while data proximity, ultra-low latency, high data rates and control are permitted (Lopez, Montresor, Epema, Datta, Higashino, Iamnitchi, Barcellos, Felber, & Riviere, 2015; Taleb, Samdanis, Mada, Flinck, Dutta, & Sabella, 2017).

Among numerous edge computing realization, one of the most popular is Multi-access Edge Computing (MEC) (ETSI, 2016). MEC deals with stand-alone platforms. Here, the focus is on RANs, as well as fixed access networks, while the main object is development of an interface which will enable the process of controlling various services from the network edge.

The second one of the most popular realization is fog computing. From a systematic perspective, fog environment provides a distributed computing system with a hierarchical topology. Fog environment is characterized by stringent latency requirements, reduced power consumption of end devices, while providing real-time data processing and control with localized computing resources. The fog available resources are typically much more heterogeneous, and the criticality of some applications requires special treatment. The complementarity between fog and cloud has traditionally been seen as a mandatory feature. However, Lingen et al. (2017) advocate for a different approach. Instead of specifying an architecture where fog and cloud are complementary by design, focus is on a service management architecture that fuses fog and cloud. With this solution, an infrastructure composed of network, cloud, and fog nodes, can be provided as promising paradigm to service administrators as a unified resource pool. Administrators can then define where to allocate resources according to the service requirements. The requirements to host and manage NFV, MEC, and fog computing services are doubtless similar. It is only a matter of time until fog computing becomes part of the convergence that is already existing between NFV and MEC, driven by operators, enterprises, and third party investments. The goal is to conceal underlying complexity, and to promote service management into simple and intuitive operations. Overlapping needs and challenges, when considering MEC and fog integration, can be identified as: virtualization, management of network, services, billing and pricing, security, etc., multi-tenancy, as well as, detailed analytics.

NETWORK SLICING IN RADIO ACCESS NETWORK DOMAIN

In order to manage limited frequency spectrum resources, for network slicing in RAN domain, the following characteristics are required: dynamic resource management, resource isolation and sharing, as well as functional requirements. Dynamic resource management implies that the procedures are flexible and programmable enabling the advantages of application programming interfaces (APIs). In this case resource sharing is provided with the help of KPIs for each slice. For example, eMBBC slice seeks high throughput, while on the other hand, uRLLC slice needs very low latency (Bojkovic, Bakmaz & Bakmaz, 2017; Sachs, Wikstrom, Dudda, Baldemair, & Kittichokechai, 2018).

RAN domain is characterized by the fact that E2E slices are in a position to provide self-contained networks together with the corresponding degree of isolation. In the RAN domain, spectrum isolation leads to be a bottleneck because of limitations in multiplexing gains. The goal of functional requirements is that each network slice need a different control/user plane functional split and corresponding VNF placement to provide an optimal performance. The interface between network slicing and RAN is shown in Figure 4.

The network slice selection function introduced by Rost et al. (2017) is significant for selecting the corresponding slice per user. As for RAN-CN interface, it is carried out in a way that the control and user plane traffic are routed to functional entities in the CN slice. User plane anchor (UP-anchor) is responsible for traffic distribution to the slice policy and for encryption. The radio flow in slice #1 is configured with two connections, while slice #2 is configured with only one connection according to the provided policy configuration.

Radio resource management and control in the base station is responsible for configuring the RAN protocol stack and QoS according to the slice requirements (Rost, Mannweiler, Michalopoulos, Sartori, Sciancalepore, Sastry… Bakker, 2017). For example, in eMBBC slice with high capacity requirements, radio bearers are configured to support multi-connectivity, similar to the dual connectivity in 4G systems. On the other hand, in uRLLC slices, lower frame error rates as well as multi-point diversity techniques need to be utilized. Also, flexible architecture incorporating millimeter wave support is required to meet different slice requirements. This support is essential for multi-connectivity realization covering both eMBBC and mMTC slices.

Figure 4. Interface between network slicing and RAN

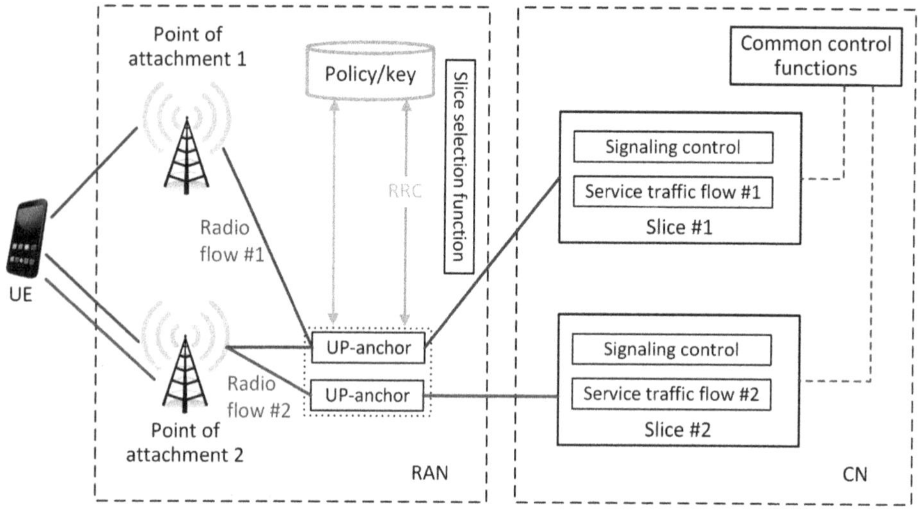

SLICING IN CORE NETWORK DOMAIN

During the last decade, evolution of mobile CN was evident. Going back to the past, the beginning was with Long Term Evolution (LTE) and a full IP CN, then passing through softwarization and utilization of the CN elements, until 5G and network slicing. Today, SDN and NFV have become the key enablers in order to obtain more dynamic evolved packet core (EPC) networks. It should be noted that the main EPC entities have been divided into granular NFs. Moreover, 3GPP has completely reshaped the CN with modular architecture and network slicing concept for many types of different services.

From the introduction of network slicing on the mobile network layer, new requirements as the result of autonomous and dynamic network configuration are raised for the transport layer. As a consequence, there is a need for a new interface between the RANs and CNs (3GPP TR 28.801, 2018). The purpose of introducing such an interface is to connect the management system of the RAN with the transport network controller. One of the main objectives is to provide and facilitate the mapping of RAN slice to CN resources (Afolabi, Taleb, Samdanis, Ksentini & Flinck, 2018).

The fact that building blocks of the CN can be deployed as virtual instances, brings more flexibility, elasticity and QoS guaranties to the service provisioning techniques. The flexibility and elasticity in service provisioning implies that operators can deploy multiple instances of the EPC at the same time, to serve different categories of users based on their service requirements. Moreover, while some services may need all the components that constitute the EPC, others do not (e.g., mMTC services with limited mobility does not need mobility management). Therefore, the notion of CN slicing is centered around the possibility to deploy multiple instances of virtual EPC (vEPC) running in parallel in order for each to fulfill different service demands, e.g. uRLLC services may require a distributed vEPC closer to the users.

In fact, vEPC can be orchestrated and managed over cloud platforms. Here, different orchestration schemes can be utilized offering an efficient management and operation of the EPC entities. Control plane and user plane entities can be provided as a service on commodity servers, bringing new dimension to the operating models of the vertical markets. The EPC as a Service (EPCaaS) model proposed by Taleb et al. (2015) leverages the cost efficiency offered by deploying vEPCs on the cloud to introduce two major virtualization approaches and different architectural implementation scenarios. First one is a full virtualization approach where both the control and user plane entities are virtualized. For the second approach, only the control plane entities are virtualized, while the data plane components are deployed on proprietary hardware in order to ensure high throughputs and to enable the implementation of traffic policies. The suggested implementation scenarios include: 1:1 mapping (where each functional component runs on a virtual machine), 1:N mapping (where each functional component runs on multiple virtual machines, N:1 mapping (where all functional components run on virtual machine), and N:2 mapping (where the control plane and data plane components run on a virtual machine each).

OPEN RESEARCH CHALLEGES

Although network slicing is currently undergoing a standardization phase, its realization is full of different challenges. The main challenges and future researches are arising from slicing implementation. First of all, since 5G systems provide ubiquitous connectivity for everything, both the RAN and CN domains need evolution and reconstruction to support E2E network slicing. In ultra-dense heterogeneous networks, the cooperation of multiple RANs should be considered to provide seamless mobility and high throughput. Secondly, although service providers and operators have started to pave the way for slicing solutions, the management remains at an early stage in terms of its development. Many technologies are introduced to create, activate, maintain and deactivate network slice at the service level.

Next, the open research challenge refers to the cooperation with traditional and emerging technologies such as broadband transmission, mobile cloud engineering, SDN and NFV. The cloud of access networks and CN have advantages of physical resource pooling, distribution of software architectures and centralization of management. It should be noted that there is still a problem with integration of network slicing with centralized RAN, SDN, and NFV to provide point-to-point connection between physical radio equipment and controller.

The open interfaces that support the programmability of the network, introduce security treats to softwarized architecture. The number of VNFs that network slices have in common or share is in direct relation to the number of security vulnerabilities between them. In the case of network sharing, when slices of different tenants are used, the security vulnerabilities are even more evident. The fact that each of the operating tenants can share resources alongside their isolated slices with different security parameters, exposes their individual slices to the intra-tenant security threats. Thus, a multi-level security including policies and mechanisms for software integrity, user authentication and management are welcome.

Finally, new charging models together with new foundations for cost sharing and standardized solutions have to be taken into consideration when dealing with open research challenges. In that way, interoperability in multi-technology environment will be supported, too.

CONCLUSION

This chapter focuses on some most relevant problems concerning principles and enabling technologies of slicing in 5G networks. Strong potential of network slicing was revealed for addressing the diverse requirements of future services. Having in mind the current maturity state, a role of network slicing in the development of the 5G architecture is provided. At the same time the enablers of network slicing such as the NFV, SDN and cloud technologies are introduced considering various scenarios, resource allocation principles for QoS requirements and interface constraints. These technologies can offer strong support in developing 5G sliced architecture. Significant efforts are still needed to develop optimal architectural configuration in order to realize cooperative environment between enablers technologies and take benefits from their advantages. With regards to E2E architecture, this chapter survey how network slicing can be achieved by slicing the RAN and CN. Finally, open research challenges in line with the realization of network slicing in 5G mobile networks are observed and discussed.

ACKNOWLEDGMENT

This research was supported in part by the Ministry of Education, Science and Technological Development, Republic of Serbia [grant number TR-32025].

REFERENCES

3GPP TR 28.801 V.15.1.0. (2018). Study on management and orchestration of network slicing for next generation network.

3GPP TS 22.261 V.16.3.0. (2018). Service requirements for the 5G system.

Abdelwahab, S., Hamdaoui, B., Guizani, M., & Znati, T. (2016). Network function virtualization in 5G. *IEEE Communications Magazine*, *54*(4), 84–91. doi:10.1109/MCOM.2016.7452271

Afolabi, I., Taleb, T., Samdanis, K., Ksentini, A., & Flinck, H. (2018). Network slicing & softwarization: A survey on principles, enabling technologies & solutions. *IEEE Communications Surveys and Tutorials*, *20*(3), 2429–2453. doi:10.1109/COMST.2018.2815638

Alliance, N. G. M. N. (2016). *Description of network slicing concept*. Retrieved from http:// www.ngmn.org

Arfaoui, G., Bisson, P., Blom, R., Borgaonkar, R., Englund, H., Félix, E., ... Zahariev, A. (2018). A security architecture for 5G networks. *IEEE Access: Practical Innovations, Open Solutions*, *6*, 22466–22479. doi:10.1109/ACCESS.2018.2827419

Berman, M., Chase, J. S., Landweber, L., Nakao, A., Ott, M., Raychaudhuri, D., ... Seskar, I. (2014). GENI: A federated testbed for innovative network experiments. *Computer Networks*, *61*, 5–23. doi:10.1016/j.bjp.2013.12.037

Bojkovic, Z., Bakmaz, B., & Bakmaz, M. (2017). Vision and enabling technologies of tactile Internet realization. In *Proceedings of 13th International Conference on Advanced Technologies, Systems and Services in Telecommunications* (pp. 113-118). Nis, Serbia: Academic Press.

ETSI GS MEC 003. (2016). Mobile Edge Computing (MEC); Framework and Reference Architecture, v. 1.1.1

ETSI GS NFV-EVE 005 (2015). *Network Functions Virtualization (NFV); Ecosystem; Report on SDN Usage in NFV Architectural Framework*, v. 1.1.1.

Foukas, X., Patounas, G., Elmokashfi, A., & Marina, M. K. (2017). Network slicing in 5G: Survey and challenges. *IEEE Communications Magazine*, *55*(5), 94–100. doi:10.1109/MCOM.2017.1600951

Goldberg, R. (1974). Survey of virtual machine research. *IEEE Computer*, *7*(6), 34–45. doi:10.1109/MC.1974.6323581

Hucheng, W., Shanzhi, C., Ming, A., & Yan, S. (2017). Mobility driven network slicing: An enabler of on demand mobility management for 5G. *Journal of China Universities of Posts and Telecommunications*, *24*(4), 16–26. doi:10.1016/S1005-8885(17)60219-X

ITU-R Rec. M.2083-0. (2015). IMT Vision – Framework and overall objectives of the future development of IMT for 2020 and beyond.

Katsalis, K., Nikaein, N., Schiller, E., Ksentini, A., & Braun, T. (2017). Network slices toward 5G communications: Slicing the LTE network. *IEEE Communications Magazine*, *55*(8), 146–154. doi:10.1109/MCOM.2017.1600936

Lopez, G. P., Montresor, A., Epema, D., Datta, A., Higashino, T., Iamnitchi, A., ... Riviere, E. (2015). Edge-centric computing: Vision and challenges. *Computer Communication Review*, *45*(5), 37–42. doi:10.1145/2831347.2831354

Manzalini, A., & Crespi, N. (2016). An edge operating system enabling anything-as-a-service. *IEEE Communications Magazine*, *54*(3), 62–67. doi:10.1109/MCOM.2016.7432173

Nikaein, N., Schiller, E., Favraud, R., Katsalis, K., Stavropoulos, D., Alyafawi, I., ... Korakis, T. (2015). Network store: Exploring slicing in future 5G networks. In *Proceedings of 10th International Workshop on Mobility in the Evolving Internet Architecture* (pp. 8-13). Paris, France: ACM.

ONF TR-526. (2016). Applying SDN architecture to 5G slicing.

Ordonez-Lucena, J., Ameigeiras, P., Lopez, D., Ramos-Munoz, J. J., Lorca, J., & Folgueira, J. (2017). Network slicing for 5G with SDN/NFV: Concepts, architectures, and challenges. *IEEE Communications Magazine*, *55*(5), 80–87. doi:10.1109/MCOM.2017.1600935

Raza, M. R., Fiorani, M., Rostami, A., Öhlen, P., Wosinska, L., & Monti, P. (2018). Dynamic slicing approach for multi-tenant 5G transport networks. *Journal of Optical Communications and Networking*, *10*(1), A77–A90. doi:10.1364/JOCN.10.000A77

Rost, P., Mannweiler, C., Michalopoulos, D. S., Sartori, C., Sciancalepore, V., Sastry, N., ... Bakker, H. (2017). Network slicing to enable scalability and flexibility in 5G mobile networks. *IEEE Communications Magazine*, *55*(5), 72–79. doi:10.1109/MCOM.2017.1600920

Sachs, J., Wikstrom, G., Dudda, T., Baldemair, R., & Kittichokechai, K. (2018). 5G radio network design for ultra-reliable low-latency communication. *IEEE Network*, *32*(2), 24–31. doi:10.1109/MNET.2018.1700232

Samdanis, K., Costa-Perez, X., & Sciancalepore, V. (2016). From network sharing to multi-tenancy: The 5G network slice broker. *IEEE Communications Magazine*, *54*(7), 32–39. doi:10.1109/MCOM.2016.7514161

Taleb, T., Corici, M., Parada, C., Jamakovic, A., Ruffino, S., Karagiannis, G., & Magedanz, T. (2015). EASE: EPC as a service to ease mobile core network deployment over cloud. *IEEE Network*, *29*(2), 78–88. doi:10.1109/MNET.2015.7064907

Taleb, T., Samdanis, K., Mada, B., Flinck, H., Dutta, S., & Sabella, D. (2017). On multi-access edge computing: A survey of the emerging 5G network edge cloud architecture and orchestration. *IEEE Communications Surveys and Tutorials*, *19*(3), 1657–1681. doi:10.1109/COMST.2017.2705720

van Lingen, F., Yannuzzi, M., Jain, A., Irons-Mclean, R., Lluch, O., Carrera, D., ... Rodríguez, J. P. (2017). The unavoidable convergence of NFV, 5G, and fog: A model-driven approach to bridge cloud and edge. *IEEE Communications Magazine*, *55*(8), 28–35. doi:10.1109/MCOM.2017.1600907

Zhang, H., Liu, N., Chu, X., Long, K., Aghvami, A. H., & Leung, V. C. M. (2017). Network slicing based 5G and future mobile networks: Mobility, resource management, and challenges. *IEEE Communications Magazine*, *55*(8), 138–145. doi:10.1109/MCOM.2017.1600940

KEY TERMS AND DEFINITIONS

Cloud Computing: Technology that allows accessing a set of shared and configurable computing resources (e.g., servers, storage facilities, infrastructure, applications) offered as services.

Edge Computing: Highly distributed computing environment that can be used to deploy services and applications, as well as storage and processing resources in close proximity to the mobile users.

Network Functions Virtualization: Software abstraction of network functions in next generation communications systems, which otherwise require dedicated hardware concerning traditional infrastructure.

Network Slice: A bundle of services functions, applications, resources and corresponding equipment.

Network Slicing: A solution for running multiple dedicated logical networks as mutually isolated according to the service requirements of different use cases.

Software-Defined Networking: Technology which simplifies network management by decoupling control and data plane and by utilizing logically centralized intelligence in form of programmability and open access.

Virtual Network Functions: Software components deployed in servers or cloud infrastructure instead of network hardware.

Chapter 12
An Improved Modeling for Network Selection Based on Graph Theory and Cost Function in Heterogeneous Wireless Systems

Mohamed Lahby
University Hassan II of Casabalanca, Morocco

Ayoub Essouiri
University Hassan II of Casabalanca, Morocco

Abderrahim Sekkaki
University Hassan II of Casabalanca, Morocco

ABSTRACT

The next generation of mobile wireless communications represents a heterogeneous environment which integrates variety of network generation like third generation (3G), fourth generation (4G), and fifth generation (5G). The major challenge in this heterogeneous environment is to decide which access point to use when multiple networks are available. Process of roaming mobile user from one technology to anther different is called vertical handover. In this chapter, the authors propose a new mechanism based on graph theory and cost function in order to determine the best path for the end user in terms of quality of service (QoS) when the vertical handover process is needed. Then, they investigate the impact of some existing weighting methods in order to determine the suitable method which can be coupled with the cost function. The experiments evaluation by using Mininet emulator demonstrate that the proposed approach can achieve a significant improvement concerning four QoS metrics: throughput, packet lost, packet delay, and packer jitter for two services FTP and video streaming.

DOI: 10.4018/978-1-5225-7570-2.ch012

1. INTRODUCTION

The future networked world have integrated multiple wireless network technologies like the third generation (3G), the fourth generation (4G) and the fifth generation (5G). The 3G mobile system is designed to support multimedia services and video teleconferencing with data rate of 2 Mbps to 11 Mbps. The 3G can be divided into two groups. The first group represents cellular networks designed by the International Telecommunication Unit's (UIT). The universal mobile telecommunication system (UMTS) is one of the most popular 3G network. The standardization of UMTS network (Javier, & Mamadou, 2001) is carried out by Third Generation Partnership Project (3GPP). In addition the UMTS network can provide wireless data with data rate of 2Mbps. The second group of 3G systems consists of wireless access networks specified by IEEE 802.11 committee (IEEE 802.11. 2007). These networks include WiFi (IEEE 802.11 a) and its extended systems such as IEEE 802.11b, IEEE 802.11n, etc. For example, the IEEE 802.11 a technology can ensure data rate of 11 Mbps. Furthermore, the 4G technology is on all IP system characterized with high data rate and high coverage area. The 4G system can integrate cellular networks such as Long Term Evolution (LTE) and wireless networks such as Interoperability for Microwave Access (WIMAX). The LTE network (Barth, 2006) is specified by 3GPP in December 2008, this standard allows supporting a variety of services such as interactive and streaming. The WiMax network (IEEE 802.16) is specified by the IEEE 802.16 committee (IEEE 802.16. 2004). This technology have been deployed in order to provide high data and to support multimedia applications. Actually both 3G and 4G have been increasing demand of utilization in spite of their limitations in term of coverage, bandwidth and mobility. In this context, in order to enhance the quality of service (QoS), to ensure high mobility and to increase the bandwidth, the companies of telecommunications are currently driving the development of the fifth generation network (5G). This new generation will provide the foundational infrastructure for building many applications such as Internet of Energy (IoE), Internet of Things (IoT), and Internet of Vehicles (IoV).

In parallel, the development of new mobile devices equipped with multiple network interfaces allow users to roam seamlessly between different technologies by using the vertical handover process (Lahby, 2013). Consequently, the users have the privilege to use different services like voice, data, web browsing, and video streaming by using different technologies at anytime and anywhere. The vertical handover can be divided into three parts namely: handover initiation, handover decision (it is also called network selection) and handover execution. The present chapter focuses on the second part which is considered the principle key of the vertical handover. However, how to choose the most suitable access network by satisfying Always Best Connected (ABC) Paradigm (Gustafsson & Jonsson, 2003), becomes a significant challenge during the network selection process, in this heterogeneous environment. To deal with this issue, several algorithms and strategies have been proposed in the literature in the recent years

Up to our knowledge, most of these existing references, have studied the network selection in the context of selecting only one access technology. In this chapter, we propose new mechanism based on graph theory and cost function in order to determine the best path for the end user in terms of quality of service (QoS). Firstly we introduce graph theory for modeling the network selection problem as graph. Then we use the cost function to calculate the weight of each edge and they apply the Dijkstra's algorithm to determine the best path in terms of QoS. Finally, we investigate the impact of some existing weighting methods in order to determine the suitable method which can be coupled with the cost function.

This chapter is organized as follows, in section 2 we outline some of existing works related to vertical handover algorithms which proposed during the last decade. In section 3 we discuss several existing weighting algorithms based handover. In section 4 we describe our new approach based on graph theory and cost function to deal with vertical handover problem. Section 5 comprises the testbed and the experimental results. Section 6 presents conclusion of this chapter, open issues and future works.

2. RELATED WORK AND PROBLEM STATEMENT

The network selection problem is one of the major trends in the heterogeneous wireless networks, to cope with this issue, several network selection algorithms and policies have been proposed in the literature in the recent years. According to reference (WANG & Eng-Sheng 2013), various mathematical theories can be used to model and to optimize the network selection problem. We can categorize them into eight main groups: the utility theory, the cost function, fuzzy logic, game theory, combinatorial optimization, Markov chain, multiple attribute decision making (MADM) and hybrids algorithms.

According to (VUONG, Thinh, DOUDANE & Nazim, 2008) the utility functions models are commonly used to describe the level of satisfaction of the mobile terminal according different services offered by the network technology. The authors Lee, Han and Hwang (2013), have proposed an effective utility function for vertical handover decision algorithm. The utility function is composed of several parameters such as signal to interference plus noise ratio (SINR), bandwidth, traffic load and velocity of terminal mobile. The proposed utility function adopts Shannon's formula as QoS function in order to increase throughput.

In the cost function based algorithms (Yan, Mani, & Sekercioglu, 2008), the goal is to find the network which minimize the output of the cost function. The authors in Hasswa et al, (2005) have proposed a new cost function in order to select the best access network. This cost function is based on the following metric attributes: cost of service, power requirements, security, proactive handoff (user preference), network conditions, network performance and velocity.

Moreover, fuzzy logic approach has been widely utilized to model and to design an intelligent system for vertical handoff process. The authors Hou and O'Brien, (2006), have proposed a new scheme for network selection based on fuzzy logic. In this scheme three input fuzzy variables are considered: the probability of a short interruption, the failure probability of handover to radio, and the size of unsent messages.

The authors Trestian, Ormond, and Muntean, (2012), have classified and reviewed the most significant game theory algorithms that have been used to deal with the network selection problem. The authors E. KHLOUSSY, et al (2015), have tackled the network selection between the cellular and wireless local area network by using the Markov decision process. This technique is applied to maximize the revenue and user satisfaction.

Another popular category of handover algorithms proposed in the recent years is MADM methods. According to recent survey (Lahby, Silki. & Sekkaki, 2015), a lot of vertical handover schemes based on MADM methods are proposed to select the best radio access network. The first category of MADM algorithms such as simple additive weighting (SAW), multiplicative exponent weighting (MEW), technique for order preference by similarity to ideal solution (TOPSIS), Grey relational analysis (GRA), distance to ideal alternative (DIA), and the novel method (NMMD) based on Mahalanobis distance are used for ranking. While the second category includes analytic hierarchy process (AHP), fuzzy analytic hierarchy process (FAHP), analytic network process (ANP) and fuzzy analytic network process (FANP) are used to calculate the corresponding weighting of criteria.

The solution based on SAW and AHP has been extensively applied for making network selection decision. For example, in Sheng-Mei et al. (2010) and De Oliveira Filho and Madeira (2011), the AHP method is used to weigh each criterion involved in the network selection. While the SAW method is used to rank all alternatives.

The authors in (Karam, 2012) and (Escobar, 2010) have modeled the network selection problem by coupling two methods AHP and MEW. The authors in (Fu, 2010) and (Yafang, Huimin, & Jinyan, 2010) have proposed an algorithm for network selection based on two MADM methods AHP and GRA. Firstly the AHP method is used to find the weights of the available networks. In the second step the GRA method is applied to rank the alternatives, in (Sgora, 2010) and (Lahby, Leghris, & Adib, 2011) the authors have proposed to combine the AHP and TOPSIS methods for the network selection problem. The AHP method is used to weigh different criteria while TOPSIS method is applied to rank the available networks.

The authors (Lahby, 2013), have proposed a new ranking algorithm NMMD based on Mahalanobis distance. The proposed algorithm allows the selection of the best available access network. In addition, according to simulation results provided in this work, the NMMD algorithm can provide the better performance than the classical MADM methods such as SAW, MEW, TOPSIS and DIA.

Reference Lahby et al. (2013), an Enhanced TOPSIS (E-TOPSIS) is proposed for the vertical handover decision using the ANP for the weighing of the criteria. The authors have shown that the number of handoffs and ranking abnormality phenomenon are reduced, and E-TOPSIS provides better results than some existing methods.

Finally, based on reviewed literature, many hybrids algorithms have been proposed to model and to optimize the vertical handover problem. The authors in Phuoc Tran and Boukhatem (2009), have developed a new scheme for network selection based on utility function and DIA algorithm. In this scheme, each interface network is associated by utility function which takes into account the application requirements and the energy consumption of the terminal. Then the DiA algorithm is used to rank the interfaces considering the interface utility values and the network side attributes. The authors in Goyal, Kaushal, and Sangaiah (2017) have proposed an optimal model for network selection based on the utility function and novel fuzzy-Analytic Hierarchy Process (AHP). Firstly, the utility function is used to model different criteria bandwidth, delay, jitter, bit error rate, and user preferences. Then a new non-linear fuzzy optimization model based AHP method is applied to weigh each criterion. Finally, scores are calculated for each network by three MADM algorithms SAW, MEW and TOPSIS.

In order to guarantee the selection of the optimal network, the authors in Lahby, Sekkaki (2017) and Lahby, Attioui, and Sekkaki (2017), combined the utility function respectively with TOPSIS algorithm and E-TOPSIS algorithm. The simulation results show that, both two hybrids algorithms reduce the reversal phenomenon, the ping-pong effect and the handoff failures better than MADM algorithms and utility function.

The authors S. Goudarzi, et al. (2017) focused on the optimization of vertical handover by combining two algorithms Markov decision process (MDP) with genetic algorithm (GA). The genetic algorithm is used to find a set of optimal decisions that ensures the best trade-off between QoS based on their priority level.

However, the major weakness of these previous approaches is that the network selection decision is based only on selecting of one access technology. In reality, the infrastructure of each operator is a heterogeneous environment which can have the coexistence of different networks technologies. In addition, each network access we can find various access points. In this context, when the mobile user communicates with other mobile user, his message can cross different access points and not only one access point.

To cope with this issue, in the literature review we find only two references (Naeem, Ngah, Hashim, & Maqbool, 2014) and (Lahby & Sekkaki 2018) that allow to select the most appropriate path during vertical handover process. The authors (Naeem, Ngah, Hashim, & Maqbool, 2014) have proposed a new history-based communication graph scheme to perform vertical handover decision. The proposed scheme is based on the graph theory in order to determine the best path between any heterogeneous nodes among existing traversed paths. In addition, the proposed scheme has shown better performance in terms of minimizing the number of unsuccessful handovers and reducing the ping-pong effect.

However, there are two drawbacks of this work, the first one lies on the cost of each edge between two nodes that use only RSS criterion. The second one lies in the performance results which are given by using simulation based on Matlab simulator. In our previous work (Lahby & Sekkaki 2018) we have coupled two approaches graph theory and cost function. Firstly the graph theory approach is used to model the vertical handover problem. Then, the cost function is used to quantify the importance degree of each edge. The Dijkstra's algorithm is applied to choose the best path for vertical handover decision. The experiments evaluation on a testbed by using Mininet emulator showed that for both services FTP and streaming, the proposed mechanism for selecting the best path provides higher throughput, exhibits smaller packer lost, decreases the delay, and the jitter better than better than handover based RSS, handover based bandwidth and handover based cost function.

This previous solution, however, consider the same importance for different criterion because the cost function is combined with the equal weighting method. To deal with this issue, in this chapter, we propose an improved modeling for vertical handover based on graph theory and cost function, we investigate the impact of these weighting techniques combined with cost function when Dijkstra's algorithm is used to select the best path in terms of QoS.

3. AN OVERVIEW ON WEIGHTING METHODS

One of the most critical steps of making decision problems is determining the importance of different criteria. In this step, the decision-makers determine the weight of each of the criteria. In the context of the vertical handover, several weighting methods are proposed such as fuzzy logic (Kaleem, Mehbodniyay, Yen, & Adachi, 2012), the genetic algorithm (Alkhawlani & Ayesh, 2008), entropy method (Sheng-mei, 2010), random weighting (Lahby, Leghris, & Adib 2012), equal weighting (Lahby, & Sekkaki, 2018), AHP method and ANP method.

In this chapter we focus on four weighting algorithms which are equal weighting, random weighting, AHP method and ANP method. We investigate the impact of these weighting techniques combined with the cost function when Dijkstra's algorithm is used to select the best path in terms of QoS.

3.1 Analytic Hierarchy Process (AHP) Method

The AHP is one of the extensive multi-attribute decision making developed by Saaty (Saaty 1998) and has been widely used in network selection process to assign weights for different criteria. The AHP approach is based on five steps:

1. **Construct of the Structuring Hierarchy:** A problem is decomposed into a hierarchy which contains three levels: the overall objective is placed at the topmost level of the hierarchy, the subsequent level presents the decision factors and the alternative solution are located at the bottom level.

2. **Construct of the Pairwise Comparisons:** To establish a decision, AHP builds the pairwise matrix comparison such as:

$$A = \begin{bmatrix} x_{11} & x_{12} & \cdots & \cdots & x_{1n} \\ x_{21} & x_{22} & \cdots & \cdots & x_{2n} \\ \vdots & \vdots & \ddots & \vdots & \vdots \\ \vdots & \vdots & \vdots & \ddots & \vdots \\ x_{n1} & x_{n2} & \cdots & \cdots & x_{nn} \end{bmatrix}, where \begin{cases} x_{ii} = 1 \\ x_{ji} = \dfrac{1}{x_{ij}} \end{cases}$$

Elements x_{ij} are obtained from the table 1 which contains the 1-9 preference scales.

3. **Construct the Normalized Decision Matrix:** A_{norm} is the normalized matrix of A, where $A(x_{ij})$ is given by, $A_{norm}(a_{ij})$ such:

$$A_{norm} = \begin{bmatrix} r_{11} & r_{12} & \cdots & \cdots & r_{1n} \\ r_{21} & r_{22} & \cdots & \cdots & r_{2n} \\ \vdots & \vdots & \ddots & \vdots & \vdots \\ \vdots & \vdots & \vdots & \ddots & \vdots \\ r_{n1} & r_{n2} & \cdots & \cdots & r_{nn} \end{bmatrix}, where\ r_{ij} = \frac{x_{ij}}{\sum_{i=1}^{n} x_{ij}}$$

4. **Calculating the Weights of Criterion:** The weight of each criterion i can be calculated by:

$$W_i = \frac{\sum_{j=1}^{n} a_{ij}}{n}\ and \sum_{j=1}^{n} W_i = 1$$ where n is the number of the compared elements.

5. **Calculating the Coherence Ratio (CR):** To test consistency of a pairwise comparison, a consistency ratio (CR) can be introduced with consistency index (CI) and random index (RI).
   - Let define consistency index $CI = \dfrac{\lambda_{max} - n}{n-1}$
   - Also, we need to calculate the λ_{max} by the following formula:

$$\lambda_{max} = \frac{\sum_{j=1}^{n} b_i}{n}\ such\ b_i = \frac{\sum_{j=1}^{n} W_i * a_{ij}}{W_i}$$

   - We calculate the coherence ratio CR by the following formula $\mathrm{CR} = \dfrac{\mathrm{CI}}{\mathrm{RI}}$

The various values of RI are displayed in table 2. If the CR is less than 0.1, the pairwise comparison is considered acceptable.

Table 1. Saaty's scale for pair-wise comparison

Saaty's Scale	The Relative Importance of the Two Sub-Elements
1	Equally important
3	Moderately important with one over another
5	Strongly important
7	Very Strongly important
9	Extermely important
2,4,6,8	Intermediate values

Table 2. Value of random consistency index RI

Criteria	3	4	5	6	7	8	9	10
RI	0.58	0.90	1.12	1.24	1.32	1.41	1.45	1.49

3.2 Analytic Network Process (ANP) Method

The ANP is a MADM method proposed by Saaty (Lee, & Kim 2000), which extends the AHP approach to overcome the problems with dependence and feed beck within clusters (inner dependence) and between clusters (outer dependence). The ANP approach is based on six steps:

1. **Model Construction:** A problem is decomposed into a network in which nodes corresponds to components. The elements in a component can interact with some or all of the elements of another component. Also, relationships among elements in the same component can exist. These relationships are represented by arcs with directions.
2. **Construct of the Pairwise Comparisons between Component and of Inter-Dependencies:** The way of conducting pairwise comparisons is the same as in the second step of AHP approach.
3. **Construct the Normalized Decision Matrix:** Is the same as in the third step of AHP approach.
4. **Calculating the Weights of Criterion:** Is the same as in the fourth step of AHP approach.
5. **Calculating the Coherence Ratio (CR):** Is the same as in the fifth step of AHP approach.
6. **Construct the Super-Matrix Formation:** The local priority vectors are entered into the appropriate columns of a super-matrix, which is a partitioned matrix where each segment represents a relationship between two components.

3.3 Equal Weighting Method

The equal weighting method is a very simple type of weighting techniques that gives the same weight, to each criterion for vertical handover. For each criterion C_j (j=1,...,n), we associate the same importance w_i with $\sum_{i=1}^{n} w_i = 1$.

3.4 Random Weighting Method

Equal weighting is a very simple type of weighting methods that generates the random weight, to each criterion for vertical handover. For each criterion C_j (j=1,...,n), we associate the random importance w_i with $\sum_{i=1}^{n} w_i = 1$.

4. OUR IMPROVED STRATEGY FOR NETWORK SELECTION

In this section, we describe how can introduce the graph theory for modeling network selection problem. Then, we investigate the impact of on four weighting techniques which are equal weighting, random weighting, AHP method and ANP method. Our goal is to determine the suitable method which should be used with the cost function when Dijkstra's algorithm is applied to find the best path in terms of QoS.

4.1 System Model

Any geographical area of each operator of telecommunication contains a set of heterogeneous networks such as Wifi, UMTS, WIMAX and LTE. The Figure 1 illustrates how can deploy different access points in order to build this heterogeneous geographical area.

Figure 1. Illustration of network selection in heterogeneous geographical area

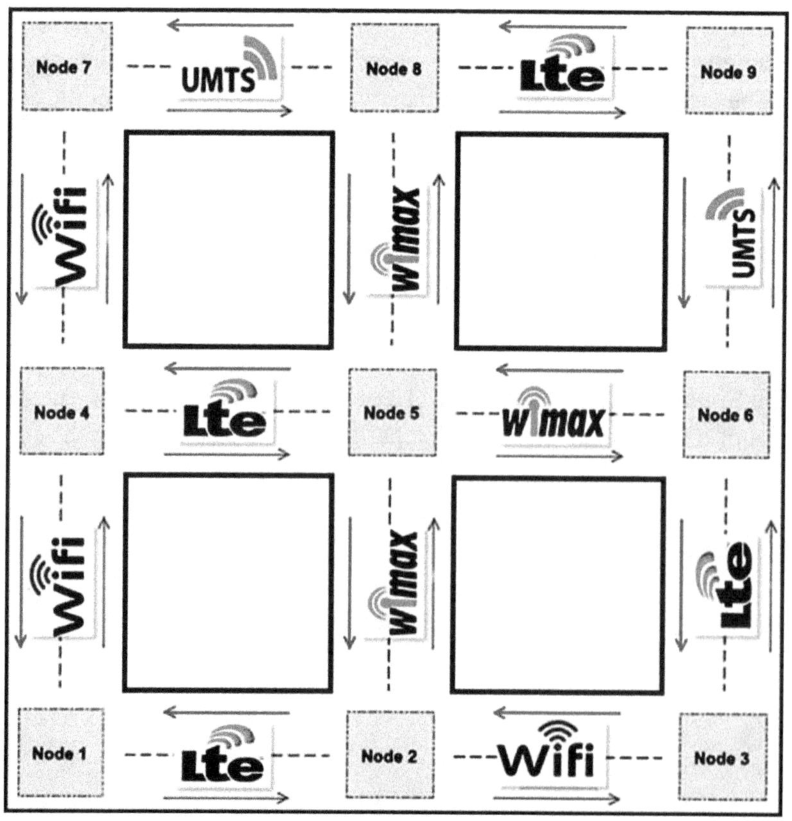

In order to model this area, we introduce the graph theory to build the graph. Let G = (V,E) be a directed graph defined by nodes set V and an edges set E. Here the nodes refer to different mobile devices (iPad, Smart phone,, ..) or other connected objects (vehicle, camera,..).. While the edge set E corresponds to the set of mobility links between nodes. In fact, the edge weight can be measured as Distance, RSS, Bandwidth, or combination between different parameters concerning each both nodes.

4.2 Our Improved Strategy for Network Selection

The network selection strategy is based on Dijkstra's algorithm in order to select the best path. In fact, for each edge, the cost function C(e) is calculated by addition of weighted metrics. The formula of C(e) is expressed as equation below:

$$C\left(e\right) = w1CB + w2\frac{1}{S} + w3\frac{1}{AB} + w4PD + w5PJ + w6PL$$

The different metrics introduced in this formula are cost per byte (CB), security (S), available bandwidth (AB), packet delay (PD), packet jitter (PJ) and packet loss (PL). With w1, w2, w3, w4, w5 and w6 represent respectively the weight of each criterion. In this chapter, we compare the performance of four weighting techniques combined with the cost function in order to determine the most weight of each attribute.

The pseudo code of our Algorithm 1 that allows choosing the best path is shown in Figure 2.

Figure 2. The pseudo code of our proposed strategy for network selection

Algorithm 1 Our mechanism for selecting the best path

Input:

- Source node S;
- Destination node D
- The adjacency matrix D representing the network selection area

Output:

- The best access path from S to D

```
N'= {u}
for all nodes v do
    if ( v neighbor) then
        D(v) = c(u; v)
    else
        D(v) = ∞
    end if
end for
repeat
    Select w ∉ N' such that D(w) minimal
    N' ← N'∪ {w}
    for all w's neighbors z ∉ N' do
        D(z) = min (D(z);D(w) + c(w; z))
    end for
until all nodes ∈ N'
return the best access path
```

5. EXPERIMENTAL RESULTS

5.1 Scenario and Testbed Setup

In order to illustrate the effectiveness of our mechanism based on graph theory and cost function, we conducted two testbed experiments, according to ftp service and streaming service by using Mininet emulator (Team, 2012). In the first experimental, we compare the performance of our policy with three other algorithms, namely Handover based RSS, Handover based bandwidth and cost function. While the second experimental consist to choose the most appropriate weighing algorithm which should be coupled with our mechanism. For that, we perform the performance comparison between four weighting techniques which are equal method, random method, AHP method and ANP method.

For each testbed we analyze four QoS metrics which are throughput, packet loss, packet delay and jitter. The network topology concerning our testbed is depicted in Figure 1.

5.2 Performance Evaluation

5.2.1 Experimental 1

In this experimental, we present performance comparison between four algorithms namely:

- **Our Mechanism:** This technique represents our proposed algorithm 1 for selecting the best path by using Dijkstra's algorithm.
- **Handover Based RSS:** In this technique, the weight of each edge is the Received Signal Strength (RSS) of the technology installed on the edge.
- **Handover Based Bandwidth:** The weight of each edge is the Received Signal Strength (RSS) of the technology installed on the edge.
- **Handover Based Cost Function:** The cost for each edge e is calculated as:

$$C\left(e\right) = \alpha\left(1 - \frac{PJ_{Te}}{D_{Te}}\right) + \beta\left(1 - \frac{AB_{Te}}{D_{Te}}\right)$$

where $\alpha=\beta=0{,}5$, when D_{Te} and PJ_{Te} are the distance (between mobile and access point) and the jitter of technology Te respectively. We notice that this cost function of each edge e is used in [15] to choose the best path for vehicular networks.

Figure 3 shows the average throughput of FTP traffic and streaming traffic for all algorithms. We remark that our policy increases the throughput with a value of 3.8 Mbps and 11.5 Mbps for FTP and streaming respectively. The all values provides by our policy are higher than the other values of the other algorithms. So, our proposed mechanism can improve the FTP throughput and streaming throughput better than handover based RSS, handover based bandwidth and handover based cost function for each service.

Furthermore, the throughput for handover based cost function is 2.8 Mbps for FTP service and 1Mbps for streaming service. Both these values are higher than the values provided by handover based RSS and Handover based Bandwidth. On the other hand, the handover based on RSS provides the smaller value

Figure 3. Throughput (Mb/s) for both services FTP and Streaming

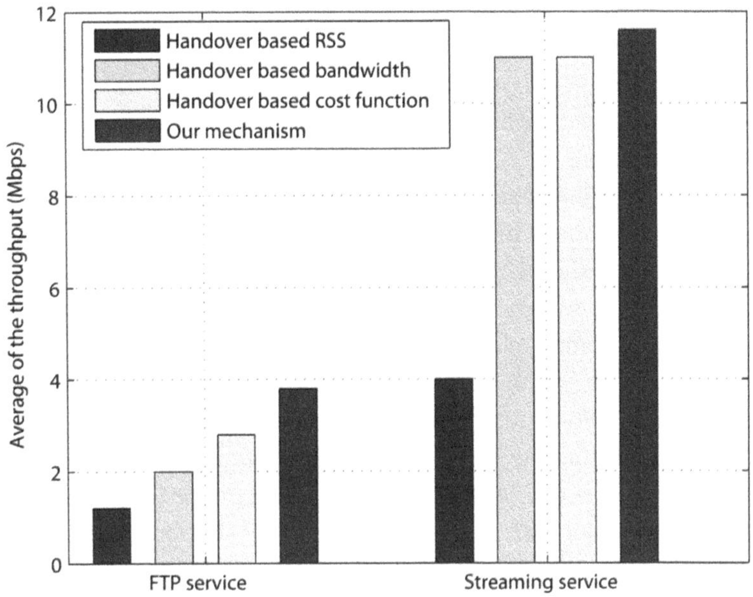

of the throughput 1.2Mbps for FTP service and 4Mbps for streaming service. For that, this algorithm is bad mechanism for selecting the best path for network selection.

Figure 4 shows the average value of the packet lost for all algorithms concerning FTP service and Streaming service. We notice that our policy diminishes the packet lost with a value of 0.42% and 0.85% for FTP service and streaming service respectively. For that, our mechanism can help to reduce the packet lost better than other algorithms because all values provides by our mechanism are smaller than

Figure 4. Average of Packet lost (Kb/s) for both services FTP and Streaming

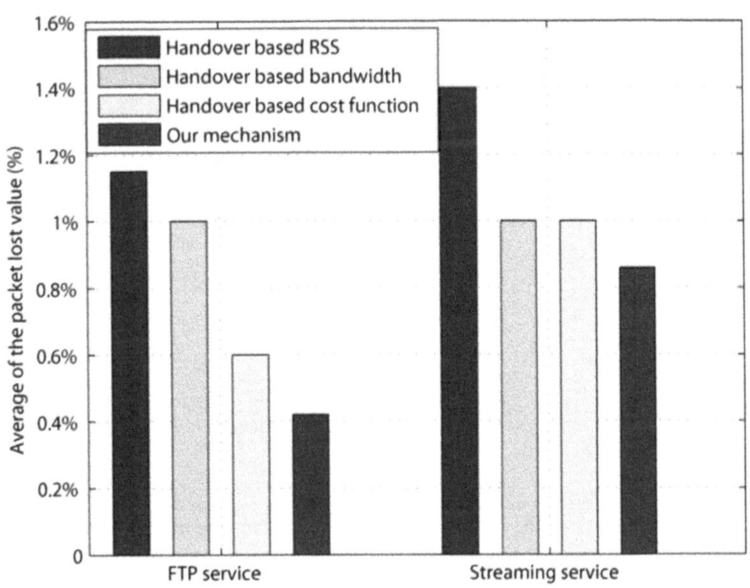

the other values of the other algorithms. While the handover based cost function can reduces the value of the packet lost with 0.6% for FTP service and 1% for streaming service, both these values are smaller than the values of handover based RSS and handover based Bandwidth. Moreover, Figure 3 shows that higher values of this parameter are provided by handover based RSS, with 1.25% for FTP service and 1.4% for streaming service. For that, the handover based RSS is not appropriate mechanism for selecting the best path for network selection.

Figure 5 shows the average value of end-to-end delay for all algorithms concerning FTP service and streaming service. We remark that our mechanism decreases the delay with 39ms for FTP service and with 24ms for streaming service. Based on these values, our mechanism exhibits smaller delay than the other algorithms. In addition, we remark that the handover based cost function provides better performance than conventional algorithms such as handover based RSS and handover based bandwidth.

Finally the Figure 5 shows that higher values of this parameter are provided by handover based RSS, with 84ms for FTP service and 96ms for streaming service. For that, the handover based RSS is not appropriate mechanism for selecting the best path for network selection.

Figure 6 shows the average jitter of FTP traffic and streaming traffic for all algorithms. We notice that our mechanism diminishes the jitter with a value of 6ms for FTP service and 7ms for streaming service. Both these values are smaller than the other values of the other algorithms. For that our proposed mechanism can improve the jitter better than handover based RSS, handover based bandwidth and handover based cost function for all traffic classes. Furthermore, the handover based cost function is 12.5 ms for FTP service and 8.5 ms for streaming service. Both these values are smaller than the values provides by handover based RSS and Handover based Bandwidth. Finally, the handover based on RSS provides the higher value of the jitter with 35ms for FTP service and 33ms for streaming service. For this reason, the handover based RSS is the bad mechanism for selecting the best path for network selection.

Figure 5. Average of Packet delay (ms) for both services FTP and Streaming

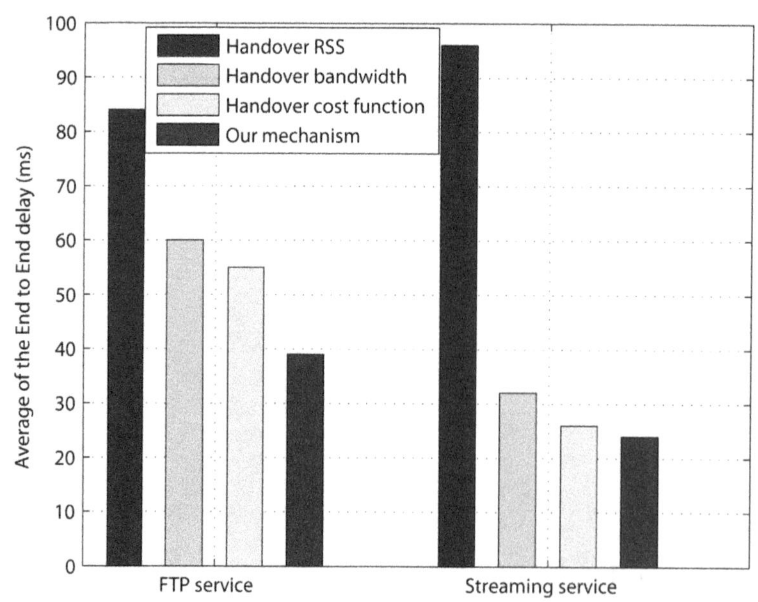

Figure 6. Average of Packet jitter (ms) for both services FTP and Streaming

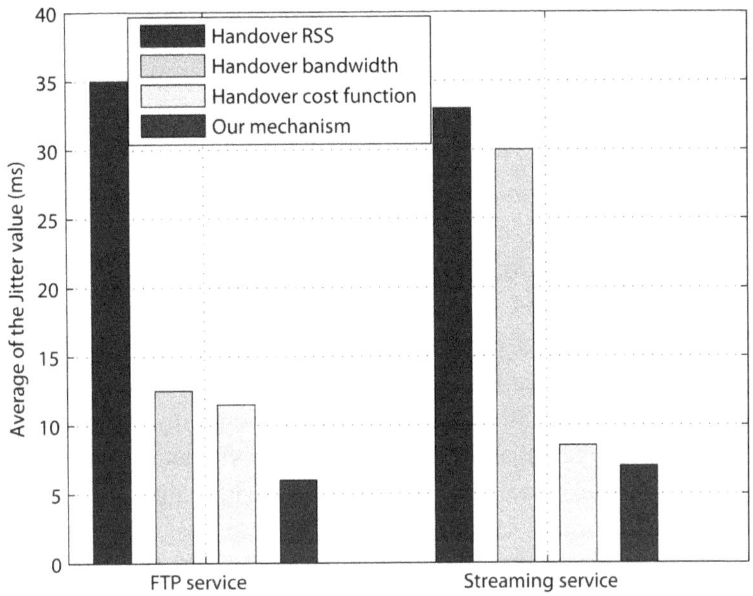

5.2.2 Experimental 2

In this experimental, we investigate the impact of four weighting techniques combined with the cost function when Dijkstra's algorithm is used to select the best path in terms of QoS. These four weighting methods are equal method, random method, AHP method and ANP method.

Figure 7 shows the average throughput of the selected path for all weighting algorithms. We remark that, for ftp service, the ANP method increases the throughput with a value of 5.3 Mbps, the AHP method increases the throughput with a value of 4 Mbps, the equal method increases the throughput with a value

Figure 7. Throughput (Mb/s) for different weighting methods based handover

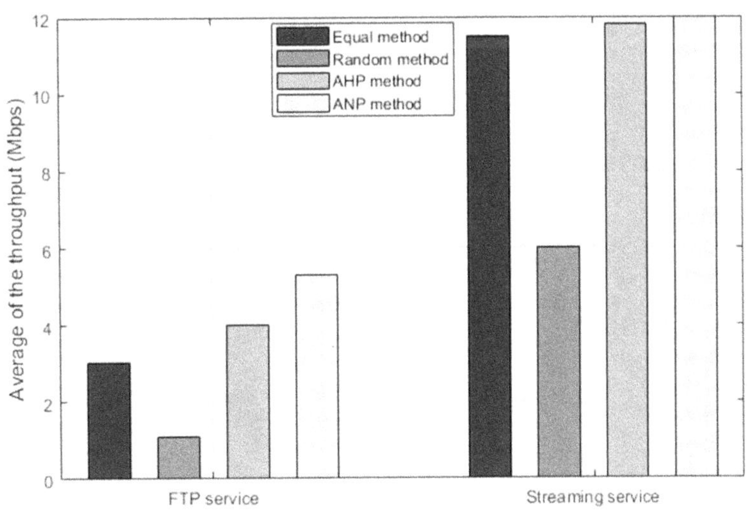

of 3.8 Mbps and the random method increases the throughput with a value of 1.09Mbps. Concerning streaming service, the ANP method increases the throughput with a value of 12 Mbps, the AHP method increases the throughput with a value of 11.8 Mbps, the equal method increases the throughput with a value of 11.5 Mbps and the random method increases the throughput with a value of 6 Mbps.

The all values provides by our mechanism based on ANP method coupled with the cost function are higher than the other values of the other methods for both services. So, our proposed mechanism based on ANP method can improve the ftp throughput and streaming throughput better than all other weighting methods which can be coupled with the cost function.

Furthermore, the AHP method can exhibit significant improvement of the ftp throughput and streaming throughput than the random method and the equal method. Finally, the random method is the only method that provides smaller results of throughput according both services, which represent the bad method that coupled with our mechanism.

Figure 8 shows the average value of the packet lost for all weighting techniques concerning ftp service and streaming service. We notice that for ftp service, the ANP method, the AHP method, the equal method and the random method can decrease the packet lost with values of 0.5%, 0.8% 0.9% and 2% respectively. In addition, for streaming service, the ANP method, the AHP method, the equal method and the random method can diminish the packet lost with values of 0.5%, 0.6% and 0.99% and 1.01% respectively.

We deduce that, for both services, our mechanism based on ANP method can be used to reduce the packet lost. On the other hand, the AHP method can diminishes the value of the packet lost better the random method and the equal method. The random method is the bad method that coupled with our mechanism.

Figure 9 shows the average of end-to-end delay for all weighting techniques concerning ftp service and streaming service. We remark that, for ftp traffic, the four algorithms Equal method, Random

Figure 8. Average of Packet lost (Ms) for different weighting methods based handover

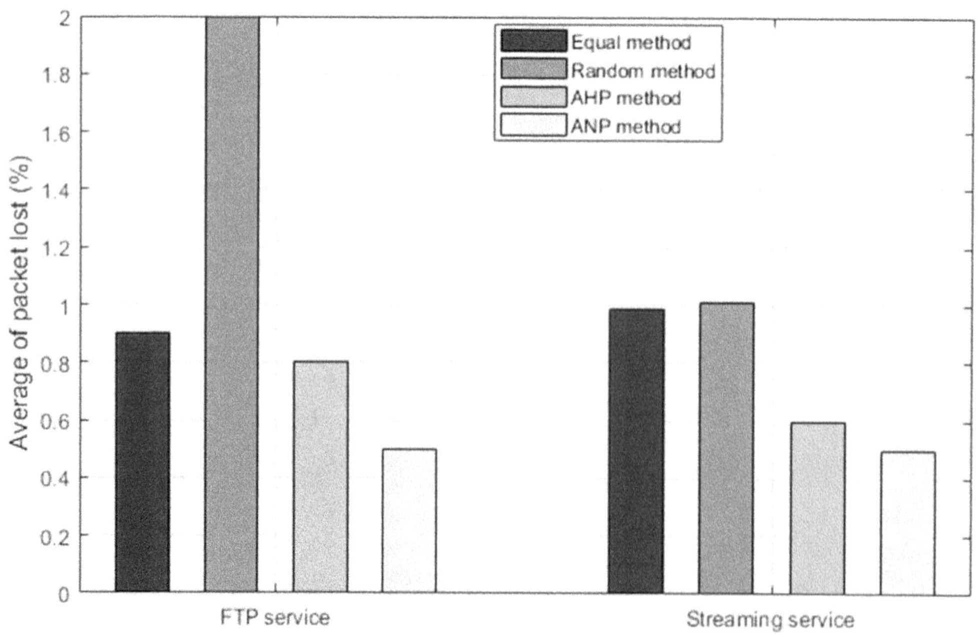

Figure 9. Average of Packet delay (ms) for different weighting methods based handover

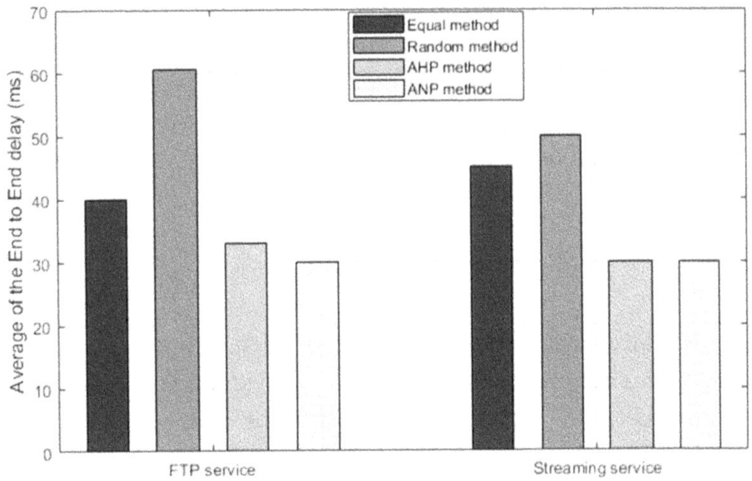

method, AHP method and ANP method decrease the packet delay with 39 ms, 60.6 ms 33 ms and 30 ms respectively. In addition, for streaming traffic, the four algorithms Equal method, Random method, AHP method and ANP method decrease the packet delay with 45 ms, 50 ms, 30 ms and 30 ms respectively.

Based on these values, for both services, the ANP method exhibits smaller delay than the other weighting algorithms. On the other hand, the higher values of this parameter are provided by random method, for that, this method is not appropriate algorithm to be combined with our mechanism for selecting the best path.

Figure 10 shows the average jitter for all weighting algorithms according to ftp service and streaming service. Note that in this figure, for ftp service, the four algorithms Equal method, Random method AHP method and ANP method decrease the packet jitter with 6 ms, 8 ms, 6 ms and 5.86 ms respectively. In addition, for streaming service, the four algorithms, Equal method, Random method, AHP method and ANP method decrease the packet delay with 10.99 ms, 11.63 ms 10.05 ms and 9.37 ms respectively.

Figure 10. Average of Packet jitter (ms) for different weighting methods based handover

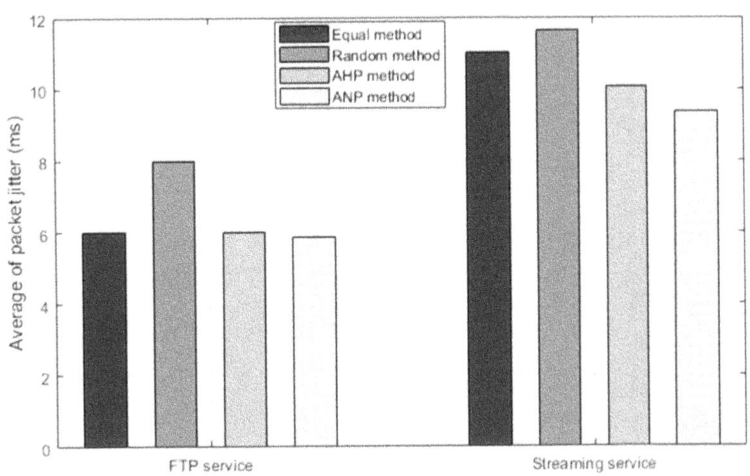

Based on these values, we remark that for both traffic classes, the ANP method provides the smaller value of the average jitter than the AHP method, the equal method and the random method. For that, the ANP method can improve the jitter better than all weighting techniques.

Furthermore, the AHP method provides better performance than equal method and random method. Finally, the random method provides the higher value of the jitter. For this reason, the mechanism coupled with this technique is the bad solution for selecting the best path for vertical handover.

6. CONCLUSION

In this chapter, we proposed an efficient mechanism for network selection decision based on graph theory and cost function in order to determine the best path during the network selection decision. Firstly, the graph theory approach is applied to model and build graph concerning the network selection problem. Then, six criteria: cost per byte (CB), security (S), available bandwidth (AB), packet delay (PD), packet jitter (PJ) and packet loss (PL) are introduced to develop a new cost function which used by Dijkstra's algorithm for selecting the best path for network selection decision. Moreover, we improved our mechanism by investigating the impact of four weighting techniques: equal method, random method, AHP method and ANP method on cost function in order to determine the most suitable weights for different criteria.

The experiments evaluation on a testbed by using Mininet emulator showed, that our efficient mechanism for selecting the best path provides higher throughput, exhibits smaller packer lost, decreases the packet delay, and the packet jitter better than handover based RSS, handover based bandwidth and handover based cost function. In addition, we notice that the vertical handover based on one criterion such as RSS or bandwidth is not efficient to choose the best path for network selection in heterogeneous wireless networks. Moreover, the cost function based on ANP method provided better performance in terms of QoS for both services than equal method, random method and AHP method.

Finally, in spite of our mechanism has achieved promising results in light of modeling vertical handover as graph, our proposed solution suffered two limitations. The first one is that, we presumed that the access technology installed along the road edges; consequently, the proposed mechanism does not consider dynamic users. The second limitation is that the proposed mechanism does not take into account what is the most appropriate time to start the network selection algorithm. For that, our on-going work focuses to deal with both limitations, we will propose k-partite graph to model the network selection problem in order to deal with the first limitation. In addition, to deal with the second limitation, we will introduce machine learning in order to predict the most appropriate time to start the network selection algorithm.

REFERENCES

Alkhawlani, M., & Ayesh, A. (2008). Access network selection based on fuzzy logic and genetic algorithms. *Advances in Artificial Intelligence*, *8*(1), 1–12. doi:10.1155/2008/793058

Barth, U. (2006). *3GPP Long-Term Evolution / System Architecture Evolution Overview*. Alcatel, Technical white paper.

Escobar, L. (2010). Vertical handoff algorithms a new approach for performance evaluation. *IEEE Globecom Workshop on Ubiquitous Computing and Networks*, 1724-1728.

Fu, J. (2010). Novel AHP and GRA Based Handover Decision Mechanism in Heterogeneous Wireless Networks. Proc. CICLing, (2), 213-220.

Goudarzi, S., Hassan, W. H., Anisi, M. H., & Soleymani, S. A. (2017). MDP-Based Network Selection Scheme by Genetic Algorithm and Simulated Annealing for Vertical-Handover in Heterogeneous Wireless Networks. *Wireless Personal Communications*, *92*(2), 399–436. doi:10.100711277-016-3549-5

Goyal, R. K., Kaushal, S., & Sangaiah, A. K. (2017). *The utility based non-linear fuzzy AHP optimization model for network selection in heterogeneous wireless networks. In Applied Soft Computing*. Elsevier.

Gustafsson, E., & Jonsson, A. (2003). Always best connected. *IEEE Wireless Communications Magazine*, *10*(1), 49–55. doi:10.1109/MWC.2003.1182111

Hasswa, A., Nasser, N., & Hassanein, H. (2005). Generic vertical handoff decision function for heterogeneous wireless. *Second IFIP International Conference on Wireless and Optical Communications Networks, (WOCN)*, 239-243.

Hou, J., & O'Brien, D. C. (2006). Vertical handover decision making algorithm using fuzzy logic for the integrated Radio-and-OW system. *IEEE Transactions on Wireless Communications*, *5*(1), 176–185. doi:10.1109/TWC.2006.1576541

IEEE802.11. (2007) Standard for information technology-telecommunications and information exchange between systems-local and metropolitan area networks-specific requirements-part 11: Wireless lan medium access control (mac) and physical layer (phy) specifications. (Revision of IEEE Std 802.11-1999), 1–1184.

IEEE 802.16. (2004). Air interface for fixed broadband wireless access systems. Local and Metropolitan Area Networks, Part 16.

Javier, S., & Mamadou, T. (2001). *UMTS: services, architecture et WCDMA*. Academic Press.

Kaleem, F., Mehbodniyay, Z., Yen, K., & Adachi, F. (2012). A dynamic weighting of attributes in heterogeneous wireless networks using fuzzy linguistic variables. *Proceedings of the First IEEE International Conference on Communications in China Wireless Networking and Applications (WNA)*, 698-703

Karam, F. (2012). Performance analysis of ranking for QoS handover algorithm for selection of access network in heterogeneous wireless networks. *21st International Conference on Computer Communications and Networks (ICCCN)*, 1-6.

Khloussy, E., Gelabert, X., & Jiang, Y. (2015). Investigation on MDP-based radio access technology selection in heterogeneous wireless networks. *Computer Networks*, *91*, 57–67. doi:10.1016/j.comnet.2015.08.005

Lahby. (2013a). A Novel Ranking Algorithm Based Network Selection For Heterogeneous Wireless Access. *Journal of Networks*, *8*(2), 263–272.

Lahby. (2013b). An Enhanced-TOPSIS Based Network Selection Technique for Next Generation Wireless Networks. *Proceedings of the 20th International Conference on Telecommunications (ICT)*, 1-5.

Lahby, Attioui, & Sekkaki. (2017). An improved policy for network selection decision based on enhanced TOPSIS and utility function. *Proceedings of the 13th International Wireless Communications ans Mobile Computing Conference (IWCMC)*, 2175-2180.

Lahby & Sekkaki. (2017). Optimal Vertical Handover based on TOPSIS algorithm and Utility function In Heterogeneous Wireless Networks. *Proceedings of the 4th International Symposium on Networks, Computers and Communications (ISNCC)*, 1-5.

Lahby, M., Leghris, C., & Adib, A. (2011). A Hybrid Approach for Network Selection in Heterogeneous Multi-Access Environments. *Proceedings of the 4th IFIP International Conference on New Technologies, Mobility and Security (NTMS)*, 1-5.

Lahby, M., Leghris, C., & Adib, A. (2012). A Survey and Comparison Study on Weighting Algorithms for Access Network Selection. *Proceedings of the 9th IFIP Annual Conference on Wireless On-Demand Network Systems and Services (WONS)*, 35-38.

Lahby, M., & Sekkaki, A. (2018). A Graph Theory Based Network Selection Algorithm in Heterogeneous Wireless Networks. *9th IFIP International Conference on New Technologies, Mobility and Security (NTMS)*, 1-4.

Lahby, M., Silki, B., & Sekkaki, A. (2015). Survey and comparison of MADM methods for network selection access in heterogeneous networks. *7th IFIP International Conference on New Technologies, Mobility and Security (NTMS)*, 1-6.

Lee, D., Han, Y., & Hwang, J. (2013). QoS Based Vertical Handoff Decision Algorithm in Heterogeneous Systems. *Sixth Joint IFIP Wireless and Mobile Networking Conference (WMNC)*, 1-6.

Naeem, B., Ngah, R., Hashim, S. Z. M., & Maqbool, W. (2014). Vertical handover decision using history-based communication graph for heterogeneous networks. *Open Systems (ICOS), 2014 IEEE Conference on. IEEE*, 7-11.

Oliveira Filho, J. L. D. E., & Madeira, E. (2011). A mechanism for vertical handover based on SAW using IEEE 802.21. In *Mobile Networks and Management* (pp. 96–108). Springer. doi:10.1007/978-3-642-21444-8_10

Phuoc Tran & Boukhatem. (2009). An utility based interface selection scheme for multi-homed mobile terminals. *Personal, Indoor and Mobile Radio Communications*, 767-772.

Sgora, A. (2010). An access network selection algorithm for heterogeneous wireless environments. *Proceedings ofThe IEEE symposium on Computers and Communications*, 890-892. 10.1109/ISCC.2010.5546556

Sheng-Mei, L. (2010). A simple additive weighting vertical handoff algorithm based on SINR and AHP for heterogeneous wireless networks. *Proceedings of IEEE International Conference on Intelligent Computation Technology and Automation (ICICTA)*, 1, 347-350.

Sheng-mei, L., Su, P., & Ming-hai, X. (2010). An improved topsis vertical handoff algorithm for heterogeneous wireless networks. *Proceedings of the 12th IEEE International Conference on Communication Technology (ICCT)*, 750-754.

Team, T. M. (2012). *Mininet: An Instant Virtual Network on Your Laptop (or Other PC)*. Retrieved from http://www.mininet.org

Trestian, R., Ormond, O., & Muntean, G. (2012). Game theory-based network selection: Solutions and challenges. *IEEE Communications Surveys and Tutorials*, *14*(4), 1212–1231. doi:10.1109/SURV.2012.010912.00081

Vuong, Thinh, Doudane, & Nazim. (2008). On utility models for access network selection in wireless heterogeneous networks. *IEEE Network Operations and Management Symposium (NOMS)*, 144-151.

Wang, L., & Kuo, G.-S. (2013). Mathematical modeling for network selection in heterogeneous wireless networks-A tutorial. *IEEE Communications Surveys and Tutorials*, *15*(1), 271–292. doi:10.1109/SURV.2012.010912.00044

Yafang, W., Huimin, C., & Jinyan, Z. (2010). Network access selection algorithm based on the Analytic Hierarehy Process and Gray Relation Analysis. *Proc New Trends in Information Science and Service Science*.

Yan, X., Mani, N., & Sekercioglu, Y. A. (2008). A travelling distance prediction based method to minimize unnecessary handovers from cellular networks to WLANs. *IEEE Communications Letters*, *12*(1), 14–16. doi:10.1109/LCOMM.2008.071430

ADDITIONAL READING

Buiati, F., Villalba, L. J. G., Corujo, D., Sargento, S., & Aguiar, R. L. (2011). IEEE 802.21 information services deployment for heterogeneous mobile environments. *IET Communications*, *5*(18), 2721–2729. doi:10.1049/iet-com.2010.1012

Keshavarz-Haddad, A., Aryafar, E., Wang, M., & Chiang, M. (2017). HetNets selection by clients: Convergence, efficiency, and practicality. *IEEE/ACM Transactions on Networking*, *25*(1), 406–419. doi:10.1109/TNET.2016.2587622

Kobayashi, H., Kameda, E., Terashima, Y., & Shinomiya, N. (2018). Towards sustainable heterogeneous wireless networks: A decision strategy for AP selection with dynamic graphs. *Computer Networks*, *132*, 99–107. doi:10.1016/j.comnet.2018.01.008

Lahby, M., & Sekkaki, A. (2018). An Efficient Policy for Vertical-Handover-Based Multi-Attribute Utility Theory in Heterogeneous Wireless Networks. In *Advances in Data Communications and Networking for Digital Business Transformation* (pp. 1–20). IGI Global. doi:10.4018/978-1-5225-5323-6.ch001

Obayiuwana, E., & Falowo, O. E. (2017). Network selection in heterogeneous wireless networks using multi-criteria decision-making algorithms: A review. *Wireless Networks*, *23*(8), 2617–2649. doi:10.100711276-016-1301-4

KEY TERMS AND DEFINITIONS

Always Best Connected (ABC) Paradigm: Means that the users have the possibility to connect to applications using the devices and access technologies that best suit his or her needs, thereby combining the features of access technologies such as Bluetooth, Wifi, WiMax, UMTS, and LTE. In addition, a user is able to choose the best available access networks at anywhere and at any in time by using mobile terminal equipped with multiple interfaces.

Handover: In cellular telecommunications, the terms handover or handoff refer to the process of transferring an ongoing call or data session of mobile terminal from one access network to another access network. Basically, handovers are mainly classified into horizontal and vertical handovers. Handover within same access networks is referred to as horizontal handover, while handover across heterogeneous access networks is referred to as the vertical.

Handover Decision: It's namely also network selection. In this step the mobile terminal evaluates the reachable wireless networks to make a decision according some criteria such as battery, velocity, QoS level, security level, users' preferences, perceived QoS, etc.

Handover Execution: It consists on establishing the target access network by using mobile IP protocol.

Handover Initiation: During this step, the mobile terminal discovers available networks. In addition, some initial parameters for measurement and handover are configured. Measurement related parameters may include the required link quality, measurement metrics, measurement interval, and so on.

International Telecommunication Unit (ITU): Is the leading United Nations agency for information and communication technologies. As the global focal point for governments and the private sector, ITU's role in helping the world communicate spans 3 core sectors: radio-communication, standardization and development. ITU also organizes TELECOM events and was the lead organizing agency of the World Summit on the Information Society.

Packet Delay: Is defined as difference between time taken by the transmitted packets from source to destination. It includes all possible delays in the network like buffering route discovery latency, and propagation and transmission delay.

Packet Jitter: Is defined as a variation in the delay of received packets. At the sending side, packets are sent in a continuous stream with the packets spaced evenly apart.

Packet Loss: Is the ratio of the number of the failure of packets and the number of packets originated by the source.

Quality of Service (QoS): Refers to the description or measurement of the overall performance of a service, such as access network or a cloud computing service, particularly the performance seen by the users of the network. To quantitatively measure quality of service, several related aspects of the network service are often considered, such as throughput, packet loss, packet delay, packet jitter, etc.

Throughput: For each access network, this parameter represents the ratio of total successful data delivered over a communication channel per unit time.

Compilation of References

3GPP TR 28.801 V.15.1.0. (2018). Study on management and orchestration of network slicing for next generation network.

3GPP TS 22.261 V.16.3.0. (2018). Service requirements for the 5G system.

3GPP. (n.d.). *Release 14*. Retrieved from http://www.3gpp.org/release-14

3GPP. Policy and Charging Control Framework for the 5G System (Release 15). TR 23.503.

3GPP. Procedures for the 5G System (Release 15). TS 23.502.

3GPP. Study on Access Traffic Steering, Switch and Splitting support in the 5G system architecture (Release 15). TR 23.793.

3GPP. System Architecture for the 5G System (Release 15). TS 23.501.

5G Americas Whitepaper. (2018). *Cellular V2XCommunications Towards 5G*. Retrieved from http://www.5gamericas.org/files/9615/2096/4441/2018_5G_Americas_White_Paper_Cellular_V2X_Communications_Towards_5G__Final_for_Distribution.pdf

Abbas, H. S., & Gregory, M. A. (2016). The next generation of passive optical networks: A review. *Journal of Network and Computer Applications*, *67*(5), 53–74. doi:10.1016/j.jnca.2016.02.015

Abbas, N., Zhang, Y., Taherkordi, A., & Skeie, T. (2018). Mobile Edge Computing: A Survey. *IEEE Internet of Things Journal*, *5*(1), 450–465. doi:10.1109/JIOT.2017.2750180

Abdelwahab, S., Hamdaoui, B., Guizani, M., & Znati, T. (2016). Network function virtualization in 5G. *IEEE Communications Magazine*, *54*(4), 84–91. doi:10.1109/MCOM.2016.7452271

Abou-Rjeily, C., & Haddad, S. (2014). Inter-relay cooperation: A new paradigm for enhanced relay-assisted FSO communications. *IEEE Transactions on Communications*, *62*(6), 1970–1982. doi:10.1109/TCOMM.2014.2316262

Afolabi, I., Taleb, T., Samdanis, K., Ksentini, A., & Flinck, H. (2018). Network slicing & softwarization: A survey on principles, enabling technologies & solutions. *IEEE Communications Surveys and Tutorials*, *20*(3), 2429–2453. doi:10.1109/COMST.2018.2815638

Ahlgren, B., D'ambrosio, M., Dannewitz, C., Eriksson, A., & Golic, M. J. (2010). *Second netinf architecture description*. 4WARD EU FP7 Project, Deliverable D-6.2 v2. 0. Retrieved from http://www.4ward-project.eu/

Ahlgren, B., Dannewitz, C., Imbrenda, C., Kutscher, D., & Ohlman, B. (2012). A survey of information-centric networking. *IEEE Communications Magazine*, *50*(7), 26–36. doi:10.1109/MCOM.2012.6231276

Ahmad, I., Noor, R. M., Ahmedy, I., Shah, S. A. A., Yaqoob, I., Ahmed, E., & Imran, M. (2018). VANET–LTE based Heterogeneous Vehicular Clustering for Driving Assistance and Route Planning Applications. *Computer Networks*, *145*, 128–140. doi:10.1016/j.comnet.2018.08.018

Ain, M., Trossen, D., Nikander, P., Tarkoma, S., Visala & Kjällman, J. (2009). *D2. 3–architecture definition, component descriptions, and requirements.* Deliverable, PSIRP 7th FP EU-funded project, 11.

Akiyama, Oodaira, Ishikawa, Hisamoto, & Watanabe. (2003). The Umbrella Cell: A Logic-Process-Compatible 2T Cell for SOC Applications. In *VLSI Circuits, 2003. Digest of Technical Papers. 2003 Symposium On* (pp. 89–92). IEEE.

Akkarajitsakul, K., Hossain, E., Niyato, D., & Kim, D. I. (2011). Game Theoretic Approaches for Multiple Access in Wireless Networks: A Survey. *IEEE Communications Surveys and Tutorials*, *13*(3), 372–395. doi:10.1109/SURV.2011.122310.000119

Alimi, R., Rahman, A., & Yang, Y. (2011). *A survey of in-network storage systems* (No. RFC 6392).

Alimi, R., Rahman, A., Kutscher, D., Yang, Y., Song, H., & Pentikousis, K. (2013). *DECoupled Application Data Enroute (DECADE)* (No. RFC 7069).

Al-Kashoash, H. A. A., Hafeez, M., & Kemp, A. H. (2017). Congestion Control for 6LoWPAN Networks: A Game Theoretic Framework. *IEEE Internet of Things Journal*, *4*(3), 760–771. doi:10.1109/JIOT.2017.2666269

Alkhawlani, M. & Ayesh, A. (2008). Access network selection based on fuzzy logic and genetic algorithm. *Advances in Artificial Intelligence, 2008*(1), 1-12.

Alkhawlani, M., & Ayesh, A. (2008). Access network selection based on fuzzy logic and genetic algorithms. *Advances in Artificial Intelligence*, *8*(1), 1–12. doi:10.1155/2008/793058

Alleven, M. (2018). *AT&T Plans to Launch Mobile 5G in a Dozen Cities by Late 2018.* Retrieved from https://www.fiercewireless.com/wireless/at-t-plans-to-launch-mobile-5g-dozen-cities-by-late-2018

Alliance, N. G. M. N. (2016). *Description of network slicing concept.* Retrieved from http:// www.ngmn.org

Almalag, M. S., & Weigle, M. C. (2010, October). Using traffic flow for cluster formation in vehicular ad-hoc networks. In *Local Computer Networks (LCN), 2010 IEEE 35th Conference on* (pp. 631-636). IEEE.

Alpcan, T., Basar, T., Srikant, R., & Atman, E. (2001). CDMA uplink power control as a noncooperative game. In *Proceedings of the 40th IEEE Conference on Decision and Control (Cat. No.01CH37228)* (Vol. 1, pp. 197–202). IEEE. 10.1109/CDC.2001.980097

Altman, E., Boulogne, T., El-Azouzi, R., Jiménez, T., & Wynter, L. (2006). A Survey on Networking Games in Telecommunications. *Computers & Operations Research*, *33*(2), 286–311. doi:10.1016/j.cor.2004.06.005

Altman, E., El-Azouzi, R., Hayel, Y., & Tembine, H. (2009). The evolution of transport protocols: An evolutionary game perspective. *Computer Networks*, *53*(10), 1751–1759. doi:10.1016/j.comnet.2008.12.023

Alves, H., Silva, L. M., Neto Marinheiro, R., & Moura, J. (2018). PMIPv6 Integrated with MIH for Flow Mobility Management: A Real Testbed with Simultaneous Multi-Access in Heterogeneous Mobile Networks. *Wireless Personal Communications*, *98*(1), 1055–1082. doi:10.100711277-017-4908-6

Anastasopoulos, M., Arapoglou, P.-D., Kannan, R., & Cottis, P. (2008). Adaptive Routing Strategies in IEEE 802.16 Multi-Hop Wireless Backhaul Networks Based On Evolutionary Game Theory. *IEEE Journal on Selected Areas in Communications*, *26*(7), 1218–1225. doi:10.1109/JSAC.2008.080918

Anderegg, L., & Eidenbenz, S. (2003). Ad hoc-VCG Ad hoc-VCG: a truthful and cost-efficient routing protocol for mobile ad hoc networks with selfish agents. In *Proceedings of the 9th annual international conference on Mobile computing and networking - MobiCom '03* (p. 245). New York: ACM Press. 10.1145/938985.939011

Andreev, S., Galinina, O., Pyattaev, A., Johnsson, K., & Koucheryavy, Y. (2015). Analyzing assisted offloading of cellular user sessions onto D2D links in unlicensed bands. *IEEE Journal on Selected Areas in Communications*, *33*(1), 67–80. doi:10.1109/JSAC.2014.2369616

Andrews, J. G. (2005). Interference cancellation for cellular systems: A contemporary overview. *IEEE Wireless Communications*, *12*(2), 19–29. doi:10.1109/MWC.2005.1421925

Andrews, J. G., Baccelli, F., & Ganti, R. K. (2011). A tractable approach to coverage and rate in cellular networks. *IEEE Transactions on Communications*, *59*(11), 3122–3134. doi:10.1109/TCOMM.2011.100411.100541

Andrews, J. G., Buzzi, S., Choi, W., Hanly, S. V., Lozano, A., Soong, A. C. K., & Zhang, J. C. (2014). What Will 5G Be? *IEEE Journal on Selected Areas in Communications*, *23*(6), 1065–1082. doi:10.1109/JSAC.2014.2328098

Andrews, J. G., Claussen, H., Dohler, M., Rangan, S., & Reed, M. C. (2012). Femtocells: Past, Present, and Future. *IEEE Journal on Selected Areas in Communications*, *30*(3), 497–508. doi:10.1109/JSAC.2012.120401

Apicella, C. L., Marlowe, F. W., Fowler, J. H., & Christakis, N. A. (2012). Social networks and cooperation in hunter-gatherers. *Nature*, *481*(7382), 497–501. doi:10.1038/nature10736 PMID:22281599

Aptilo Networks. (2013). *Seamless Wi-Fi Offload: A Huge Business Opportunity for Mobile Carriers.* White paper.

Arevalo, G. V., Hincapié, R. C., & Gaudino, R. (2017). Optimization of multiple PON deployment costs and comparison between GPON, XGPON, NGPON2 and UDWDM PON. *Optical Switching and Networking*, *25*(7), 80–90. doi:10.1016/j.osn.2017.03.003

Arfaoui, G., Bisson, P., Blom, R., Borgaonkar, R., Englund, H., Félix, E., ... Zahariev, A. (2018). A security architecture for 5G networks. *IEEE Access: Practical Innovations, Open Solutions*, *6*, 22466–22479. doi:10.1109/ACCESS.2018.2827419

Arianfar, S., Nikander, P., & Ott, J. (2010, June). Packet-level caching for information-centric networking. In *ACM SIGCOMM*. ReArch Workshop.

Arianfar, S., Nikander, P., & Ott, J. (2010, November). On content-centric router design and implications. In *Proceedings of the Re-Architecting the Internet Workshop* (p. 5). ACM.

Arkian, H. R., Atani, R. E., Diyanat, A., & Pourkhalili, A. (2015). A cluster-based vehicular cloud architecture with learning-based resource management. *The Journal of Supercomputing*, *71*(4), 1401–1426. doi:10.100711227-014-1370-z

Arnold, O., Richter, F., Fettweis, G., & Blume, O. (2010, June). Power consumption modeling of different base station types in heterogeneous cellular networks. In *Future Network and Mobile Summit, 2010* (pp. 1–8). IEEE.

Arsenyan, J., Büyüközkan, G., & Feyzioğlu, O. (2015). Modeling collaboration formation with a game theory approach. *Expert Systems with Applications*, *42*(4), 2073–2085. doi:10.1016/j.eswa.2014.10.010

Atlam, H. F., Walters, R. J., & Wills, G. B. (2018). Fog Computing and the Internet of Things: A Review. *Big Data and Cognitive Computing, 2*(2).

Awada, A., Wegmann, B., Viering, I., & Klein, A. (2013). A SON-based algorithm for the optimization of inter-RAT handover parameters. *IEEE Transactions on Vehicular Technology*, *62*(5), 1906–1923. doi:10.1109/TVT.2013.2251923

Axelrod, R. (1981). The Emergence of Cooperation among Egoists. *The American Political Science Review*, *75*(02), 306–318. doi:10.2307/1961366

Babaioff, M., Dobzinski, S., Oren, S., & Zohar, A. (2012). On bitcoin and red balloons. In *Proceedings of the 13th ACM Conference on Electronic Commerce - EC '12* (p. 56). New York: ACM Press. 10.1145/2229012.2229022

Babu, A. V., & Joshy, S. (2012). Maximizing the data transmission rate of a cooperative relay system in an underwater acoustic channel. *International Journal of Communication Systems*, *25*(2), 231–253. doi:10.1002/dac.1266

Badri, T. F. Z., Saadane, R., Wahbi, M., & Samir, M. (2013). Handover Management Scheme in LTE Femtocell Networks. *International Journal of Computer Science & Information Technology*, *5*(3), 89–100. doi:10.5121/ijcsit.2013.5307

Bae, Ryu, & Park. (2011). *Analysis* of Handover Failures in LTE Femtocell Systems. In *Telecommunication Networks and Applications Conference (ATNAC), 2011 Australasian* (pp. 1–5). IEEE.

Baek, S. Y., & Hong, D. Y. J. (2009). Adaptive Transmission Scheme for Mixed Multicast and Unicast Traffic in Cellular Systems. *IEEE Transactions on Vehicular Technology*, *58*(6), 2899–2908. doi:10.1109/TVT.2008.2010515

Bai, Wang, Liu, & Zhang. (2011). A Policy-Based Handover Mechanism between Femtocell and Macrocell for LTE Based Networks. In *Communication Technology (ICCT), 2011 IEEE 13th International Conference On* (pp. 916–920). IEEE.

Bakmaz, B. M. (2013). Network selection equilibrium in heterogeneous wireless environment. *Electronics and Electrical Engineering*, *19*(4), 91–96.

Balachandran, Kang, Karakayali, & Rege (2012). Virtual Soft Handoff Enabled Dominant Interference Cancellation for Enhanced Uplink Performance in Heterogeneous Cellular Networks. In *Wireless Communications and Networking Conference Workshops (WCNCW), 2012 IEEE* (pp. 109–114). IEEE.

Balevi, E., & Gitlin, R. D. (2018). Optimizing the Number of Fog Nodes for Cloud-Fog-Thing Networks. *IEEE Access: Practical Innovations, Open Solutions*, *6*, 11173–11183. doi:10.1109/ACCESS.2018.2808598

Bansal, S., & Baker, M. (2003). Observation-based Cooperation Enforcement in Ad Hoc Networks. *CoRR,* cs.NI/0307.

Barth, U. (2006). *3GPP Long-Term Evolution / System Architecture Evolution Overview*. Alcatel, Technical white paper.

Basagni, S. (1999). Distributed clustering for ad hoc networks. In Parallel Architectures, Algorithms, and Networks, 1999. (I-SPAN'99) Proceedings. Fourth InternationalSymposium on (pp. 310-315). IEEE. doi:10.1109/ISPAN.1999.778957

BBF. Hybrid Access Broadband Network Architecture. TR-348.

Becvar & Mach. (2011). On Enhancement of Handover Decision in Femtocells. *In Wireless Days (WD), 2011 IFIP* (pp. 1–3). IEEE.

Bennis, M., Guruacharya, S., & Niyato, D. (2011). Distributed Learning Strategies for Interference Mitigation in Femtocell Networks. In *Global Telecommunications Conference (GLOBECOM 2011)*, 2011 *IEEE* (pp. 1–5). IEEE. 10.1109/GLOCOM.2011.6134218

Benslimane, A., Taleb, T., & Sivaraj, R. (2011). Dynamic clustering-based adaptive mobile gateway management in integrated VANET—3G Heterogeneous Wireless Networks. *Selected Areas in Communications. IEEE Journal on*, *29*(3), 559–570.

Berman, M., Chase, J. S., Landweber, L., Nakao, A., Ott, M., Raychaudhuri, D., ... Seskar, I. (2014). GENI: A federated testbed for innovative network experiments. *Computer Networks*, *61*, 5–23. doi:10.1016/j.bjp.2013.12.037

Bhaskaran, S. R., & Krishnan, V. (2009). Effort, Revenue, and Cost Sharing Mechanisms for Collaborative New Product Development. *Management Science*, *55*(7), 1152–1169. doi:10.1287/mnsc.1090.1010

Blum, J., Eskandarian, A., & Hoffman, L. (2003). Mobility management in IVC networks. In *Intelligent Vehicles Symposium, 2003. Proceedings. IEEE* (pp. 150-155). IEEE.

Böge, W., & Eisele, T. (1979). On solutions of Bayesian games. *International Journal of Game Theory*, *8*(4), 193–215. doi:10.1007/BF01766706

Bojkovic, Z., Bakmaz, B., & Bakmaz, M. (2017). Vision and enabling technologies of tactile Internet realization. In *Proceedings of 13th International Conference on Advanced Technologies, Systems and Services in Telecommunications* (pp. 113-118). Nis, Serbia: Academic Press.

Bond, R. M., Fariss, C. J., Jones, J. J., Kramer, A. D. I., Marlow, C., Settle, J. E., & Fowler, J. H. (2012). A 61-million-person experiment in social influence and political mobilization. *Nature*, *489*(7415), 295–298. doi:10.1038/nature11421 PMID:22972300

Borah, S. J., Dhurandher, S. K., Woungang, I., & Kumar, V. (2017). A game theoretic context-based routing protocol for opportunistic networks in an IoT scenario. *Computer Networks*, *129*, 572–584. doi:10.1016/j.comnet.2017.07.005

Bouhaddi, M., Radjef, M. S., & Adi, K. (2018). An efficient intrusion detection in resource-constrained mobile ad-hoc networks. *Computers & Security*, *76*, 156–177. doi:10.1016/j.cose.2018.02.018

Bucciol, A., & Zarri, L. (2013). *Financial Risk Aversion and Personal Life History*. Retrieved June 19, 2018, from: http://leonardo3.dse.univr.it/home/workingpapers/risk_events_BZ.pdf

Buchegger, S., & Le Boudec, J.-Y. (2002). Performance analysis of the CONFIDANT protocol. In *Proceedings of the 3rd ACM international symposium on Mobile ad hoc networking & computing - MobiHoc '02* (p. 226). New York: ACM Press. 10.1145/513800.513828

Bu, S., Yu, F. R., & Yanikomeroglu, H. (2015). Interference-Aware Energy-Efficient Resource Allocation for OFDMA-Based Heterogeneous Networks With Incomplete Channel State Information. *IEEE Transactions on Vehicular Technology*, *64*(3), 1036–1050. doi:10.1109/TVT.2014.2325823

Buttyan, L., & Hubaux, J.-P. (2000). Enforcing service availability in mobile ad-hoc WANs. In *2000 First Annual Workshop on Mobile and Ad Hoc Networking and Computing. MobiHOC (Cat. No.00EX444)* (pp. 87–96). IEEE. 10.1109/MOBHOC.2000.869216

Buttyan, L., & Hubaux, J.-P. (2001). Nuglets: A Virtual Currency to Stimulate Cooperation in Self-Organized Mobile. *Ad Hoc Networks*.

Buttyán, L., & Hubaux, J.-P. (2003). Stimulating Cooperation in Self-Organizing Mobile Ad Hoc Networks. *Mobile Networks and Applications*, *8*(5), 579–592. doi:10.1023/A:1025146013151

Cao, R., Yang, L., & Qu, F. (2010, April). On the capacity and system design of relay-aided underwater acoustic communications. In *Wireless Communications and Networking Conference (WCNC)*, 2010 *IEEE* (pp. 1-6). IEEE.

Cao, Y., Wang, N., Kamel, G., & Kim, Y. J. (2015). An electric vehicle charging management scheme based on publish/subscribe communication framework. *IEEE Systems Journal*.

Chaganti, Sathya, Ahammed, Rex, & Tamma. (2013). Efficient Son Handover Scheme for Enterprise Femtocell Networks. In *Advanced Networks and Telecommuncations Systems (ANTS), 2013 IEEE International Conference On* (pp. 1–6). IEEE.

Chan, V., Chapin, J., Elson, P., Fisher, D., Frost, V., Jones, K., & Miller, G. (2011, March). Workshop on Future Heterogeneous Networks. In *FutureHetNets 2011*. Mountain View.

Chandrasekhar, V., Andrews, J. G., & Gatherer, A. (2008a). Femtocell Networks: A Survey. *IEEE Communications Magazine*, *46*(9).

Chang, B. J., Tsai, Y. L., & Liang, Y. H. (2017). Platoon-based cooperative adaptive cruise control for achieving active safe driving through mobile vehicular cloud computing. *Wireless Personal Communications*, *97*(4), 5455–5481. doi:10.100711277-017-4789-8

Chang, J., Li, Y., Feng, S., Wang, H., Sun, C., & Zhang, P. (2009,). A fractional soft handover scheme for 3GPP LTE-advanced system. *IEEE International Conference on Communications Workshops*, 1-5.

Charilas, D. E., & Panagopoulos, A. D. (2010). A Survey on Game Theory Applications in Wireless Networks. *Computer Networks*, *54*(18), 3421–3430. doi:10.1016/j.comnet.2010.06.020

Charilas, D. E., Panagopoulos, A. D., Vlacheas, P., Markaki, O. I., & Constantinou, P. (2009). *Mobile Lightweight Wireless Systems Congestion Avoidance Control through Non-cooperative Games between Customers and Service Providers*. Berlin: Springer.

Charilas, D., & Panagopoulous, A. (2010). Multiaccess radio network enviroments. *IEEE Vehicular Technology Magazine*, *5*(4), 40–49. doi:10.1109/MVT.2010.939107

Chawla, S., & Sivan, B. (2014). Bayesian Algorithmic Mechanism Design. *SIGecom Exch.*, *13*(1), 5–49. doi:10.1145/2692375.2692378

Cheikh, Ayari, Langar, Pujolle, & Saidane. (2013). Optimized Handover Algorithm for Two-Tier Macro-Femto Cellular LTE Networks. In *Wireless and Mobile Computing, Networking and Communications (WiMob), 2013 IEEE 9th International Conference On* (pp. 608–613). IEEE.

Chen, J., Chiang, M., Erman, J., Li, G., Ramakrishnan, K. K., & Sinha, R. K. (2015). *Fair and optimal resource allocation for LTE multicast (eMBMS): Group partitioning and dynamics*. Paper presented at the IEEE Conference on Computer Communications.

Chen, C., Hu, J., Zhang, J., Sun, C., Zhao, L., & Ren, Z. (2016). Information Congestion Control on Intersections in VANETs: A Bargaining Game Approach. In *Vehicular Technology Conference (VTC Spring), 2016 IEEE 83rd* (pp. 1-5). IEEE.

Chen, K., & Nahrstedt, K. (2004). iPass: an incentive compatible auction scheme to enable packet forwarding service in MANET. In *24th International Conference on Distributed Computing Systems, 2004. Proceedings.* (pp. 534–542). IEEE. 10.1109/ICDCS.2004.1281620

Chen, K., Yang, Z., Wagener, C., & Nahrstedt, K. (2005). Market models and pricing mechanisms in a multihop wireless hotspot network. In *The Second Annual International Conference on Mobile and Ubiquitous Systems: Networking and Services* (pp. 73–82). IEEE. 10.1109/MOBIQUITOUS.2005.36

Cherif, M. O., Senouci, S. M., & Ducourthial, B. (2009). A new framework of self-organization of vehicular networks. In *Information Infrastructure Symposium, 2009. GIIS'09. Global* (pp. 1-6). IEEE.

Cheriton, D. R., & Gritter, M. (2000). *TRIAD: A New Next-Generation Internet Architecture*. Academic Press.

Chiang, M., & Zhang, T. (2016). Fog and IoT: An Overview of Research Opportunities. *IEEE Internet of Things Journal*, *3*(6), 854–864. doi:10.1109/JIOT.2016.2584538

Cho, K., Lee, M., Park, K., Kwon, T. T., Choi, Y., & Pack, S. (2012, March). Wave: Popularity-based and collaborative in-network caching for content-oriented networks. In *Computer Communications Workshops (INFOCOM WKSHPS), 2012 IEEE Conference on* (pp. 316-321). IEEE.

Choi, Seo, Nam, Park, & Cho. (2012). Handover Decision Algorithm Based on Available *Data* Volume in Hierarchical Macro/Femto-Cell Networks. In *Communications and Electronics (ICCE),* 2012 *Fourth International Conference On* (pp. 145–149). IEEE.

Chowdhury & Jang. (2010). *Handover Control for WCDMA Femtocell Networks.* ArXiv Preprint ArXiv:1009.3779.

Chowdhury, Bui, & Jang. (2011). Neighbor Cell List Optimization for Femtocell-to-Femtocell Handover in Dense Femtocellular Networks. In *Ubiquitous and Future Networks (ICUFN), 2011 Third International Conference On* (pp. 241–245). IEEE.

Chowdhury, Jang, & Haas. (2010). *Network Evolution and QoS Provisioning for Integrated Femtocell/Macrocell Networks.* ArXiv Preprint ArXiv:1009.2368.

Chowdhury, Jang, & Haas. (2011). Cost-Effective Frequency Planning for Capacity Enhancement of Femtocellular Networks. *Wireless Personal Communications, 60*(1), 83–104.

Christidis, K., & Devetsikiotis, M. (2016). Blockchains and Smart Contracts for the Internet of Things. *IEEE Access: Practical Innovations, Open Solutions, 4,* 2292–2303. doi:10.1109/ACCESS.2016.2566339

Cisco Visual Networking Index. (2017, March). *Global Mobile Data Traffic Forecast Update, 2016-2021.* White Paper.

CISCO. (2015). *Cisco Fog Computing Solutions: Unleash the Power of the Internet of Things.* Retrieved July 26, 2018, from https://www.cisco.com/c/dam/en_us/solutions/trends/iot/docs/computing-solutions.pdf

Cisco. (2016). *Cisco Virtual Networking Index: Forecast and Methodology, 2015-2020.* Retrieved from http://www.webvideomarketing.org/pdf/Cisco_Video_and_Visual_Networking_Index_Report_8.10.16.pdf

CISCO. (2016). *VNI Global Fixed and Mobile Internet Traffic Forecasts (2016-2021).* Retrieved June 2, 2018, from https://www.cisco.com/c/en/us/solutions/service-provider/visual-networking-index-vni/index.html

Cisco. (2017). *Cisco Visual Networking Index: Global Mobile Data Traffic Forecast Update, 2016–2021.* Author.

Cisco. (2017). *Cisco Visual Networking Index: Global Mobile Data Traffic Forecast Update.* Cisco.

Cisco. (2017). *The Zettabyte Era: Trends and Analysis.* Cisco White Paper.

Cisco. (2017b). *Cisco Visual Networking Index: Forecast and Methodology, 2016-2021.* Retrieved from https://www.cisco.com/c/en/us/solutions/collateral/service-provider/visual-networking-index-vni/complete-white-paper-c11-481360.html

Cisco. (2017b). *The Zettabyte Era: Trends and Analysis.* Retrieved from https://www.cisco.com/c/en/us/solutions/collateral/service-provider/visual-networking-index-vni/vni-hyperconnectivity-wp.html

Ciubotaru, B., Ghinea, G., & Muntean, G. M. (2014). Subjective assessment of region of interest-aware adaptive multimedia streaming quality. *IEEE Transactions on Broadcasting, 60*(1), 50–60. doi:10.1109/TBC.2013.2290238

Ciubotaru, B., Muntean, C. H., & Muntean, G. M. (2017). Mobile multi-source high quality multimedia delivery scheme. *IEEE Transactions on Broadcasting, 63*(2), 391–403. doi:10.1109/TBC.2016.2635079

Cohen & Levin. (2012). Handovers with Forward Admission Control for Adaptive TCP Streaming in LTE-Advanced with Small Cells. In *Computer Communications and Networks (ICCCN), 2012 21st International Conference On* (pp. 1–7). IEEE.

Coninck, Q. D., Baerts, M., Hesmans, B., & Bonaventure, O. (2016). Observing real smartphone applications over multipath TCP. *IEEE Communications Magazine*, *54*(3), 88–93. doi:10.1109/MCOM.2016.7432153

Cooper, C., Franklin, D., Ros, M., Safaei, F., & Abolhasan, M. (2017). A comparative survey of VANET clustering techniques. *IEEE Communications Surveys and Tutorials*, *19*(1), 657–681. doi:10.1109/COMST.2016.2611524

Corcoran, P. (2016). The Internet of Things: Why now, and what's next? *IEEE Consumer Electronics Magazine*, *5*(1), 63–68. doi:10.1109/MCE.2015.2484659

Čuchran, J. & Róka, R. (2006). *Optocommunication systems and networks*. Bratislava, Slovakia: Slovak University of Technology Publishing house.

Dahlman, E., Parkvall, S., & Skold, J. (2008). *3G evolution: HSPA and LTE for mobile broadband*. Burlington: Academic Press.

Darehshoorzadeh, A., & Boukerche, A. (2015). Underwater sensor networks: A new challenge for opportunistic routing protocols. *IEEE Communications Magazine*, *53*(11), 98–107. doi:10.1109/MCOM.2015.7321977

de Lima, Ghaboosi, Bennis, MacKenzie, & Latva-aho. (2010). A Stochastic Association Mechanism for Macro-to-Femtocell Handover. In *Signals, Systems and Computers (ASILOMAR), 2010 Conference Record of the Forty Fourth ASILOMAR Conference On* (pp. 1570–1574). IEEE.

Deb, S., Jaiswal, S., & Nagaraj, K. (2008). *Real-Time Video Multicast in WiMAX Networks*. Paper presented at the IEEE Conference on Computer Communications (INFOCOM).

Demir, C., & Comaniciu, C. (2007). An Auction based AODV Protocol for Mobile Ad Hoc Networks with Selfish Nodes. In *2007 IEEE International Conference on Communications* (pp. 3351–3356). IEEE. 10.1109/ICC.2007.555

Deng, D.-J., Lien, S.-Y., Lin, C.-C., Hung, S.-C., & Chen, W.-B. (2017). Latency Control in Software-Defined Mobile-Edge Vehicular Networking. *IEEE Communications Magazine*, *55*(8), 87–93. doi:10.1109/MCOM.2017.1601165

Deng, Tao, & Lu. (2012). Qos-aware resource allocation for mixed multicast and unicast traffic in OFDMA networks. *EURASIP Journal on Wireless Communications and Networking*.

Desai, S. (2014). *Electronic Health Information Exchange, Switching Costs, and Network Effects*. SSRN Electronic Journal. doi:10.2139srn.2525084

Dharmaraja, S., Trivedi, K. S., & Logothetis, D. (2003). Performance Modeling of Wireless Networks with Generally Distributed Handoff Interarrival Times. *Computer Communications*, *26*(15), 1747–1755. doi:10.1016/S0140-3664(03)00044-6

Dimou, K., Wang, M., & Yang, Y. (2009). Handover within 3GPP LTE: Design Principles and Performance. In Vehicular Technology Conference Fall (VTC 2009-Fall), 2009 IEEE 70th (pp. 1–5). IEEE.

Ding, X., Hartog, J., & Sun, Y. (2010). *Can we measure individual risk attitudes in a survey?* IZA Discussion Paper No 4807. Retrieved June 19, 2018, from http://ftp.iza.org/dp4807.pdf

Dixit, A. (2012). Wavelength switched hybrid TDMA/WDM-PON: A flexible next-generation optical access solution. *International Conference on Transparent Optical Network*, 14, 1-5.

Djahel, S., Salehie, M., Tal, I., & Jamshidi, P. (2013). Adaptive Traffic Management for Secure and Efficient Emergency Services in Smart Cities. *Pervasive Computing and Communications (PerCom WIP track)*, 340-343.

Dohmen, T., & Falk, A. (2011). Performance pay and multidimensional sorting - productivity, preferences and gender. *The American Economic Review*, *101*(2), 556–590. doi:10.1257/aer.101.2.556

Dohmen, T., Falk, A., Huffman, D., & Sunde, U. (2011a). The intergenerational transmission of risk and trust attitudes. *The Review of Economic Studies*, *79*(2), 645–677. doi:10.1093/restud/rdr027

Dohmen, T., Falk, A., Huffman, D., Sunde, U., Schupp, J., & Wagner, G. (2011b). Individual risk attitudes: Measurement, determinants and behavioural consequences. *Journal of the European Economic Association*, *9*(3), 522–550. doi:10.1111/j.1542-4774.2011.01015.x

Doolan, R., & Muntean, G. M. (2017). EcoTrec—a novel VANET-based approach to reducing vehicle emissions. *IEEE Transactions on Intelligent Transportation Systems*, *18*(3), 608–620. doi:10.1109/TITS.2016.2585925

Doosti-Aref, A., & Ebrahimzadeh, A. (2017). Adaptive Relay Selection and Power Allocation for OFDM Cooperative Underwater Acoustic Systems. *IEEE Transactions on Mobile Computing*.

Dror, E., Avin, C., & Lotker, Z. (2011). Fast randomized algorithm for hierarchical clustering in vehicular ad-hoc networks. In *Ad Hoc Networking Workshop (Med-Hoc-Net), 2011 The 10th IFIP Annual Mediterranean* (pp. 1-8). IEEE.

Dror, E., Avin, C., & Lotker, Z. (2013). Fast randomized algorithm for 2-hops clustering in vehicular ad-hoc networks. *Ad Hoc Networks*, *11*(7), 2002–2015. doi:10.1016/j.adhoc.2012.02.006

Duan, X., Liu, Y., & Wang, X. (2017). SDN enabled 5G-VANET: Adaptive vehicle clustering and beamformed transmission for aggregated traffic. *IEEE Communications Magazine*, *55*(7), 120–127. doi:10.1109/MCOM.2017.1601160

Duong, N. D., Madhukumar, A. S., & Niyato, D. (2016). Stackelberg Bayesian Game for Power Allocation in Two-Tier Networks. *IEEE Transactions on Vehicular Technology*, *65*(4), 2341–2354. doi:10.1109/TVT.2015.2418297

Dutta, P. K. (1999). *Strategies and games: Theory and practice*. MIT Press.

Effenberger, F., Mukai, H., Kani, J., & Rasztovits-Wiech, M. (2009). Next-generation PON - Part III: System specification for XG-PON. *IEEE Communications Magazine*, *47*(11), 58–64. doi:10.1109/MCOM.2009.5307467

Effenberger, F., Mukai, H., Soojin Park, & Pfeiffer, T. (2009). Next-generation PON - Part II: Candidate systems for next-generation PON. *IEEE Communications Magazine*, *47*(11), 50–57. doi:10.1109/MCOM.2009.5307466

Eidenbenz, S., Resta, G., & Santi, P. (2008). The COMMIT Protocol for Truthful and Cost-Efficient Routing in Ad Hoc Networks with Selfish Nodes. *IEEE Transactions on Mobile Computing*, *7*(1), 19–33. doi:10.1109/TMC.2007.1069

El Mouna Zhioua, G., Tabbane, N., Labiod, H., & Tabbane, S. (2015). A fuzzy multi-metric QoS-balancing gateway selection algorithm in a clustered VANET to LTE advanced hybrid cellular network. *IEEE Transactions on Vehicular Technology*, *64*(2), 804–817. doi:10.1109/TVT.2014.2323693

Ellouze, R., Gueroui, M., & Alimi, A. M. (2011). MacroFemto Cell Handover with Enhanced QoS in Mobile WiMAX. In *Wireless Telecommunications Symposium (WTS)*, 2011 (pp. 1–6). IEEE.

ElSawy, H., Hossain, E., & Haenggi, M. (2013). Stochastic geometry for modeling, analysis, and design of multi-tier and cognitive cellular wireless networks: A survey. *IEEE Communications Surveys and Tutorials*, *15*(3), 996–1019. doi:10.1109/SURV.2013.052213.00000

Ericsson. (2018). *Mobility report*. Author.

Escobar, L. (2010). Vertical handoff algorithms a new approach for performance evaluation. *IEEE Globecom Workshop on Ubiquitous Computing and Networks*, 1724-1728.

ETSI GS MEC 003. (2016). Mobile Edge Computing (MEC); Framework and Reference Architecture, v. 1.1.1

ETSI GS NFV-EVE 005 (2015). *Network Functions Virtualization (NFV); Ecosystem; Report on SDN Usage in NFV Architectural Framework*, v. 1.1.1.

EXFO. (2012). *FTTH PON guide - Testing passive optical networks.* Quebec City, Canada: EXFO Electro-Optical Engineering Inc.

Eyal, I. (2015). The Miner's Dilemma. In *2015 IEEE Symposium on Security and Privacy* (pp. 89–103). IEEE. 10.1109/SP.2015.13

Fadlullah, Z. M., Nozaki, Y., Takeuchi, A., & Kato, N. (2011). A survey of game theoretic approaches in smart grid. In *Wireless Communications and Signal Processing (WCSP), 2011 International Conference on* (pp. 1–4). Academic Press. 10.1109/WCSP.2011.6096962

Fan & Sun. (2010). Access and Handover Management for Femtocell Systems. In *Vehicular Technology Conference (VTC 2010-Spring), 2010 IEEE 71st* (pp. 1–5). IEEE.

Fan, P., Haran, J. G., Dillenburg, J., & Nelson, P. C. (2005). Cluster-based framework in vehicular ad-hoc networks. In Ad-Hoc, Mobile, and Wireless Networks (pp. 32-42). Springer Berlin Heidelberg. doi:10.1007/11561354_5

Feki, Capdevielle, Roullet, & Sanchez. (2013). Handover Aware Interference Management in LTE Small Cells Networks. In *Modeling & Optimization in Mobile, Ad Hoc & Wireless Networks (WiOpt), 2013 11th International Symposium On* (pp. 49–53). IEEE.

Felegyhazi, M., & Hubaux, J.-P. (2006). *Game Theory in Wireless Networks: A Tutorial.* Academic Press.

Fernandes, S., & Karmouch, A. (2012). Vertical Mobility Management Architectures in Wireless Networks: A Comprehensive Survey and Future Directions. *IEEE Communications Surveys and Tutorials*, *14*(1), 45–63. doi:10.1109/SURV.2011.082010.00099

Fettweis, G., & Zimmermann, E. (2008, September). ICT energy consumption-trends and challenges. *Proceedings of the 11th international symposium on wireless personal multimedia communications,* 2, 6.

Fodor, G., Dahlman, E., Mildh, G., Parkvall, S., Reider, N., Miklós, G., & Turányi, Z. (2012). Design aspects of network assisted device-to-device communications. *IEEE Communications Magazine*, *50*(3), 170–177. doi:10.1109/MCOM.2012.6163598

Foukas, X., Patounas, G., Elmokashfi, A., & Marina, M. K. (2017). Network slicing in 5G: Survey and challenges. *IEEE Communications Magazine*, *55*(5), 94–100. doi:10.1109/MCOM.2017.1600951

Frey, B. J., & Dueck, D. (2007). Clustering by passing messages between data points. *Science, 315*(5814), 972-976.

Fu, J. (2010). Novel AHP and GRA Based Handover Decision Mechanism in Heterogeneous Wireless Networks. Proc. CICLing, (2), 213-220.

Fudenberg, D., & Maskin, E. (1986). The Folk Theorem in Repeated Games with Discounting or with Incomplete Information. *Econometrica*, *54*(3), 533–554. doi:10.2307/1911307

Furukawa & Akaiwa. (1994). A Microcell Overlaid with Umbrella Cell System. In *Vehicular Technology Conference, 1994 IEEE 44th* (pp. 1455–1459). IEEE.

Gabry, F., Bioglio, V., & Land, I. (2016). On Energy-Efficient Edge Caching in Heterogeneous Networks. *IEEE Journal on Selected Areas in Communications*, *34*(12), 3288–3298. doi:10.1109/JSAC.2016.2611845

Gazdar, T., Belghith, A., & Benslimane, A. (2010). A cluster based secure architecture for vehicular ad hoc networks. In *Computer Systems and Applications (AICCSA), 2010 IEEE/ACS International Conference on* (pp. 1-8). IEEE.

Gazdar, T., Benslimane, A., & Belghith, A. (2011). Secure clustering scheme based keys management in VANETs. In *Vehicular Technology Conference (VTC Spring), 2011 IEEE 73rd* (pp. 1-5). IEEE.

Ge, X., Pan, L., Tu, P. S., Chen, H. H., & Wang, C. X. (2016). Wireless Backhaul Capacity of 5G Ultra-Dense Cellular Networks. In *IEEE Vehicular Technology Conference* (pp. 1-6). Montreal, Canada: IEEE.

Gerhardt, H., Biele, G. P., Heekeren, H. R., & Uhlig, H. (2016). *Cognitive load increases risk aversion* (No. 2016-011). SFB 649 Discussion Paper.

Ghazvini, M., Movahedinia, N., Jamshidi, K., & Moghim, N. (2013). Game Theory Applications in CSMA Methods. *IEEE Communications Surveys and Tutorials*, *15*(3), 1062–1087. doi:10.1109/SURV.2012.111412.00167

Ghergulescu, I., Moldovan, A. N., & Muntean, C. H. (2014, June). Energy-aware Adaptive Multimedia for Game-based E-learning. In *Broadband Multimedia Systems and Broadcasting (BMSB), 2014 IEEE International Symposium on* (pp. 1-6). IEEE. 10.1109/BMSB.2014.6873573

Gibbons, R. (1992). *A primer in game theory*. Harvester.

Girinath, D. R., & Selvan, S. (2010). A novel cluster based routing algorithm for hybrid mobility model in VANET. *International Journal of Computers and Applications*, *1*(15), 35–42. doi:10.5120/326-495

Gluhak, A., Vermesan, O., Bahr, R., Clari, F., MacchiaMaria, T., Delgado, T., … Barchetti, V. (2016). *Report on IoT platform activities*. Retrieved July 26, 2018, from http://www.internet-of-things-research.eu/pdf/D03_01_WP03_H2020_UNIFY-IoT_Final.pdf

Goldberg, R. (1974). Survey of virtual machine research. *IEEE Computer*, *7*(6), 34–45. doi:10.1109/MC.1974.6323581

Goldsmith, A. (2012). Beyond 4G: What Lies Ahead for Cellular System Design. In *Wireless Communications and Networking Conference*. IEEE.

Golrezaei, N., Shanmugam, K., Dimakis, A. G., Molisch, A. F., & Caire, G. (2012, March). Femtocaching: Wireless video content delivery through distributed caching helpers. In INFOCOM, 2012 Proceedings IEEE (pp. 1107-1115). IEEE.

Gonzalez-Brevis, P., Gondzio, J., Fan, Y., Poor, H. V., Thompson, J., Krikidis, I., & Chung, P. J. (2011, May). Base station location optimization for minimal energy consumption in wireless networks. In *Vehicular Technology Conference (VTC Spring), 2011 IEEE 73rd* (pp. 1-5). IEEE.

Goonewardene, R. T., Ali, F. H., & Stipidis, E. (2009). Robust mobility adaptive clustering scheme with support for geographic routing for vehicular ad hoc networks. *Intelligent Transport Systems, IET*, *3*(2), 148–158. doi:10.1049/iet-its:20070052

Goudarzi, S., Hassan, W. H., Anisi, M. H., & Soleymani, S. A. (2017). MDP-Based Network Selection Scheme by Genetic Algorithm and Simulated Annealing for Vertical-Handover in Heterogeneous Wireless Networks. *Wireless Personal Communications*, *92*(2), 399–436. doi:10.100711277-016-3549-5

Goyal, R. K., Kaushal, S., & Sangaiah, A. K. (2017). *The utility based non-linear fuzzy AHP optimization model for network selection in heterogeneous wireless networks. In Applied Soft Computing*. Elsevier.

Goyal, S., & Joshi, S. (2003). Networks of collaboration in oligopoly. *Games and Economic Behavior*, *43*(1), 57–85. doi:10.1016/S0899-8256(02)00562-6

Grinnemo, K. J., & Brunstrom, A. (2015). A first study on using MPTCP to reduce latency for cloud based mobile applications. *IEEE Symposium on Computers and Communication (ISCC)*.

Grobe, K., & Elbers, J. P. (2008). PON in adolescence: From TDMA to WDM-PON. *IEEE Communications Magazine, 46*(1), 26–34. doi:10.1109/MCOM.2008.4427227

Grossman, P. J. (2013). Holding fast: The persistence and dominance of gender stereotypes. *Economic Inquiry, 51*(1), 747–763. doi:10.1111/j.1465-7295.2012.00479.x

Guiso, L., Sapienza, P., & Zingales, L. (2018). Time varying risk aversion. *The National Bureau of Economic Research. Journal of Financial Economics, 128*(3), 403–421. doi:10.1016/j.jfineco.2018.02.007

Gunter, Y., Wiegel, B., & Großmann, H. P. (2007). Medium access concept for VANETs based on clustering. In *Vehicular Technology Conference, 2007. VTC-2007 Fall. 2007 IEEE 66th* (pp. 2189-2193). IEEE.

Guo, Quddus, & Tafazolli. (2012). Seamless Handover for LTE Macro-Femto Networks Based on Reactive Data Bicasting. *IEEE Communications Letters, 16*(11).

Guo, C., Guo, Z., Zhang, Q., & Wenwu, Z. (2004). A seamless and proactive end-to-end mobility solution for roaming across heterogeneous wireless networks. *IEEE Journal on Selected Areas in Communications, 22*(5), 834–848. doi:10.1109/JSAC.2004.826921

Guo, S., Xie, H., & Shi, G. (2012, May). Collaborative forwarding and caching in content centric networks. In *International Conference on Research in Networking* (pp. 41-55). Springer.

Gupta & Deswal. (2015). Survey of Handover Schemes in Femtocellular Networks. *International Journal of Science, Engineering and Technology Research, 4*(4).

Guruacharya, S., Niyato, D., Hossain, E., & Kim, D. I. (2010). Hierarchical Competition in Femtocell-Based Cellular Networks. In *2010 IEEE Global Telecommunications Conference GLOBECOM 2010* (pp. 1–5). IEEE. 10.1109/GLOCOM.2010.5683278

Guruprasad, K. R. (2011). Generalized Vornoi partition: A new tool for optimal placement of base stations. In *IEEE International Conference on Advanced Networks and Telecommunication Systems* (pp. 1-3). Bangalore: IEEE.

Gustafsson, E., & Jonsson, A. (2003). Always best connected. *IEEE Wireless Communications Magazine, 10*(1), 49–55. doi:10.1109/MWC.2003.1182111

Gutierrez, L. (2010). Next-generation optical access networks: from TDM to WDM. In *Trends in Telecommunication Technologies*. London, UK: Intechopen Limited. doi:10.5772/8473

Habak, K., Ammar, M., Harras, K. A., & Zegura, E. (2015). Femto Clouds: Leveraging Mobile Devices to Provide Cloud Service at the Edge. In *2015 IEEE 8th International Conference on Cloud Computing* (pp. 9–16). IEEE. Retrieved from https://www.cc.gatech.edu/%7B~%7Dkhabak3/papers/FemtoCloud-CLOUD%7B%25%7D2715.pdf

Habeeb & Qadeer. (2009). Interference Evaluation and MS Controlled Handoff Technique for Femtocell. In *Internet, 2009. AH-ICI 2009. First Asian Himalayas International Conference On* (pp. 1–5). IEEE.

Hafeez, K. A., Zhao, L., Liao, Z., & Ma, B. N. W. (2011). Clustering and OFDMA-based MAC protocol (COMAC) for vehicular ad hoc networks. *EURASIP Journal on Wireless Communications and Networking*, (1): 1–16.

Hajir, M., Langar, R., & Gagnon, F. (2016). Coalitional Games for Joint Co-Tier and Cross-Tier Cooperative Spectrum Sharing in Dense Heterogeneous Networks. *IEEE Access: Practical Innovations, Open Solutions, 4*, 2450–2464. doi:10.1109/ACCESS.2016.2562498

Halgamuge, Vu, Rarnamohanarao, & Zukerman. (2005). Signal-Based Evaluation of Handoff Algorithms. *IEEE Communications Letters, 9*(9), 790–792.

Hammitt, J., & Haninger, K. (2010). Valuing fatal risks to children and adults: Effects of disease, latency, and risk aversion. *Journal of Risk and Uncertainty*, *40*(1), 57–83. doi:10.100711166-009-9086-9

Han, B., Hui, P., Kumar, V. A., Marathe, M. V., Shao, J., & Srinivasan, A. (2012). Mobile data offloading through opportunistic communications and social participation. *IEEE Transactions on Mobile Computing*, *11*(5), 821–834. doi:10.1109/TMC.2011.101

Han, Z., Niyato, D., Saad, W., Baar, T., & Hjrungnes, A. (2012). *Game Theory in Wireless and Communication Networks: Theory, Models, and Applications*. Cambridge University Press.

Harikrishnan, Y., & He, J. (2013). Clustering algorithm based on minimal path loss ratio for vehicular communication. In *Computing, Networking and Communications (ICNC), 2013 International Conference on* (pp. 745-749). IEEE.

Harsanyi, J. C. (2004). Games with Incomplete Information Played by "Bayesian" Players, I–III: Part I. The Basic Model. *Management Science, 50*(12S), 1804–1817. doi:10.1287/mnsc.1040.0270

Hasrouny, H., Samhat, A. E., Bassil, C., & Laouiti, A. (2017). VANet security challenges and solutions: A survey. *Vehicular Communications*, *7*, 7–20. doi:10.1016/j.vehcom.2017.01.002

Hassan, M. Z., Leung, V. C., Hossain, M. J., & Cheng, J. (2017, December). Delay-QoS Aware Adaptive Resource Allocations for Free Space Optical Fronthaul Networks. In *GLOBECOM 2017-2017 IEEE Global Communications Conference* (pp. 1-6). IEEE.

Hasswa, A., Nasser, N., & Hassanein, H. (2005). Generic vertical handoff decision function for heterogeneous wireless. *Second IFIP International Conference on Wireless and Optical Communications Networks, (WOCN)*, 239-243.

He, Q., Wu, D., & Khosla, P. (2004). SORI: a secure and objective reputation-based incentive scheme for ad-hoc networks. In 2004 IEEE Wireless Communications and Networking Conference (IEEE Cat. No.04TH8733) (p. 825–830 Vol.2). IEEE. doi:10.1109/WCNC.2004.1311293

Hou, F., Cai, L. X., Ho, Pin-Han, Shen, X., & Zhang, J. (2009). A cooperative multicast scheduling scheme for multimedia services in IEEE 802.16 networks. *IEEE Trans. on Wireless Comm.*, *8*(3), 1508-1519.

Hou, J., & O'Brien, D. C. (2006). Vertical handover decision making algorithm using fuzzy logic for the integrated Radio-and-OW system. *IEEE Transactions on Wireless Communications*, *5*(1), 176–185. doi:10.1109/TWC.2006.1576541

Hsieh, P.-J., Lin, P.-H., & Lee, Y.-C. (2014). CoPS: Context Prefetching Handover Scheme on 4G Outdoor Small Cell Testbed. In *Modeling and Optimization in Mobile, Ad Hoc, and Wireless Networks (WiOpt), 2014 12th International Symposium On* (pp. 50–56). IEEE.

Hu, Y., Zhu, M., Xia, Y., Chen, K., & Luo, Y. (2012). GARDEN: Generic Addressing and Routing for Data Center Networks. In *Cloud Computing (CLOUD), 2012 IEEE 5th International Conference on* (pp. 107–114). IEEE. 10.1109/CLOUD.2012.9

Huang, J., & Han, Z. (2010). Game {Theory} for {Spectrum} {Sharing}. In Cognitive {Radio} {Networks} (pp. 291–317). CRC Press. Retrieved from doi:10.1201/EBK1420077759-c10

Huang. (2012). A Handover Scheme for TD-SCDMA Femtocell Networks. In *Computer and Information Technology (CIT), 2012 IEEE 12th International Conference On* (pp. 948–952). IEEE.

Huang, C. J., Lin, C. F., Li, C. Y., Lee, C. Y., Chen, H. M., Shen, H. Y., & Chen, I. F. (2011). Service-oriented routing and clustering strategies for vehicle infotainment dissemination. *International Journal of Innovative Computing, Information, & Control*, *7*(3), 1467–1480.

Huang, H. C., Wang, T. Y., & Hsieh, F. M. (2012). Constructing an adaptive mobile learning system for the support of personalized learning and device adaptation. *Procedia: Social and Behavioral Sciences*, *64*, 332–341. doi:10.1016/j.sbspro.2012.11.040

Huang, J. (2013). Economic viability of dynamic spectrum management. In T. Alpcan, H. Boche, M. L. Honig, & H. V. Poor (Eds.), *Mechanisms and Games for Dynamic Spectrum Allocation* (pp. 396–432). Cambridge University Press; doi:10.1017/CBO9781139524421.018

Huang, S., & Safari, M. (2017). Free-Space Optical Communication Impaired by Angular Fluctuations. *IEEE Transactions on Wireless Communications*, *16*(11), 7475–7487. doi:10.1109/TWC.2017.2749219

Hucheng, W., Shanzhi, C., Ming, A., & Yan, S. (2017). Mobility driven network slicing: An enabler of on demand mobility management for 5G. *Journal of China Universities of Posts and Telecommunications*, *24*(4), 16–26. doi:10.1016/S1005-8885(17)60219-X

Hügelschäfer, S., & Achtziger, A. (2014). On confident men and rational women: It's all on your mind (set). *Journal of Economic Psychology*, *41*, 31–44. doi:10.1016/j.joep.2013.04.001

Hung, S.-C., Hsu, H., Lien, S.-Y., & Chen, K.-C. (2015). Architecture Harmonization Between Cloud Radio Access Networks and Fog Networks. *IEEE Access: Practical Innovations, Open Solutions*, *3*, 3019–3034. doi:10.1109/ACCESS.2015.2509638

Hussein, A., Elhajj, I. H., Chehab, A., & Kayssi, A. (2017, May). SDN VANETs in 5G: An architecture for resilient security services. In *Software Defined Systems (SDS), 2017 Fourth International Conference on* (pp. 67-74). IEEE.

Hwang, J., Shin, A., & Yoon, H. (2008). Dynamic Reputation-based Incentive Mechanism Considering Heterogeneous Networks. In *Proceedings of the 3Nd ACM Workshop on Performance Monitoring and Measurement of Heterogeneous Wireless and Wired Networks* (pp. 137–144). New York, NY: ACM. 10.1145/1454630.1454651

IEEE 802.16. (2004). Air interface for fixed broadband wireless access systems. Local and Metropolitan Area Networks, Part 16.

IEEE Standard for Local and Metropolitan Area Networks-Part 16. (2005). Air Interface for Fixed Broadband Wireless Access Systems (Incorporated into IEEE Standard 802.16e-2005 and IEEE Standard 802.16-2004/Cor 1-2005 E). *IEEE Standard P802.16/Cor1/D5.*

IEEE802.11. (2007) Standard for information technology-telecommunications and information exchange between systems-local and metropolitan area networks-specific requirements-part 11: Wireless lan medium access control (mac) and physical layer (phy) specifications. (Revision of IEEE Std 802.11-1999), 1–1184.

IETF. Use Cases and Operational Experience with Multipath TCP. RFC 8041.

Ileri, O., Siun-Chuon Mau, & Mandayam, N. B. (2005). Pricing for enabling forwarding in self-configuring ad hoc networks. *IEEE Journal on Selected Areas in Communications*, *23*(1), 151–162. doi:10.1109/JSAC.2004.837356

INRIX. (n.d.). *INRIX Global Traffic Scorecard.* Retrieved from http://inrix.com/scorecard/

Islam, M. H., Dziong, Z., Sohraby, K., Daneshmand, M. F., & Jana, R. (2012). Capacity-Optimal Relay and Base Station Placement in Wireless Networks. *International Conference on Information Networking*, 358-363.

ITU-R Rec. M.2083-0. (2015). IMT Vision – Framework and overall objectives of the future development of IMT for 2020 and beyond.

ITU-T Telecommunication Standardization Sector. (2009). *Characteristics of a bending-loss insensitive single-mode optical fiber and cable for the access network*. Recommendation G.657.

ITU-T Telecommunication Standardization Sector. (2009). *Characteristics of a single-mode optical fiber and cable*. Recommendation G.652.

ITU-T Telecommunication Standardization Sector. (2010). *Characteristics of a fiber and cable with non-zero dispersion for wideband optical transport*. Recommendation G.656.

Jacobson, V., Smetters, D. K., Thornton, J. D., Plass, M. F., Briggs, N. H., & Braynard, R. L. (2009). Networking Named Content. In *Proceedings of the 5th International Conference on Emerging Networking Experiments and Technologies* (pp. 1–12). New York: ACM. 10.1145/1658939.1658941

Jacobson, V., Smetters, D. K., Thornton, J. D., Plass, M. F., Briggs, N. H., & Braynard, R. L. (2009, December). Networking named content. In *Proceedings of the 5th international conference on Emerging networking experiments and technologies* (pp. 1-12). ACM.

Jang, J.-S. R. (2002). ANFIS: Adaptive-network-based fuzzy inference system. *IEEE Transactions on Systems, Man, and Cybernetics*, *23*(3), 665–685. doi:10.1109/21.256541

Jaramillo, J. J. & Srikant, R. (2007). DARWIN: Distributed and adaptive reputation mechanism for wireless ad-hoc networks. In *Proceedings of the 13th annual ACM international conference on Mobile computing and networking - MobiCom '07* (p. 87). New York: ACM Press. 10.1145/1287853.1287865

Jararweh, Y., Tawalbeh, L., Ababneh, F., Khreishah, A., & Dosari, F. (2014). Scalable Cloudlet-based Mobile Computing Model. *Procedia Computer Science*, *34*, 434–441. doi:10.1016/j.procs.2014.07.051

Javier, S., & Mamadou, T. (2001). *UMTS: services, architecture et WCDMA*. Academic Press.

Jin, J., He, R., Gong, H., Xu, X., & He, C. (2017). Farmers' Risk Preferences in Rural China: Measurements and Determinants. *International Journal of Environmental Research and Public Health*, *14*(7), 713. doi:10.3390/ijerph14070713 PMID:28665332

Jochen, M., & Le, B. J. (2005). Analysis of a robust reputation system for self-organised networks. *European Transactions on Telecommunications*, *16*(5), 375–384. doi:10.1002/ett.1066

Jordan, S. (2017). A critical survey of the literature on broadband data caps. *Telecommunications Policy*, *41*(9), 813–837. doi:10.1016/j.telpol.2017.09.001

Kaleem, F., Mehbodniyay, Z., Yen, K., & Adachi, F. (2012). A dynamic weighting of attributes in heterogeneous wireless networks using fuzzy linguistic variables. *Proceedings of the First IEEE International Conference on Communications in China Wireless Networking and Applications (WNA)*, 698-703

Kam, C., Kompella, S., Nguyen, G. D., Ephremides, A., & Jiang, Z. (2014). Frequency selection and relay placement for energy efficiency in underwater acoustic networks. *IEEE Journal of Oceanic Engineering*, *39*(2), 331–342. doi:10.1109/JOE.2013.2245057

Kani, J., Bourgart, F., Cui, A., Rafel, A., Campbell, M., Davey, R., & Rodrigues, S. (2009). Next-generation PON - Part I: Technology roadmap and general requirements. *IEEE Communications Magazine*, *47*(11), 43–49. doi:10.1109/MCOM.2009.5307465

Karadeniz, S. (2011). Effects of gender and test anxiety on student achievement in mobile based assessment. *Procedia: Social and Behavioral Sciences*, *15*, 3173–3178. doi:10.1016/j.sbspro.2011.04.267

Karagiannis, G., Altintas, O., Ekici, E., Heijenk, G., Jarupan, B., Lin, K., & Weil, T. (2011). Vehicular networking: A survey and tutorial on requirements, architectures, challenges, standards and solutions. *IEEE Communications Surveys and Tutorials*, *13*(4), 584–616. doi:10.1109/SURV.2011.061411.00019

Karam, F. (2012). Performance analysis of ranking for QoS handover algorithm for selection of access network in heterogeneous wireless networks. *21st International Conference on Computer Communications and Networks (ICCCN)*, 1-6.

Karhunen, P., & Ledyaeva, S. (2010). Determinants of entrepreneurial interest and risk tolerance among Russian university students: Empirical study. *Journal of Enterprising Culture*, *18*(3), 229–263. doi:10.1142/S0218495810000574

Katsalis, K., Nikaein, N., Schiller, E., Ksentini, A., & Braun, T. (2017). Network slices toward 5G communications: Slicing the LTE network. *IEEE Communications Magazine*, *55*(8), 146–154. doi:10.1109/MCOM.2017.1600936

Katsaros, K., Xylomenos, G., & Polyzos, G. C. (2011). MultiCache: An overlay architecture for information-centric networking. *Computer Networks*, *55*(4), 936–947. doi:10.1016/j.comnet.2010.12.012

Kayis, O., & Acarman, T. (2007). Clustering formation for inter-vehicle communication. In *Intelligent Transportation Systems Conference, 2007. ITSC 2007. IEEE* (pp. 636-641). IEEE. 10.1109/ITSC.2007.4357779

Kazovsky, L. G. (2011). *Broadband optical access networks*. Wiley Interscience, John Wiley & Sons. doi:10.1002/9780470910931

Kebriaei, H., Maham, B., & Niyato, D. (2015). Double Sided Bandwidth-Auction Game for Cognitive Device-to-Device Communication in Cellular Networks. *IEEE Transactions on Vehicular Technology*, (99), 1–1. doi:10.1109/TVT.2015.2485304

Khalid & Kwak. (2013). Handover Optimization in Femtocell Networks. In *ICT Convergence (ICTC), 2013 International Conference On* (pp. 122–127). IEEE.

Khan, M. A. & Sivrikaya, F. & Albayrak, S. & Mengaly, K. Q. (2009). Auction based interface selection in heterogeneous wireless networks. *Second IFIP Wireless Days*, 1-6.

Khan, A. A., Abolhasan, M., & Ni, W. (2018). An Evolutionary Game Theoretic Approach for Stable and Optimized Clustering in VANETs. *IEEE Transactions on Vehicular Technology*, *67*(5), 4501–4513. doi:10.1109/TVT.2018.2790391

Khan, M. A., Tembine, H., & Vasilakos, A. V. (2012a). Evolutionary coalitional games: Design and challenges in wireless networks. *IEEE Wireless Communications*, *19*(2), 50–56. doi:10.1109/MWC.2012.6189413

Khan, M. A., Tembine, H., & Vasilakos, A. V. (2012b). Game Dynamics and Cost of Learning in Heterogeneous 4G Networks. *IEEE Journal on Selected Areas in Communications*, *30*(1), 198–213. doi:10.1109/JSAC.2012.120118

Khan, M. A., Toseef, U., Marx, S., & Goerg, C. (2010). Auction based interface selection with Media Independent Handover services and flow management. *European Wireless Conference*, 429-436.

Khan, M. A., Toseef, U., Marx, S., & Goerg, C. (2010). Game-Theory Based User Centric Network Selection with Media Independent Handover Services and Flow Management. *Eighth Annual Communication Networks and Services Research Conference*, 248-255. 10.1109/CNSR.2010.40

Khan, Z., Lehtomaki, J. J., DaSilva, L. A., Hossain, E., & Latva-aho, M. (2016). Opportunistic Channel Selection by Cognitive Wireless Nodes Under Imperfect Observations and Limited Memory: A Repeated Game Model. *IEEE Transactions on Mobile Computing*, *15*(1), 173–187. doi:10.1109/TMC.2015.2412940

Khloussy, E., Gelabert, X., & Jiang, Y. (2015). Investigation on MDP-based radio access technology selection in heterogeneous wireless networks. *Computer Networks*, *91*, 57–67. doi:10.1016/j.comnet.2015.08.005

Kiayias, A., Koutsoupias, E., Kyropoulou, M., & Tselekounis, Y. (2016). Blockchain Mining Games. In *Proceedings of the 2016 ACM Conference on Economics and Computation - EC '16* (pp. 365–382). New York: ACM Press. 10.1145/2940716.2940773

Kim & Lee. (2010). Handover in UMTS Networks with Hybrid Access Femtocells. In *Advanced Communication Technology (ICACT), 2010 The 12th International Conference On* (pp. 904–908). IEEE.

Kim, S. (2016). Cognitive hierarchy thinking based behavioral game model for IoT power control algorithm. *Computer Networks, 110*, 79–90. doi:10.1016/j.comnet.2016.09.020

Konrad, M., Koch, W., & Huschke, J. (2007). *Throughput Based Switching Threshold for Multicast and Unicast in Cellular OFDM Networks.* Paper presented at the IEE VTC-2007 Fall Conference.

Kopeinik, S., Lex, E., Seitlinger, P., Albert, D., & Ley, T. (2017, March). Supporting collaborative learning with tag recommendations: a real-world study in an inquiry-based classroom project. In *Proceedings of the Seventh International Learning Analytics & Knowledge Conference* (pp. 409-418). ACM. 10.1145/3027385.3027421

Koponen, T., Chawla, M., Chun, B. G., Ermolinskiy, A., Kim, K. H., Shenker, S., & Stoica, I. (2007, August). A data-oriented (and beyond) network architecture. *Computer Communication Review, 37*(4), 181–192. doi:10.1145/1282427.1282402

Kroll, J. A., Davey, I. C., & Felten, E. W. (2013). The Economics of Bitcoin Mining, or Bitcoin in the Presence of Adversaries. In *Proceedings of WEIS* (p. 21). Academic Press.

Kuklinski, S., & Wolny, G. (2009). Density based clustering algorithm for VANETs. In *Testbeds and Research Infrastructures for the Development of Networks & Communities and Workshops, 2009. TridentCom 2009. 5th International Conference on* (pp. 1-6). IEEE. 10.1109/TRIDENTCOM.2009.4976256

Kumar, N., Chilamkurti, N., & Misra, S. (2015). Bayesian coalition game for the internet of things: An ambient intelligence-based evaluation. *IEEE Communications Magazine, 53*(1), 48–55. doi:10.1109/MCOM.2015.7010515

Kumar, N., Misra, S., Rodrigues, J. J. P. C., & Obaidat, M. S. (2015). Coalition Games for Spatio-Temporal Big Data in Internet of Vehicles Environment: A Comparative Analysis. *IEEE Internet of Things Journal, 2*(4), 310–320. doi:10.1109/JIOT.2015.2388588

Kumar, N., Zeadally, S., Chilamkurti, N., & Vinel, A. (2015). Performance analysis of Bayesian coalition game-based energy-aware virtual machine migration in vehicular mobile cloud. *IEEE Network, 29*(2), 62–69. doi:10.1109/MNET.2015.7064905

Kuo, W. H., Liu, T., & Liao, W. (2007). *Utility-Based Resource Allocation for Layer-Encoded IPTV Multicast in IEEE 802.16 (WiMAX) Wireless Networks.* Paper presented at the IEEE International Conference on Communications (ICC).

Kuo, Y.-L., Wu, E. H.-K., & Chen, G.-H. (2004). Noncooperative admission control for differentiated services in IEEE 802.11 WLANs. In *IEEE Global Telecommunications Conference, 2004. GLOBECOM '04* (Vol. 5, pp. 2981–2986). IEEE. 10.1109/GLOCOM.2004.1378899

Kwon & Cho. (2011). Load Based Cell Selection Algorithm for Faulted Handover in Indoor Femtocell Network. In *Vehicular Technology Conference (VTC Spring), 2011 IEEE 73rd* (pp. 1–5). IEEE.

Kwon, T., Lim, S., Choi, S., & Hong, D. (2010). Optimal Duplex Mode for DF Relay in Terms of the Outage Probability. *IEEE Transactions on Vehicular Technology, 59*(7), 3628–3634. doi:10.1109/TVT.2010.2050503

Lahby & Sekkaki. (2017). Optimal Vertical Handover based on TOPSIS algorithm and Utility function In Heterogeneous Wireless Networks. *Proceedings of the 4th International Symposium on Networks, Computers and Communications (ISNCC)*, 1-5.

Lahby, Attioui, & Sekkaki. (2017). An improved policy for network selection decision based on enhanced TOPSIS and utility function. *Proceedings of the 13th International Wireless Communications ans Mobile Computing Conference (IWCMC)*, 2175-2180.

Lahby, M., Leghris, C., & Adib, A. (2011). A Hybrid Approach for Network Selection in Heterogeneous Multi-Access Environments. *Proceedings of the 4th IFIP International Conference on New Technologies, Mobility and Security (NTMS)*, 1-5.

Lahby, M., Leghris, C., & Adib, A. (2012). A Survey and Comparison Study on Weighting Algorithms for Access Network Selection. *Proceedings of the 9th IFIP Annual Conference on Wireless On-Demand Network Systems and Services (WONS)*, 35-38.

Lahby. (2013a). A Novel Ranking Algorithm Based Network Selection For Heterogeneous Wireless Access. *Journal of Networks*, *8*(2), 263–272.

Lahby. (2013b). An Enhanced-TOPSIS Based Network Selection Technique for Next Generation Wireless Networks. *Proceedings of the 20th International Conference on Telecommunications (ICT)*, 1-5.

Lahby, M., & Sekkaki, A. (2018). A Graph Theory Based Network Selection Algorithm in Heterogeneous Wireless Networks. *9th IFIP International Conference on New Technologies, Mobility and Security (NTMS)*, 1-4.

Lahby, M., Silki, B., & Sekkaki, A. (2015). Survey and comparison of MADM methods for network selection access in heterogeneous networks. *7th IFIP International Conference on New Technologies, Mobility and Security (NTMS)*, 1-6.

Lai, L., & El Gamal, H. (2008). The Water-Filling Game in Fading Multiple-Access Channels. *IEEE Transactions on Information Theory*, *54*(5), 2110–2122. doi:10.1109/TIT.2008.920340

La, Q. D., Quek, T. Q. S., Lee, J., Jin, S., & Zhu, H. (2016). Deceptive Attack and Defense Game in Honeypot-Enabled Networks for the Internet of Things. *IEEE Internet of Things Journal*, *3*(6), 1025–1035. doi:10.1109/JIOT.2016.2547994

Larsson, E. G., Jorswieck, E. A., Lindblom, J., & Mochaourab, R. (2009). Game theory and the flat-fading gaussian interference channel. *IEEE Signal Processing Magazine*, *26*(5), 18–27. doi:10.1109/MSP.2009.933370

Latouche, M., Rauschen, C., Oszabó, O., Creusat, J.-B., & Belmans, W. (2013). *Mobile Data Explosion How Mobile Service Providers Can Monetize the Growth in Mobile Data Through Value-Added Services.* Cisco White Paper: Cisco Internet Business Solutions Group (IBSG).

Lauriola, M., Panno, A., Levin, I. P., & Lejuez, C. W. (2013). Individual differences in risky decision making: A meta-analysis of sensation seeking and impulsivity with the balloon analogue risk task. *Journal of Behavioral Decision Making*. Doi:10.1002/bdm.1784

Lecompte, D., & Gabin, F. (2012). Evolved Multimedia Broadcast/Multicast Service (eMBMS) in LTE-Advanced: Overview and Rel-11 Enhancements. *IEEE Communications Magazine*, *50*(11), 68–74. doi:10.1109/MCOM.2012.6353684

Lee, D., Han, Y., & Hwang, J. (2013). QoS Based Vertical Handoff Decision Algorithm in Heterogeneous Systems. *Sixth Joint IFIP Wireless and Mobile Networking Conference (WMNC)*, 1-6.

Lee, Huang, Lin, Su, & Wang. (2013). Reducing Handover Cost for LTE Femtocell/Macrocell Network. In *Globecom Workshops (GC Wkshps), 2013 IEEE* (pp. 4988–4993). IEEE.

Lee, K. C., Lee, U., & Gerla, M. (2009). Survey of routing protocols in vehicular ad hoc networks. Advances in Vehicular Ad-Hoc Networks: Developments and Challenges, 21.

Lee, Shin, Lim, & Hong. (2010). Effects of Time-to-Trigger Parameter on Handover Performance in SON-Based LTE Systems. In *Communications (APCC), 2010 16th Asia-Pacific Conference On* (pp. 492–496). IEEE.

Lee, C. C., & Steele, R. (1998). Effect of soft and softer handoffs on CDMA system capacity. *IEEE Transactions on Vehicular Technology*, *47*(3), 830–841. doi:10.1109/25.704838

Lee, C.-W., Chuang, M.-C., Meng, C. C., & Sun, Y. S. (2014). Seamless Handover for High-Speed Trains Using Femtocell-Based Multiple Egress Network Interfaces. *IEEE Transactions on Wireless Communications*, *13*(12), 6619–6628. doi:10.1109/TWC.2014.2364179

Lee, D., Choi, J., Kim, J. H., Noh, S. H., Min, S. L., Cho, Y., & Kim, C. S. (2001). LRFU: A spectrum of policies that subsumes the least recently used and least frequently used policies. *IEEE Transactions on Computers*, *50*(12), 1352–1361. doi:10.1109/TC.2001.970573

Lee, H., Son, H., & Lee, S. (2009). Semisoft handover gain analysis over OFDM-based broadband systems. *IEEE Transactions on Vehicular Technology*, *58*(3), 1443–1453. doi:10.1109/TVT.2008.927041

Leitao, F., Ros, R. D. C., & Riu, J. R. (2016). Fixed-mobile convergence towards the 5G era: Convergence 2.0: The past, present and future of FMC standardization. *IEEE Conference on Standards for Communications and Networking (CSCN).*

Leung, V. C. M. (1992). Internetworking Wireless Terminals to Local Area Networks via Radio Bridges. In *Wireless Communications, 1992. Conference Proceedings., 1992 IEEE International Conference on Selected Topics In* (pp. 126–129). IEEE.

Lewenberg, Y., Bachrach, Y., Sompolinsky, Y., Zohar, A., & Rosenschein, J. S. (2015). Bitcoin Mining Pools: A Cooperative Game Theoretic Analysis. In *Proceedings of the 2015 International Conference on Autonomous Agents and Multiagent Systems* (pp. 919–927). Richland, SC: International Foundation for Autonomous Agents and Multiagent Systems. Retrieved from http://dl.acm.org/citation.cfm?id=2772879.2773270

Li, P., Zhang, H., Zhao, B., & Rangarajan, S. (2009). *Scalable Video Multicast in Multi-carrier Wireless Data Systems.* Paper presented at the IEEE International Conference on Network Protocols (ICNP).

Li, Q., Zhu, S., & Cao, G. (2010). Routing in Socially Selfish Delay Tolerant Networks. In 2010 Proceedings IEEE INFOCOM (pp. 1–9). IEEE. doi:10.1109/INFCOM.2010.5462138

Li, C., Xue, Y., Wang, J., Zhang, W., & Li, T. (2018). Edge-Oriented Computing Paradigms: A Survey on Architecture Design and System Management. *ACM Computing Surveys*, *51*(2), 1–34. doi:10.1145/3154815

Lin, S., Ni, W., Tian, H., & Liu, R. P. (2015). An Evolutionary Game Theoretic Framework for Femtocell Radio Resource Management. *IEEE Transactions on Wireless Communications*, *14*(11), 6365–6376. doi:10.1109/TWC.2015.2453170

Lin, X., Andrews, J., Ghosh, A., & Ratasuk, R. (2014). An overview of 3GPP device-to-device proximity services. *IEEE Communications Magazine*, *52*(4), 40–48. doi:10.1109/MCOM.2014.6807945

Liu, C. H., & Wang, L. C. (2015, June). Random cell association and void probability in poisson-distributed cellular networks. In *Communications (ICC), 2015 IEEE International Conference on* (pp. 2816-2821). IEEE.

Liu, H., & Krishnamachari, B. (2006). A price-based reliable routing game in wireless networks. In *Proceeding from the 2006 workshop on Game theory for communications and networks - GameNets '06* (p. 7). New York: ACM Press. 10.1145/1190195.1190201

Liu, Hong, Xue, & Peng. (2010). Conflict Avoidance between Mobility Robustness Optimization and Mobility Load Balancing. In *Global Telecommunications Conference (GLOBECOM 2010), 2010 IEEE* (pp. 1–5). IEEE.

Liu, H., De Foy, X., & Zhang, D. (2012, August). A multi-level DHT routing framework with aggregation. In *Proceedings of the second edition of the ICN workshop on Information-centric networking* (pp. 43-48). ACM. 10.1145/2342488.2342498

Liu, H., Eldarrat, F., Alqahtani, H., Reznik, A., de Foy, X., & Zhang, Y. (2017). Mobile Edge Cloud System: Architectures, Challenges, and Approaches. *IEEE Systems Journal*, 1–14. doi:10.1109/JSYST.2017.2654119

Liu, L., Ma, M., Liu, C., & Shu, Y. (2017). Optimal Relay Node Placement and Flow Allocation in Underwater Acoustic Sensor Networks. *IEEE Transactions on Communications*, *65*(5), 2141–2152. doi:10.1109/TCOMM.2017.2677448

Lobato, W., Rosario, D., Gerla, M., & Villas, L. A. (2017, December). Platoon-based driving protocol based on game theory for multimedia transmission over vanet. In *GLOBECOM 2017-2017 IEEE Global Communications Conference* (pp. 1-6). IEEE.

Lopez, G. P., Montresor, A., Epema, D., Datta, A., Higashino, T., Iamnitchi, A., ... Riviere, E. (2015). Edge-centric computing: Vision and challenges. *Computer Communication Review*, *45*(5), 37–42. doi:10.1145/2831347.2831354

López-Pérez, Valcarce, Ladányi, de la Roche, & Zhang. (2010). Intracell Handover for Interference and Handover Mitigation in OFDMA Two-Tier Macrocell-Femtocell Networks. *EURASIP Journal on Wireless Communications and Networking*, (1).

Louhab, F. E., Bahnasse, A., & Talea, M. (2018). Considering mobile device constraints and context-awareness in adaptive mobile learning for flipped classroom. *Education and Information Technologies*, 1–26.

Louta, M., Zournatzis, P., Kraounakis, S., Sarigiannidis, P., & Demetropoulos, I. (2011). Towards realization of the ABC vision: A comparative survey of Access Network Selection. In *2011 IEEE Symposium on Computers and Communications (ISCC)* (pp. 472–477). IEEE. 10.1109/ISCC.2011.5983882

Low, T. P., Pun, M. O., Hong, Y. W. P., & Kuo, C. C. J. (2009). Optimized opportunistic multicast scheduling (OMS) over wireless cellular networks. *IEEE Transactions on Wireless Communications*, *9*(2), 791–801. doi:10.1109/TWC.2010.02.090387

Lu, R., Lin, X., Zhu, H., & Shen, X. (2009). SPARK: a new VANET-based smart parking scheme for large parking lots. In INFOCOM 2009, IEEE (pp. 1413-1421). IEEE.

Luo, Y., Zhang, W., & Hu, Y. (2010). A new cluster based routing protocol for VANET. In *Networks Security Wireless Communications and Trusted Computing (NSWCTC), 2010 Second International Conference on* (Vol. 1, pp. 176-180). IEEE.

Lu, Y., Zhou, B., Tung, L.-C., Gerla, M., Ramesh, A., & Nagaraja, L. (2013). Energy-efficient content retrieval in mobile cloud. In *Proceedings of the second ACM SIGCOMM workshop on Mobile cloud computing - MCC '13* (p. 21). New York: ACM Press. 10.1145/2491266.2491271

Machado, R., & Tekinay, S. (2008). A survey of game-theoretic approaches in wireless sensor networks. *Computer Networks*, *52*(16), 3047–3061. doi:10.1016/j.gaceta.2008.07.003

Mach, P., & Becvar, Z. (2017). Mobile Edge Computing: A Survey on Architecture and Computation Offloading. *IEEE Communications Surveys and Tutorials*, *19*(3), 1628–1656. doi:10.1109/COMST.2017.2682318

MacKenzie, A. B., & DaSilva, L. A. (2006). Game Theory for Wireless Engineers. *Synthesis Lectures on Communications*, *1*(1), 1–86. doi:10.2200/S00014ED1V01Y200508COM001

Mahloo, M., Jiajia Chen, Wosinska, L., Dixit, A., Lannoo, B., Colle, D., & Mas Machuca, C. (2014). Toward reliable hybrid WDM/TDM passive optical networks. *IEEE Communications Magazine*, *52*(2), S14–S23. doi:10.1109/MCOM.2014.6736740

Mahmud, R., Kotagiri, R., & Buyya, R. (2018). Fog Computing: A Taxonomy, Survey and Future Directions. In B. Di Martino, K.-C. Li, L. T. Yang, & A. Esposito (Eds.), *Internet of Everything: Algorithms, Methodologies, Technologies and Perspectives* (pp. 103–130). Singapore: Springer Singapore. doi:10.1007/978-981-10-5861-5_5

Manzalini, A., & Crespi, N. (2016). An edge operating system enabling anything-as-a-service. *IEEE Communications Magazine*, *54*(3), 62–67. doi:10.1109/MCOM.2016.7432173

Mao, Y., You, C., Zhang, J., Huang, K., & Letaief, K. B. (2017). A Survey on Mobile Edge Computing: The Communication Perspective. *IEEE Communications Surveys and Tutorials*, *19*(4), 2322–2358. doi:10.1109/COMST.2017.2745201

Marina, N., Saad, W., Han, Z., & Hjørungnes, A. (2011). Modeling malicious behavior in cooperative cellular wireless networks. In E. Hossain, D. I. Kim, & V. K. Bhargava (Eds.), *Cooperative Cellular Wireless Networks* (pp. 382–422). Cambridge University Press. doi:10.1017/CBO9780511667008.015

Marinelli, E. E. (2009). *Hyrax: Cloud Computing on Mobile Devices using MapReduce* (MSc thesis). CMU. Retrieved from https://pdfs.semanticscholar.org/8cd2/11cc816952f036ed65a7022adba063486008.pdf

Marsal-Llacuna, M.-L. (2018). Future living framework: Is blockchain the next enabling network? *Technological Forecasting and Social Change*, *128*, 226–234. doi:10.1016/j.techfore.2017.12.005

Martinez, D. A., Feijoo, F., Zayas-Castro, J. L., Levin, S., & Das, T. K. (2018). A strategic gaming model for health information exchange markets. *Health Care Management Science*, *21*(1), 119–130. doi:10.100710729-016-9382-2 PMID:27600378

Martinez, F. J., Toh, C. K., Cano, J. C., Calafate, C. T., & Manzoni, P. (2010). Emergency services in future intelligent transportation systems based on vehicular communication networks. *Intelligent Transportation Systems Magazine, IEEE*, *2*(2), 6–20. doi:10.1109/MITS.2010.938166

Mas Machuca, C., Jiajia Chen, & Wosinska, L. (2012). Cost-efficient protection in TDM PONs. *IEEE Communications Magazine*, *50*(8), 110–117. doi:10.1109/MCOM.2012.6257535

Maslekar, N., Boussedjra, M., Mouzna, J., & Houda, L. (2009). Direction based clustering algorithm for data dissemination in vehicular networks. In *Vehicular Networking Conference (VNC)*, 2009 *IEEE* (pp. 1-6). IEEE.

Maslekar, N., Boussedjra, M., Mouzna, J., & Labiod, H. (2011). A stable clustering algorithm for efficiency applications in VANETs. In *Wireless Communications and Mobile Computing Conference (IWCMC), 2011 7th International* (pp. 1188-1193). IEEE.

Maslekar, N., Mouzna, J., Labiod, H., Devisetty, M., & Pai, M. (2011). Modified C-DRIVE: Clustering based on direction in vehicular environment. In *Intelligent Vehicles Symposium (IV)*, 2011 *IEEE* (pp. 845-850). IEEE.

Massin, R., Le Martret, C. J., & Ciblat, P. (2017). A Coalition Formation Game for Distributed Node Clustering in Mobile Ad Hoc Networks. *IEEE Transactions on Wireless Communications*, *16*(6), 3940–3952. doi:10.1109/TWC.2017.2690419

Mateus, A., & Marinheiro, R. N. (2010). A Media Independent Information Service Integration Architecture for Media Independent Handover. In *2010 Ninth International Conference on Networks* (pp. 173–178). IEEE. 10.1109/ICN.2010.71

Mekikis, P. V., Kartsakli, E., Lalos, A. S., Antonopoulos, A., Alonso, L., & Verikoukis, C. (2015, June). Connectivity of large-scale WSNs in fading environments under different routing mechanisms. In *Communications (ICC), 2015 IEEE International Conference on* (pp. 6553-6558). IEEE.

Milan, F., Jaramillo, J. J., & Srikant, R. (2006). Achieving cooperation in multihop wireless networks of selfish nodes. In *Proceeding from the 2006 workshop on Game theory for communications and networks - GameNets '06* (p. 3). New York: ACM Press. 10.1145/1190195.1190197

Ming, Z., Xu, M., & Wang, D. (2014, August). Age-based cooperative caching in information-centric networking. In *Computer Communication and Networks (ICCCN), 2014 23rd International Conference on* (pp. 1-8). IEEE. 10.1109/ICCCN.2014.6911725

Mingxiao, D., Xiaofeng, M., Zhe, Z., Xiangwei, W., & Qijun, C. (2017). A review on consensus algorithm of blockchain. In *2017 IEEE International Conference on Systems, Man, and Cybernetics (SMC)* (pp. 2567–2572). IEEE. 10.1109/SMC.2017.8123011

Mishra, D. & De, S. (2016, April). Energy harvesting and sustainable M2M communication in 5G mobile technologies. *Internet of Things (IoT) in 5G Mobile Technologies, Springer-Verlag Handbook*, 99–125.

Mishra, D., & Alexandropoulos, G. C. (2018). Transmit Precoding and Receive Power Splitting for Harvested Power Maximization in MIMO SWIPT Systems. *IEEE Transactions on Green Communications and Networking*, *2*(3), 774–786. doi:10.1109/TGCN.2018.2835409

Mishra, D., & De, S. (2017). i^{2}RES: Integrated Information Relay and Energy Supply Assisted RF Harvesting Communication. *IEEE Transactions on Communications*, *65*(3), 1274–1288. doi:10.1109/TCOMM.2016.2637915

Mishra, D., De, S., & Krishnaswamy, D. (2017). Dilemma at RF Energy Harvesting Relay: Downlink Energy Relaying or Uplink Information Transfer? *IEEE Transactions on Wireless Communications*, *16*(8), 4939–4955. doi:10.1109/TWC.2017.2704084

Mohammed, A., Kamal, H., & AbdelWahab, S. (2007). 2G/3G inter-RAT handover performance analysis. *The Second European Conference on Antennas and Propagation*, 1-8.

Molnar, A. (2011). *Cost Efficient Educational Multimedia Delivery* (Doctoral thesis). National College of Ireland. Retrieved August 26, 2013, from http://trap.ncirl.ie/762/

Molnar, A. (2016b). Factors affecting user preference for mobile video quality. In *Multimedia and Expo (ICME), 2016 IEEE International Conference on* (pp. 1-6). IEEE.

Molnar, A., & Muntean, C. H. (2012). Mobile learning: An economic approach. Intelligent and Adaptive Learning Systems: Technology Enhanced Support for Learners and Teachers. IGI Global.

Molnar, A., & Muntean, C. H. (2014). Economical Aspects of Heterogeneous Networks: When Money, Quality, and Context Matter. In Convergence of Broadband, Broadcast, and Cellular Network Technologies (pp. 1-15). IGI Global.

Molnar, A. (2016a). Video quality vs. mobile data billing plans. *Telematics and Informatics*, *33*(1), 227–231. doi:10.1016/j.tele.2015.07.009

Molnar, A., & Muntean, C. H. (2010). Educational content delivery: An experimental study assessing student preference for multimedia content when monetary cost is involved. *10th International Conference on Intelligent Systems Design and Applications*, 871-876.

Molnar, A., & Muntean, C. H. (2013a). Cost oriented adaptive multimedia delivery. *IEEE Transactions on Broadcasting*, *59*(3), 484–499. doi:10.1109/TBC.2013.2244786

Molnar, A., & Muntean, C. H. (2013b). COMEDY: Viewer trade-off between multimedia quality and monetary benefits. *IEEE International Symposium on Broadband Multimedia Systems and Broadcasting*, 1-7.

Molnar, A., & Muntean, C. H. (2015). What can people risk attitude tell us about people preference for multimedia quality? *Information Technology & People*, *28*(2), 383–404. doi:10.1108/ITP-11-2013-0199

Molnar, A., Virseda, J., & Frias-Martinez, V. (2015). Insights from EducaMovil: Involving teachers in creating educational content for mobile learning games. *Journal of Interactive Learning Research*, *26*(2), 209–221.

Molnar, A., & Weerakkody, V. (2016). Clinicians' view of tele-glaucoma. *Health Policy and Technology*, *5*(2), 123–130. doi:10.1016/j.hlpt.2016.02.001

Monserrat, J. F., Calabuig, J., Fernandez-Aguilella, A., & Gomez-Barquero, D. (2012). Joint Delivery of Unicast and E-MBMS Services in LTE Networks. *IEEE Transactions on Broadcasting*, *58*(2), 157–167. doi:10.1109/TBC.2012.2191030

Monteiro, R., Silva, J., Lourenço, J., & Paulino, H. (2015). Decentralized Storage for Networks of Hand-held Devices. In *Proceedings of the 12th EAI International Conference on Mobile and Ubiquitous Systems: Computing, Networking and Services* (pp. 299–300). ACM. 10.4108/eai.22-7-2015.2260263

Moon & Cho. (2010). Novel Handoff Decision Algorithm in Hierarchical Macro/Femto-Cell Networks. In *Wireless Communications and Networking Conference (WCNC), 2010 IEEE* (pp. 1–6). IEEE.

Moon, J.-M., & Cho, D.-H. (2009). Efficient Handoff Algorithm for Inbound Mobility in Hierarchical Macro/Femto Cell Networks. *IEEE Communications Letters*, *13*(10).

Moreira, Lage, Gomes, Radwan, & Rodriguez. (2013). Context-Aware Macro-Femto Handover for Energy Efficiency of Outdoor Users. In *Future Network and Mobile Summit (FutureNetworkSummit), 2013* (pp. 1–10). IEEE.

Mougayar, W. (2016). *The Business Blockchain: Promise, Practice, and Application of the Next Internet Technology*. John Wiley & Sons, Inc.

Moura, J., & Edwards, C. (2015). Future Trends and Challenges for Mobile and Convergent Networks. In A. Yarali (Ed.), *4G & Beyond: The Convergence of Networks, Devices and Services* (pp. 35–80). Nova Science Publishers. Retrieved from https://arxiv.org/abs/1601.06202v1

Moura, J., & Hutchison, D. (2018). Survey of Game Theory and Future Trends with Application to Emerging Wireless Data Communication Networks. *ArXiv E-Prints*, 46. Retrieved from https://arxiv.org/abs/1704.00323v1

Moura, J., & Edwards, C. (2016). Efficient access of mobile flows to heterogeneous networks under flash crowds. *Computer Networks*, *107*(2), 163–177. doi:10.1016/j.comnet.2016.04.010

Moustafa, H., Senouci, S. M., & Jerbi, M. (2009). Introduction to Vehicular Networks. In H. Moustafa & Y. Zhang (Eds.), *Vehicular Networks: Techniques, Standards and Applications*. CRC Press. doi:10.1201/9781420085723

Mugume, E., So, D. K., & Alsusa, E. (2015, December). Energy efficient deployment of dense heterogeneous cellular networks. In *Global Communications Conference (GLOBECOM)*, 2015 *IEEE* (pp. 1-6). IEEE.

Munjal, M., & Singh, N. P. (2018). Utility aware network selection in small cell. *Wireless Networks*, 1–14. doi:10.100711276-018-1676-5

Naeem, B., Ngah, R., Hashim, S. Z. M., & Maqbool, W. (2014). Vertical handover decision using history-based communication graph for heterogeneous networks. *Open Systems (ICOS), 2014 IEEE Conference on. IEEE*, 7-11.

Nakamoto, S. (2008). *Bitcoin: A peer-to-peer electronic cash system*. Retrieved July 27, 2018, from https://bitcoin.org/bitcoin.pdf

Nakamura, H. (2013). *NG-PON2 technologies*. NTT Access Network Service Systems Laboratories, NTT Corporation. doi:10.1364/NFOEC.2013.NTh4F.5

Nasser, N., Hasswa, A., & Hassanein, H. (2006). Handoffs in fourth generation heterogeneous networks. *IEEE Communications Magazine*, *44*(10), 96–103. doi:10.1109/MCOM.2006.1710420

Nawab, F., Agrawal, D., & El Abbadi, A. (2018). Nomadic Datacenters at the Network Edge: Data Management Challenges for the Cloud with Mobile Infrastructure. In M. H. Böhlen, R. Pichler, N. May, E. Rahm, S.-H. Wu, & K. Hose (Eds.), *Proceedings of the 21th International Conference on Extending Database Technology, {EDBT} 2018, Vienna, Austria, March 26-29, 2018* (pp. 497–500). OpenProceedings.org. 10.5441/002/edbt.2018.56

Nazir, M., Bennis, M., Ghaboosi, K., MacKenzie, A. B., & Latva-aho, M. (2010). Learning based mechanisms for interference mitigation in self-organized femtocell networks. In *2010 Conference Record of the Forty Fourth Asilomar Conference on Signals, Systems and Computers* (pp. 1886–1890). Academic Press. 10.1109/ACSSC.2010.5757866

Nikaein, N., Schiller, E., Favraud, R., Katsalis, K., Stavropoulos, D., Alyafawi, I., ... Korakis, T. (2015). Network store: Exploring slicing in future 5G networks. In *Proceedings of 10th International Workshop on Mobility in the Evolving Internet Architecture* (pp. 8-13). Paris, France: ACM.

Niyato, D., & Hossain, E. (2007). Radio resource management games in wireless networks: An approach to bandwidth allocation and admission control for polling service in IEEE 802.16. *IEEE Wireless Communications*, *14*(1), 27–35. doi:10.1109/MWC.2007.314548

Niyato, D., & Hossain, E. (2008). A Noncooperative Game-Theoretic Framework for Radio Resource Management in 4G Heterogeneous Wireless Access Networks. *IEEE Transactions on Mobile Computing*, *7*(3), 332–345. doi:10.1109/TMC.2007.70727

Nouri, H., Uysal, M., & Panayirci, E. (2016). Information theoretical performance analysis and optimisation of cooperative underwater acoustic communication systems. *IET Communications*, *10*(7), 812–823. doi:10.1049/iet-com.2015.0640

Ntemos, K., Plata-Chaves, J., Kolokotronis, N., Kalouptsidis, N., & Moonen, M. (2018). Secure Information Sharing in Adversarial Adaptive Diffusion Networks. *IEEE Transactions on Signal and Information Processing over Networks*, *4*(1), 111–124. doi:10.1109/TSIPN.2017.2787910

Nurmi, P., & Nurmi, P. (2006). A Bayesian framework for online reputation systems. In *Advanced International Conference on Telecommunications and International Conference on Internet and Web Applications and Services. IEEE.*

O'Sullivan, A., Sheffrin, S., & Perez, S. (2003). *Economics : principles in action*. Prentice Hall.

Oeldorf-Hirsch, A., Donner, J., & Cutrell, E. (2012). How bad is good enough? Exploring mobile video quality trade-offs for bandwidth-constrained consumers. *ACM 7th Nordic Conference on Human-Computer Interaction: Making Sense Through Design*, 49-58.

Ofcom. (2015). *Measuring mobile broadband performance in the UK. 4G and 3G network performance*. White paper.

Ofcom. (2017). *UK Home broadband performance: The performance of fixed line broadband delivered to UK residential consumers*. White paper.

Oliveira Filho, J. L. D. E., & Madeira, E. (2011). A mechanism for vertical handover based on SAW using IEEE 802.21. In *Mobile Networks and Management* (pp. 96–108). Springer. doi:10.1007/978-3-642-21444-8_10

ONF TR-526. (2016). Applying SDN architecture to 5G slicing.

Ong, E. H., & Khan, J. (2008). Dynamic access network selection with QoS parameters estimation: A step closer to ABC. In *IEEE Vehicular Technology Conference* (pp. 2671-2676). Singapore: IEEE.

Ordonez-Lucena, J., Ameigeiras, P., Lopez, D., Ramos-Munoz, J. J., Lorca, J., & Folgueira, J. (2017). Network slicing for 5G with SDN/NFV: Concepts, architectures, and challenges. *IEEE Communications Magazine*, *55*(5), 80–87. doi:10.1109/MCOM.2017.1600935

Ormond, O., Murphy, J., & Muntean, G. M. (2006). Utility-based intelligent network selection in beyond 3G systems. *IEEE International Conference on Communications- ICC'06,* 4, 1831-1836.

Ormond, O., Murphy, J., & Muntean, G. M. (2006). Utility-based intelligent network selection in beyond 3G systems. *IEEE International Conference on Communications*, 1831-1836.

Page, M., Molina, M., & Jones, G. (2013). The mobile economy. *GSMA*. Retrieved June 18, 2018, from http://www.atkearney.com/documents/10192/760890/The_Mobile_Economy_2013.pdf/

Panta, R. K., Jana, R., Cheng, F., Chen, Y.-F. R., & Vaishampayan, V. A. (2013). Phoenix: Storage Using an Autonomous Mobile Infrastructure. *IEEE Transactions on Parallel and Distributed Systems*, *24*(9), 1863–1873. doi:10.1109/TPDS.2013.84

Pantisano, F., Bennis, M., Saad, W., Debbah, M., & Latva-Aho, M. (2014). Improving Macrocell-Small Cell Coexistence through Adaptive Interference Draining. *IEEE Transactions on Wireless Communications*, *13*(2), 942–955. doi:10.1109/TWC.2013.120613.130617

Peng, J., Hong, P., & Xue, K. (2015). Energy-aware cellular deployment strategy under coverage performance constraints. *IEEE Transactions on Wireless Communications*, *14*(1), 69–80. doi:10.1109/TWC.2014.2329858

Peng, M., Liu, Y., Wei, D., Wang, W., & Chen, H.-H. (2011). Hierarchical cooperative relay based heterogeneous networks. *IEEE Wireless Communications Magazine*, *18*(3), 48–56. doi:10.1109/MWC.2011.5876500

Pervaiz, H., & Bigham, J. (2009). Game theoretical formulation of network selection in competing wireless networks: An analytic hierarchy process model. *Third International Conference on Next Generation Mobile Applications, Services and Technologies*, 292–297.

Phuoc Tran & Boukhatem. (2009). An utility based interface selection scheme for multi-homed mobile terminals. *Personal, Indoor and Mobile Radio Communications*, 767-772.

Prasad, G., Mishra, D., & Hossain, A. (2018). Joint Optimal Design for Outage Minimization in DF Relay-assisted Underwater Acoustic Networks. *IEEE Communications Letters*, *12*(8), 1724–1727. doi:10.1109/LCOMM.2018.2837019

Prasad, G., Mishra, D., & Hossain, A. (2018). Joint Optimization Framework for Operational Cost Minimization in Green Coverage-Constrained Wireless Networks. *IEEE Transactions on Green Communications and Networking*, *2*(3), 693–706. doi:10.1109/TGCN.2018.2828092

Prat, J. (2008). Next-generation FTTH passive optical networks: Research towards unlimited bandwidth access. New York: Springer Science+Business Media. doi:10.1007/978-1-4020-8470-6

Prat, J. (2016). Roadmap on optical communications: Access networks. *Journal of Optics*, *18*(6), 27–28.

Prince, K., Gibbon, T. B., Rodes, R., Hviid, E., Mikkelsen, C. I., Neumeyr, C., ... Monroy, I. T. (2012). GigaWaM - Next-generation WDM-PON enabling gigabit per-user data bandwidth. *Journal of Lightwave Technology*, *30*(10), 1444–1454. doi:10.1109/JLT.2012.2185039

Psaras, I., Chai, W. K., & Pavlou, G. (2012, August). Probabilistic in-network caching for information-centric networks. In *Proceedings of the second edition of the ICN workshop on Information-centric networking* (pp. 55-60). ACM. 10.1145/2342488.2342501

Qiu & Wang. (2012). A Novel Handover Strategy in LTE Networks with Hybrid Access Femtocells. In *Wireless Communications, Networking and Mobile Computing (WiCOM), 2012 8th International Conference On* (pp. 1–5). IEEE.

Racz, Temesvary, & Reider. (2007). Handover Performance in 3GPP Long Term Evolution (LTE) Systems. In *Mobile and Wireless Communications Summit, 2007. 16th IST* (pp. 1–5). IEEE.

Ragheb, A. M., & Fathallah, H. (2011). Performance analysis of next generation PON architectures. *International Conference on High-capacity Optical Networks and Emerging Technologies*, 8, 339-345. 10.1109/HONET.2011.6149766

Raj, H., Saroiu, S., Wolman, A., & Padhye, J. (2013). Splitting the bill for mobile data with SIMlets. *ACM Workshop on Mobile Computing Systems and Applications*, 1-6.

Raj, M., Narayan, A., Datta, S., Das, S. K., & Pathak, J. K. (2010). Fixed mobile convergence: Challenges and solutions. *IEEE Communications Magazine*, *48*(12), 26–34. doi:10.1109/MCOM.2010.5673069

Rakha, H., & Kamalanathsharma, R. K. (2011). Eco-driving at signalized intersections using V2I communication. In *Intelligent Transportation Systems (ITSC), 2011 14th International IEEE Conference on* (pp. 341-346). IEEE.

Ramakrishnan, B., Rajesh, R. S., & Shaji, R. S. (2011). CBVANET: A cluster based vehicular adhoc network model for simple highway communication. *J. Advanced Networking and Applications*, *2*(04), 755–761.

Ramanathan, Subramanian (2002). Self-Organizing Mobile Wireless Station Network.

Ramaswami, R., & Sivarajan, K. N. (2001). *Optical networks - A practical perspective.* San Francisco, CA: Morgan Kaufmann Publishers.

Rangan, S., & Madan, R. (2012). Belief propagation methods for intercell interference coordination in femtocell networks. *IEEE Journal on Selected Areas in Communications*, *30*(3), 631–640. doi:10.1109/JSAC.2012.120412

Rappaport, T. S. (1996). Wireless Communications: Principles and Practice (vol. 2). Prentice Hall.

Rath & Panwar. (2012). Fast Handover in Cellular Networks with Femtocells. In *Communications (ICC), 2012 IEEE International Conference On* (pp. 2752–2757). IEEE.

Raza, M. R., Fiorani, M., Rostami, A., Öhlen, P., Wosinska, L., & Monti, P. (2018). Dynamic slicing approach for multi-tenant 5G transport networks. *Journal of Optical Communications and Networking*, *10*(1), A77–A90. doi:10.1364/JOCN.10.000A77

Razzaque, M. A., Dobson, S., & Nixon, P. (2010). *Enhancement of Self-organisation in Wireless Networking through a Cross-layer Approach*. Berlin: Springer.

Reardon, M. (2010, February 9). *Cisco Predicts Wireless Data Explosion* [Press Release]. Cisco.

Reumerman, H. J., Roggero, M., & Ruffini, M. (2005). The application-based clustering concept and requirements for intervehicle networks. *Communications Magazine, IEEE*, *43*(4), 108–113. doi:10.1109/MCOM.2005.1421913

RFC 793. (1981). *Transmission Control Protocol.* Retrieved June 19, 2018, from ietf: http://www.ietf.org/rfc/rfc793.txt

Richard, A., Dadlani, A., & Kim, A. (2013). Multicast scheduling and resource allocation algorithms for OFDMA-based systems: A survey. *IEEE Communications Surveys and Tutorials*, *15*(1), 240–254. doi:10.1109/SURV.2012.013012.00074

Rich, E. (1979). User modelling via stereotypes. *Cognitive Science*, *3*(4), 329–354. doi:10.120715516709cog0304_3

Róka, R. (2003). The utilization of the DWDM/CWDM combination in the metro/access networks. *Joint 1st Workshop on Mobile Future & Symposium on Trends in Communications, 1*, 160-162.

Róka, R. (2010). The utilization of the HPON Network Configurator at designing of passive optical networks, *International Conference on Telecommunication and Signal Processing*, 33, 444-448.

Róka, R. (2011). The extension of the HPON Network Configurator at designing of NG-PON networks, *International Conference on Telecommunication and Signal Processing*, 34, 79-84.

Róka, R. (2012). The designing of NG-PON networks using the HPON Network Configurator. *Journal of Communication and Computer*, *9*(6), 669–678.

Róka, R. (2013a). The analysis of SUCCESS HPON networks using the HPON Network Configurator. *Advances in Electrical and Electronic Engineering*, *1*(5-SI), 420–425.

Róka, R. (2013b). The analysis of SARDANA HPON networks using the HPON Network Configurator. *Advances in Electrical and Electronic Engineering*, *11*(6), 522–527. doi:10.15598/aeee.v11i6.866

Róka, R. (2014). Broadband NG-PON networks and their designing using the HPON Network Configurator. In *Convergence of Broadband, Broadcast, and Cellular Network Technologies*. Hershey, PA: IGI Global. doi:10.4018/978-1-4666-5978-0.ch012

Róka, R. (2015). *Hybrid PON networks - features, architectures and configuration*. Saarbrücken, Germany: LAP Lambert Academic Publishing.

Róka, R. (2016). Optimization of traffic protection schemes for utilization in hybrid passive optical networks. *International Journal of Application or Innovation in Engineering & Management*, *5*(9), 107–116.

Rosensweig, E. J., & Kurose, J. (2009, April). Breadcrumbs: Efficient, best-effort content location in cache networks. In INFOCOM 2009, IEEE (pp. 2631-2635). IEEE.

Rossini, G., & Rossi, D. (2012, September). A dive into the caching performance of content centric networking. In *Computer Aided Modeling and Design of Communication Links and Networks (CAMAD), 2012 IEEE 17th International Workshop on* (pp. 105-109). IEEE.

Rost, P., Mannweiler, C., Michalopoulos, D. S., Sartori, C., Sciancalepore, V., Sastry, N., ... Bakker, H. (2017). Network slicing to enable scalability and flexibility in 5G mobile networks. *IEEE Communications Magazine*, *55*(5), 72–79. doi:10.1109/MCOM.2017.1600920

Roy, A., Shin, J., & Saxena, N. (2012). Multi-Objective Handover in LTE Macro/Femto-Cell Networks. *Journal of Communications and Networks (Seoul)*, *14*(5), 578–587. doi:10.1109/JCN.2012.00016

Saad, W., Han, Z., & Hjørungnes, A. (2011). Coalitional games for cooperative cellular wireless networks. In E. Hossain, D. I. Kim, & V. K. Bhargava (Eds.), *Cooperative Cellular Wireless Networks* (pp. 347–381). Cambridge University Press. doi:10.1017/CBO9780511667008.014

Saad, W., Han, Z., Poor, H. V., & Basar, T. (2012). Game-Theoretic Methods for the Smart Grid: An Overview of Microgrid Systems, Demand-Side Management, and Smart Grid Communications. *IEEE Signal Processing Magazine*, *29*(5), 86–105. doi:10.1109/MSP.2012.2186410

Sachs, J., Wikstrom, G., Dudda, T., Baldemair, R., & Kittichokechai, K. (2018). 5G radio network design for ultra-reliable low-latency communication. *IEEE Network*, *32*(2), 24–31. doi:10.1109/MNET.2018.1700232

Saidi, H., Tourki, K., & Hamdi, N. (2016, November). Performance analysis of PSK modulation in DF dual-hop hybrid RF/FSO system over gamma gamma channel. In *Signal, Image, Video and Communications (ISIVC), International Symposium on* (pp. 213-216). IEEE.

Salem & Reguiga. (2013). Context-Aware Handoff Selection Algorithm in Femtocell Networks. In *Wireless Communications and Mobile Computing Conference (IWCMC), 2013 9th International* (pp. 243–247). IEEE.

Samdanis, K., Costa-Perez, X., & Sciancalepore, V. (2016). From network sharing to multi-tenancy: The 5G network slice broker. *IEEE Communications Magazine*, *54*(7), 32–39. doi:10.1109/MCOM.2016.7514161

Samdanis, K., Leitao, F., Oechsner, S., Riu, J. R. I., Ros, R. D. C., & Fabregas, G. (2017). From Interworking to Hybrid Access Systems and the Road toward the Next-Generation of Fixed-Mobile Convergence. *IEEE Communications Standards Magazine*, *1*(1), 36–43. doi:10.1109/MCOMSTD.2017.1600536ST

Santi, P. (2012). *Mobility Models for Next Generation Wireless Networks*. Sussex, UK: Wiley. doi:10.1002/9781118344774

Sapirshtein, A., Sompolinsky, Y., & Zohar, A. (2017). *Optimal Selfish Mining Strategies in Bitcoin*. Berlin: Springer; doi:10.1007/978-3-662-54970-4_30

Saraydar, C. U., Mandayam, N. B., & Goodman, D. J. (2002). Efficient power control via pricing in wireless data networks. *IEEE Transactions on Communications*, *50*(2), 291–303. doi:10.1109/26.983324

Schneir, J. R., & Xiong, Y. (2014). Cost analysis of network sharing in FTTH/PONs. *IEEE Communications Magazine*, *52*(8), 126–134. doi:10.1109/MCOM.2014.6871680

Scutari, G., Palomar, D. P., & Barbarossa, S. (2008). Competitive Design of Multiuser MIMO Systems Based on Game Theory: A Unified View. *IEEE Journal on Selected Areas in Communications*, *26*(7), 1089–1103. doi:10.1109/JSAC.2008.080907

Sedjelmaci, H., Senouci, S. M., & Al-Bahri, M. (2016). A lightweight anomaly detection technique for low-resource IoT devices: A game-theoretic methodology. In *2016 IEEE International Conference on Communications (ICC)* (pp. 1–6). IEEE. 10.1109/ICC.2016.7510811

Selimi, M., Kabbinale, A. R., Ali, A., Navarro, L., & Sathiaseelan, A. (2018). Towards Blockchain-enabled Wireless Mesh Networks. In *Proceedings of the 1st Workshop on Cryptocurrencies and Blockchains for Distributed Systems - CryBlock'18* (pp. 13–18). New York: ACM Press. 10.1145/3211933.3211936

Semasinghe, P., Hossain, E., & Zhu, K. (2015). An Evolutionary Game for Distributed Resource Allocation in Self-Organizing Small Cells. *IEEE Transactions on Mobile Computing*, *14*(2), 274–287. doi:10.1109/TMC.2014.2318700

Semasinghe, P., Maghsudi, S., & Hossain, E. (2017). Game Theoretic Mechanisms for Resource Management in Massive Wireless IoT Systems. *IEEE Communications Magazine*, *55*(2), 121–127. doi:10.1109/MCOM.2017.1600568CM

Seo, H., Kwack, S., & Lee, B. G. (2007). *Channel Structuring and Subchannel Allocation for Efficient Multicast and Unicast Services Integration in Wireless OFDM Systems*. Paper presented at the IEEE Globecom Conference.

Sgora, A. (2010). An access network selection algorithm for heterogeneous wireless environments. *Proceedings of The IEEE symposium on Computers and Communications*, 890-892. 10.1109/ISCC.2010.5546556

Sgora, A., Chatzimisios, P., & Vergados, D. D. (2010). *Mobile Lightweight Wireless Systems: Access Network Selection in a Heterogeneous Environment Using the AHP and Fuzzy TOPSIS Methods*. Berlin: Springer.

Sharangi, S., Krishnamurti, R., & Hefeeda, M. (2011). Energy-Efficient Multicasting of Scalable Video Streams Over WiMAX Networks. *IEEE Transactions on Multimedia*, *13*(1), 102–115. doi:10.1109/TMM.2010.2076799

Shastry, N., & Adve, R. (2006). Stimulating Cooperative Diversity in Wireless Ad Hoc Networks through Pricing. In *2006 IEEE International Conference on Communications* (pp. 3747–3752). IEEE. 10.1109/ICC.2006.255655

Shbat & Tuzlukov. (2012). Handover Technique between Femtocells in LTE Network Using Collaborative Approach. In *Communications (APCC), 2012 18th Asia-Pacific Conference On* (pp. 61–66). IEEE.

Shea, C., Hassanabadi, B., & Valaee, S. (2009). Mobility-based clustering in VANETs using affinity propagation. In *Global Telecommunications Conference, 2009. GLOBECOM 2009. IEEE* (pp. 1-6). IEEE.

Shen, J., Yi, N., Wu, B., Jiang, W., & Xiang, H. (2009). *A greedy-based resource allocation algorithm for multicast and unicast services in OFDM system*. Paper presented at the WCSP Conference.

Shen, D. M. (2010) The QoE-oriented Heterogeneous Network Selection Based on Fuzzy AHP Methodology. *The Fourth International Conference on Mobile Ubiquitous Computing, Systems, Services and Technologies*, 275-280.

Sheng-mei, L., Su, P., & Ming-hai, X. (2010). An improved topsis vertical handoff algorithm for heterogeneous wireless networks. *Proceedings of the 12th IEEE International Conference on Communication Technology (ICCT),* 750-754.

Sheng-Mei, L. (2010). A simple additive weighting vertical handoff algorithm based on SINR and AHP for heterogeneous wireless networks. *Proceedings of IEEE International Conference on Intelligent Computation Technology and Automation (ICICTA)*, 1, 347-350.

Shen, S., Yue, G., Cao, Q., & Yu, F. (2011). A Survey of Game Theory in Wireless Sensor Networks Security. *JNW*, *6*(3), 521–532. doi:10.4304/jnw.6.3.521-532

Shen, W., & Zeng, Q.-A. (2007). Cost-function-based network selection strategy in integrated wireless and mobile networks. *21st International Conference on Advanced Information Networking and Applications Workshops*, 2, 314-319. 10.1109/AINAW.2007.136

Shi, H.-Y., Wang, W.-L., Kwok, N.-M., & Chen, S.-Y. (2012). Game Theory for Wireless Sensor Networks: A Survey. *Sensors (Basel)*, *12*(7), 9055–9097. doi:10.3390120709055 PMID:23012533

Shivshankar, S., & Jamalipour, A. (2015). An Evolutionary Game Theory-Based Approach to Cooperation in VANETs Under Different Network Conditions. *IEEE Transactions on Vehicular Technology*, *64*(5), 2015–2022. doi:10.1109/TVT.2014.2334655

Shoaib, M., Song, W. C., & Kim, K. H. (2012). Cluster based data aggregation in vehicular adhoc network. In *Communication Technologies for Vehicles* (pp. 91–102). Springer Berlin Heidelberg. doi:10.1007/978-3-642-29667-3_8

Silva, J., Marinheiro, R., Moura, J., & Almeida, J. (2013). Differentiated classes of service and flow management using an hybrid broker. *ACEEE International Journal on Communication, 4*(2), 13–22.

Singhal, C., Kumar, S., De, S., Panwar, N., Tonde, R., & De, P. (2014). Class-based shared resource allocation for cell-edge users in OFDMA networks. *IEEE Transactions on Mobile Computing*, *13*(1), 48–60. doi:10.1109/TMC.2012.210

Singoria, Oliveira, & Agrawal. (2011). Reducing Unnecessary Handovers: Call Admission Control Mechanism between WiMAX and Femtocells. In *Global Telecommunications Conference (GLOBECOM 2011), 2011 IEEE* (pp. 1–5). IEEE.

Sivaraj, R., Gopalakrishna, A. K., Chandra, M. G., & Balamuralidhar, P. (2011). QoS-enabled group communication in integrated VANET-LTE heterogeneous wireless networks. In *Wireless and Mobile Computing, Networking and Communications (WiMob), 2011 IEEE 7th International Conference on* (pp. 17-24). IEEE.

Skraparlis, D., Sakarellos, V., Panagopoulos, A., & Kanellopoulos, J. (2009). Outage Performance Analysis of Cooperative Diversity with MRC and SC in Correlated Lognormal Channels. *EURASIP Journal on Wireless Communications and Networking*, *707839*(1), 707839. doi:10.1155/2009/707839

Smolorz, S. (2009). Next generation access networks: PIEMAN and beyond. *International Conference on Photonics in Switching*, 1-4.

Solomon, M. R., Bamossy, G., Askegaard, S., & Hogg, M. K. (2010). *Consumer Behaviour: A European Perspective*. Financial Times/Prentice Hall.

Son, I. K., & Mao, S. (2017). A survey of free space optical networks. *Digital Communications and Networks, 3*(2), 67-77.

Srinivasa, S., & Haenggi, M. (2010). Distance distributions in finite uniformly random networks: Theory and applications. *IEEE Transactions on Vehicular Technology*, *59*(2), 940–949. doi:10.1109/TVT.2009.2035044

Srirama, S., & Paniagua, C. (2013). Mobile web service provisioning and discovery in android days. In *Conference on Mobile Services* (pp. 15–22). Washington, DC: IEEE Computer Society. 10.1109/MS.2013.14

Srivastava, V., Neel, J., Mackenzie, A. B., Menon, R., Dasilva, L. A., Hicks, J. E., ... Gilles, R. P. (2005). Using game theory to analyze wireless ad hoc networks. *IEEE Communications Surveys and Tutorials*, *7*(4), 46–56. doi:10.1109/COMST.2005.1593279

Stojanovic, M. (2007). On the relationship between capacity and distance in an underwater acoustic communication channel. *Mobile Computing and Communications Review*, *11*(4), 34–43. doi:10.1145/1347364.1347373

Su, G., Li, L., Lin, X., & Wang, H. (2014, July). On the optimal small cell deployment for energy-efficient heterogeneous cellular networks. In *Proceedings Ubiquitous and Future Networks (ICUFN),* 2014 Sixth International Conf (pp. 172-175). IEEE.

Su, H., & Zhang, X. (2007). Clustering-Based Multichannel MAC Protocols for QoS Provisionings Over Vehicular. *Vehicular Technology. IEEE Transactions on*, *56*(6), 3309–3323.

Suh, D., Ko, H., & Pack, S. (2016). Efficiency analysis of WiFi offloading techniques. *IEEE Transactions on Vehicular Technology*, *65*(5), 3813–3817. doi:10.1109/TVT.2015.2437325

Sun, Z. G., Yu, H., & Zhu, Y. J. (2017). A Superimposed Relaying Strategy and Power Allocation for Outdoor Visible Light Communications. *IEEE Access: Practical Innovations, Open Solutions*, *5*, 9555–9561. doi:10.1109/ACCESS.2017.2700880

Tal, I., & Muntean, G. M. (2012). User-oriented cluster-based solution for multimedia content delivery over VANETs. In *Broadband Multimedia Systems and Broadcasting (BMSB), 2012 IEEE International Symposium on* (pp. 1-5). IEEE.

Tal, I., & Muntean, G. M. (2014). Towards smarter cities and roads: A survey of clustering algorithms in vanets. In Convergence of Broadband, Broadcast, and Cellular Network Technologies (pp. 16-50). IGI Global.

Taleb, T., Samdanis, K., Mada, B., Flinck, H., Dutta, S., & Sabella, D. (2017). On Multi-Access Edge Computing: A Survey of the Emerging 5G Network Edge Architecture & Orchestration. *IEEE Communications Surveys & Tutorials,* (99), 1–1. doi:10.1109/COMST.2017.2705720

Taleb, T., & Benslimane, A. (2010). Design guidelines for a network architecture integrating vanet with 3g & beyond networks. In *Global Telecommunications Conference (GLOBECOM 2010)*, 2010 *IEEE* (pp. 1-5). IEEE.

Taleb, T., Corici, M., Parada, C., Jamakovic, A., Ruffino, S., Karagiannis, G., & Magedanz, T. (2015). EASE: EPC as a service to ease mobile core network deployment over cloud. *IEEE Network*, *29*(2), 78–88. doi:10.1109/MNET.2015.7064907

Tal, I., Ciubotaru, B., & Muntean, G. M. (2016). Vehicular-Communications-Based Speed Advisory System for Electric Bicycles. *IEEE Transactions on Vehicular Technology*, *65*(6), 4129–4143. doi:10.1109/TVT.2015.2442338

Tal, I., & Muntean, G. M. (2013a). User-oriented Fuzzy Logic-based Clustering Scheme for Vehicular Ad-hoc Networks. In *Vehicular Technology Conference*, (pp. 1-5). IEEE.

Tal, I., & Muntean, G. M. (2017). Towards Reasoning Vehicles: A Survey of Fuzzy Logic-Based Solutions in Vehicular Networks. *ACM Computing Surveys*, *50*(6), 80. doi:10.1145/3125640

Tal, I., & Muntean, G.-M. (2013b). V2X communication-based power saving strategy for electric bicycles. In *Global Telecommunications Conference Workshops, 2013. GlobeCom Workshops 2013.* (pp 1-6).

Tal, I., Tianhua, Z., & Muntean, G.-M. (2013). On the Potential of V2X Communications in Helping Electric Bicycles Save Energy. *Vehicular Networking Conference*, 1-4.

Talli, G. (2007). Integrated metro and access network: PIEMAN. *European Conference on Networks and Optical Communications*, 12, 493-500.

Tanaka, K., Agata, A., & Horiuchi, Y. (2010). IEEE 802.3av 10G-EPON standardization and its research and development status. *Journal of Lightwave Technology*, *28*(4), 651–661. doi:10.1109/JLT.2009.2038722

Tang, H., Hong, P., & Xue, K. (2013). HeNB-Aided Virtual-Handover for Range Expansion in LTE Femtocell Networks. *Journal of Communications and Networks (Seoul)*, *15*(3), 312–320. doi:10.1109/JCN.2013.000054

Tawil, Pujolle, & Demerjian. (2008). Distributed Handoff Decision Scheme Using MIH Function for the Fourth Generation Wireless Networks. In *Information and Communication Technologies: From Theory to Applications, 2008. ICTTA 2008. 3rd International Conference On* (pp. 1–6). IEEE.

Taylor, J. W. (1974). The role of risk in consumer behaviour. *Journal of Marketing*, *38*(2), 54–60. doi:10.2307/1250198

Taylor, P. D., & Jonker, L. B. (1978). Evolutionary stable strategies and game dynamics. *Mathematical Biosciences*, *40*(1–2), 145–156. doi:10.1016/0025-5564(78)90077-9

Team, T. M. (2012). *Mininet: An Instant Virtual Network on Your Laptop (or Other PC)*. Retrieved from http://www.mininet.org

Tehrani, M. N., Uysal, M., & Yanikomeroglu, H. (2014). Device-to-device communication in 5G cellular networks: Challenges, solutions, and future directions. *IEEE Communications Magazine*, *52*(5), 86–92. doi:10.1109/MCOM.2014.6815897

Tembine, H., Altman, E., El-Azouzi, R., & Hayel, Y. (2010). Evolutionary Games in Wireless Networks. *IEEE Transactions on Systems, Man, and Cybernetics. Part B, Cybernetics*, *40*(3), 634–646. doi:10.1109/TSMCB.2009.2034631 PMID:19963703

Teshima, S., Ohta, T., Kohno, E., & Kakuda, Y. (2011). A data transfer scheme using autonomous clustering in VANETs environment. In *Autonomous Decentralized Systems (ISADS), 2011 10th International Symposium on* (pp. 477-482). IEEE.

The European Commission. (n.d.). *The 5G Infrastructure Public Private Partnership: The Next Generation of Communication Networks and Services*. Retrieved from https://5g-ppp.eu/wp-content/uploads/2015/02/5G-Vision-Brochure-v1.pdf

Third Generation Partnership Project. (2012, September). TS 36.300, Evolved Universal Terrestrial Radio Access (E-UTRA) and Evolved Universal Terrestrial Radio Access Network (E-UTRAN), Release 11.

Tian, D., Wang, Y., Lu, G., & Yu, G. (2010). A VANETs routing algorithm based on Euclidean distance clustering. In *Future Computer and Communication (ICFCC), 2010 2nd International Conference on* (Vol. 1, pp. V1-183). IEEE.

Tielert, T., Rieger, D., Hartenstein, H., Luz, R., & Hausberger, S. (2012). Can V2X communication help electric vehicles save energy? In *ITS Telecommunications (ITST), 2012 12th International Conference on* (pp. 232-237). IEEE.

Tordera, E., Masip-Bruin, X., Garcia-Almiana, J., Jukan, A., Ren, G.-J., Zhu, J., & Farre, J. (2016). *What is a Fog Node? A Tutorial on Current Concepts towards a Common Definition*. Retrieved from http://arxiv.org/abs/1611.09193

Trestian, R., Ormond, O., & Muntean, G.-M. (2011). Reputation-based network selection mechanism using game theory. *Physical Communication*, *4*(3), 156–171. doi:10.1016/j.phycom.2011.06.004

Trestian, R., Ormond, O., & Muntean, G.-M. (2012). Game theory-based network selection: Solutions and challenges. *IEEE Communications Surveys and Tutorials*, *14*(4), 1212–1231. doi:10.1109/SURV.2012.010912.00081

Trivedi, H., Veeraraghavan, P., Loke, S., Desai, A., & Singh, J. (2011). SmartVANET: The Case for a Cross-Layer Vehicular Network Architecture. In *Advanced Information Networking and Applications (WAINA), 2011 IEEE Workshops of International Conference on* (pp. 362-368). IEEE.

Tung, L.-C., Mena, J., Gerla, M., & Sommer, C. (2013). A Cluster Based Architecture for Intersection Collision Avoidance Using Heterogeneous Networks. In *Annual Mediterranean Ad Hoc Networking Workshop (Med-Hoc-Net 2013)*. IFIP/IEEE.

Ucar, S., Ergen, S. C., & Ozkasap, O. (2013). VMaSC: Vehicular multi-hop algorithm for stable clustering in Vehicular Ad Hoc Networks. In *Wireless Communications and Networking Conference (WCNC)*, 2013 *IEEE* (pp. 2381-2386). IEEE.

Ungureanu, V. (2018). Strategic Form Games on Digraphs. In *Pareto-Nash-Stackelberg Game and Control Theory. Smart Innovation, Systems and Technologies* (pp. 167–197). Cham: Springer; doi:10.1007/978-3-319-75151-1_8

Uzcategui, R. A., & Acosta-Marum, G. (2009). Wave: A tutorial. *IEEE Communications Magazine*, *47*(5), 126–133. doi:10.1109/MCOM.2009.4939288

van Lingen, F., Yannuzzi, M., Jain, A., Irons-Mclean, R., Lluch, O., Carrera, D., ... Rodríguez, J. P. (2017). The unavoidable convergence of NFV, 5G, and fog: A model-driven approach to bridge cloud and edge. *IEEE Communications Magazine*, *55*(8), 28–35. doi:10.1109/MCOM.2017.1600907

Varela, G., Paz-Lopez, A., Becerra, J. A., & Duro, R. J. (2015). Autonomous evaluation of interaction resource adequateness for ambient intelligence scenarios. In *International Conference on Ubiquitous Computing and Ambient Intelligence* (pp. 174-186). Springer.

Vassaki, S., Panagopoulos, A. D., Constantinou, P., & Vázquez-Castro, M. A. (2010). Market-Based Bandwidth Allocation for Broadband Satellite Communication Networks. In *2010 Second International Conference on Advances in Satellite and Space Communications* (pp. 110–115). IEEE. 10.1109/SPACOMM.2010.22

Vasudeva, K., López-Pérez, D., & Güven, I. (2017). Analysis of Handover Failures in Heterogeneous Networks With Fading. *IEEE Transactions on Vehicular Technology*, *66*(7), 6060–6074. doi:10.1109/TVT.2016.2640310

Vedhavathy, T. R., & Manikandan, M. S. K. (2018). Triple referee incentive mechanism for secure mobile adhoc networks. *Cluster Computing*, 1–10. doi:10.100710586-017-1631-8

Verenzuela, D., Björnson, E., & Sanguinetti, L. (2016, May). Optimal design of wireless networks for broadband access with minimum power consumption. In *Communications (ICC), 2016 IEEE International Conference on* (pp. 1-6). IEEE.

Vermesan, O., Bröring, A., Tragos, E., Serrano, M., Bacciu, D., Chessa, S. … Bahr, R. (2017). Internet of robotic things : converging sensing/actuating, hypoconnectivity, artificial intelligence and IoT Platforms. In Cognitive Hyperconnected Digital Transformation : Internet of Things Intelligence Evolution (pp. 97–155). Academic Press.

Vetter, P., Suvakovic, D., Chow, H., Anthapadmanabhan, P., Kanonakis, K., Lee, K., ... Lannoo, B. (2014). Energy-efficiency improvements for optical access. *IEEE Communications Magazine*, *52*(4), 136–144. doi:10.1109/MCOM.2014.6807958

Vodopivec, S., Bester, J., & Kos, A. (2012). A survey on clustering algorithms for vehicular ad-hoc networks. In *Telecommunications and Signal Processing (TSP), 2012 35th International Conference on* (pp. 52-56). IEEE.

Vrzic & Fong. (2004). *Semi-Distributed Scheduling Scheme for the Reverse Link of Wireless Systems*. Academic Press.

Vuong, Thinh, Doudane, & Nazim. (2008). On utility models for access network selection in wireless heterogeneous networks. *IEEE Network Operations and Management Symposium (NOMS)*, 144-151.

Wahab, O. A., Otrok, H., & Mourad, A. (2013). VANET QoS-OLSR: QoS-based clustering protocol for Vehicular Ad hoc Networks. *Computer Communications*, *36*(13), 1422–1435. doi:10.1016/j.comcom.2013.07.003

Wang, Katz, & Giese. (1999). Policy-Enabled Handoffs across Heterogeneous Wireless Networks. In *Mobile Computing Systems and Applications, 1999. Proceedings. WMCSA'99. Second IEEE Workshop On* (pp. 51–60). IEEE.

Wang, Li, Yu, & Qiu. (2012). Automated Handover Optimization Mechanism for LTE Femtocells. In *Communications and Networking in China (CHINACOM), 2012 7th International ICST Conference On* (pp. 601–605). IEEE.

Wang, B., Wu, Y., & Liu, K. J. R. (2010). Game Theory for Cognitive Radio Networks: An Overview. *Computer Networks*, *54*(14), 2537–2561. doi:10.1016/j.comnet.2010.04.004

Wang, B., Wu, Y., Liu, K. J. R., & Clancy, T. C. (2011). An anti-jamming stochastic game for cognitive radio networks. *IEEE Journal on Selected Areas in Communications*, *29*(4), 877–889. doi:10.1109/JSAC.2011.110418

Wang, J. M., Zhang, J., & Bensaou, B. (2013, August). Intra-AS cooperative caching for content-centric networks. In *Proceedings of the 3rd ACM SIGCOMM workshop on Information-centric networking* (pp. 61-66). ACM. 10.1145/2491224.2491234

Wang, L., & Kuo, G.-S. (2013). Mathematical modeling for network selection in heterogeneous wireless networks-A tutorial. *IEEE Communications Surveys and Tutorials*, *15*(1), 271–292. doi:10.1109/SURV.2012.010912.00044

Wang, S., Zhang, X., Zhang, Y., Wang, L., Yang, J., & Wang, W. (2017). A Survey on Mobile Edge Networks: Convergence of Computing, Caching and Communications. *IEEE Access: Practical Innovations, Open Solutions*, *5*, 6757–6779. doi:10.1109/ACCESS.2017.2685434

Wang, Z., Liu, L., Zhou, M., & Ansari, N. (2008). A position-based clustering technique for ad hoc intervehicle communication. *Systems, Man, and Cybernetics, Part C: Applications and Reviews. IEEE Transactions on*, *38*(2), 201–208.

Wanner, L. (2004). *Introduction to Clustering Techniques*. International Union of Local Authorities.

Watanabe, E. H., Menasche, D. S., de Souza e Silva, E., & Leao, R. M. M. (2008). Modeling Resource Sharing Dynamics of VoIP Users over a WLAN Using a Game-Theoretic Approach. In *IEEE INFOCOM 2008 - The 27th Conference on Computer Communications* (pp. 915–923). IEEE. 10.1109/INFOCOM.2008.144

Watanabe, E. H., Nenasche, D. S., de Souza, S., & Leao, R. M. M. (2008). Modeling Resource Sharing Dynamics of VoIP Users over a WLAN Using a Game-Theoretic Approach. *IEEE International Conference on Computer Communications*.

Weerakkody, V., Molnar, A., Irani, Z., & El-Haddadeh, R. (2013). A research proposition for using high definition video in emergency medical services. *Health Policy and Technology*, *2*(3), 131–138. doi:10.1016/j.hlpt.2013.04.001

Wen, Y. F. (2009). Heterogeneous Base Station Placement for Wireless Networks. *IEEE Mobile WiMAX Symposium*, 87-92.

Wolny, G. (2008, October). Modified DMAC clustering algorithm for vanets. In *Systems and Networks Communications, 2008. ICSNC'08. 3rd International Conference on* (pp. 268-273). IEEE. 10.1109/ICSNC.2008.28

Wongchokprasitti, C., Peltonen, J., Ruotsalo, T., Bandyopadhyay, P., Jacucci, G., & Brusilovsky, P. (2015). User model in a box: Cross-system user model transfer for resolving cold start problems. In *International Conference on User Modeling, Adaptation, and Personalization* (pp. 289-301). Springer. 10.1007/978-3-319-20267-9_24

Wu, C., Sun, X., & Zhang, T. (2012). *A power-saving scheduling algorithm for mixed multicast and unicast traffic in MBSFN.* Paper presented at the ComComAp Conference.

Wu, Chu, & Fang. (2013). The Periodic Scan and Velocity Decision Handover Scheme for next Generation Femtocell/ Macrocell Overlay Networks. In *ICT Convergence (ICTC), 2013 International Conference On* (pp. 201–206). IEEE.

Wu. (2011). A New Handover Strategy between Femtocell and Macrocell for LTE-Based Network. In *Ubi-Media Computing (U-Media), 2011 4th International Conference On* (pp. 203–208). IEEE.

Wu, J., Yuen, C., Cheng, B., Shang, Y., & Chen, J. (2015). Goodput-Aware Load Distribution for Real-Time Traffic over Multipath Networks. *IEEE Transactions on Parallel and Distributed Systems, 26*(8), 2286–2299. doi:10.1109/ TPDS.2014.2347031

Xenakis, Passas, Merakos, & Verikoukis. (2013). Energy-Efficient and Interference-Aware Handover Decision for the LTE-Advanced Femtocell Network. In *Communications (ICC), 2013 IEEE International Conference On* (pp. 2464–2468). IEEE.

Xiao, Y., Chen, K. C., Yuen, C., Han, Z., & DaSilva, L. A. (2015). A Bayesian Overlapping Coalition Formation Game for Device-to-Device Spectrum Sharing in Cellular Networks. *IEEE Transactions on Wireless Communications, 14*(7), 4034–4051. doi:10.1109/TWC.2015.2416178

Yafang, W., Huimin, C., & Jinyan, Z. (2010). Network access selection algorithm based on the Analytic Hierarehy Process and Gray Relation Analysis. *Proc New Trends in Information Science and Service Science.*

Yang, Q., Lim, A., & Agrawal, P. (2008). Connectivity aware routing in vehicular networks. In *Wireless Communications and Networking Conference, 2008. WCNC 2008. IEEE* (pp. 2218-2223). IEEE. 10.1109/WCNC.2008.392

Yan, X., Mani, N., & Sekercioglu, Y. A. (2008). A travelling distance prediction based method to minimize unnecessary handovers from cellular networks to WLANs. *IEEE Communications Letters, 12*(1), 14–16. doi:10.1109/ LCOMM.2008.071430

Yan, Y., Huang, J., & Wang, J. (2013). Dynamic Bargaining for Relay-Based Cooperative Spectrum Sharing. *IEEE Journal on Selected Areas in Communications, 31*(8), 1480–1493. doi:10.1109/JSAC.2013.130812

Yap, K.-K., Huang, T.-Y., Kobayashi, M., Yiakoumis, Y., McKeown, N., Katti, S., & Parulkar, G. (2012). Making use of all the networks around us. In *Proceedings of the 2012 ACM SIGCOMM workshop on Cellular networks: operations, challenges, and future design - CellNet '12* (p. 19). New York: ACM Press. 10.1145/2342468.2342474

Ydenberg, A., Heir, N., & Gill, B. (2018, January). Security, SDN, and VANET technology of driver-less cars. In *Computing and Communication Workshop and Conference (CCWC), 2018 IEEE 8th Annual* (pp. 313-316). IEEE.

Yeh, Talwar, Lee, & Kim. (2008). WiMAX Femtocells: A Perspective on Network Architecture, Capacity, and Coverage. *IEEE Communications Magazine, 46*(10).

Ye, H., Liang, L., Li, G. Y., Kim, J., Lu, L., & Wu, M. (2018). Machine Learning for Vehicular Networks: Recent Advances and Application Examples. *IEEE Vehicular Technology Magazine, 13*(2), 94–101. doi:10.1109/MVT.2018.2811185

Yeh, S., Talwar, S., Wu, G., Himayat, N., & Johnsson, K. (2011). Capacity and coverage enhancement in heterogeneous networks. *IEEE Wireless Communications, 18*(3), 32–38. doi:10.1109/MWC.2011.5876498

Yin, C., Ogata, H., Tabata, Y., & Yano, Y. (2010). JAPELAS2: Japanese polite expressions learning assisting system in ubiquitous environments. *International Journal of Mobile Learning and Organisation, 4*(2), 214–234. doi:10.1504/ IJMLO.2010.032637

Yi, S., Li, C., & Li, Q. (2015). A Survey of Fog Computing: Concepts, Applications and Issues. In *Proceedings of the 2015 Workshop on Mobile Big Data - Mobidata '15* (pp. 37–42). New York: ACM Press. 10.1145/2757384.2757397

Yu, J. Y., & Chong, P. H. J. (2005). A survey of clustering schemes for mobile ad hoc networks. *IEEE Communications Surveys and Tutorials*, *7*(1-4), 32–48. doi:10.1109/COMST.2005.1423333

Yusof, Zainali, Nasir, & Ya'Acob. (2014). Handover Adaptation for Load Balancing Scheme in Femtocell Long Term Evolution (LTE) Network. In *Control and System Graduate Research Colloquium (ICSGRC), 2014 IEEE 5th* (pp. 242–246). IEEE.

Zaydoun, Z. Y., & Mahmud, S. M. (2009). Toward Strongly Connected Clustering Structure in Vehicular Ad Hoc Networks. In *Conference on Vehicular Technology* (pp. 1-5). IEEE.

Zeydan, E., Tan, A. S., Mester, Y., & Özcan, G. (2016, April). Integration and management of Wi-Fi offloading in service provider infrastructures. In *Network Operations and Management Symposium (NOMS), 2016 IEEE/IFIP* (pp. 1153-1156). IEEE. 10.1109/NOMS.2016.7502979

Zhang, D., Ileri, O., & Mandayam, N. (2008). Bandwidth exchange as an incentive for relaying. In *2008 42nd Annual Conference on Information Sciences and Systems* (pp. 749–754). IEEE. 10.1109/CISS.2008.4558621

Zhang, Wu, & Zhang. (2005). Optimization and Performance Analysis of Handover Process in Micro-Cell Systems. In *Wireless Communications, Networking and Mobile Computing, 2005. Proceedings. 2005 International Conference On* (pp. 1047–1051). IEEE.

Zhang, Y., Lou, W., & Fang, Y. (2004). SIP: a secure incentive protocol against selfishness in mobile ad hoc networks. In 2004 IEEE Wireless Communications and Networking Conference (IEEE Cat. No.04TH8733) (pp. 1679–1684). IEEE. doi:10.1109/WCNC.2004.1311805

Zhang, H., Liu, N., Chu, X., Long, K., Aghvami, A. H., & Leung, V. C. M. (2017). Network slicing based 5G and future mobile networks: Mobility, resource management, and challenges. *IEEE Communications Magazine*, *55*(8), 138–145. doi:10.1109/MCOM.2017.1600940

Zhang, J., & De la Roche, G. (2011). *Femtocells: Technologies and Deployment*. John Wiley & Sons.

Zhang, L., Afanasyev, A., Burke, J., Jacobson, V., Claffy, K., Crowley, P., ... Zhang, B. (2014). Named Data Networking. *Computer Communication Review*, *44*(3), 66–73. doi:10.1145/2656877.2656887

Zhang, W. (2004). Handover decision using fuzzy MADM in heterogeneous networks. *IEEE Wireless Communications and Networking Conference*, 2, 653-658.

Zhang, X., & Cao, Y. (2017). A Cooperation-Driven ICN-based Caching Scheme for Mobile Content chunk Delivery at RAN. *Wireless Communications and Mobile Computing Conference (IWCMC), 13th International.*

Zhang, X., Wang, N., Vassilakis, V. G., & Howarth, M. P. (2015). A distributed in-network caching scheme for P2P-like content chunk delivery. *Computer Networks*, *91*, 577–592. doi:10.1016/j.comnet.2015.08.035

Zhang, Y., & Guizani, M. (2011). *Game theory for wireless communications and networking*. CRC Press.

Zhang, Z., Long, K., Vasilakos, A. V., & Hanzo, L. (2016). Full-Duplex Wireless Communications: Challenges, Solutions, and Future Research Directions. *Proceedings of the IEEE*, *104*(7), 1369–1409. doi:10.1109/JPROC.2015.2497203

Zheng, G., Liu, S., & Qi, X. (2012). Clustering routing algorithm of wireless sensor networks based on Bayesian game. *Journal of Systems Engineering and Electronics*, *23*(1), 154–159. doi:10.1109/JSEE.2012.00019

Zhong, S., Chen, J., & Yang, Y. R. (2003). Sprite: a simple, cheat-proof, credit-based system for mobile ad-hoc networks. In *IEEE INFOCOM 2003. Twenty-second Annual Joint Conference of the IEEE Computer and Communications Societies (IEEE Cat. No.03CH37428)* (Vol. 3, pp. 1987–1997). IEEE. 10.1109/INFCOM.2003.1209220

Zhou, Hu, Mao, Agrawal, & Reddy. (2013). Cell Association and Handover Management in Femtocell Networks. In *Wireless Communications and Networking Conference (WCNC), 2013 IEEE* (pp. 661–666). IEEE.

Zhu, K., Niyato, D., & Wang, P. (2010). Network Selection in Heterogeneous Wireless Networks: Evolution with Incomplete Information. In 2010 IEEE Wireless Communication and Networking Conference (pp. 1–6). IEEE. doi:10.1109/WCNC.2010.5506371

Zhu, K., & Hossain, E. (2015). Joint Mode Selection and Spectrum Partitioning for Device-to-Device Communication: A Dynamic Stackelberg Game. *IEEE Transactions on Wireless Communications*, *14*(3), 1406–1420. doi:10.1109/TWC.2014.2366136

Zhu, K., Niyato, D., & Wang, P. (2010). Network Selection in Heterogeneous Wireless Networks: Evolution with Incomplete Information. *IEEE Wireless Communications and Networking Conference*, 1-6.

Zoubir, A., & Boashash, B. (1998). The bootstrap and its application in signal processing. *IEEE Signal Processing Magazine*, *15*(1), 56–76. doi:10.1109/79.647043

Zou, L., Trestian, R., & Muntean, G. M. (2018). E 3 DOAS: Balancing QoE and Energy-Saving for Multi-Device Adaptation in Future Mobile Wireless Video Delivery. *IEEE Transactions on Broadcasting*, *64*(1), 26–40. doi:10.1109/TBC.2017.2722221

About the Contributors

Ramona Trestian received the B. Eng. degree in Telecommunications from the Electronics, Telecommunications and the Technology of Information Department, Technical University of Cluj-Napoca, Romania in 2007, and the Ph.D. degree from the School of Electronic Engineering, Dublin City University (DCU), Dublin, Ireland in 2012 for her research in adaptive multimedia systems and network selection mechanisms. She was with Dublin City University as an IBM/Irish Research Council Exascale Postdoctoral Researcher, from December 2011 to August 2013. She is currently a Senior Lecturer with the Design Engineering and Mathematics Department, Faculty of Science and Technology, Middlesex University, London, UK. She has published in prestigious international conferences and journals and has three edited books. Her research interests include mobile and wireless communications, multimedia streaming, user quality of experience, handover and network selection strategies, and software-defined networks. She is a reviewer for international journals and conferences and a member of the IEEE Young Professionals, IEEE Communications Society and IEEE Broadcast Technology Society.

Gabriel-Miro Muntean is an Associate Professor with the School of Electronic Engineering, Dublin City University (DCU), Ireland, and the Co-Director of the DCU Performance Engineering Laboratory. He has published over 300 papers in top-level international journals and conferences, authored three books and 17 book chapters, and edited six additional books. His research interests include quality, performance, and energy saving issues related to multimedia and multiple sensorial media delivery, technology-enhanced learning, and other data communications over heterogeneous networks. He is an Associate Editor of the IEEE Transactions on Broadcasting, the Multimedia Communications Area Editor of the IEEE Communications Surveys and Tutorials, and a Reviewer for important international journals, conferences, and funding agencies. He is a Project Coordinator for the EU-funded project NEWTON http://www.newtonproject.eu.

***

Ayaz Ahmad is currently serving as Assistant Professor in the Department of Electrical and Computer Engineering, COMSATS University Islamabad - Wah Canpus, Pakistan. He obtained his MS and Ph.D. degrees both in Telecommunication from Ecole Superieure d'Electricite (Supelec), Gif-sur-Yvette, France, in 2008 and 2011, respectively. He obtained his B.Sc. in Electrical Engineering from University of Engineering and Technology, Peshawar, Pakistan in 2006. From 2006 to 2007, he was a faculty member at the Department of Electrical Engineering, FAST-NUCES, Peshawar, Pakistan. He has several years of research experience and has authored/coauthored several scientific publications in various refereed international journals and conferences. He has also published several book chapters and is the leading co-editor of the book entitled, "Smart Grid as a Solution for Renewable and Efficient Energy" published in 2016. He is currently an Associate Editor with IEEE Access, and Springer Human-centric Computing and Information Sciences. He has also served as the lead guest editor of a special issue on optimization for emerging wireless networks and co-guest editor of a special issue on energy harvesting both for IEEE Access. He is regularly serving as TPC member for several international conferences including IEEE GLOBECOM, IEEE ICC, and IEEE PIMRC, and as reviewer for several renowned international journals. He is Senior Member of IEEE. His research interests include resource allocation in wireless communication systems, energy management in smart grid, and application of optimization methods to engineering problems.

Raheel Ahmed is pursuing his master degree in Electrical Engineering from Department of Electrical Engineering, COMSATS University Islamabad, Wah campus.

Giuseppe Araniti received the Laurea and Ph.D. degrees in electronic engineering from the University Mediterranea of Reggio Calabria, Italy, in 2000 and 2004, respectively. He is currently an Assistant Professor of telecommunications with the University Mediterranea of Reggio Calabria. His research activities are conducted by cooperating with several national and international research groups, from academia and industries, and the results of such researches are reported in more than 130 scientific publications. His major area of research is on 5G networks and it includes personal communications systems, enhanced wireless and satellite systems, traffic and radio resource management, multicast and broadcast services, device-to-device and machine-type communications. He is the Vice-Chair of the IEEE BTS Italian Chapter and the Special Interest Group on Social Behavior Driven Cognitive Radio Networks - IEEE Comsoc.

Bojan Bakmaz received his B.Sc., M.Sc. and Ph.D. degrees in telecommunications traffic and networks from the University of Belgrade in 2004, 2007, and 2011 respectively. He is an Associate Professor at the Faculty of Transport and Traffic Engineering, University of Belgrade, Serbia. His research interests include the field of information distribution, handover algorithms in heterogeneous wireless networks, and teletraffic theory.

Miodrag Bakmaz received his B.Sc., M.Sc. and Ph.D. degrees in electrical engineering from the University of Belgrade in 1975, 1980 and 1983, respectively. He is a Full Professor at the Faculty of Transport and Traffic Engineering, University of Belgrade, Serbia. His areas of expertise include teletraffic theory, telecommunications networks, and switching techniques.

Zoran Bojkovic (Life Senior Member IEEE) received the BSc., MSc., and Ph.D. degrees, all from the Faculty of Electrical Engineering, University of Belgrade, Serbia. He is a Full Professor of Electrical Engineering, University of Belgrade. His areas of expertise include information theory and coding, signal processing, wireless multimedia communications. He has coauthored 6 International Books as well as 7 International Books Chapters. Also, he has published extensively in peer-reviewed journals.

Yue Cao received the PhD degree from the Institute for Communication Systems (ICS), at University of Surrey, Guildford, UK in 2013. He was a Research Fellow at the ICS until September 2016, and Lecturer in Department of Computer and Information Sciences, at Northumbria University, Newcastle upon Tyne, UK until July 2017, and the Senior Lecturer since August 2017. His research interests focus on DTNs, E Mobility, QoS/QoE in 5G. He is the Associate Editor of IEEE Access.

Massimo Condoluci received the B.S. and M.S. degrees in telecommunications engineering in 2008 and 2011, respectively, from the University Mediterranea of Reggio Calabria, Italy. From the same university, he received the Ph.D. degree in information technology in 2016 with focus on random access optimization for machine-type communications over 4G. From 2016 to 2017, he was a Research Associate at the Centre for Telecommunications Research (CTR), King's College London, UK, where he worked on H2020 (VirtuWind, 5GCAR) and industry (Ericsson, BT) projects with focus on fixed-mobile convergence and network optimization. Since 2018, he is with Ericsson Research, Kista, Sweden. He serves as Editor for Mobile Information Systems and Wireless Communications and Mobile Computing. His current research interests include 5G network and protocol design, V2X, QoS management, and machine-type communications.

Swades De (S'02-M'04-SM'14) received his BTech in radiophysics and electronics from University of Calcutta, India, in 1993, MTech in optoelectronics and optical communication from IIT Delhi, India, in 1998, and PhD in electrical engineering from State University of New York at Buffalo, NY, USA, in 2004. He is currently a Professor with the Department of Electrical Engineering, IIT Delhi. From 2004 to 2007, he was a tenure-track Assistant Professor with the Department of Electrical and Computer Engineering, New Jersey Institute of Technology, NJ, USA. He was an ERCIM Researcher with ISTI-CNR, Pisa, Italy, in 2004, and has nearly five years industry experience in India on telecom hardware and software development, from 1993 to 1997, and in 1999. His research interests are in communication networks, with emphasis on performance modeling and analysis. Current directions include energy harvesting sensor networks, broadband wireless access and routing, cognitive/white space access networks, smart grid, and IoT communications. Dr. De currently serves as a Senior Editor of IEEE Communications Letters, and an Associate Editor of the IEEE Transactions on Vehicular Technology, IEEE Wireless Communications Letters, IEEE Networking Letters, and IETE Technical Review Journal.

Mischa Dohler is full Professor in Wireless Communications at Kings College London, driving cross-disciplinary research and innovation in technology, sciences and arts. He is a Fellow of the IEEE, the Royal Academy of Engineering, the Royal Society of Arts (RSA), the Institution of Engineering and Technology (IET); and a Distinguished Member of Harvard Square Leaders Excellence. He is a serial entrepreneur; composer & pianist with 5 albums on Spotify/iTunes; and fluent in 6 languages. He acts as policy advisor on issues related to digital, skills and education. He has had ample coverage by national and international press and media. He is a frequent keynote, panel and tutorial speaker, and has received numerous awards. He has pioneered several research fields, contributed to numerous wireless broadband, IoT/M2M and cyber security standards, holds a dozen patents, organized and chaired numerous conferences, was the Editor-in-Chief of two journals, has more than 200 highly-cited publications, and authored several books. He was the Director of the Centre for Telecommunications Research at Kings from 2014-2018. He is the Cofounder of the Smart Cities pioneering company Worldsensing, where he was the CTO from 2008-2014. He also worked as a Senior Researcher at Orange/France Telecom from 2005-2008.

Ayoub Essouiri is a male graduate student and his main research interests are: wireless communication and networking.

Ashraf Hossain earned the Ph.D degree in Electronics & Electrical Communication Engg., from Indian Institute of Technology (IIT), Kharagpur, India in 2011. He received the M.Tech. and B.Tech. degrees in Radio Physics & Electronics from the Institute of Radio Physics & Electronics, University of Calcutta, Kolkata, India in 2004 and 2002, respectively. He served Dept. of Electronics & Communication Engg., Aliah University, Kolkata and Haldia Institute of Technology as Asst. Professor. He is currently working as an Assistant Professor in the Dept. of Electronics & Communication Engg., National Institute of Technology (NIT), Silchar, Assam, India. His research interests include wireless sensor network, communication theory and systems. He is a Senior Member of IEEE and Member of IE (India).

Junhuai Li received the B.S. degree in electrical automation from Xi'an University of Technology of China, Xi'an in 1992, M.S. degree in computer application technology from Xi'an University of Technology of China, Xi'an in 1999, and Ph.D. degree in computer software and theory from Northwest University of China, Xi'an in 2002. He is currently a professor with School of Computer Science and Engineering, Xi'an University of Technology, China. His research interests include Internet of Things technology and network computing.

Toktam Mahmoodi is Associate Professor in Telecommunications and Director of the Centre for Telecommunications Research (CTR) in King's College London. Previously, she was visiting research scientist at F5 Networks, research associate in Imperial College London, Mobile VCE researcher, and telecom R&D engineer. Toktam is and has been involved in number of European FP7 and H2020 projects on 5G mobile networks, vehicular communications, and Software-defined Networking. She is executive editor of Transactions on Emerging Telecommunications Technologies, series guest editor of JSAC on network softwarization, and area editor of IEEE Communications Standard Magazine on management & orchestration. Her research focuses on the areas of mobile and cloud networking, and includes ultra-low latency networking, network virtualisation, mobility management, network modelling and optimisation. She also works on applications of mobile communications in the healthcare, smart cities, industrial networks and intelligent transportation.

Rui Marinheiro is a tenured Assistant Professor, at ISCTE—Lisbon University Institute, Portugal and a researcher at Instituto de Telecomunicações. He obtained a Ph.D. in the field of Multimedia Information Systems, at the University of Southampton, United Kingdom, under the supervision of Professor Dame Wendy Wall. He was graduated with a M.Eng. in Electrical and Computer Science Engineering with a minor in Telecommunications from the Faculty of Engineering of University of Porto, Portugal. Dr. Marinheiro has an extensive experience in teaching and researching in the fields of Telecommunications, Computer Networks, and Hypermedia. Dr Marinheiro has coordinated and participated in many national and international research projects.

Deepak Mishra received the B.Tech. degree in electronics and communication engineering from the Guru Gobind Singh Indraprastha University, New Delhi, India, in 2012, and the Ph.D. degree in electrical engineering from the Indian Institutes of Technology Delhi, New Delhi, in 2017. He is currently a Post-Doctoral Researcher with the Department of Electrical Engineering, Linköping University, Linköping, Sweden. His current research interests include energy harvesting cooperative communication networks, massive MIMO, physical layer security, as well as signal processing and energy optimization schemes for uninterrupted operation of wireless networks. Dr. Mishra was a recipient of the IBM Ph.D. Fellowship Award in 2016, the Raman–Charpak Fellowship Award in 2017, and the Endeavour Research Fellowship Award in 2018. He was selected as an Exemplary Reviewer for the IEEE Transactions on Wireless Communications in 2017.

Lahby Mohamed received his PhD in Computer Science from Faculty of Sciences and Technology of Mohammedia, University Hassan II of Casablanca, in 2013. He is currently working as an Associate Professor at the Ecole Normale Superieur (ENS) University Hassan II of Casablanca. The current research interest of Dr. Mohamed Lahby includes Heterogeneous Wireless Networks, Mobility Management and Mobile Computing, QoS/QoE, Optimization and Machine learning for communications. He already published more than 30 papers in international journals, conferences and book chapters. Dr. Lahby has an extensive experience in teaching in the fields of Networking, Telecommunications and Internet Programming. Dr. Lahby has coordinated and participated in many national and international conferences.

Antonella Molinaro is an Associate Professor of Telecommunications at University Mediterranea of Reggio Calabria, Italy. She was with the University of Messina (1998-2001) and the University of Calabria (2001-04) as an Assistant Professor, and with the Polytechnic of Milano (1997-98) with a research contract. She worked at Telesoft, Rome (1992-93), and at Siemens, Munich (DE) (1994-95) with an European fellowship contract. Her current research interests focus on vehicular networking and future Internet architectures. She is a Member of IEEE.

Andreea Molnar is a Senior Lecturer at Swinburne University of Technology. She has a PhD from National College of Ireland. Her research interests include video adaptation, quality of experience and technology enhance learning. She is in the editorial board of several journals such as International Journal of Game-Based Learning.

Jose Moura received in 1989 the BsC for Electronics and Telecommunications from Universidade de Aveiro, Portugal. From 1989 to 2000, he worked as a project manager at EFACEC Sistemas Electrónica. He holds a MSc since 2001 in computer networks from Universidade do Porto. Since 2001, he has been teaching in computer networks at Instituto Universitário de Lisboa (ISCTE-IUL), Lisboa. He is an assistant professor in the Department of Information Science and Technology at ISCTE-IUL. Jose holds since 2011 a PhD in managing the mobile access in heterogeneous wireless networks from Lancaster University, UK. His research interests include Wireless Networks, Game Theory, Software-Defined Networking, Virtualization, Cloud Computing and Big Data.

Cristina Hava Muntean received the B.Eng. degree in computer science from the "Politehnica" University of Timisoara, Romania, in 2000, and the Ph.D. degree from Dublin City University, Ireland, in 2005. She is a Senior Lecturer with the School of Computing, National College of Ireland. She has been constantly involved in various research-related activities over the past 16 years fostering and promoting research, leading research projects and supervising Ph.D. and M.Sc. students. She has published over 100 papers in international peer-reviewed books, journals, and conferences. Her major research areas include adaptive multimedia and energy saving solutions, adaptive and personalized e/m-learning, user modeling, consumer behavior, end-user quality of experience assessment wireless communication and STEM education. She chaired or served as a Technical Program Committee Member for top international conferences and acted as a Reviewer for several journals, including the IEEE Transactions on Broadcasting.

Muhammad Faheem Mustafa completed his MS degree in Electrical Engineering from COMSATS University Islamabad, Wah campus in 2016. He is currently working in telecommunication & electronics design field with NRTC Pakistan.

Linyu Peng received the Ph.D. degree in mathematics from the University of Surrey, UK, in 2013. He is currently an Assistant Professor at Waseda Institute for Advanced Study, Waseda University, Japan. His research interests concern a wide range of fields of pure and applied mathematics, including mathematical optimization, algebraic and geometric theories of differential and difference equations, and information geometry.

Sara Pizzi is a Research Associate in Telecommunications at University Mediterranea of Reggio Calabria, Italy. From the same University she received the 1st (2002) and 2nd (2005) level Laurea Degree, both cum laude, in Telecommunication Engineering and the Ph.D. degree (2009) in Computer, Biomedical and Telecommunication Engineering. Her current research interests focus on QoS provisioning in wireless networks, multicast support in broadband wireless networks, and RFID inclusion in the Internet of Things.

Ganesh Prasad received the M. Tech degree in Electronics and Communication Engineering from Indian Institute of Technology (BHU), Varanasi, India, in 2012, and pursuing the Ph.D. degree in the same domain from National Institute of Technology (NIT), Silchar, India. He is currently an Assistant Professor in the Department of Electronics and Communication Engineering, NIT, Silchar. His research interests include cooperative wireless communications, underwater networks, free space optical communications, and green communications.

Rastislav Róka was born in Šaľa, Slovakia on January 27, 1972. He received his MSc. and PhD. degrees in Telecommunications from the Slovak University of Technology, Bratislava, in 1995 and 2002. Since 1997, he has been working as a senior lecturer at the Institute of MICT, FEI STU, Bratislava. Since 2009, he is working as an associated professor at this institute. His teaching and educational activities are realized in areas of fixed transmission media, digital and optocommunication transmission systems and network. At present, his research activity is focused on the signal transmission through optical transport, metropolitan and access networks by means of new WDM and TDM technologies using advanced optical signal processing included various modulation and coding techniques. His main effort is dedicated to effective utilization of the optical fiber's transmission capacity of the broadband passive optical networks by means of DBA and DWA algorithms applied in various advanced hybrid network infrastructures.

Abderrahim Sekkaki received a D.Sc. in Network Management domain from the Paul Sabatier University, France, in 1991: and a Dr. of State Degree from Hassan II University, Morocco, in 2002. Presently, he is a Professor in Computer Science at the Hassan II University, Casablanca, Morocco. His research interests include Distributed Systems, Policies based network management and security, Heterogeneous Wireless Networks, QoS/QoE and Mobile Computing, Cloud Services and Big Data. He has authored more than 100 papers in the above areas. and security.

Joao Silva received the BSc and MSc degree for Aerospace Engineering from Instituto Superior Técnico (IST) – Lisbon Technical University, (1995-2000). From 2000-2002 he worked as a business consultant in strategic management in McKinsey & Company. From 2002 to 2006 he undertook his PhD in telecommunications at IST, integrated in 3 EU projects (Seacorn, B-Bone e C-Mobile), having been the representative of the university. Since 2006, he has been working on computer networks, as an assistant professor in Instituto Superior das Ciências do Trabalho e da Empresa (ISCTE). Recently, he finished his Masters on Business Management at ISEG (Instituto Superior de Economia e Gestão), and aims to work on technological management in the near future.

Chetna Singhal is working as an Assistant Professor at Indian Institute of Technology (IIT) Kharagpur. I have completed my Ph.D from IIT Delhi in May, 2015 and have worked as a Postdoctoral Researcher in Department of Electrical Engineering, IIT Delhi till October 2015. I received my M.Tech. in Computer Technology from Electrical Engineering Department, IIT Delhi in 2010 and B.Eng. in Electronics and Telecommunications from University of Pune in 2008. I worked in IBM Software Lab, New Delhi, as a Software Engineer, from June 2010 to July 2011. My research interests are in heterogeneous wireless networks, multimedia transmissions, resource allocation, and wireless handovers.

Irina Tal is a Lecturer in Computing with National College of Ireland since 2017. She was previously a Postdoctoral Researcher with the Performance Engineering Laboratory (PEL) at Dublin City University (DCU), Ireland since December 2016. In March 2016 she was successfully awarded the PhD from Dublin City University, Ireland.

Xu Zhang received the B.Eng. degree in telecommunication engineering from the Xidian University, Xi'an, China, in 2007, and the M.Sc. and Ph.D. degrees from the University of Science and Technology, Beijing, China, and the University of Surrey, Surrey, U.K., in 2010 and 2015, respectively. She is currently a Lecturer with the Department of Computer Science, Xi'an University of Technology, Xi'an, China. Her research interests include networking and network management, including aspects such as content delivery networks, information-centric networking and transportation & energy, e.g., Electro-Mobility.

Index

I

L

M

N

O

P

Q

R

S

T

U

V

W

www.ingramcontent.com/pod-product-compliance
Lightning Source LLC
LaVergne TN
LVHW062134300926
849161LV00010B/2532